Survival Strategies in Cold-adapted Microorganisms

Reeta Goel • Ravindra Soni •
Deep Chandra Suyal • Mahejibin Khan
Editors

Survival Strategies in Cold-adapted Microorganisms

Springer

Editors
Reeta Goel
Institute of Applied Sciences and Humanities
G L A University
Mathura, Uttar Pradesh, India

Ravindra Soni
Department of Agricultural Microbiology
Indira Gandhi Krishi Vishwa Vidyalaya
Raipur, Chhattisgarh, India

Deep Chandra Suyal
Department of Microbiology, Akal College of Basic Sciences
Eternal University, Baru Sahib
Sirmaur, Himachal Pradesh, India

Mahejibin Khan
Microbiology and Fermentation Technology
Central Food Technological Research Institute
Mysore, Karnataka, India

ISBN 978-981-16-2624-1 ISBN 978-981-16-2625-8 (eBook)
https://doi.org/10.1007/978-981-16-2625-8

This Springer imprint is published by the registered company Springer Nature Singapore Pte Ltd.
The registered company address is: 152 Beach Road, #21-01/04 Gateway East, Singapore 189721, Singapore

Preface

This book describes my journey from Csp to CowN. After getting trained in protein structure–function studies, the big challenge in front of me was the selection of the right candidate. While going through a published review article on Antarctic microbiota, Csps attracted me a lot. Luckily, DST financed the first project for the development of a cold-adapted/tolerant mutant of PGPR. Later on in this journey, M. Sc. and Ph.D. students accompanied me to different distances and selected destinations. Surprisingly, the last financial support was again from DST/SERB as a young scientist project to one of our associate editors, wherein CowN was. Moreover, these low-temperature warriors were isolated from high altitude agroecosystems, which gave immense satisfaction to my concept thought.

This book consists of 19 chapters based on four major themes: diversity of cold-adapted microorganisms; their adaptation and survival strategies; use of omics approaches in the respective field; and their application in agriculture and biotechnology. We trust the readers will find the precise and latest update on microbial cold adaptation in this book besides our small contribution in the field.

We gratefully acknowledge the help, support, and suggestions provided by the colleagues and well-wishers. Moreover, the contributing authors are greatly acknowledged for their cooperation and patience. Any suggestion for the improvement of this book will be highly appreciated and these will be incorporated in the subsequent editions.

Mathura, India — Reeta Goel
Raipur, India — Ravindra Soni
Sirmaur, India — Deep Chandra Suyal
Mysore, India — Mahejibin Khan

Contents

About the Editors

Reeta Goel is a Professor and former Head of the Department of Microbiology, G. B. Pant University of Agriculture & Technology, Uttarakhand, India. She has also served as visiting scientist at various institutes in the USA and has more than 35 years of teaching and research experience in the area of plant–microbe interactions, microbial ecology, biodegradation, and metagenomics, especially with reference to higher altitudes. Moreover, she has worked extensively in the field of genomics and proteomics of cold-adapted microorganisms. She has published over 100 research articles, book chapters, and edited books. She has six patents in her credit, including one US Patent, and received numerous awards for her contributions to science and technology. Presently, she is serving as a distinguished Professor at GLA University, Mathura, UP, India after her retirement.

Ravindra Soni is an Assistant Professor at the Department of Agricultural Microbiology, I.G.K.V., Raipur, India. He has more than 11 years of teaching and research experience in the field of agricultural microbiology, microbial ecology, molecular biology, and biotechnology, especially with reference to higher altitudes. He was the recipient of DST-Young Scientist award from SERB, Department of Science & Technology, India. He was also selected for UGC-DS Kothari Postdoc Fellowship and INSA summer Fellowship. He has published edited books, lab manual, several research articles, and chapters in reputed international journals and books.

Deep Chandra Suyal is an Assistant Professor at the Department of Microbiology, Akal College of Basic Sciences, Eternal University, Baru Sahib, Himachal Pradesh, India. He has been actively engaged in research for 10 years, and has experience in the fields of agricultural microbiology, molecular biology and biotechnology, and metagenomics. He is currently investigating the genomics and proteomics of cold-adapted microorganisms. He has published numerous articles and books with reputed international journals and publishers. He has three patents in his credit.

Mahejibin Khan is a Senior Scientist at CSIR-Central Food Technological Research Institute, Mysore, India. She has more than 15 years of experience in

Microbiology, Food Microbiology, Microbial Biotechnology, and Proteomics. She visited several countries and has been awarded United Nations University fellowship and worked at National Food Research Institute, Tsukuba, Japan. She is the recipient of several national and international awards and author of various international publications including research papers, books, and chapters. She has good working experience on cold-adapted microorganisms and their associated mechanisms.

Cyanobacteria in Cold Ecosystem: Tolerance and Adaptation

1

Khushboo Dasauni, Divya, and Tapan K. Nailwal

Abstract

Cyanobacteria are dominant primary producers and near ubiquitous, inhabiting diverse range of ecosystems. They are found in freshwater, marine system to extreme environment such as the Antarctic, Arctic, and alpine regions. It seems paradoxical that most polar cyanobacteria are psychrotolerant rather than psychrophilic. Fluctuation in temperature, availability of nutrients and liquid water, and irradiance can alter the microbial community dynamics. Cyanobacteria use screening compounds, antifreeze proteins, antioxidants, membrane proteins, ion regulation, etc. to thrive in extremely challenging conditions. This chapter reviews various mechanisms of acclimatization of cyanobacteria at low temperatures of the Arctic and Antarctic.

Keywords

Acclimatization · Cyanobacteria · Psychrotolerant · Antifreeze protein · Arctic region · Antarctic region

1.1 Introduction

Cyanobacteria (blue-green algae) are the only known photosynthetic prokaryotes capable of fixing both carbon dioxide and nitrogen. They are considered as among the first microorganisms to inhabit the earth, therefore playing a key role in nutrient cycling and energy flow on earth. Unicellular, surface-attached, multicellular, filamentous colony and mat-forming are the various morph types of cyanobacteria.

K. Dasauni · Divya · T. K. Nailwal (✉)
Department of Biotechnology, Sir J. C. Bose Technical Campus, Bhimtal, Kumaun University, Nainital, Uttarakhand, India

R. Goel et al. (eds.), *Survival Strategies in Cold-adapted Microorganisms*,
https://doi.org/10.1007/978-981-16-2625-8_1

Polar or low-temperature ecosystems are the largest untouched or unlocked biological resources of our planet. In these regions, temperature remains below 0 °C during most part of the year. Cyanobacteria are frequently found in cryospheric ecosystems such as polar deserts, permafrost, cryoconites, snow lake ice, glacial ice, etc. Swedish-Finnish explorer Adolf Erik and his team are one of the first to discover cyanobacteria thriving in cryoconite of cryosphere during their expedition to green ice cap in 1870. Commonly found polar cyanobacteria belong to the orders *Chroococcales* (*Gloeocapsa* (several species), *Chroococcidiopsis*, *Aphanocapsa*, *Hormathonema*), *Oscillatoriales* (*Lyngbya* and *Leptolyngbya*), and *Nostocales* (*Anabaena* (several species) and *Microchaete*). Some of these assemblages are conspicuously pigmented, such as the red *Gloeocapsa* community. Cryospheric cyanobacteria are required to deal with the challenges of fluctuating low temperature, high irradiance (PAR and UVR), desiccation, availability of nutrient and liquid water, and, now due to global warming, elevated temperature. Though cyanobacteria are dominant organisms in cryo-ecosystems, it seems paradoxical that most polar cyanobacteria are psychrotolerant rather than psychrophilic. It has been reported that cyanobacteria are able to survive in extreme conditions of polar region and are hypothetically also capable of acclimatization under hard environmental conditions and could contribute to the field of astrobiology research. Here we review current literature on ecology and the functional role played by cyanobacteria in various Arctic and Antarctic environments. We will focus on the ecological importance of cyanobacterial communities in polar regions and assess what is known regarding their general mechanism of adaptation (Nienow and Friedmann 1993).

1.2 Significance of Cold Ecosystem

Arctic and Antarctic polar regions together have attention gathering impact on global biogeochemical cycle and share a large proportion of the earth's surface area. Various different habitats in polar ecosystem include soil, permafrost, cryptic niche (biological soil crust, hypoliths and endoliths), ice, snow, and a number of aquatic habitats. Low water temperature in cryo-ecosystems shows increased carbon dioxide solubility accounting for 30% of global uptake that helps sequester a huge amount of carbon from atmosphere enriched with greenhouse gases. Mobilizing frozen methane deposits into ocean water and atmosphere as a result of rising Arctic temperature ultimately exacerbates the global warming. Sulfur particles formed via conversion of polar phytoplankton metabolite dimethylsulfoniopropionate (DMSP) to dimethylsulfide (DMS), a volatile gas, help seed clouds which mitigate climate warming. On the other hand, polar microbial population plays a primarily key role in various biogeochemical cycles, food webs, and re-mineralization processes. These microbial communities are dominated by various orders of cyanobacteria which form the base of food webs and account for large proportion of polar biomass despite extremely low temperature and other stress factors prevailing in polar ecosystems. Poor ideal level of nutrient availability for biological activity restricts the diversity of microbial communities in polar environment. Physiological and phenotypic characteristic of species determines its vulnerability to extreme weather events resulting in some species being more susceptible to challenging stress than others (Nienow and Friedmann 1993).

1.3 Ecology and Biogeochemistry of Cyanobacteria

Cryo-ecosystem active cyanobacteria are challenged with extreme environmental parameters of high altitude and latitude. These environmental parameters include high UV radiation, desiccation tolerance, and freeze and thaw damage. Cyanobacteria within microbial communities are the only dominant and most widely distributed photoautotrophs contributing to structural stability, moisture retention, and fertility of surrounding microbes (Belnap and Gardner 1993). They are considered a dominant contributor of essential ecosystem services (carbon and nitrogen cycle). Classical and modern microbiological techniques have revealed the presence of cyanobacteria as primary colonizer in a wide range of polar niche such as permafrost, ice shelves, rocks, ponds, lakes, and glaciers (Vincent 2000). Variability in environmental factors restricts development of other photoautotrophic clades, while domination of cyanobacterial autotrophy supports a diverse number of heterotrophic microorganisms (such as *Actinobacteria*, *Proteobacteria*, *Firmicutes*, and *Bacteroidetes*) as well as a small number of organisms at higher trophic level (De los Rios et al. 2014; Yung et al. 2014). Besides polar cyanobacterial cohorts, polar ecosystems are also inhabited by a number of eukaryotic photoautotrophs which include diatoms, green algae, and mixotrophic organisms. Studying and understanding the mechanism of adaptation of these dominant cyanobacteria helps understand the whole ecosystem dynamics in cryo-ecosystems (Curtis 2006). Cyanobacteria exhibit biogeographical patterns; as such, their diversity is often linked to latitudinal gradients, with highest diversity and density found between 70 and 80 °S (south) (Namsaraev et al. 2010) (Table 1.1).

Table 1.1 Types of lithic environment

Lithic environment	Definition	References
Epilithic	Rock surface-inhabiting organisms	Omelon (2008)
Endolithic	Organisms inhabiting the interior of rock	Nienow and Friedmann (1993)
Hypolithic	Communities inhabiting the soil-rock interface	Cockell and Stokes (2004)
Euendolithic	Microorganisms can bore actively into the rock and inhabit the resultant hole	Cockell and Herrera (2007)
Cryptoendolithic	Microorganisms inhabit the space between grains of porous rocks	Omelon et al. (2007)
Chasmoendolithic	Microorganisms inhabiting rocks cracks and fissures	Budel et al. (2008)

1.3.1 Cryptic Niches

Cryptic niches include biological soil crust (BSC), hypoliths, and endoliths that provide physical stability, allowing slow-growing cyanobacterial and other specialized microbial communities to develop (Agawin and Agusti 1997). Antarctic BSCs are composed of unique filamentous cyanobacteria such as *Nostoc commune* and *Tolypothrix*, *Calothrix*, and *Leptolyngbya* species (Budel and Colesie 2014), while BSC communities in high Arctic favor members of *Nostocales*, *Chroococcales*, and *Oscillatoria*.

1.3.2 Hypoliths

In an extreme cold ecosystem, the underside of rocks acts as a "refuge" for photoautotrophic microorganisms. The community performs photosynthesis with irradiance levels less than 0.1% of incident light. Hypolithons are protected from UV radiation, wind scouring, and trapped water. This provides bioavailable liquid water for hypolithons (Makhalanyane et al. 2014; Ramond et al. 2015). Among hypolithons cyanobacteria forms the base of community structure and functional processes, and a clonal-based analysis of hypolith samples showed close homology to *Nostocales* and *Oscillatoriales* dominance. Arctic hypoliths are also dominated by cyanobacteria, and species found include *Gloeocapsa atrata* Kützing, *Gloeocapsa punctata* Nägeli, *Gloeocapsa kuetzingiana* Nägeli and *Chroococcidiopsis*-like cells; unicellular algal chlorophytes are also present (Cockell and Stokes 2004).

In another study by Pointing et al. (2009), colonized quartz rock examination was done, and observations indicated dominance of oscillatorian cyanobacterial morphotypes, belonging to genus *Leptolyngbya*. De los Rios et al. (2014) demonstrated the unique spatial organization by hypoliths dominated by cyanobacteria and found that most were structured by filamentous cyanobacteria and associated with extracellular polymeric components, largely forming a biofilm. The presence of extracellular polymeric substances has a major impact on survival of these cyanobacteria especially in water-limited environment as EPS plays a key role in water retention. On the other hand, Arctic hypoliths are dominated by a diverse range of cyanobacterial species like *Gloeocapsa*, *Chroococcidiopsis*-like cells, and unicellular algal chlorophytes (Cockell and stokes 2004).

1.3.3 Endoliths

Study on gypsum crusts on Alexander Island, West Antarctic Peninsula, revealed the predominance of *Cyanobacterium chlorogloea* sp. (Makhalanyane et al. 2014). Cyanobacteria are dominant colonists in endolith niches, and their colonization mode depends on micro-morphological and structural properties of the rock (Hughes and Lawley 2003). In a similar study conducted in Canadian High Arctic, analysis of endolthic communities of gypsum crust indicated predominance of phototrophic cyanobacteria which included *Nostoc* species, *Loriellopsis* sp., and *Chroococcidiopsis* sp. (Ziolkowski et al. 2013). In Mcmurdo Dry Valley, Antartica, endolithic communities were dominated by three communities: *Gloeocapsa*, *Hormathonema*, and *Chroococcidiopsis*. Bu del et al. (2008) through their studies of endoliths from Antarctica reported presence of different genera of cyanobacteria such as *Chroococcidiopsis*, *Cyanothece*, and *Nostoc* species.

1.3.4 Cryoconites

Cryoconites play a key role in glacial ecosystems exhibiting different boundaries, nutrient cycling, and energy flow. Cyroconites act as refugia from harsh environment for a diverse range of microbes as well as different species of cyanobacteria. These are cylindrical cavities of thin layer of sediment present in the ice surfaces. These layers of sediments provide a favorable environment for microbial colonization and growth, thereby acting as biomass seeding sources in polar regions (Mueller et al. 2001). Dominating species of cyanobacteria in cryoconites are *Phormidium*, *Nostoc*, and various species of the genus *Leptolyngbya* (Cameron et al. 2012).

1.3.5 Aquatic Habitats

Studies have suggested presence of cyanobacterial species in inland aquatic systems of Antarctica to be low in diversity. However, 20 years later (Taton et al. 2003), in their study using 16S rRNA gene and internal transcribed spaces (ITS), regions came up with 15 phylotypes present on microbial mats of Lake Fryxell in McMurdo Dry Valley, Antartica, and belonged to genera *Geitlerinema*, *Nostoc*, *Hydrocoryne*, *Leptolyngbya*, *Lyngbya*, *Pseudonabaena*, *Phormidium*, *Oscillatoria*, *Schizothrix*, and *Nodularia* (Table 1.2).

1.4 Ecophysiology of Polar Cyanobacteria and Functional Role of Arctic and Antarctic Cyanobacteria

Current scientific exploration shows that both Arctic and Antarctic cyanobacterial species are essential for ecosystem services (Wagner and Adrian 2009). Fluctuation in temperature and availability of nutrients and liquid water can alter microbial community dynamics (Chan et al. 2013). For example, global warming leading to warmer water in polar lakes and ponds and nutrient accumulation from catchment often results in cyanobacterial blooms (Wagner and Adrian 2009). Application of modern molecular phylogenetic techniques can enhance the understanding of functionality of cryo-microbial communities. Experimental studies by various scientific communities have made it clear that cyanobacteria are central to such ecosystems (Fernandez-Valiente et al. 2001). Recent evidence suggests that even moderate change or reduction in microbial community will affect the whole functional

Table 1.2 Genera of dominant cyanobacteria with their cryospheric habitat

Genus	Habitat	References
Anabaena	Permafrost, soil, endolithic, cryoconite	Friedmann et al. (1987), Vishnivetskaya (2009), Mataloni et al. (2000), Mueller and Pollard (2004)
Aphanocapsa	Soil, endolithic	Friedmann et al. (1987), Fermani et al. (2007)
Calothrix	Soil, hypolithic, chasmoendolithic, epilithic	Broady (1981, 1986, 1989), Cavacini (2001)
Chroococcidiopsis	Soil, hypolithic, chasmoendolithic, endolithic, epilithic	Friedmann et al. (1987), Ryan et al. (1989), Cockell and Stokes (2004)
Eucapsis	Cryptoendolithic	Friedmann et al. (1987)
Gloeocapsa	Soil, hypolithic, cryptoendolithic, chasmoendolithic, cryoconite	Broady (1981, 1986, 1989), Friedmann et al. (1987), Cockell and Stokes (2004)
Leptolyngbya	Permafrost, soil cryptoendolithic, cryoconite, glacial ice, lake ice	Vishnivetskaya (2009), Friedmann et al. (1987), Cavacini (2001), Mataloni et al. (2000)
Lyngbya	Soil, hypolithic, chasmoendolithic, cryoconite	Broady (1981)
Microchaete	Soil, endolithic	Friedmann et al. (1987)
Microcoleus	Soil, cryoconite	Komárek et al. (2008), Mueller and Pollard (2004)
Myxosarcina	Epilithic	Broady (1981)
Nodularia	Soil, hypolithic, chasmoendolithic	Broady (1981)
Nostoc	Permafrost, soil, hypolithic, chasmoendolithic, cryoconite	Budel et al. (2008), Mueller and Pollard (2004)
Oscillatoria	Soil, endolithics, cryoconite	Cameron (1972), Mataloni et al. (2000), Mueller and Pollard (2004)

dynamics of cold ecosystem (Philippot et al. 2013; Singh 2014). Recent metagenomic analysis of genes responsible for stress response to various environmental stress parameters in microbial mat communities which are dominated by cyanobacteria (in Arctic and Antarctic) includes sigma B, EPS, cold shock proteins, and membrane modifications (Varin et al. 2012). Under low temperatures, proteins are more susceptible to tertiary and quaternary structural damage, while nucleic acid content becomes more stable, unable to support basic processes like replication, transcription, and translation (D'Amico et al. 2006). Thus, cyanobacteria for survival produce specialized proteins and other biomolecules such as cold shock proteins, antifreeze protein, and glycine betaine. Cyanobacteria survival is also accompanied by higher production of unsaturated fatty acid to retain membrane fluidity as loss of fluidity highly affects the nutrient transport (Los and Mironov 2013). DEAD-box RNA helicases are also expressed in abundance which help maintain cellular processes even under thermodynamic constraints (Rocak and Linder 2004). To cope up with the challenges of freezing and desiccation, cyanobacteria have adapted the strategy of entering to dormant state followed up by low metabolic activity (Vincent et al. 2004).

Cyanobacteria-driven biofilms provide exopolymer matrix which supports biological as well as biogeochemical interaction. These cyanobacterial biofilms secrete a large number of exopolysaccharides (EPS). EPS matrixes also play a structural role and are responsible for creation and maintenance of microenvironment to support growth and metabolism of cyanobacterial consortia. De los Rios et al. (2014) showed that EPS also provides cryoprotection and desiccation protection for a diverse range of microorganisms.

1.5 Polar Region: Extreme Environmental Parameters and Stress Factors

Generally, cyanobacteria surviving in high latitude and altitude are cold tolerant known as psychrotrophs, and they show suboptimal growth at low temperature as against psychrophiles that show optimal growth at low temperature (Tang and Vincent 1999). Psychrotrophs exhibit a variety of mechanisms to tolerate and grow at extremely low temperatures. They have adapted to freeze-thaw conditions by enhancing the production of polyunsaturated fatty acids accompanied with short chain length to be incorporated in their membrane. Cyanobacteria also produce compatible solutes which reduce freezing point of intracellular solutes such as trehalose, which also reduces cell desiccation. Cyanobacteria also produce and secrete extracellular polymeric substances which help minimize ice nucleation around the cells. Through these ways, cyanobacteria consortiums are able to withstand long-term seasonal dormancy phases in frozen environment. Later, these freeze-dried cyanobacterial mats resume photosynthesis and high metabolics upon rethawing (Vincent 2007). Temperature, liquid water availability, and irradiance are crucial stress factors in cryo-ecosystem. Different adaptation and acclimatization responses are due to various combinations of seasonal and diurnal variation, range of values, and periodicity of these major stress factors. Studies have reported dynamic fluctuation of temperature in polar regions. For example, average temperature during vegetation season ranges from 0 to 5 °C, where temperature can be much higher or lower than air temperature particularly at localized microenvironments. Elster et al. (2012) demonstrated that the temperature of dry hummock top ranges from 2 to 10 °C in the beginning of vegetative season, and with time when hummock bottom is completely submerged in water, the temperature becomes constant around 0 °C. Therefore, decrease in liquid water availability from aquatic to dry habitats is a characteristic of seasonal variation. These seasonal variations are responsible for causing precipitation events. Polar streams, rock surfaces, or wet hummock tundra meadows are exposed to variable wet and dry periods during vegetation season. Another stress encountered by microbial consortia is when temperature for a short period of time rapidly increases to 0 °C and liquid water prevails for 1 or 2 weeks. This event of rain on snow takes place from January to February and results in significant reduction in survival rate of cyanobacteria. Irradiance is yet another key factor and includes PAR (400–700 nm) and UVR (280–400 nm). Polar nights as well as continuous irradiance during polar day cause stress to thriving life. PAR is the main known source of energy for photoautotrophic microorganism. UVR is known to have greater impact on polar aquatic and terrestrial ecosystems. Microclimate can be very different from macroclimate; such dynamic differences can be observed in thermal springs or within cryoendolithic communities (Friedmann et al. 1987).

1.6 Polar Cyanobacteria: Response to Various Stress Factors

1.6.1 General Mechanism of Adaptation

Duration of stress and stress intensity determines the stress effect and course of stress reaction (Schulze 2005). Stress response also depends on adaptation capabilities of microbial communities. These cryo-microbial communities in nature are exposed to combinations of different types of stresses at the same time. For example, high PAR is followed by UVR. As a result of multiple stresses, stress responses in turn are highly complex involving various protection mechanisms. Basic stress response is categorized as active (stress tolerance) and passive (stress avoidance) according to Schulze (2005). For algae and cyanobacteria, the following terms are defined: adaptation, genetically fixed responses to outer environmental conditions (Elster 1999), and acclimation, response to sporadic extremes of environment that is not genetically fixed, but biochemical (e.g., synthesis of screening pigments), morphological (e.g., cell wall modification), or physiological (e.g., state transitions) changes occur. Elster (1999) defined this type of response as "acclimatization." Stress reaction usually triggers specific processes, for example, state transitions (Allen 2003), metabolic changes like increased synthesis of polyunsaturated fatty acids (Shivaji et al. 2007), or modification of cell ultrastructure, viz., increased number of photosystem units in membranes.

1.6.2 Stress Avoidance

Stress avoidance is done by microbial communities using escape strategies to avoid harsh conditions or insulate cells from the surroundings. Escape mechanisms involve migration to acceptable (favorable) conditions or suitable habitats. For migration, unicellular and filamentous forms use gliding movement (Hoiczyk 2000; McBride 2001), and planktonic species use buoyancy mechanisms involving gas vacuoles to control and monitor their position in the water column (Oliver 1994). Habitat

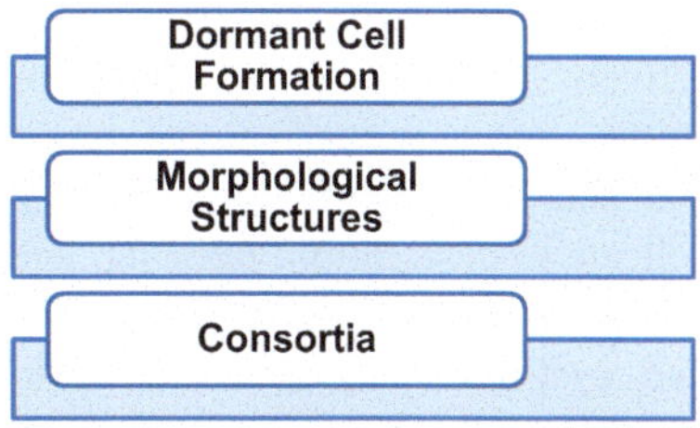

Fig. 1.1 Strategies of polar cyanobacteria use for stress avoidance

selection is a favorable option for non-motile living forms. Habitat selection is advantageous as it provides favorable conditions than outside such as rock interior is warmer and wetter, protecting against PAR and UVR) (Friedmann et al. 1987). Gilichinsky et al. (2008) observed that permafrost samples were dominated by viable non-heterocystous filamentous cyanobacteria of the family *Oscillatoriales* (Fig. 1.1).

1.6.3 Stress Tolerance

When stress factors damage the cell, stress tolerance mechanism comes into play. Stress response events that help restore the cell's function are known as stress reactions. The point when cell structure as well as function is disturbed, signaling molecules trigger appropriate receptor to carry a cascade of reactions, resistance of organism increases up to a maximal level, and resistance need not to last constantly as it can decrease again during long-term and intensive stress (Schulze 2005).

1.6.4 Dormant Cell Formation

One of the mechanisms of protection from harsh environment is to produce akinetes whose formation has not been observed in polar cyanobacteria (Lubzens et al. 2010), and it can rarely be seen in genera *Anabaena* or *Hydrocoryne*. Nonpolar cyanobacteria do exhibit the ability to form akinetes, which is defined as a resting stage where the cell wall is thick and large amounts of intracellular storage compounds are present. This allows survival under extremely harsh conditions. Trigger mechanisms for akinete formation from vegetative cells are probably species specific (Van den Hoek et al. 1995). It is reported that hormogonia and hormocyst are polar akinete in polar regions. Hormogonia and hormocyst are formed via fragmentation of vegetative or mother filaments and possess ability to migrate to favorable conditions.

Tashyreva and Elster (2016) in their study on polar *Microcoleus* (a filamentous species of cyanobacteria) observed that at the end of vegetative season, a dense sheet starts to develop, and later in spring, a large number of hormogonia is released. They also stated that at the end of vegetative season, freezing, desiccation, and nitrogen starvation trigger the acclimatization process. This promotes development of resistance to winter (Tashyreva and Elster 2016). Under stress conditions of vegetative season, hormogonia and hormocysts may still form in large number and spread actively or passively via gliding or by water, respectively, to a suitable habitat. Despite very low survival rate, few hormogonia manage to outlast severe conditions and later act as inoculums for colonization of habitat. Since filaments are more delicate to desiccation, freezing, and biochemical modifications, therefore, production of osmotically active compounds, or formation of other types of dormant cells, like cyst-like cells in chroococcidiopsis (Caiola et al. 1996), could facilitate continued cell existence (Tashyreva and Elster 2012, 2016).

1.6.5 Morphological Structures

To cope with hostile external conditions, cyanobacteria promote differential development of morphological structure. These morphological structures can help survive in times when migration toward suitable environment is restricted. Against desiccation and freezing, polar cyanobacteria develop various types of extracellular mucilaginous envelopes and sheaths (De los Ríos et al. 2004; Makhalanyane et al. 2014; Tashyreva and Elster 2016; Deming and Young 2017). The surrounding envelops protect via different compounds that screen the cell against PAR and UVR. Colonial forms are more resistant to stress (Xiong et al. 1996), probably because surface cells protect those inside the colony. Formation of a colony of cells or filaments was indeed observed after exposure to high irradiance and nutrient depletion (Callieri et al. 2012).

1.6.6 Consortia

In polar environments, the consortium of different species develops into several structures like firm layered structures, microbial mats, and biofilms. These colonial patterns are often characterized by an outer or upper layer of cell containing screening pigments. A sheath pigment, scytonemin, protects subsurface layers. This shows that the condition is more favorable in the interior of the mat than outside (Vincent et al. 1993). Such microbial mats are found at aerial-liquid-solid interfaces as periphyton in seepages or at the aerial-solid interface on the soil surface as a soil crust. These microbial mats and biofilms show more resistance response to a number of stress factors (Elster and Benson 2004). Tashyreva and Elster et al. (2012) demonstrated an extreme case of avoidance strategy by lichen formation where fungal filaments play a role in structural and chemical protection and provide water to phycobiont, and in return cyanobactreium offers organic nutrients to the mycobiont. Here lichen system provides an advantage of good water management in polar regions, which are often exposed to water stress.

1.6.7 Low Temperature

In a study, oscillatorian cyanobacteria isolated from melt water ponds which are psychrophilic (temperature range from 0 to 8 °C). Polar cyanobacteria also exhibit longer doubling time than psychrophilic eukaryotic algae and heterotrophic bacteria (Vincent 2007) (Fig. 1.2).

1.6.8 Temperature Perception

Cold adapted microbial plus cyanobacterial consortia possess temperature sensors which trigger signal transduction in cells. The scheme of acclimatization reaction in synechocystis PCC6803 was studied by Murata and Wada (1995) and Nishida and Murata (1996), and they reported that the fall in external temperature decreases fluidity of cellular membrane which is sensed by membrane-bound sensors. Receptors of cold shocks could possibly be membrane proteins. Temperature sensors trigger a cascade of reaction that ultimately leads to expression of desaturase genes leading to restoration of membrane fluidity with accumulation of desaturated fatty acids.

1.6.9 Lipids

Based on fatty acid composition, cyanobacteria belong to four different groups. Alteration in fatty acid composition during exposure to low polar temperature in cyanobacterial membranes depends on the group to which cyanobacterium belongs. When temperature shifts to extreme low, fatty acid composition of membrane changes and results in an increased ratio of unsaturated fatty acids, increase in cis-double bonds, methyl branching, and shortening of fatty acid chain (Gounot and Russell 1999). It involves changes in level of desaturation together with acyl chain length (Shivaji et al. 2007). In a study conducted on polar *Calothrix* sp. Samples, it was shown that membrane phospholipid containing branched long chain fatty acids improves cold tolerance (Řezanka et al. 2009). In another study, it was shown that in *Synechocystis* sp., level of unsaturated fatty acid changes due to unsaturation of existing fatty acid and not because of de novo synthesis.

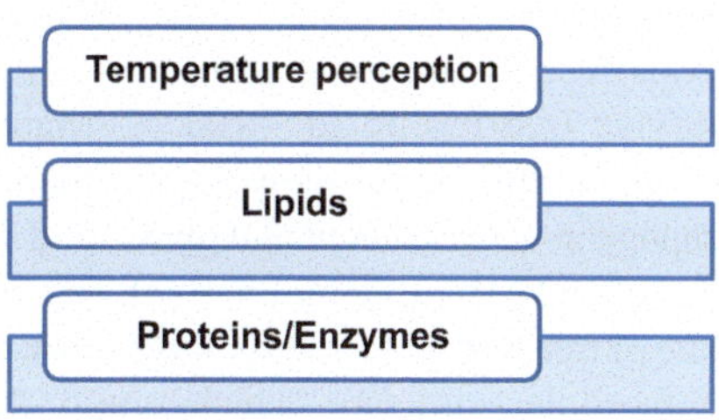

Fig. 1.2 Survival strategies of cyanobacteria at low temperature

1.6.10 Proteins/Enzymes

Conservation of structural and functional integrity of cellular proteins at extreme low temperature is challenging for polar cyanobacteria and other microbial communities. At low temperature, decaying of compact protein structure can be observed, further resulting in individual subunit dissociation (Primalov 1979). Catalytic activity of enzymes too gets affected at low temperature. Protein folding must take play in such a way as to maintain structural and functional integrity (Walles et al. 1999). Sensitivity to low temperature could be different for various enzymes even when isolated from the same species, and different enzymes have developed different responses to low temperature (Loppes et al. 1996). When compared to mesophilic enzymes, enzymes from cold adapted organisms are more active at lower temperature, and their catalytic activity is more temperature independent, though their active site is less stable (Feller and Gerday 2003).

Still, few enzymes follow opposite scenario, for example, the highly conserved enzyme ribulose-1,5-bisphosphate carboxylase/oxygenase (RuBisCO) catalytic activity was observed to be low in psychrophilic algae in comparison with mesophiles. Lower enzymatic activity in RuBisCO is compensated with synthesis of greater amount of enzyme subunits in psychrophilic alga (Devos et al. 1998). Many microbial species together with diverse cyanobacterial communities undergo dormancy or hibernation to survive in polar environment. Cold shock proteins are synthesized as they are essential for restoration of growth under lower temperature, and they function at transcriptional and translational levels (Raymond-Bouchard and Whyte 2017). Cold shock proteins act differently in psychrophiles and mesophiles. The number of cold shock protein is higher and is directly proportional to the intensity of stress, i.e., the higher the cold stress, the greater the production of cold shock proteins. Another group of proteins called cold acclimation proteins (CAPs) are also considered to play a role in promoting growth in polar environments along with cold shock proteins (Berger et al. 1997).

1.6.11 Freeze/Melting Cycles

In polar regions especially polar hydroterrestrial and terrestrial ecosystems, cyanobacteria are more frequently subjected to freeze-thaw cycle. There are many factors on which sensitivity of cell to freezing and melting depends (Elster et al. 2012), like the physiological state of cell, rate of cooling, freezing and melting, and also the chemical composition of freezing medium. It was observed that polar cyanobacteria are more resistant to freezing than polar algae and could possibly survive freezing up to −100 °C (Šabacká and Elster 2006). Polar freezing and melting in turn are responsible for inducing many stresses like temperature change, change in water content and phase, and change in concentration of many compounds that often leads to alteration in pH, salt precipitation, and intracellular ice crystal. The rate of freezing and melting is very slow (Elster et al. 2012). Ice crystals penetrate the intracellular structure which could damage cells (Tanghe et al. 2003). ROS

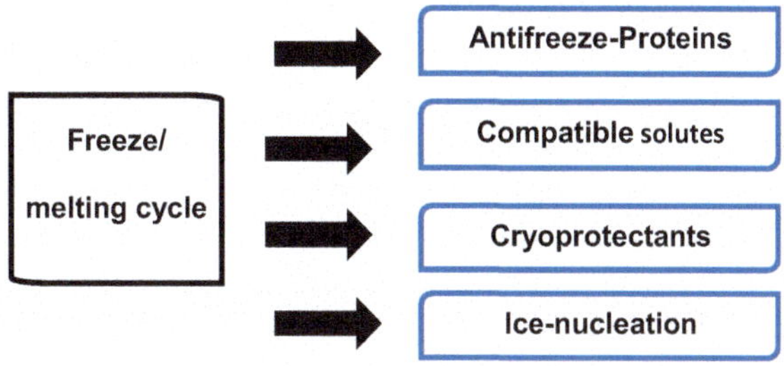

Fig. 1.3 Survival strategies against freeze/melting cycle

production together with in–/outflow of water with osmotic gradient also damages the cell. Synthesis of AFPs and dimethylsulfoxide (DMSO) prevents ice crystal formation and could account for freeze avoidance strategy, while freeze tolerance mechanism involves phenomenon of ice nucleation (Cockell et al. 2000) (Fig. 1.3).

1.6.12 Antifreeze Proteins

AFPs function by depressing the freezing point and avoiding recrystallization of ice (Chao et al. 1996). In a study on Antarctic cyanobacterial mats (dominated by *Nostoc* sp. and *Phormidium*), presence of AFPs was observed (Raymond and Fritsen 2000). These AFPs include a diverse range of proteins and are present at low concentration in cells.

1.6.13 Compatible Solutes and Cryoprotectants

Psychrophilic and psychrotolerant cyanobacteria are expected to synthesize and accumulate compatible solutes, though their accumulation in them is not well studied (Klähn and Hagemann 2011). However, marine algae have been known to produce cryoprotectant or compatible solutes, for example, DMSO which provides protection against freezing injury (Day et al. 2005; Day and Brand 2005). In halotolerant cyanobacteria, a similar compound was found, namely, dimethylsulfoniopropionate (DMSP) which can also serve as a cryoprotectant like DMSO (Kirst et al. 1991; Karsten et al. 1992). EPS (extracellular polysaccharides) present in thick mucilaginous envelopes and sheaths are also observed to slow down water movement during freeze and thaw (to prevent sudden alteration in cell volume) and, therefore, can also be considered as a sort of cryoprotectant (Deming and Young 2017). EPS are known to consist of glucose, xylose, galactose, and uronic acid (Helm et al. 2000). Chain-forming carbohydrates contain hydrophilic solutes. So, the extracellular matrix provides cryoprotection to cells during encapsulation. This effectively protective encapsulation has been demonstrated in coccal algae and yet not sufficiently studied in polar cyanobacteria (Elster et al. 2008; Lukešová et al. 2008).

1.6.14 Ice Nucleation Proteins

In bacteria, proteins present in the outer cell membrane act as template for ice nucleation which helps prevent desiccation. Ice nucleation separates water source present near the cell surface, and ice particles spreading outward from the cell can be observed. In this way, damage to the cell via ice formation reduces to minimum (Lee et al. 1995). Freezing temperatures required for ice nucleation activity are −13 °C, −15 °C, and −18 °C for Antarctic soil *Phormidium scotii*, *Pseudophormidium* sp., and *Phormidium attenuatum*, respectively, indicating that the cyanobacteria do not use ice nucleation mechanism for freezing protection (Worland and Lukešová 2000).

1.6.15 Dessication

Poikilohydric organisms such as microalgae, lichens, and mosses are defined as organisms that can tolerate desiccation (Alpert 2000, 2005). Cyanobacteria are also poikilohydric. Desiccation causes failure of various physiological processes, therefore impairing essential cellular process. In a study, long-term survival of desiccation in cyanobacterial *Nostoc* sp. was observed where *Nostoc* sp. was successfully revived after 55 (Shirkey et al. 2003), 87 (Lipman 1941), and more than 100 years (Cameron 1962). Cyanobacteria follow diverse strategies to minimize osmotic and mechanical stresses which are also common in poikilohydric organisms, indicating early evolutionary relatedness of these photosynthetic organisms (Fig. 1.4).

These strategies include complex interaction networks and processes at several cellular levels. In cyanobacterial consortia, desiccation leads to inhibition of nitrogen fixation and decline in photosynthesis and also affects respiration (Potts 2000; Qiu and Gao 2001). Even upon rehydration, it takes hours to recover photosynthetic function and days for nitrogenase activity recovery. Cyanobacterial tolerance to extreme desiccation is mainly due to its ability to tolerate extreme low water potentials. In a study by Potts and Friedmann (1981), it was shown that cryptoendolithic *Chroococcus* and *Chroococcidiopsis* were able to fix carbon dioxide at unusually low potential (Potts and Friedmann 1981; Palmer and Friedmann 1990). In another study, it was observed that *Nostoc* sp. carry out usual photochemical and nitrogen fixation activity even after losing 50% of its original weight when present at fully hydrated state. Stress response during desiccation is reported in *Calothrix* sp. and *Nostoc* sp., respectively. These responses include enhanced rate of lipid production and accumulation of polyunsaturated membrane lipids. Superoxide

Fig. 1.4 Survival strategies in response to desiccation

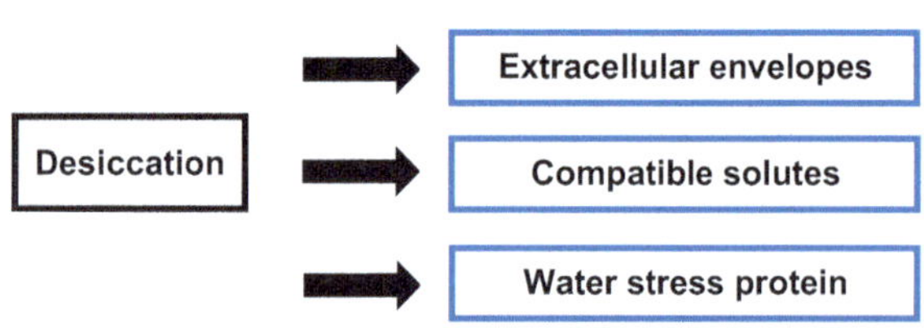

dismutase and catalase are the oxygen scavenging enzymes produced to reduce or prevent membrane damage due to ROS generation. As a desiccation tolerance strategy, cyanobacteria modify protein structure by acetylation, phosphorylation, and glycosylation. Photoreactivation, excision repair, and post-replication repair are the DNA protection mechanisms adapted by polar cyanobacteria.

1.6.16 Extracellular Envelopes

In polar regions, microbial species are usually desiccation resistant as they develop thick mucilage envelopes and sheaths (Tamaru et al. 2005). The major component of this envelop is a mucopolysaccharide that helps delay desiccation and retains water (Caiola et al. 1996; Pereira 2009). In *Nostoc commune* colonies, it has been demonstrated that EPS are responsible for possible water storage function and, therefore, keep cells in hydrated state (Kvíderová et al. 2011). Prolonged desiccation does affect the size and biochemical composition, resulting in larger cellular envelops in *Chrococcidiopsis* sp. (Caiola et al. 1996).

1.6.17 Water Stress Proteins

Another challenging parameter in the polar region is availability of water. Many microbial species including cyanobacterial species have water stress proteins (WSPs), late embryogenesis abundant (LEA) proteins, and dehydrin among them. LEA proteins have mechanism to stabilize other proteins and cellular membrane during drying (in presence of trehalose) (Close 1997). Dehydrins exhibit a functional role of inhibition of coagulation of macromolecules. Both proteins are found to be present in cyanobacteria (Close and Lammers 1993). Disaccharides such as trehalose, sucrose, and glucosylglycerol are considered compatible solutes especially in water-stressed mesophilic cyanobacteria (Reed et al. 1984; Klähn and Hagemann 2011). Some WSPs have been detected in *Nostoc commune* colonies that perform modification in extracellular envelop (or extracellular glycan) (Potts 1999).

1.6.18 Salinity

Osmotic and ionic components are required to combat with salinity stress through their specific response (Epstein 1985; Läuchli and Epstein 1990). Osmotic component is linked to water efflux, and response mechanism helps minimize or reduce water loss. Ionic component acts against toxic effects caused by higher concentration of ions and nutritional misbalance by performing selective or specific ion transport (Elster 1999). Cyanobacterial response to increase tolerance to salinity includes reduction in total lipid content and undergoes desaturation of membrane lipids. Salt tolerance mechanism also involves regulation of ion and water membrane channels (Singh et al. 2002). Salt stress often generates ROS, and cyanobacteria

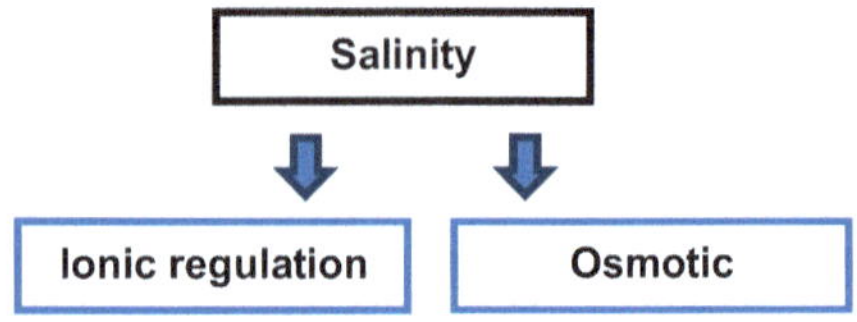

Fig. 1.5 Survival strategies against salinity in cold ecosystems

deals with it by activating quenching mechanisms (Latifi et al. 2009). Response involves several steps: (1) active Na^+ extrusion (ionic regulation), (2) uptake/synthesis of compatible solutes (osmotic regulations), (3) modification of membrane lipid composition, and (4) increased energetic capacity due to increased photosynthesis and respiration (Joset et al. 1996) (Fig. 1.5).

1.6.19 Ionic Regulation

Ion regulation is a crucial process that takes place in every microbial as well as eukaryotic cell. It can be achieved by active and passive ion transport. Active ion transport provides rapid response to changes in intracellular Na^+ concentration. When extracellular concentration of Na^+ increases, to maintain osmotic balance as a result, intracellular Na^+ concentration increases steeply and later declines slowly to attain physiological value (Packer et al. 1987; Blumwald et al. 1983). Active ion transport eliminates excess Na^+ to maintain an intracellular range of 10–30 mM. K^+ uptake too helps in adjustment of turgor (Apte 2001). These active ion transport systems such as antiport Na^+/H^+ are powered by P-ATPase (Joset et al. 1996).

1.6.20 Osmotic Regulation

In *Synechocystis* sp., the processes of uptake/synthesis of compatible compounds takes place simultaneously along with ion transport to maintain positive turgor (Reed et al. 1984; Joset et al. 1996). However, very less has been explored about the osmotic regulation in polar cyanobacteria. Cyanobacteria surviving under combined stress of desiccation and salinity in hypersaline deserts do synthesize necessary compatible solutes, for example, glycinebetaine, the most commonly synthesized compatible solute (Klähn and Hagemann 2011; Oren 2007).

1.6.21 Irradiance (PAR) and Ultraviolet Radiation (UVR)

Orthodoxically, cyanobacteria in polar regions have the ability to tolerate high doses of UVR and are less sensitive to it (Seckbach and Oren 2007). They exhibit different morphological and community organization patterns such as bigger cells or grow in large colonies, coenobia, or mats which provides protection to internal cells or filaments (Pattanaik et al. 2007). Screening strategies are species specific. For

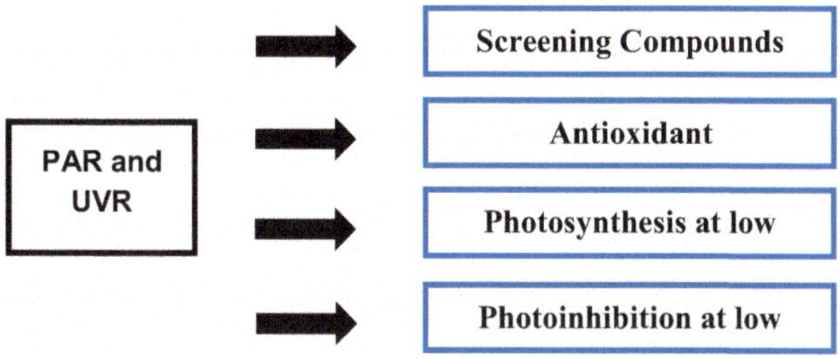

Fig. 1.6 Survival strategies of polar cyanobacteria against PAR and UVR

example, outer layers of *Scytonema* mats are characterized by actively moving filaments, whereas the upper layer cells in oscillatoracean mats contain low pigment concentration thereby protecting internal cells (Quesada et al. 1999) (Fig. 1.6).

High level of UVR exposure to cyanobacteria may cause photo-inhibition, photochemical damage, and cellular component degradation; as a consequence of which, generation of ROS takes place. Primary target of ROS is DNA leading to harmful mutations (Quesada and Vincent 1997; Xue et al. 2005). Consequences of high-dose exposure of UVR include slow growth rate and damage to photosynthetic apparatus, nitrogenase complex, and cellular membranes. As an adaptive mechanism, cyanobacteria possess four lines of defense which also favor cyanobacteria acclimatization under high light (Vincent 2000).

- Selection of habitat
- Production of screening compounds
- Production of antioxidants
- Repair mechanisms

Sunlight is the main source of energy for phototrophic clades including cyanobacteria. Variations in the availability of PAR together with its spectral composition and intensity impose challenges to cyanobacterial communities (Falkowski and Raven 2007; Markager and Vincent 2000). In water, blue and blue-green light prevails more, while red light diminishes in very short time. Its intensity can vary from minute intensities in cryptoendolithic communities of 0.1 μmol m^{-2} s^{-1} up to full sunlight of 1500 μmol m^{-2} s^{-1} (Nienow et al. 1988). To survive in cold habitats, cyanobacteria are required to acclimatize to a broad range of PAR values (Vincent 2000).

1.6.22 Photosynthesis and Photoinhibition at Low Temperature

Rate of photosynthetic reaction depends mainly on factors like temperature, sunlight, water, and CO_2. Photosynthesis occurs at temperature ranges of −7 to +75 °C (Castenholz 1969). It is observed that rate of photosynthesis first increases to an optimum temperature and then declines (Davison 1991). Generally, at cold temperature, rate of individual steps of electron transport, carbon dioxide diffusion, enzymatic activity, and turnover of membrane protein is slow. This limits photosynthesis in phototrophic clades when exposed to higher irradiance resulting in photoinhibition (called as photo-inhibition of low temperature) (Powels and Berry 1983; Smillie and Hetherington 1988). In a study, it was shown that *Synechocystis* sp. are resistant to photoinhibition due to lipid desaturation, through light regulation of desaturase genes expression, resulting in higher photosynthetic productivity (Sakurai 2003). Therefore, cryo-cyanobacteria attain higher photosynthetic rates even at low temperatures considering their upper temperature limits of photosynthesis are still lower than algae of warmer region (Kuebler et al. 1991).

1.6.23 Screening Compounds

Cyanobacterial consortia often exhibit colored colonies. Cyanobacteria produce pigments that absorb in UV and blue regions of electromagnetic spectrum (Quesada and Vincent 1997). The major screening compound mycosporine-like amino acid (MAA) accumulates in cytoplasm of cyanobacteria and shows absorbance maxima at 310–360 nm (Oren and Seckbach 2001; Pattanaik et al. 2007). Another screening pigment scytonemin (yellowish-brown) shows maximal absorbance at 370–390 nm (Oren and Seckbach 2001; Pattanaik et al. 2007). Scytonemin is a stable pigment and provides prolonged protection. Cyanobacterial mats exhibit black and dark coloration due to the presence of scytonemin in extracellular sheath. *Gloeocapsa* species of cyanobacteria lacks scytonemin and exhibits brown coloration due to the presence of a different screening pigment in the envelope, the gloeocapsin. Colored *Gloeocapsa* sp. were reported by Nováček (1930, 1934), and the observation of Jaag (1945) may imply possible effects of environmental conditions, namely, pH, on sheath pigment color in *Gloeocapsa*. Presence of gloeocapsin is usually connected with absence of scytonemin in some cyanobacterial species, thus suggesting independent development of scytonemin and gloeocapsin-based protective strategies (Storme 2015). It has been reported that even common cellular compounds have screening property against excess UVR. There is an observation by Araoz and Häder (1999) that synthesis of phycoerytherin increases with increased exposure to UVR due to faster rate of repair of photosynthetic apparatus, whereas phycobiliprotein (PBP) do show absorbance in UVR region but still can be easily damaged by it (Häder 2001). In cyanobacterial mats, mucilaginous sheath performs a dual function to provide resistance against UVR, firstly by providing matrix for extracellular screening compound and secondly EPS showing UVR absorption (Ehling-Schulz et al. 1997). Dark red gloeocapsin in the envelopes of *Gloeocapsa* sp. is responsible for the brown coloration of crusts on rocks (Sheath et al. 1996); however, further data on its properties are lacking.

1.6.24 Antioxidants

Cryo-cyanobacteria have adapted a system of enzymatic and non-enzymatic antioxidant to scavenge ROS. These ROS are formed in cyanobacteria when exposed to high PAR and UVR (Pattanaik et al. 2007). In cryo-cyanobacteria, non-enzymatic oxidants include carotinoids, tocopherol, ascorbic acid (act rapidly to URV induced damage (Ehling-Schulz et al. 1997), and reduced glutathione though not considered as very effective ROS quenchers (Wolfe-Simon et al. 2005). SOD, catalase, and glutathione reductase as well as enzymes of ascorbate-glutathione cycle however account for enzymatic antioxidant systems of cyanobacteria growing in the polar regions (Pattanaik et al. 2007).

1.6.25 Survival Strategies: Insight from Metagenomics

A study tested whether increasing age and associated stress challenges drive adaptive changes in diversity of microbial community and their function. Pleistocene permafrost chronosequence from 19,000 to 33,000 (kyr) was performed with deep metagenomics and 16S rRNA gene sequencing. It was concluded that age affected microbial community composition and also reduced its diversity. Consistent shifts were observed with long-term strategies in cryo-ecosystem; this includes increased reliance of community on scavenging detritus biomass, horizontal gene transfer, chemotaxis, dormancy, environmental sensing, and stress response.

1.6.26 Subzero Temperature Effect

Under subzero temperature, nucleic acid and its secondary structure as well as structural flexibility of proteins are stabilized that inhibits replication, transcription, and translation. At extremely low temperature, concentration of ROS increases and damages DNA, RNA, proteins, and lipids. Stressors on the microbial community accumulate over time, demanding a counteractive adaptation for long-term survival in extremely harsh polar conditions. However, we know little about the ecological strategies utilized by microbial communities in response to the challenges presented by spending millennia in permafrost. The study showed older permafrost samples were enriched with pathways involved in synthesis of cell envelope component, amino acid, peptide, carbohydrate metabolism, environmental sensing, membrane transport, and degradation of recalcitrant biomass as compared to younger permafrost samples.

Along the permafrost chronosequence, the ability to respond to harsh environmental conditions increased, for example, ancient cryo-environment includes nutrients and other resource limitation, low temperature, and high osmolarity (Raivio 2014). Older permafrost samples are also enriched with diverse sensor system genes involved in temperature sensing, protein misfolding, H^+ regulation, salt stress, osmolarity, oxygen limitation, and cell membrane stress. Expression of genes responsible for nutrient and resource sensing ion, trace metal, nitrogen, acetoacetate, malate, and glucose amplified in chronosequence. Chemotaxis pathway also showed significant increase in ancient permafrost.

ATP-binding cassette transporter pathway genes, importer genes responsible for transfer of amino acid, peptides, osmoprotectant, and stress compounds together with exporter genes involved in transfer of LPS layer and cell wall component, increased in abundance along with chronosequence. The study mapped high expression of genes accounting from four of the six classes of secretion machinery and two membrane-spanning systems in older permafrost. These abundantly expressed secretion systems include type IV secretion system which encodes conjugation machinery, type I secretion system that secretes product in extracellular milieu, and type III secretion system involved in interaction with eukaryotic domain of life. Abundance of pathways involving biosynthesis of three cell envelope components—fatty acids, lipopolysaccharides (LPS), and peptidoglycan—increased with age. Fatty acid chains in phospho- and glycolipids form the membrane and are altered to increase membrane fluidity in response to cold (Phadtare 2004).

1.7 Impact of Rise in Global Temperature on Polar Cyanobacteria

Ancient literature and ongoing scientific studies suggest how important Arctic and Antarctic cyanobacterial communities are, for a number of ecosystem processes. The rise in global temperature has resulted in global climate change, which is accompanied with melting of polar glaciers. This causes harmful alteration in community dynamics in polar habitats (Wynn-Williams 1996; Strauss et al. 2012; Chan et al. 2013). One can observe the rise in cyanobacterial blooms in lakes and ponds having warmer water and enriched with nurients due to discharge of untreated industrial and community waste. It is crucial to understand cyanobacterial functionality along with its interaction networks with other microbial species under changing global climate. This can reveal the acclimatization strategies and also tell us in what way global changes are affecting the biogeochemical processes derived by microbes. Studies on soil systems suggest that moderate losses in microbial diversity definitely affect the functionality of the whole ecosystem (Philippot et al. 2013; Singh 2014).

1.7.1 Nitrogen Cycling

Climate change does have impact on microbial communities, for example, it affects nitrogen-fixing microorganisms. Yergeau (2008) did an environmental microarray analysis of Antarctic soil communities, where they observed a rise in cyanobacterial population with increasing latitude. This study revealed a strong connection between community structure and functional gene distribution in Antarctic soils. Results of Principal Coordinate Analysis (PCA) exhibited a relationship between cyanobacteria and genes for nitric oxide reductase (*nor*B). Surprisingly, the study showed a strong relationship between cyanobacteria and genes implicated in nitrogen fixation (*nif*H), ammonium oxidizing bacteria (*amo*A), and other genes associated with nitrogen cycling (nar, nos, nas). However, in a similar study undertaken on Arctic permafrost, cyanobacteria-mediated nitrogen fixation was suggested through the identification of a single type of nitrogenase (*nif*H) (Yergeau et al. 2010). Cyanobacteria also appear to drive nitrogen fixation in moist soil communities (Niederberger et al. 2012). For instance, it has been shown that N_2-fixing activity of cyanobacteria in Arctic regions is primarily governed by moisture gradients associated with topography that determines nutrient availability. The energy demand leads to high photosynthetic and respiration rates (Joset et al. 1996). Since respiration rates are low, photosynthesis seems to be the primary energy source (Apte 2001). Due to the allocation of energy to ion regulation and inhibition of nitrogenase activity, nitrogen fixation may be depressed in N_2-fixing species at high salt concentrations (Apte 2001) as observed in nonpolar *Anabaena* sp. (Apte et al. 1987). On the contrary, optimum nitrogen fixation in marine tropic and subtropic *Trichodesmium* sp. occurs in a salinity range from 3.3% to 3.7%, indicating adaptation of this cyanobacterium to the marine environment.

1.7.2 Carbon Cycling

Cyanobacteria, the globally dominant photoautotrophic lineage, are considered to be particularly important in Antarctic carbon cycling. In a study, using metagenomics, Pearce et al. (2012) reported cyanobacteria being underrepresented in southern maritime Antarctic soil, with only 3.4% of total sequences belonging to this phylum, although 1% of the genes identified were involved in CO_2 fixation. This result is surprising and may be a localized phenomenon, rather than representative of all southern maritime Antarctic soils. In any event, cyanobacteria have been shown to use a number of methods in order to increase photosynthesis (Rae et al. 2013). For example, they are able to produce carboxysomes, which together with CO_2-concentrating mechanisms (CCM) augment chemical conditions in the locality of the primary CO_2-fixing enzyme (RuBisCO), resulting in increased photosynthesis (Rae et al. 2013). In Antarctic habitats, such mechanisms are essential for nutrient input. Interestingly, Functional Forms II and III of RubisCO were assigned to other groups including *Archaea*, *Actinobacteria*, and *Proteobacteria*, the known chemolithotrophs. Cyanobacterium *Nostoc commune* is a prominent primary producer in continental Antarctica and has been used as a model for elucidating the ecological constraints on total carbon fixation particularly for ice-free areas (Novis 2007).

1.8 Conclusion

The knowledge of these adaptation or acclimation processes is believed to determine several important survival strategies. The scientific data collected from numerous studies have the potential to improve current models which can impact global climate change on polar ecosystems and determine the future distribution of invasive species, especially in the Arctic. Cyanobacteria do not seem to be particularly adaptable to low temperatures because they cannot maintain a rapid growth rate in a cold environment. On the other hand, studies in the modern cryosphere have shown that they have a wide range of adaptive mechanisms that can make them grow under irradiance such as freezing, ice cover, and regular exposure to ultraviolet rays and bright PAR. These mechanisms include concentrating pigments, screening pigments, ROS quenching compound (such as carotenoids), and alteration in membrane fluidity at low temperatures and production of cold stable proteins. Despite the cold, polar and alpine cyanobacteria can maintain a slow but steady growth. This cold tolerance strategy has been very successful in many cold ecosystems. The fitness test of extreme microorganisms should also help determine the possible habitability to extreme environments on earth and other regions. Detailed biochemical and molecular research may help discover new specific biotechnologically important compounds. These compounds can be used as food supplements (polyunsaturated fatty acids), cosmetics (colors), cryoprotectants (antifreeze), drugs (antifreeze and antibiotics), etc. Low-temperature biotechnology applications using polar cyanobacteria will benefit from the continuous light in the polar summer and the tolerance range of selected species. Cyanobacteria could play an important role in regenerative system for waste disposal in future polar stations or small settlements in distant areas as is proposed for space stations (Gòdia et al. 2002, 2004).

References

Agawin NS, Agusti S (1997) Abundance, frequency of dividing cells and growth rates of Synechococcus sp. (cyanobacteria) in the stratified Northwest Mediterranean Sea. J Plankton Res 19:1599–1615

Allen JF (2003) State transitions—a question of balance. Science 299:1530–1532

Alpert P (2000) The discovery, scope, and puzzle of desiccation tolerance in plants. Plant Ecol 151 (1):5–17

Alpert P (2005) The limits and frontiers of desiccation-tolerant life. Integr Comp Biol 45 (5):685–695

Apte SK (2001) Copying with salinity/water stress: cyanobacteria show the way. Proc Indian Natl Sci Acad 67(5):285–310

Apte SK, Reddy BR, Thomas J (1987) Relationship between sodium influx and salt tolerance of nitrogen-fixing cyanobacteria. American Society of Civil Engineers, New York, NY, pp 113–137

Araoz R, Häder DP (1999) Phycoerythrin synthesis is induces by solar UV-B in the cyanobacterium *Nostoc*. Plant Physiol Biochem 37:223–229

Belnap J, Gardner JS (1993) Soil microstructure in soils of the Colorado plateau: the role of the cyanobacterium Microcoleus vaginatus. West N Am Nat 53:40–47

Berger F, Morellet N, Menu F, Potier C (1997) Cold shock and cold acclimation proteins in the psychrotrophic bacterium Arthrobacter globiformis SI55. J Bacteriol 178:2999–3007
Blumwald E, Mehlhorn RJ, Packer L (1983) Ionic osmoregulation during salt adaptation of the cyanobacterium *Synechococcus* 6311. Plant Physiol 73(2):377–380
Broady PA (1981) Ecological and taxonomic observations on subaerial epilithic algae from Princess Elisabeth and MacRobertson land, Antarctica. Br Phycol J 16:257–266
Broady PA (1986) Ecology and taxonomy of the Vestfold Hills. In: Pickard J (ed) Antarctic oasis: terrestrial environments and history of the Vestfold Hills. Academic, Sydney, pp 165–202
Broady PA (1989) Survey of algae and other terrestrial biota from Edward VII peninsula, Marie Byrd Land. Antarct Sci 1:215–224
Budel B, Bendix J, Bicker FR, Allan Green T (2008) Dewfall as a water source frequently activates the endolithic cyanobacterial communities in the granites of Taylor Valley, Antarctica. J Phycol 44:1415–1424
Budel B, Colesie C (2014) Biological soil crusts. In: Antarctic terrestrial microbiology. Springer, Berlin, pp 131–161
Caiola MG, Billi D, Friedmann EI (1996) Effect of desiccation on envelopes of the cyanobacterium *Chroococcidiopsis* sp. (Chroococcales). Eur J Phycol 31(1):97–105
Callieri C, Cronberg G, Stockner J (2012) Freshwater picocyanobacteria: single cells, microcolonies and colonial forms. In: Whitton BA (ed) Ecology of cyanobacteria II. Springer, Dordrecht, pp 229–269
Cameron KA, Hodson AJ, Osborn AM (2012) Structure and diversity of bacterial, eukaryotic and archaeal communities in glacial cryoconite holes from the Arctic and the Antarctic. FEMS Microbiol Ecol 82:254–267
Cameron RE (1962) Species of *Nostoc* Vaucher occurring in the Sonoran Desert in Arizona. Trans Am Microsc Soc 81(4):379–384
Cameron RE (1972) Microbial and ecological investigations in Victoria Dry Valley, southern Victoria land, Antarctica. In: Llano GA (ed) Antarctic terrestrial biology. American Geophysical Union, Washington, DC, pp 195–260
Castenholz RW (1969) Thermophilic blue-green algae and the thermal environment. Bacteriol Rev 33:476–504
Cavacini P (2001) Soil algae from northern Victoria Land (Antarctica). Polar Biosci 14:45–60
Chan Y, Van Nostrand JD, Zhou J, Pointing SB, Farrell RL (2013) Functional ecology of an Antarctic Dry Valley. Proc Natl Acad Sci 110(22):8990–8995
Chao HM, Davies PL, Carpenter JF (1996) Effect of antifreeze proteins on red blood cell survival during cryopreservation. J Exp Biol 199:2071–2076
Close TJ (1997) Dehydrins: a commonalty in the response of plants to dehydration and low temperature. Physiol Plant 100(2):291–296
Close TJ, Lammers PJ (1993) An osmotic stress protein of cyanobacteria is immunologically related to lantdehydrins. Plant Physiol 101(3):773–779
Cockell CS, Herrera A (2007) Why are some microorganisms boring? Trends Microbiol 16:101–106
Cockell CS, Stokes MD (2004) Widespread colonization by polar hypoliths. Nature 431:414
Cockell CS, Stokes MD, Korsmeyer KE (2000) Overwintering strategies of Antarctic organisms. Environ Rev 8:1–19
Curtis T (2006) Microbial ecologists: it's time to go large. Nat Rev Microbiol 4:488
D'Amico S, Collins T, Marx JC, Feller G, Gerday C (2006) Psychrophilic microorganisms: challenges for life. EMBO Rep 7:385–389
Davison IR (1991) Environmental effects on algal photosynthesis: temperature. J Phycol 27:2–8
Day JG et al (2005) Cryopreservation and conservation of microalgae: the development of a pan-European scientific and biotechnological resource (the COBRA project). Cryo-Letters 26 (4):231–238
Day JG, Brand JJ (2005) Cryopreservation methods for maintaining microalgal cultures. In: Algal culturing techniques, pp 165–187

De los Ríos A, Ascaso C, Wierzchos J, Fernández-Valiente E, Quesada A (2004) Microstructural characterization of cyanobacterial mats from the McMurdo ice shelf, Antarctica. Appl Environ Microbiol 70:569–558
De los Rios A, Cary C, Cowan D (2014) The spatial structures of hypolithic communities in the dry valleys of East Antarctica. Polar Biol 37(12):1823–1833
Deming JW, Young JN (2017) The role of exopolysaccharides in microbial adaptation to cold habitats. In: Margesin R (ed) Psychrophiles: from biodiversity to biotechnology. Springer, Cham, pp 259–284
Devos N, Ingouff M, Loppes R, Matagne RF (1998) RUBISCO adaptation to low temperatures: a comparative study in psychrophilic and mesophilic unicellular algae. J Phycol 34:655–660
Ehling-Schulz M, Bilger W, Scherer S (1997) UV-B-induced synthesis of photoprotective pigments and extracellular polysaccharides in the terrestrial cyanobacterium Nostoc commune. J Bacteriol 179(6):1940–1945. https://doi.org/10.1128/jb.179.6.1940-1945.1997
Elster J (1999) Algal versatility in various extreme environments. In: Seckbach J (ed) Enigmatic microorganisms and life in extreme environments. Springer, Dordrecht
Elster J, Benson EE (2004) Life in the polar terrestrial environment with a focus on algae and cyanobacteria. In: Fuller BJ, Lane N, Benson EE (eds) Life in the frozen state. CRC Press, Boca Raton, pp 111–150
Elster J, Kvíderová J, Hájek T, Láska K, Šimek M (2012) Impact of warming on Nostoc colonies (Cyanobacteria) in a wet hummock meadow. Polish Polar Res 33(4):395–420
Elster J, Lukavský J, Harding K, Benson EE, Day JG (2008) Development of the encapsulation/dehydration protocol to cryopreserve polarmicroalgae held at the Czech Republic Academy of Science Institute of Botany. Cryo-Letters 29(1):27–28
Epstein E (1985) Salt-tolerant corps—origin, development and prospects. Plant and Soil 89 (1–3):187–198
Falkowski P, Raven JA (2007) Aquatic photosynthesis, 2nd edn. Princeton University Press, Princeton, pp 655–663
Feller G, Gerday C (2003) Psychrophilic enzymes: hot topics in cold adaptation. Nat Rev Microbiol 1(3):200–208
Fermani P, Mataloni G, Van de Vijver B (2007) Soil microalgal communities on an active volcano (Deception Island, South Shetlands). Polar Biol 30:1381–1393
Fernandez-Valiente E, Quesada A, Howard-Williams C, Hawes I (2001) N2-fixation in cyanobacterial mats from ponds on the McMurdo ice shelf, Antarctica. Microb Ecol 42:338–349
Friedmann EI, McKay CP, Nienow JA (1987) The cryptoendolithic microbial environment in the Ross Desert of Antarctica: satellite-transmitted continuous nanoclimate data, 1984 to 1986. Polar Biol 7:273–287
Gilichinsky D et al (2008) Bacteria in permafrost. In: Margesin R et al (eds) Psychrophiles: from biodiversity to biotechnology. Springer, Berlin/Heidelberg, pp 83–102
Gòdia F et al (2002) MELISSA: a loop of interconnected bioreactors to develop life support in space. J Biotechnol 99(3):319–330
Gòdia F et al (2004) The MELISSA pilot plant facility as an integration test-bed for advanced life support systems. Adv Space Res 34(7):1483–1493
Gounot AM, Russell NJ (1999) Physiology of cold-adapted microorganisms. In: Margesin R, Schinner F (eds) Cold-adapted organisms. Springer, Berlin, pp 33–55
Häder D (2001) Adaptation to UV stress in algae. https://doi.org/10.1007/978-3-642-59491-5_7
Helm RF et al (2000) Structural characterization of the released polysaccharide of desiccation-tolerant *Nostoc commune* DRH-1. J Bacteriol 182(4):974–982
Hoiczyk E (2000) Gliding motility in cyanobacteria: observations and possible explanations. Arch Microbiol 174(1–2):11–17
Hughes KA, Lawley B (2003) A novel Antarctic microbial endolithic community within gypsum crusts. Environ Microbiol 5:555–565

Jaag O (1945) Untersuchungen über die Vegetation und Biologie der Algen des nackten Gesteins in den Alpen, im Jura und im schweizerischen Mittelland. Kommisionverlag Buchdruckerei Büchler & Co., Bern

Joset F, Jeanjean R, Hagemann M (1996) Dynamics of the response of cyanobacteria to salt stress: deciphering the molecular events. Physiol Plant 96:738–744

Karsten U, Wiencke C, Kirst GO (1992) Dimethylsulphoniopropionate (DMSP) accumulation in green macroalgae from polar to temperate regions: interactive effects of light versus salinity and light versus temperature. Polar Biol 12:603–607

Kirst GO et al (1991) Dimethylsulfoniopropionate (DMSP) in ice algae and its possible biological role. Mar Chem 35(1–4):381–388

Klähn S, Hagemann M (2011) Compatible solute biosynthesis in cyanobacteria. Environ Microbiol 13(3):551–562

Komárek J, Elster J, Komárek O (2008) Diversity of the cyanobacterial microflora of the northern part of James Ross island, NW Weddell Sea, Antarctica. Polar Biol 31:853–865

Kuebler JE, Davison IR, Yarish C (1991) Photosynthetic temperature adaptation in the red algae *Lomentaria baileyana* and *Lomentaria orchadensis*. Br Phycol J 26:9–19

Kvíderová J, Elster J, Šimek M (2011) *In situ* response of *Nostoc commune* s.l. colonies to desiccation in Central Svalbard, Norwegian high Arctic. Fottea 11(1):87–97

Latifi A, Ruiz M, Zhang CC (2009) Oxidative stress in cyanobacteria. FEMS Microbiol Rev 33 (2):258–278

Läuchli A, Epstein E (1990) Plant responses to saline and sodic conditions. In: Tanji KK (ed) Agricultural salinity assessment and management. United Book Prints, Jodhpur

Lee RE Jr, Warren GJ, Gusta LV (1995) Biological ice nucleation and its application. APS Press, St Paul

Lipman CB (1941) The successful revival of *Nostoc commune* from a herbarium specimen eighty-seven years old. Bull Torrey Bot Club 68(9):664–666

Loppes R, Devos N, Willem S, Barthélemy P, Matagne RF (1996) Effect of temperature on two enzymes from a psychrophilic Chloromonas (Chlorophyta). J Phycol 32:276–278

Los D, Mironov K (2013) Membrane properties and cold stress responses in cyanobacteria and plants. Wiley, Hoboken, p 224

Lubzens E et al (2010) Akinetes: dormant cells of cyanobacteria. In: Lubzens E, Cerda J, Clark M (eds) Dormancy and resistance in harsh environments. Springer, Berlin/Heidelberg, pp 5–27

Lukešová A, Hrouzek P, Harding K, Benson EE, Day JG (2008) Deployment of the encapsulation/dehydration protocol to cryopreserve diverse microalgae held at the Institute of Soil Biology, Academy of Sciences of the Czech Republic. Cryo-Letters 29(1):21–26

Makhalanyane TP, Pointing SB, Cowan DA (2014) Lithobionts: cryptic and refuge niches. In: Cowan DA (ed) Antarctic terrestrial microbiology: physical and biological properties of Antarctic soils. Springer, Berlin, pp 163–179

Potts M (1999) Mechanisms of desiccation tolerance in cyanobacteria. Eur J Phycol 34(4):319–328. https://doi.org/10.1080/09670269910001736382

Markager S, Vincent WF (2000) Spectral light attenuation and the absorption of UV and blue light in natural waters. Limnol Oceanogr 45(3):642–650

Mataloni G, Tell G, Wynn-Williams DD (2000) Structure and diversity of soil algal communities from Cierva point (Antarctic Peninsula). Polar Biol 23:205–211

McBride MJ (2001) Bacterial gliding motility: multiple mechanisms for cell movement over surfaces. Annu Rev Microbiol 55:49–75

Mueller D, Pollard WH (2004) Gradient analysis of cryoconite ecosystems from two polar glaciers. Polar Biol 27:66–74

Mueller DR, Vincent WF, Pollard WH, Fritsen CH (2001) Glacial cryoconite ecosystems: a bipolar comparison of algal communities and habitats. Nova Hedwig Beih 123:173–198

Murata N, Wada H (1995) Acid lipid desaturases and their importance in the tolerance and acclimatization to cold of cyanobacteria. Biochem J 308:1–8

Namsaraev Z, Mano M-J, Fernandez R, Wilmotte A (2010) Biogeography of terrestrial cyanobacteria from Antarctic ice-free areas. Ann Glaciol 51:171–177
Niederberger TD, Sohm JA, Tirindelli J, Gunderson T, Capone DG, Carpenter EJ, Cary SC (2012) Diverse and highly active diazotrophic assemblages inhabit ephemerally wetted soils of the Antarctic dry valleys. FEMS Microbiol Ecol 82:376–390
Nienow JA, Friedmann EI (1993) Terrestrial lithophytic (rock) communities. In: Friedmann EI (ed) Antarctic microbiology. Wiley Liss, New York, pp 343, 634 pp–412
Nienow JA, McKay CP, Friedmann EI (1988) Cryptoendolithic microbial environment in the Ross Desert of Antarctica: light in the photosynthetically active region. Microb Ecol 16:271–289
Nishida I, Murata N (1996) Chilling sensitivity in plants and cyanobacteria. Annu Rev Plant Biol 47:541–568
Nováček F (1930) De statibus Gloeocapsae sanguineae Ag.notula. In: Sborník Klubu přírodovědců v Brně, vol 8. Krajské nakladatelství, Brno
Nováček F (1934) Epilithické sinice Serpentinů Mohelských [Epilithic cyanobacteria of serpentines in Mohelensko region]. In: Archiv des Verbandes für
Novis PM (2007) Annual carbon fixation in terrestrial populations of Nostoc commune (cyanobacteria) from an Antarctic dry valley is driven by temperature regime. Glob Chang Biol 13:1224–1237
Oliver RL (1994) Floating and sinking in gas-vacuolate cyanobacteria. J Phycol 30(2):161–173
Omelon CR (2008) Endolithic microbial communities in polar desert habitats. Geomicrobiol J 25:404–414
Omelon CR, Pollard WH, Ferris FG (2007) Inorganic species distribution and microbial diversity within high arctic cryptoendolithic habitats. Microb Ecol 54:740–752
Oren A (2007) Diversity of organicosmotic compounds andosmotic adaptation in cyanobacteria and algae. In: Seckbach J (ed) Algae and cyanobacteria in extreme environments. Springer, Dordrecht, pp 639–655
Oren A, Seckbach J (2001) Oxygenic photosynthesis in extreme environments. In: Elster J et al (eds) Algae and extreme environments. J. Cramerin der Gebrüder Borntraeger Verlags buchhandlung, Berlin, pp 13–31
Packer L, Spath S, Martin JB, Roby C, Bligny R (1987) 23Na and 31P NMR studies of the effects of salt stress on the freshwater cyanobacterium Synechococcus 6311. Arch Biochem Biophys 256 (1):354–361
Palmer RJ, Friedmann EI (1990) Water relations and photosynthesis in the cryptoendolithic microbial habitat of hot and cold deserts. Microb Ecol 19:111–118
Pattanaik B, Schumann R, Karsten U (2007) Effects of ultraviolet radiation on cyanobacteria and their protective mechanisms. In: Seckbach J (ed) Algae and cyanobacteria in extreme environments. Springer, Dordrecht, pp 29–45
Pearce DA, Newsham KK, Thorne MAS, Calvo-Bado L, Krsek M, Laskaris P, Hodson A, Wellington EM (2012) Metagenomic analysis of a southern maritime Antarctic soil. Front Microbiol 3:403. https://doi.org/10.3389/fmicb.2012.00403
Pereira S (2009) Complexity of cyanobacterial exopolysaccharides: composition, structures, inducing factors and putative genes involved in their biosynthesis and assembly. FEMS Microbiol Rev 33(5):917–941
Phadtare S (2004) Recent developments in bacterial cold-shock response. Curr Issues Mol Biol 6 (2):125–136. PMID: 15119823
Philippot L et al (2013) Loss in microbial diversity affects nitrogen cycling in soil. ISME J 7:1609–1619
Pointing SB, Chan Y, Lacap DC, Lau MC, Jurgens JA, Farrell RL (2009) Highly specialized microbial diversity in hyper-arid polar desert. Proc Natl Acad Sci U S A 106:19964–19969
Potts M (2000) Nostoc. In: Whitton BA, Potts M (eds) The ecology of cyanobacteria. Kluwer Academic Publishers, Dordrecht, pp 465–504
Potts M, Friedmann EI (1981) Effects of water stress on cryptoendolithic cyanobacteria from hot desert rocks. Arch Microbiol 130(4):267–271

Powels SB, Berry JA, Björkman O (1983) Interaction between light and chilling temperatures on the inhibition of photosynthesis in chilling-sensitive plants. Plant Cell Environ 6:117–123
Primalov PL (1979) Stability of proteins. Adv Protein Chem 33:167–241
Qiu B, Gao K (2001) Photosynthetic characteristics of the terrestrial blue-green alga, *Nostoc flagelliforme*. Eur J Phycol 36(02):147–156
Quesada A, Vincent WF (1997) Strategies of adaptation by Antarctic cyanobacteria to ultraviolet radiation. Eur J Phycol 32:335–342
Quesada A, Vincent WF, Lean DRS (1999) Community and pigment structure of Arctic cyanobacterial assemblages: the occurrence and distribution of UV-absorbing compounds. FEMS Microbiol Ecol 28:315–323
Raivio TL (2014) Everything old is new again: an update on current research on the Cpx envelope stress response. Biochim Biophys Acta 1843:1529–1541
Rae BD, Long BM, Badger MR, Price GD (2013) Functions, compositions, and evolution of the two types of carboxysomes: polyhedral microcompartments that facilitate CO2 fixation in cyanobacteria and some proteobacteria. Microbiol Mol Biol Rev 77:357–379
Ramond JB, Makhalanyane TP, Tuffin MI, Cowan DA (2015) Normalization of environmental metagenomic DNA enhances the discovery of under-represented microbial community members. Lett Appl Microbiol 60(4):359–366. https://doi.org/10.1111/lam.12380
Raymond JA, Fritsen CH (2000) Ice active substances associated with Antarctic freshwater and terrestrial photosynthetic organisms. Antarct Sci 12(4):418–424
Raymond-Bouchard I, Whyte LG (2017) From transcriptomes to metatranscriptomes: cold adaptation and active metabolisms of psychrophiles from cold environments. In: Margesin R (ed) Psychrophiles: from biodiversity to biotechnology. Springer, Cham, pp 437–457
Reed RH, Richardson DL, Warr SRC, Stewart WDP (1984) Carbohydrate accumulation and osmotic stress in cyanobacteria. J Gen Microbiol 130(1):1–4
Řezanka T, Nedbalová L, Elster J, Cajthaml T, Sigler K (2009) Very-long-chain iso and anteiso branched fatty acids in N-acylphosphatidylethanolamines. Phytochemistry 70(5):655–663
Rocak S, Linder P (2004) DEAD-box proteins: the driving forces behind RNA metabolism. Nat Rev Mol Cell Biol 5:232–241
Ryan PG, Watkins BP, Smith RIL, Dastych H, Eicker R, Foissner W, Heatwole H, Miller WR (1989) Biological survey of Robertskollen, western Dronning Maud land: area description and preliminary species list. S Afr J Antarct Res 19:10–20
Šabacká M, Elster J (2006) Response of cyanobacteria and algae from Antarctic wetland habitats to freezing and desiccation stress. Polar Biol 30(1):31–37
Sakurai I (2003) Requirement of phosphatidylglycerol for maintenance of photosynthetic machinery. Plant Physiol 133:1376–1384
Schulze WX (2005) Protein analysis in dissolved organic matter: what proteins from organic debris, soil leachate and surface water can tell us–a perspective. Biogeosciences 2:75–86. https://doi.org/10.5194/bg-2-75-2005
Seckbach J, Oren A (2007) Oxygenic photosynthetic microorganisms in extreme environments: possibilities and limitations. In: Seckbach J (ed) Algae and cyanobacteria in extreme environments. Springer, Dordrecht, pp 3–25
Sheath R, Vis M, Hambrook J, Cole K (1996) Tundra stream macroalgae of North America: composition, distribution and physiological adaptations. Hydrobiologia 336(1-3):67–82
Shirkey B et al (2003) Genomic DNA of Nostoc commune (Cyanobacteria) becomes covalently modified during long-term (decades) desiccation but is protected from oxidative damage and degradation. Nucleic Acids Res 31:2995–3005
Shivaji S, Kiran MD, Chintalapati S (2007) Perception and transduction of low temperature in bacteria. In: Gerday C, Glansdorff N (eds) Physiology and Biochemistry of Extremophiles. ASM Press, Washington, DC, pp 194–207
Singh JS (2014) Cyanobacteria: a vital bio-agent in eco-restoration of degraded lands and sustainable agriculture. Climate Change Environ Sustain 2:133–137

Singh SC, Sinha RP, Hader DP (2002) Role of lipids and fatty acids in stress tolerance in cyanobacteria. Acta Protozool 41:297–308. https://doi.org/10.1016/S0378-1097(02)01030-3
Smillie RM, Hetherington SE (1988) Stress tolerance and stress-induced injury in corp plants measured by chlorophyll fluorescence in vivo: chilling, freezing, ice cover, heat and high light. Plant Physiol 72:1043–1050
Storme JY (2015) Raman characterization of the UV protective pigment gloeocapsin and its role in the survival of cyanobacteria. Astrobiology 15(10):843–857
Strauss J, Schirrmeister L, Wetterich S, Borchers A, Davydov SP (2012) Grain-size properties and organic-carbon stock of Yedoma Ice Complex permafrost from the Kolyma lowland, northeastern Siberia. Glob Biogeochem Cycles 26(3). https://doi.org/10.1029/2011GB004104. Article GB3003
Tamaru Y, Takani Y, Yoshida T, Sakamoto T (2005) Crucial role of extracellular polysaccharides in desiccation and freezing tolerance in the terrestrial cyanobacterium *Nostoc commune*. Appl Environ Microbiol 71(11):7327–7333
Tang EPY, Vincent WF (1999) Studies of thermal adaptation by high-latitude cyanobacteria. New Phytol 142:315–323
Tanghe A, Van Dijck P, Thevelein JM (2003) Determinants of freeze tolerance in microorganisms, physiological importance and biotechnological applications. Adv Appl Microbiol 53:129–176
Tashyreva D, Elster J (2012) Production of dormant stages and stress resistance of polar cyanobacteria. In: Hanslmeier A, Kempe S, Seckbach J (eds) Life on earth and other planetary bodies. Springer, Dordrecht, pp 367–386
Tashyreva D, Elster J (2016) Annual cycles of two cyanobacterial mat communities in hydro-terrestrial habitats of the high Arctic. Microb Ecol 71(4):887–900
Taton A, Grubisic S, Brambilla E, De Wit R, Wilmotte A (2003) Cyanobacterial diversity in natural and artificial microbial mats of Lake Fryxell (McMurdo Dry Valleys, Antarctica): a morphological and molecular approach. Appl Environ Microbiol 69(9):5157–5169
Van den Hoek C, Mann DG, Jahns HM (1995) Algae: an introduction to phycology. Cambridge University Press, Cambridge
Varin T, Lovejoy C, Jungblut AD, Vincent WF, Corbeil J (2012) Metagenomic analysis of stress genes in microbial mat communities from Antarctica and the high Arctic. Appl Environ Microbiol 78:549–559
Vincent WF (2000) Cyanobacterial dominance in the polar regions. In: Whitton BA, Potts M (eds) The ecology of cyanobacteria. Kluwer Academic Publishers, Dordrecht, pp 321–340
Vincent WF (2007) Cold tolerance of cyanobacteria and life in the cryosphere. In: Seckbach J (ed) Algae and cyanobacteria in extreme environments. Springer, Dordrecht, pp 289–304
Vincent WF, Castenholz RW, Downes MT, Howard Williams C (1993) Antarctic cyanobacteria: light, nutrients, and photosynthesis in the microbial mat environment. J Phycol 29(6):745–755
Vincent WF, Mueller DR, Bonilla S (2004) Ecosystems on ice: the microbial ecology of Markham ice shelf in the high Arctic. Cryobiology 48(2):103–112
Vishnivetskaya TA (2009) Viable cyanobacteria and green algae from permafrost darkness. In: Margesin R (ed) Permafrost soils, soil biology, vol 16. Springer, Heidelberg, pp 73–84
Wagner C, Adrian R (2009) Cyanobacteria dominance: quantifying the effects of climate change. Limnol Oceanogr 54:2460
Walles S et al (1999) Protein adaptation to low temperatures: a comparative study of a-tubulin sequences in mesophilic and psychrophilic algae. Extremophiles 3:221–226
Wolfe-Simon F, Grzebyk D, Schofield O, Falkowski PG (2005) The role and evolution of superoxide dismutases in algae. J Phycol 41(3):453–465
Worland MR, Lukešová A (2000) The effect of feeding on specific soil algae on the cold-hardiness of two Antarctic micro-arthropods (Alaskozetes antarcticus and Cryptopygus antarcticus). Polar Biol 23:766–774
Wynn-Williams D (1996) Antarctic microbial diversity: the basis of polar ecosystem processes. Biodivers Conserv 5:1271–1293

Xiong F, Lederer F, Lukavský J, Nedbal L (1996) Screening of freshwater algae (Chlorophyta, Chromophyta) for ultraviolet-B sensitivity of the photosynthetic apparatus. J Plant Physiol 14 (8):42–48

Xue L, Zhang Y, Zhang T, An L, Wang X (2005) Effects of enhanced ultraviolet-B radiation on algae and cyanobacteria. Crit Rev Microbiol 31:79–89

Yergeau E (2008) Environmental microarray analyses of Antarctic soil microbial communities. ISME J 3:340–351

Yergeau E, Hogues H, Whyte LG, Greer CW (2010) The functional potential of high Arctic permafrost revealed by metagenomic sequencing, qPCR and microarray analyses. ISME J 4:1206–1214

Yung CC et al (2014) Characterization of chasmoendolithic community in Miers Valley, McMurdo dry valleys, Antarctica. Microb Ecol 68:351–359

Ziolkowski L, Mykytczuk N, Omelon C, Johnson H, Whyte L, Slater G (2013) Arctic gypsum endoliths: a biogeochemical characterization of a viable and active microbial community. Biogeosci Discuss 10:2269–2304

Cold-Adapted Fungi: Evaluation and Comparison of Their Habitats, Molecular Adaptations and Industrial Applications

2

Angeline Jessika Suresh and Regina Sharmila Dass

Abstract

The versatility displayed by kingdom *Fungi* in terms of physiological, genomic and metabolic complexities has ensured their presence in all major ecosystems. Given that 85% of the Earth experiences cold temperatures of below 5 °C, either seasonally or permanently, there is no shortage of cold environments resulting in global distribution of psychrophilic and psychrotrophic fungi. The cold-adapted extremophilic fungi possess molecular adaptations to persist and proliferate against harsh conditions exerted on them by their environment such as multiple freeze-thaw cycles, desiccation, low water activity, high exposure to harmful UV radiation or complete absence, high hydrostatic pressure and low nutrient availability. Cold habitats include polar regions such Antarctica and the Arctic as well as non-polar regions such as the deep seas and alpine regions. These regions offer a broad spectrum of niches for colonization of fungi including but not limited to rocks, ice sheets, snow cover, glaciers, cold soils, frozen seas, freshwater ice and permafrost, with varying levels of abundance and diversity.

Keywords

Cold-adapted fungi · Arctic · Antarctic · Extremophilic · Fungal genomes · Ecosystems

A. J. Suresh · R. S. Dass (✉)
Fungal Genetics and Mycotoxicology Laboratory, Department of Microbiology, School of Life Sciences, Pondicherry University, Pondicherry, India

R. Goel et al. (eds.), *Survival Strategies in Cold-adapted Microorganisms*,
https://doi.org/10.1007/978-981-16-2625-8_2

2.1 Introduction

As suggested by one of the pioneering and late mycologists Alexopoulos, the attempt to classify and define the boundaries of any group of living organisms is futile, given that we are in a constant state of discovery (Alexopoulos and Mims 1979). This may not be truer when it comes to the members of the kingdom *Mycota* as many resist classifications, perhaps more vehemently than any other group of *Eukaryota*. Traditionally, fungi are defined as heterotrophic, spore-bearing eukaryotes, capable of producing tubular networks made of chitin or cellulose and do not contain chlorophyll. Fungi lead an osmotrophic lifestyle, producing an arsenal of secretory enzymes and obtain nutrients through extracellular digestion and endocytosis. Their life cycles consist of both a sexual and asexual stage, and they come in many forms from the unicellular yeasts and zoosporic parasites to multicellular filamentous fungi to macroscopic fruiting bodies formed by many members of the phyla *Ascomycota* and *Basidiomycota* (Alexopoulos and Mims 1979). In recent years, this definition more or less still holds true but has become more inclusive taking into consideration the genomic, cellular and physiological characteristics. As reviewed by Naranjo-Ortiz and Gabaldón (2019), currently, nine phylum clades of fungi are recognized: *Opisthosporidia*, *Chytridiomycota*, *Neocallimastigomycota*, *Blastocladiomycota*, *Zoopagomycota*, *Mucoromycota*, *Glomeromycota*, *Basidiomycota* and *Ascomycota*.

2.2 Natural Habitats and Their Occurrence

Fungi are ubiquitously found in both terrestrial and aquatic communities throughout the globe. It is believed that there are over 1.5 million species of fungi, with 120,000 identified species, and only a handful of that number have been characterized and studied in detail. Members of the kingdom *Fungi* are often found in associations and interactions with other organisms such as plants, animals, insects, nematodes and macroalgae. In accordance with their relationship status, fungi can be classified as free-living, symbiotic or parasitic, pathogenic or predatory fungi. Fungi have also evolved their lifestyles to suit "extreme environments" and are found in the polar, alpine and deep-sea regions as well as micro-niches of increased temperature. The diversity of fungi in a particular ecological niche varies as fungal species present in the specific micromilieu can inhabit a broad range of substrata, depending on the metabolic activities they possess. Terrestrial fungi have been extensively studied. The aquatic ecosystem consists of both freshwater and marine ecosystems. Zoosporic fungi dominate aquatic environments.

2.3 Temperature Range

Thermophilic and thermotolerant fungi compromise only a tiny slice of the Earth's mycobiota, with approximately 50 species being identified in phylum *Ascomycota* and phylum *Mucoromycotina* (Salar and Aneja 2007; Maheshwari et al. 2000). de Oliveira et al. (2014) define thermophilic fungi as those with optimum growth, ranging between 40 and 50 °C. Thermotolerant fungi also exhibit this optimum growth temperature but are able to grow below 20 °C also, while thermophilic fungi cannot. However, it should be noted there is no set definition for distinguishing the boundaries of heat-tolerant fungi. Thermophilic fungi generally do not exceed growth temperature of 55 °C due to the upper limit placed on eukaryotic life forms, despite their "extremity" found throughout most temperate and tropical environments, present in transient microenvironments (Hutchinson et al. 2019).

The habitats and lifestyles of thermophilic fungi are surprisingly varied. They are found in associations with animals (for instance, in bird's nests) and plant hosts, as free-living saprotrophs in soil and more frequently as key microbial components in composts. Thermophilic fungi are believed to play an important role in plant biomass decomposition in association with other microorganisms (Hutchinson et al. 2019; de Oliveira et al. 2014; Korniłłowicz-Kowalska and Kitowski 2012). Like most filamentous fungi, thermophilic fungi synthesize a slew of secretory enzymes that act on different components present in the biomass. These include cellulases, proteases, xylanses, amylases, pectinases, phytases and amylases (Maheshwari et al. 2000). The metabolites of thermophilic fungi including enzymes and nanoparticles have been characterized for potential use in the industries (Chadha et al. 2019; Molnár et al. 2018;). Toxigenic thermophilic fungi also have severe implications in agriculture and pose a threat to global food security under the light of global warming (Paterson and Lima 2017).

2.4 Cold Adaptations in Fungi: Definition

It is not uncommon for microorganisms to be categorized according to their maximum, minimum and optimum temperature of growth. According to Gounot (1986), psychrophilic and psychrotrophic organisms are those that are able to grow at 0 °C (Robinson 2001). The two groups are differentiated on the basis of maximum growth temperature with psychrophilic unable to grow above 20 °C, whereas psychrotrophic beings can. However, as Cavicchioli (2015) eloquently explained, temperature-dependent growth rate determined by laboratory studies may not hold to describe how well an organism is adapted to cold environments and where it stands ecologically. A newer classification of isolated cold-adapted organisms divides them into eurypyschrophiles and stenopyschrophiles, with the former having a broader growth range and the latter exhibiting growth in a restricted temperature range (Raymond-Bouchard et al. 2018).

2.5 Cold-Adapted Fungi: A Background

The study of cold-adapted fungi is a field that was born out of necessity. In the early twentieth century, psychrophilic and psychrotrophic microorganisms were causing economic losses due to contamination of cold storage food (Kuehn and Gunderson 1963; Rangaswami and Venkatesan 1961). However, the collection and taxonomic identification of fungi in cold-dominated regions (alpine and polar regions) were done by several eminent botanists (as fungi were considered part of plants) and mycologists well before the effects of spoilage fungi were observed. Several sailing expeditions to the Canadian Arctic, Alaska and Greenland were taken starting from the early nineteenth century (Noffsinger et al. 2020). Notable mycologists and naturalists include Rev. Miles Joseph Berkeley, David Lyall, David Walker, Emil Rostrup, Herman George Simmons and John Dearness Jules Favre who is considered the father of alpine mycology and was one of the earlier pioneers among those who studied macrofungi in the alpine regions (Brunner et al. 2017). Among the studies that predominated in the early twentieth century was Arctic and Antarctic aero-mycology research on fungal spores. However, it wasn't until the mid-twentieth century that the physiology of psychrophilic and psychrotrophic fungi was studied. Apart from ecology and spoilage, an interest in the biotechnological use of psychrophilic organisms began to arise in the early 1990s.

2.6 Molecular Adaptations

Microorganisms dominate extreme cold environments such as the polar regions and the deep sea. However, life for these organisms is often a stressful existence, as psychrophiles and psychrotrophs must persist against unfavourable conditions. Limiting factors include low water availability, low temperatures, freezing, unpredictable and frequent fluctuations in temperature, high exposure to UV radiation, high pressure and low nutrient availability. In order for cold-adapted fungi to persevere against the extremophilic conditions presented by their surroundings, they have evolved adaptations to ensure their growth and survival.

Biological life is sustained through numerous physiological processes, in which enzymes play a central role. Psychrophilic and some psychrotrophic fungi synthesize an array of cold-adapted enzymes that are often functional over a broad temperature range, including cold-adapted proteases, amylases, pectinases, lipases, etc. (Hassan et al. 2016). These enzymes not only are essential for the growth and proliferation of cold-adapted fungi, but also contribute to ecosystem functioning and global geochemical cycles (citation required). A detailed account on fungal cold-adapted enzymes and their potential applications is covered here.

Low temperatures exert numerous changes in the structure, physiology and metabolism of an organism. One such example is the increase in rigidity in the plasma membranes, compromising its integrity (Collins and Margesin 2019). Cold-adapted fungi surmount the loss of fluidity by altering the composition, ratio and distribution of lipids throughout the plasma membrane. Fluidity is maintained by increasing the concentration of unsaturated and polyunsaturated fatty acyls as well as by decreasing the chain length of said polymers (Hassan et al. 2016). Fatty acid desaturase is an important enzyme in this adaption mechanism (Boo et al. 2012). A study done by He et al. (2015) demonstrates this relationship between polyunsaturated fatty acids and membrane fluidity in cold-adapted yeast of *Rhodotorula glutinis.* Another commonly found genus, *Mrakia*, in cold environments, was found to contain a high amount of unsaturated fatty acids in its plasma membrane (Tsuji et al. 2013).

The formation of ice crystals in the cytosol of cells under subzero temperatures can cause cryoinjury and osmotic stress (Collins and Margesin 2019). Many cold-adapted fungi are known to synthesize and/or secrete substances known as cryoprotectants that halt the freezing of cellular interiors. Antifreeze proteins (AFPs) and ice-binding proteins (IBPs) are well-recognized cryoprotectants and have been well characterized in snow moulds *Antarctomyces psychrotrophicus* and *Typhula ishikariensis* (Xiao et al. 2009). AFPs exert their antifreeze activity by attaching to minute ice crystals, preventing the formation of larger ice crystals or recrystallization (Rahman et al. 2019). AFBs lower the freezing point of the fluid without altering the melting point, a phenomenon known as thermal hysteresis. This is hypothesized to keep essential water channels responsible for physiological processes functioning, such as those involved in nutrient uptake as well as in maintaining fluidity of exterior cell environments (Alcaíno et al. 2015). A very recent study (Batista et al. 2020) detected the presence of an ice-binding protein in *Antarctomyces pellizariae*, a psychrotolerant fungi endemic to Antarctica (Villarreal et al. 2018), and identified and purified AFPs and IBPs from numerous Antarctic yeast species.

Increased accumulation of compatible solutes such as glycerol, mannitol and trehalose is another mechanism through which the fluidity of membrane structure and osmotic balance of cells is maintained in response to stresses such as desiccation and low temperatures (Hassan et al. 2016). While there is no doubt in the importance of adaptation mechanisms in response to low temperatures, Robinson (2001) attributes the survival of fungi in extreme cold environments primarily to their ability to withstand multiple freeze-thaw cycles. Villarreal et al. (2018) suggested that no single mechanism is responsible for this resistance. Instead, they stated that combating freeze-thaw cycles most likely involved the initiation of a complex stress response involving AFBs, membrane composition, accumulation of compatible solutes and other factors unknown. The ability of fungi to induce a cold response as well as the degree of the response is highly species dependent (Alcaíno et al. 2015).

Psychrophilic and psychrotrophic fungi exploit a broad range of substrates, many of which receive high levels of UV irradiation. Pigment production has been observed in several fungi taxa in the polar regions (Sajjad et al. 2020). These pigments, such as melanin, are often heavily concentrated in the cell walls of hyphae and yeasts (Wang et al. 2017). The presence of pigments provides a photo-protective mechanism by preventing DNA damage caused by prolonged UV exposure (Hassan et al. 2016). Black fungi or microcolonic fungi are members of cryptoendolithic communities present in and on the exposed surfaces of rocks and are excellent examples of organisms that utilize melanin as a sun shield. Microcolonic fungi are also known to secrete exopolysaccharides, which are gel-like polymers of mostly carbohydrate nature secreted by a diverse set of microorganisms (Mahapatra and Banerjee 2013). EPS is believed to aid in the survival of the organism by providing protection from drying and freezing (Selbmann et al. 2005). These adaptations along with their minimal life cycles slow proliferation, and their meristematic nature contributes to the highly extremophilic and resilient character of black fungi (Wang et al. 2017; Pacelli et al. 2017). Other pigments such as carotenoids, mycosporines and mycosporine like amino acids act as UV sunscreens by absorbing radiation at 310–360 nm (Vaz et al. 2011; Sajjad et al. 2020). Vaz et al. (2011) identified mycosporine production in both pigmented and non-pigmented yeast species from Antarctic soil *Microglossum* sp. and *Exophiala xenobiotica.*

Adaptations in cold-adapted fungi vary among fungi, with respect to the specific niche inhabited by the organism and by its ecological role as seen in saprobic, parasitic and mutualistic interactions (Wang et al. 2017). It is not necessary that all the adaptations discussed must be present in a single organism (Alcaíno et al. 2015). Due to the versatility of cold environments, cold-adapted fungi may exhibit polyextremophile capabilities such as high salinity and high-pressure tolerance (Turk et al. 2007; Hassan et al. 2016).

2.7 The Arctic

The Arctic is a vast circumpolar region located in the northernmost part of the Earth above latitude 60 N. The boundaries separating it from the lower non-Arctic regions remain blurry (AMAP 2009). It consists predominantly of ice-covered oceans including the Arctic Ocean, Nordic Sea and Bering Sea (two thirds of the region), enclosed by the terrestrial boreal forests, tundra regions and polar deserts of the eight Arctic nations (ACIA 2005). The Arctic tundra is considered a maritime environment (Walker et al. 2005). Cryosphere components such as sea, lake and river ice, permafrost, glaciers and continental ice shelves pervade the Arctic. Svalbard, Franz Joseph Island, Ellesmere Island and the New Siberian Islands all fall within the region of the Cold Arctic (Robinson 2001).

2.8 The Antarctic

The Antarctica continent is an isolated, dry and cold land mass present in the southern hemisphere. More than 99% of the region is permanently covered by snow or ice. It is often divided into three terrestrial biogeographic zones: the sub-Antarctica, Maritime Antarctica and Continental Antarctica (Convey 2013). Maritime Antarctica includes the Scotia Arc island archipelagos and the western coastal regions of the Antarctic Peninsula. Continental Antarctica is marked by varying substratum such as nunatalks, old glacial lakes and permafrost. It is also home to the unique ecosystem of an arid polar desert, the McMurdo Dry Valley. Ice-free zones are mostly found in continental Antarctica and host microbial communities such as endolithic communities or those present in transient freshwater communities or hypersaline ice-covered lakes (Robinson 2001). Antarctica experiences a variable climate. Continental Antarctica remains engulfed in darkness throughout the winter and experiences mean temperatures as low as −30 °C. In fact, this region of the continent only witnesses positive temperatures only during summer months (de La Torre et al. 2003). In the case of Maritime Antarctica, the region experiences mean positive temperatures of 1–2 °C for 3–4 months of the year, and its climate is influenced to a certain degree by the surrounding ocean (Convey 2013).

2.9 Nonpolar Regions

It is estimated that approximately 90% of the ocean biome experiences temperatures below 5 °C, with deep seas routinely encountering temperatures between −1 and −4 °C (Wang et al. 2017). Fungi are distributed throughout the marine ecosystem from the surface waters to the deep seas well below 10,000 m sea level (Nagano et al. 2010). Currently, there are 1112 recognized species of marine fungi amongst 472 genera (Jones et al. 2015).

2.10 Arctic Fungi

2.10.1 Plant-Associated and Free-Living Fungi of Arctic Soils

Fungi in terrestrial ecosystems often form associations with plant hosts and contribute enormously to the survival and growth of the plant, especially in harsh environments (Bjorbækmo et al. 2010). In the Arctic tundra, they also play a role in biogeochemical and mineral cycles and have the potential to influence the Earth's climate in this era of global warming and climate change (Timling and Taylor 2012). The Arctic soil mycobiota is diverse and is represented by saprophytes, mycobionts in lichens, mycorrhizal fungi and root-associated endophytic fungi (Zhang et al. 2016; Newsham et al. 2009).

The Arctic can be divided into five bioclimatic subzones in accordance with summer temperature and vegetation (Walker et al. 2005). Areas classified as subzone A prove to be the harshest and coldest, whereas areas designated subzone E are the warmest. Bioclimatic zones influence the composition of fungal communities present in the Arctic soils. For instance, the abundance of melanized forms of fungi rose in subzones A and B in comparison to the warmer zones, possibly due to the increase in UV exposure (Timling et al. 2012).

Arctic plants inhabiting endophytes predominantly belong to phylum *Basidiomycota* (Botnen et al. 2014; Bjorbækmo et al. 2010). Zhang and Yao (2015) investigated the nature of endophytic fungal communities present in the leaves and stems of vascular plants *Cassiope tetragona*, *Saxifraga cespitosa*,

Saxifraga oppositifolia and *Silene acaulis* present in the Svalbard archipelago and found them to differ from endophytes present on non-Arctic plants. Species detected frequently by the group included *Rhizosphaera macrospora*, *Phaeosphaeria triglochinicola*, *Leptosphaeria pedicularis*, *Venturia alpina*, *Phoma herbarum* and *Mrakia frigida*. Timling et al. (2012) detected mycorrhizal fungal sequences belonging to the genera *Thelephora*, *Tomentella*, *Sebacina*, *Inocybe*, *Cortinarius*, *Russula*, *Hebeloma*, *Laccaria* and *Clavulina* across the North American tundra, with the fungal class *Thelephoraceae* dominating a third of the isolates. Along with *Thelephoraceae* (Abrego et al. 2020), several mycorrhizal fungi were identified associated with Arctic plants in Northeast Greenland belonging to the family *Cortinariaceae*, *Gloniaceae*, *Sebacinaceae* and *Inocybaceae*.

Timling et al. (2012) noted that there was widespread distribution of EMF communities associated with *S. antarctica* and *D. integrifolia*, but these communities had little to no pattern. They attributed this finding to the non-specific nature of EMF and how plant hosts were not rigid in their EMF partners. Botnen et al. (2014) also reported similar findings. In contrast to mycorrhizal fungi of the Arctic, endophytic fungi and root-associated fungi show high host specificity, to the point that phylogenetically related plant hosts have similar communities of root-associated fungi (Zhang and Yao 2015; Abrego et al. 2020; Botnen et al. 2020).

Aside from bioclimatic zones, abiotic factors including glaciation history, chemistry of bedrocks, pH, C/N ratio and available P and temperature can also influence the structure and distribution patterns of Arctic fungi communities in soil (Fujimura and Egger 2012; Zhang et al. 2016). Gittel et al. (2013) reported the decrease in fungal abundance with depth in the buried soils of the Siberian Tundra, while Semenova et al. (2016) observed alterations in fungi composition with increase of snow depth, including the decline of many functional groups such as the ectomycorrhizal fungi. Abrego et al. (2020) found that root-associated fungal communities in Northeast Greenland differed with elevation and were less species rich in higher elevations.

2.10.2 Glacial Ice

Slow-forming glacial ice are reservoirs of the global freshwater supply and consist of mountain glaciers, ice caps, ice cheers and shelf ice (Boetius et al. 2015). Subglacial ice is a relatively thin layer of ice containing debris found in the basal zone of glaciers and houses rich and abundant microbial communities (Skidmore et al. 2000; Butinar et al. 2011). The genus *Penicillium* is widespread in subglacial environments (Perini et al. 2019a). Sonjak et al. (2006) isolated several species of common foodborne penicillia such as *P. crustosum*, *P. nordicum*, *P. solitum*, *P. expansum* and *P. chrysogenum*.

An elaborate investigation into the subglacial, three polythermal glaciers (Pedersenbreen, Vestre Brøggerbreen and Midtre Lovénbreen) in Kongsfjorden, Svalbard, done by Perini et al. (2019a) offered illuminating insights on the fungal diversity, abundance and species richness of this unique habitat. By using high-throughput amplicon-based screening, they demonstrated that *Basidiomycetes* (e.g. *Mrakia* sp., *Rhodotorula svalbardensi*, *Glaciozyma watsonii*) is the dominant phylum, followed by *Chytridiomycota* and *Ascomycota* (e.g. black yeast *Hortaea werneckii, Pseudogymnoascus* sp., *Penicillium bialowiezense*).

Pyschrophillic fungi of genera *Rhodotorula*, *Thelebolus*, *Cryptococcus*, *Mrakia*, *Phialophora*, *Articulospora* and *Varicosporium* were identified in cryoconite holes of Brøggerbreen, Midre Lovénbreen and Vestre Brøggerbreen glaciers in Svalbard (Singh and Singh 2011; Edwards et al. 2013; Singh et al. 2016). Cryoconite holes are protective water-filled microenvironments of differing sizes distributed among the supraglacial surfaces (Poniecka et al. 2020). Novel species *Articulospora tetracladia* were identified using DNA sequencing in cryoconite holes in Svalbard (Singh et al. 2016).

A research team analysed several substrata from the Greenland Ice Sheet, the largest ice sheet in the northern hemisphere, such as snow, dark ice (algal containing ice), clear ice, supraglacial water and cryoconite holes. They found varied community composition between substratum and isolated numerous taxa including those belonging to plant-associated fungi, filamentous fungi and basidiomycete yeasts (Perini et al. 2019b).

2.10.3 Marine Fungi from the Arctic

The marine fungi comprise a large group of phylogenetically diverse organisms (Jones et al. 2015). Much of the marine mycobiota remains unclassified, highlighting the biodiversity potential of this ecosystem (Richards et al. 2012; Amend et al. 2019). They are known to utilize a wide range of substrate such as driftwood, marine sediments, macroalgae, sea water columns, sea grasses and sponges. However, there are comparatively less studies on the nature of the Arctic marine mycobiota, possibly due to the inaccessibility of the region.

Driftwood, acting as both a shelter and an energy source, is an excelled substratum for Arctic marine fungi and is known to house diverse, species-rich, fungal communities (Blanchette et al. 2016; Rämä et al. 2016). Another study reported several genera of fungi from the marine sediments of Kongsfjorden, Svalbard, including dominant genera *Fusarium* and *Pichia* and other genera such as *Malassezia*, *Alternaria*, *Rhodotorula* and *Aspergillus* (Zhang et al. 2015). Hagestad et al. (2019) isolated poorly studied marine fungi *Lulworthiaceae* as well as genera *Mucor*, *Penicillium*, *Oidiodendron*, *Tolypocladium* and the dimorphic *Glaciozyma* from varying samples (e.g. sea ice, algae and driftwood) in Svalbard, Norway.

It is known that phylum *Ascomycota* dominate the marine world, followed by *Basidiomycetes*, *Chytridiomycota* and *Mucoromyctoa* (Jones et al. 2015). But a study done by Comeau et al. (2016) demonstrated the widespread distribution of phylum *Chytridiomycota*, with varying abundance across the Arctic ocean and related seas. Chytrids are zoosporic parasites and are participants in major ecological processes, such as modulating the flow of carbon through degradation of organic compounds (Hassett et al. 2019; Kilias et al. 2020). Apart from genes responsible for biodegradation of compounds such as lignin, a study also detected the presence of genes related to the assimilation of nitrate and denitrification, a possible indication in its role in the global nitrogen cycle (Hassett et al. 2019).

2.11 Antarctic Fungi

2.11.1 Soils

Antarctic soils are primarily of oligotrophic nature and present unfavourable conditions, such as dryness and oscillating temperatures, for organisms to survive in (Godinho et al. 2015). Yet the inhospitable soil is rich in cosmopolitan, bipolar (taxa found in both the polar regions but not elsewhere) and endemic species of fungi that vary in diversity and abundance in respect to different soil types (Gomes et al. 2018; Cox et al. 2019) and other factors like C/N ratio (Durán et al. 2019). Soil diversity can range from poorly developed soils without vegetation, moderately developed soils and mature ornithogenic soils (da Silva et al. 2020).

Like most fungi, many members of Antarctic mycota form associations with other members of the Antarctic biome including plants (Santiago et al. 2016a; Yu et al. 2018), mosses (Hirose et al. 2016; Melo et al. 2013; Rosa et al. 2020), lichens (Santiago et al. 2015), animals (Godinho et al. 2015) and macroalgae (Furbino et al. 2017). In these relationships, Antarctic fungi exist as either parasites or mutualists or live independently as saprophytes (Newsham et al. 2018). The abundance of fungi in root associated soil is much more when compared to bulk soil (Wentzel et al. 2018). In these soils, fungi prove extremely useful for plants, aiding in their survival, for example, through acquiring nitrogen and proving exposure against UV (Hill et al. 2019; Oses-Pedraza et al. 2020; Ramos et al. 2018).

Ascomycota is the dominant phylum in Antarctica. In fact, it is estimated that members of the *Basidiomycota* represent less than 3% of fungal sequences in sub-Antarctica, low maritime Antarctica and High Maritime Antarctica (Cox et al. 2016). Most of the fungi detected in Antarctic soils are in anamorphic forms, possibly indicating the loss of sexual lifestyles to ensure survival in this extreme habitat (Ruisi et al. 2006). However, Durán et al. (2019) observed that telemorphs of *Ascomycetes* such as *Cadophora*, *Antarctomyces* and *Thelebolus*, *Basidiomycetes* such as *Vishniacozyma* and *Bjerkandera* and *Mucormycetes Mortierella* and *Rhizopus* were dominant in the soils of King George Island, Maritime Antarctica.

Filamentous fungi recognized from the Antarctic soils belong mostly to phylum *Ascomycota*, whereas the detected yeasts belong mostly to *Basidiomycota* (Godinho et al. 2015). Dominant genera of fungi include *Pseudogymnoascus*, *Penicillium*, *Peniophora*, *Cryptococcus* and *Mortierella* (Gonçalves et al. 2015; Wentzel et al. 2018; Gomes et al. 2018). Vaz et al. (2011) isolated numerous genera of yeasts and yeasts like organisms from Maritime Antarctica including endemic species *Cryptococcus antarcticus*, *Cr. victoriae*, *Dioszegia hungarica* and *Leucosporidium scottii.*

Cold-tolerant strains of *Saccharomyces cerevisiae* are also present in Antarctic soils (Gomes et al. 2018).

2.11.2 Antarctic Permafrost

The Antarctic region encompasses 37% of the global permafrost (French 2007). Permafrost refers to soil that has been frozen for 2 or more consecutive years. Goordial et al. (2016) investigated the nature of microbial communities present in permafrost in the McMurdo Dry Valleys of Continental Antarctica. Their group reported low abundance and diversity of fungi, with *Dothideomycetes* being the dominantly detected fungal class. The group also isolated cryophilic yeast *Rhodotorula*, which was able to grow at temperatures as low as −10 °C.

Another research team isolated 27 taxa from the permafrost present in Maritime Antarctica as well as an additional 31 taxa from the active layer, with only 5 taxa present in both groups. *Oidiodendron*, *Penicillium* and *Pseudogymnoascus* were the most abundant taxa in permafrost samples, whereas *Bionectriaceae*, *Helotiales*, *Mortierellaceae* and *Pseudeurotium* were more abundant in the active layer (da Silva et al. 2020). *Mrakia blollopis*, a cold-tolerant yeast of research interest (Tsuji 2016), was detected and was the only taxa belonging to *Basidiomycota* that was detected in their study. Interestingly, da Silva et al. (2020) observed that while *Mortierella*, *Pseudogymnoascus* and *Penicillium* were present in all collected samples, there was no commonality between the samples when it came to species.

2.11.3 Endolithic Communities

In the ice-free zones of continental Antarctica, exposed rocky substrates such as sandstones, nunataks and mountain tops are the primary source of biomass in the region and are utilized by endolithic microbial communities (Coleine et al. 2018; Archer et al. 2016). This inhospitable polar desert is burdened with unpredictable living conditions, desiccation and high UV exposure, factors that have led its comparison to the likes of Mars (Onofri et al. 2004). Hence microbiota of this border ecosystem is fragile, with extinction events believed to be common (Selbmann et al. 2017). There are numerous reports of endolithic colonies worldwide and are known to be involved in mineral transformation and bioweathering as well as disease and biocorrosion (Muggia et al. 2015; Selbmann et al. 2005; Pernice et al. 2019; Gleason et al. 2017). Fungi occur in these communities either as mycobionts in lichen associations (citation required) or independently (Fleischhacker et al. 2015). According to Ríos et al. (2004), fungi present in these communities possess a very low metabolic rate, which may be a mechanism of adaptation for survival against extreme conditions.

2.12 Harmful Effects in Plants, Animals and Humans

While fungi are indispensable to the environment and modern society, they are agents of destruction, causing severe economic, health-related, agricultural and biodiversity losses on an international scale. In comparison with their viral and bacterial counterparts, fungal and fungal-like diseases have been largely neglected, possibly due to the non-communicable nature of human mycoses, and often have suffered from insufficient funding and lack of global surveillance especially in developing countries (Rodrigues 2016). However, there has been a growing increase in fungal and fungal-like diseases in the last few decades, starting from the latter half of the twentieth century.

This phenomenon of fungi as animal and plant pathogens in emerging infectious diseases (EIDs) has been attributed to numerous reasons: (1) a steady increase in immunocompromised patients (e.g. due to administration of immunosuppressive drugs in autoimmune and cancer treatments), (2) global networks of trade (including the "exotic" species market), (3) the chronic nature of fungal infections and (4) growing resistance to antifungal agents (Almeida et al. 2019; Fisher et al. 2012, 2016; Casadevall 2018). Furthermore, fungi are biologically suited as persistent infectious agents as they propagate by spore dispersal leading to persistence in soil and airborne environments, and many have the ability to switch between parasitic and saprophytic life stages (Fisher et al. 2012).

Fungi also produce toxic, low-molecular-weight secondary metabolites termed as mycotoxins, which have widespread implications for animals, plants and humans as they are frequent contaminants of food and animal feed (Alshannaq and Yu 2017). The Food and Agriculture Organization has estimated that 25% of the world's crops are affected by mycotoxins each year, with annual losses of around one billion metric tons of foods and food products. The genera *Aspergillus*, *Fusarium* and *Penicillium* are among the major producers of mycotoxins. Consumption of mycotoxins results in a variety of symptoms.

In humans, there has been far-reaching emergence of mycoses across the continents, with new pathogens arising and being reported in intensive care units from patients who have undergone transplantation or contracted AIDS or are suffering from leukaemia and other malignancies (Low and Rotstein 2011; Friedman 2019). Invasive candidiasis caused by non-*albicans* strains of *Candida* (e.g. *C. glabrata* and *C. auris*, multidrug-resistant opportunistic pathogens with the latter being a potent nosocomial agent) and invasive aspergillosis caused by both azole drug-resistant *Aspergillus fumigatus* and non-*fumigatus* strains of *Aspergillus* (*A. flavus*) are among the more recent EIDs. Drug resistance in fungi can be attributed to the long-term use of antifungal agents in treatment and the use of fungicides in agriculture settings, as in the case of azole-resistant *Aspergillus fumigatus*.

Oomycetes reside in kingdom Protoctista but resemble fungi in their filamentous growth pattern and production of spores (Latijnhouwers et al. 2003). *Phytophthora infestans*, *Phytophthora palmivora*, *Plasmopara viticola* and *Albugo candida* are among some of the major oomycetous plant pathogens causing an assortment of diseases (including late blight, downy mildew and white blister rust) in economically important plants such as citrus plants, cocoa, soybean, tubers, rubber plants and grapes. Much of the deadly invasive tree diseases are caused by fungi and fungi-like organisms.

A current global crisis in chytridiomycosis pandemic amphibians is caused by fungal pathogen *Batrachochytrium dendrobatidis*. *Geomyces destructans* infects brown bats and has caused several epidemics in little brown bats in various parts of the world. Finally, aquatic animal pathogen oomycetes include *Halioticida noduliformans*, *Saprolegnia parasitica* and *Aphanomyces invadans* are responsible for great economic losses in the fish industry (Kamoun et al. 2014).

2.13 Applications of Fungi in Industry

Fungi are vital to modern industrial microbiology and biotechnology. Since ancient times, fungi have been utilized in the production of fermented foods such as bread, wine, soy sauce, tempeh and miso to enhance, preserve and add flavour. In the current scenario, fungi and fungal metabolites are exploited by the food, textile, leather, detergent and pharmaceutical industries. Filamentous fungi are used extensively as they produce large quantities of industrially important metabolites and secrete them extracellularly. Filamentous fungi like *Aspergillus niger* and *A. terreus* are major producers of organic acids such as citric acid, gluconic acid, lactic acid and itaconic acid. Cephalosporins and ergot alkaloids (a natural toxin and a pharmaceutical) are produced by *Cephalosporium acremonium* and *Claviceps purpurea*, respectively. Several studies have been done on the use of white rot fungi for the decolourization of environmentally hazardous textile dyes and tannery wastewater effluents. More novel uses, namely, the potential of anaerobic gut fungi in biofuel production and the production of higher quality animal feed by addition of fungal probiotics, have also been highlighted in recent reviews (Chuang et al. 2020; Ranganathan et al. 2017). With advent of metagenomics, the untapped potential of fungi in industrial applications can be explored further.

2.13.1 Cold-Active Enzymes

The vast potential of cold-adapted enzymes has been much explored in the last decade, especially with the advent of culture-independent methods including metagenomic approaches and through genome mining (Ekkers et al. 2011; Gong et al. 2013; Vester et al. 2014). However, there is still a necessity for culture-dependent methods in order to assess and characterize metabolite production capabilities of psychrophilic and psychrotrophic organisms in order to exploit them for industrial and biotechnological applications (Zucconi et al. 2020). When compared to their mesophilic homologues, cold-active enzymes require a greater protein flexibility thus compromising the stability of the enzyme (Gerday et al. 2000). Cold-adapted enzymes have a wide range of temperature and pH optima, possibly in order to withstand sudden and frequent fluctuations in their environment (Duarte et al. 2018). Psychrophilic and some psychrotrophic fungi produce an arsenal of both intra- and extracellular cold-adapted enzymes to combat the harsh realities of their environment, ensuring their survival, growth and proliferation (Hassan et al. 2016).

The increasing interest in the biotechnological applications of cold-adapted enzymes can be attributed to several advantages that cold-adapted enzymes have over their mesophilic homologues. Firstly, employing cold-adapted enzymes in industrial processes cuts down on energy costs, creating a more efficient, environmentally conscious and economic production (Santiago et al. 2016b). Secondly, unwanted chemical side-chain reactions that occur at high temperatures can be avoided (Siddiqui 2015). Finally, due to their heat-labile nature and relatively high thermo-sensitivity, they can be easily inactivated, bypassing the need for chemical inactivation ensuring quality of product is retained (Cavicchioli et al. 2011).

2.13.1.1 Proteases

Proteases are a large class of hydrolytic enzymes that break down proteins into peptides and amino acid units by cleaving peptide bonds. Mesophilic proteases enjoyed a long history of commercial use as a therapeutic agent (e.g. for wound debridement), in the cosmetic industry (e.g. for removal of frown lines), in the food processing industry, in bioremediation and for pharmaceutical applications (Fornbacke and Clarsund 2013). Cold-adapted alkaline proteases were introduced as third-generation proteases in the detergent industry and were found to have excellent activity and stability in surfactants and bleaches (Hao and Sun 2014). In the food industry, psychrophilic enzymes have potential uses in flavour enhancement of frozen meat, casein degradation in dairy products, improving stability and solubility of health foods and other food-processing applications (Furhan 2020). Cold-adapted proteases were also reviewed for their therapeutic role in (Fornbacke and Clarsund 2013).

Alkaline peptidase and aspartic protease produced by marine cold-adapted *Penicillium chrysogenum* and Antarctic psychrophilic fungus *Geomyces pannorum*, respectively, proved to be suitable for cheesemaking (Furhan 2020). Psychrophilic fungal strains of *Saccharomyces cerevisiae*, *Candida parapsilosis*, *Candida mogii* and *Schizosaccharomyces pombe* all secrete extracellular proteases, with potential commercial applications in food, fruit and milk processing (Srilakshmi et al. 2015).

2.13.1.2 Chitinases

Chitin is a linear natural polymer, present in fungal cell walls, exoskeletons and inner parts of certain invertebrates. As chitin is the second most abundant biopolymer on Earth, the chitinase group of enzymes plays an essential ecological role in breaking down chitin. In a biotechnological perspective, some of the uses of chitinases include the manufacturing of food supplements and chito-oligosaccharides and synthesis of N-acetyl D glucosamine and in generating fungal protoplasts for biological research (Hamid et al. 2013; Oyeleye and Normi 2018). Chitin is thought to be a major contributor of coastal pollution (Kumar et al. 2018); hence, it has the potential to be used in bioremediation of chitin-rich wastes. It can also be used as a biological control agent to lessen the economic burden caused by plant diseases and post-harvest pathogens (Kumar et al. 2018; Veliz and Martínez-Hidalgo 2017). The psychrotolerant Antarctic fungus *Lecanicillium muscarium* CCFEE 5003 is a remarkable producer of chitinolytic enzymes (Fenice 2016).

2.13.1.3 Cellulases and Pectinases

Cellulose is the β-1,4-linked homopolymer of β-d-glucose. Cellulases are a class of enzymes that degrade cellulose. They can be used in the bioconversion of lignocellulosic biomass to liquid biofuel and also have applications in the paper, textile and leather industry (Payne et al. 2015; Phitsuwan et al. 2012). Novel isolates of psychrophilic basidiomycetous yeast species *Mrakia hoshinonis* sp. isolated from the Canadian Arctic showed high production of lipases and cellulases that were active even at −3 °C (Tsuji et al. 2019). Krishnan et al. (2014) recognized 13 isolates from King George Island that showed significant cellulase activity. These fungal genera included *Geomyces* sp., *Galerina fallax*, *Glomerella* sp., and four unidentified fungi. Pectinases are a heterogeneous group of enzyme hydrolases consisting of polygalacturonase, pectinesterase and pectin lyase. They act on the high-molecular-weight plant polysaccharide pectin and have applications spread across the disciplines (Satapathy et al. 2020). Cold-adapted pectinases are of enormous interest as their high specific enzymatic activity at low temperatures makes them suitable for use in the fruit processing industry to eliminate pectin in fruit juices, avoiding contamination associated with higher temperatures and retaining volatile aromatic compounds to preserve aroma and flavour. They are also ideal for efficient juice extraction and improving appearance of juice in wine fermentation which is carried out at low temperatures (Adapa et al. 2014). Apart from the fruit and food industry, pectinases can be exploited by the paper processing and textile industry as well as for biofuel production and wastewater treatment (Sharma et al. 2012). Cold-adapted pectinase producing yeasts have been isolated from geographically diverse locations such the snow-covered Europeans alps of northern Siberia, frozen environmental samples of Iceland, forest soil in Hokkaido (Japan), grape vineyard soils in San Rafael (Argentina) and the cold soils and spoiled fruits and vegetables of fruit yards in the Himalayas (Birgisson et al. 2003; Nakagawa et al. 2004; Margesin et al. 2005; Naga Padma et al. 2011; Merín et al. 2011).

A research group identified high pectinolytic activity at 15 ° C in the filamentous fungus *Geomyces* sp. Strain F09-T3–2 isolated from Antarctic marine sponges and promoted its biotechnological potential (Poveda et al. 2018). Extracellular pectinase activity was also identified in three yeast and yeast-like microorganisms (*Dioszegia* sp., *Phenoliferia glacialis* and *Tetracladium* sp.) in a study done by Carrasco et al. (2019). In this study, the best activity was exhibited by *Tetracladium* sp., and the enzyme was identified as a polygalacturonase.

2.13.1.4 Amylases

Amylases (subgroups: α-amylase, β-amylase and γ-amylase) hydrolyse starch and related polymers. Hence, they are used extensively not only in the food processing industry but also in the detergent, laundry and textile industries. Three isolates of *Geomyces*, one of *Pseudeurotium*, one of *Phialemonium* and one unidentified isolate from King George Island, showed significant amylase activity (Krishnan et al. 2014). Polyextremophilic marine fungi from deep-hypersaline anoxic basins are a rich source of extracellular enzymes like amylases, lipases and esterase (Barone et al. 2019). Carrasco et al. (2016) identified high amylase activities in psychrotolerant yeasts *Tetracladium* sp. and *Rhodotorula glacialis*; however, both yeasts had their activity maxima at different temperature ranges. *Tetracladium* sp. showed optimum amylase activity between 30 and 37 °C, whereas *Rhodotorula glacialis* showed optimum amylase production between 10 and 22 °C.

2.13.1.5 Xylanases

Xylan is a major constituent of hemicellulose. The latter, along with cellulose and lignin, contributes to the bulk of lignocellulosic biomass. As xylan is a complex polymer, several different types of xylanase enzymes are required to act in a synergistic manner for it is degraded into xylose (Juturu and Wu 2012). Some of the xylanase enzymes include endo-1,4-β-xylanase, β-D-xylosidase, acetylxylan esterase and arabinase. Xylanase production in fungi is well documented, and their applications have been reviewed (Polizeli et al. 2005). However, among other cold-adapted enzymes, xylanases are characterized much less, despite their enormous biotechnological potential (Sarmiento et al. 2015). Cold-adapted xylanases can contribute to the softening and enhancement of bread dough texture by conversion of hemicellulose present in flour. It can be used in pulp and paper biobleaching and in bioremediation and biofuel production acting in combination with cellulases and other enzymes (Sarmiento et al. 2015; Walia et al. 2017; Gottschalk et al. 2010; Watanabe et al. 2015).

In the literature, instances of xylanase production in cold-adapted fungi seem to be predominantly associated with marine-derived fungal species. In a study done by Del-Cid et al. (2013), 38 fungal isolates derived from marine sponges in King George Island, Antarctica, demonstrated xylanase activities, with the *Cladosporium* sp. isolate showing the highest activity. In a similar more recent study (Duarte et al. 2017), various fungi specimens recovered from marine samples in King George Island exhibited xylanase activity. Among them, isolates belonging to *Penicillium* sp. showed the greatest activity.

2.13.1.6 Lipolytic Enzymes

Lipolytic enzymes are a robust group of enzymes consisting of true lipases and esterases. They catalyse two different reactions on the basis of solvents: in water, lipolytic enzymes catalyse the hydrolytic cleavage of an ester bond between a carboxylic acid and an alcohol group, and in organic solvents or non-aqueous media, they are capable of catalysing the formation of an ester bond by transesterification or esterification (Lopez-Lopez et al. 2014).

Yeasts isolated from the Antarctic environment all seemed to possess a minimum of low-to-moderate lipase activity (Troncoso et al. 2016). Nine yeast isolates (genera *Metschnikowia*, *Cryptococcus* and *Leucosporidium)* from marine sediments of

Admiralty Bay (King George Island, Antarctica) showed significant lipase activity, producing lipase above 0.5 U mL $^{-1}$. *Metschnikowia* sp. CRM1589 showed the highest activity (0.88 U mL $^{-1}$) among the isolates (Wentzel et al. 2018). Tsuji et al. (2018) identified excellent lipase activity from novel psychrophilic yeast *Mrakia arctica* sp. nov. *Mrakia arctica* was isolated from ice island Disraeli Fjord (northern Ellesmere Island, Canadian High Arctic) and showed high lipase activity at an optimum temperature at −3 °C but still exhibited strong lipase activity even at 20 °C. The low optimum temperature of the cold-adapted lipase has led the authors to believe that *M. arctica* may play an essential role in the biochemical cycle of its habitat. The highly specific and efficient biocatalyst Lipase B produced by *Candida antarctica* catalyses the eco-friendly synthesis of biodegradable polymers and in the biodegradation of aliphatic polyesters (Kundys et al. 2017).

The use of esterases can be applied widely, for example, in pharmaceutical industries for synthesis of chiral drugs and in bioremediation, as esterases function to degrade both natural materials like cereal wastes and toxic chemicals like polystyrene (Panda and Gowrishankar 2005; Tahir et al. 2013). Unfortunately, to our knowledge, the production of esterases by psychrophilic and psychrotrophic fungi has not been well characterized.

2.13.2 Pharmaceutical Products

The screening of filamentous fungi and yeasts in cold environments for synthesis of novel bioactive metabolites has been a fruitful endeavour. cold-adapted fungi have developed for themselves a huge repertoire of unique both known and novel secondary metabolites to survive in their harsh environments of climate extremes. There are numerous instances in the literature of psychrophilic and psychrotrophic fungi exhibiting antibacterial, antifungal, antiviral and antiprotozoal activities (Gonçalves et al. 2015; Zain ul Arifeen et al. 2019; Zucconi et al. 2020; Bratchkova and Ivanova 2011). *Oidiodendron truncatum* isolate GW25–13, an Antarctic psychrophilic fungus, shows high antifungal and anti-mycobacterium activity. Additionally, *Oidiodendron truncatum* GW25–13 also shows extraordinary cytotoxicity activity against several cancer cell lines, possibly through production of secalonic acid and chetracin, and has the potential to be developed further as a chemotherapeutic agent (Ding et al. 2016). In another instance of anticancer activity, 30 fungal strains isolated from soils and mosses of the Schirmacher Hills region (Dronning Maud Land, East Antarctica) of genera *Aspergillus*, *Coprinopsis* and *Trichosporon* exhibited production of L-asparaginase free of glutaminase and urease, which can be used as a therapeutic against lymphoblastic leukaemia (Godinho et al. 2019).

In a study performed by Gonçalves et al. (2015), extracts of fungal isolates collected from King George's Island, Antarctica, was used in various bioassays to test for antibacterial, antiprotozoal and anticancer activities. In this study, *Purpureocillium lilacinum* UFMGCB 1510 showed broad-spectrum antimicrobial activity effectively preventing the growth of amastigote intracellular forms of *Trypanosoma cruzi*, *Pennicium brasiliensis* and *Staphylococcus aureus*.

Psychrophilic yeasts from the Antarctic Peninsula (*Cryptococcus gastricus*, *C. victoriae* (syn. *Vishniacozyma victoriae*), *C. gilvescens*, *Leucosporidium* sp. and *Rhodotorula mucilaginosa*) exhibited anti-yeast activity by secretion of a novel protein factor (Troncoso et al. 2016).

Many fungal inhabitants from cold deep-sea environments are known to produce pharmaceutically valuable novel bioactive compounds. Among them are polyketides, alkaloids, polypeptides, phenolic derivatives and terpenoid compounds with the capabilities of treating human diseases (Wang et al. 2015b). Numerous strains of deep-sea fungi from diverse environments synthesized polyketides which showed impressive antimicrobial activity against pathogens *Bacillus subtilis*, *Acinetobacter baumannii*, *Escherichia coli*, *Staphylococcus aureus*, *Enterococcus faecalis*, *Staphylococcus epidermidis* and *Propionibacterium acnes*. In this study, numerous strains also showed high cytotoxic abilities (Zain ul Arifeen et al. 2019). Scequinadoline A extracted from marine fungus *Dichotomomyces cejpii* F31–1 showed activity against dengue-virus serotype 2 and is worth further investigation as an antiviral agent against the *Dengue virus* (Wu et al. 2018). A team of workers (Hagestad et al. 2019) found ten isolates of marine fungi from the Arctic Archipelago of Svalbard that showed antibacterial activity against Gram-positive human pathogens (*S. aureus*, *E. faecalis* and *S. agalactiae*) with strains of genera *Eurotiomycetes* and *Leotiomycetes* showing the strongest antimicrobial activity.

2.13.3 Bioremediation

The polar regions and other cold environments of the Earth suffer severely from petroleum hydrocarbon contamination. This is partly due to slower ecosystem recovery and low temperatures, but the issue is exacerbated due to human activities such as resource mining, accidental spills and improper waste disposal (Mair et al. 2013; van Dorst et al. 2020; Chaudhary and Kim 2019). The conventional bioremediation methods are not as effective in cold and temperate environments as there is poor bioavailability of hydrocarbons and enzymatic activity is reduced (Aislabie et al. 2006; Camenzuli and Freidman 2015). This has led to investigation of psychrophilic and psychrotrophic organisms for assimilation and degradation of environmental pollutants such as toxic metals and petroleum-based contaminants. While fungi and psychrophilic bacteria have been extensively studied for their bioremediation potential (Wang et al. 2015a), the use of cold-adapted fungi in bioremediation is an emerging field. Several fungi from Antarctica have been assessed for their ability to degrade pollutants. Antarctic strains of *Aspergillus fumigatus* demonstrated degradation phenolic compounds (Gerginova et al. 2013). Biodegradation of aliphatic and aromatic hydrocarbons using the filamentous fungus *Penicillium* sp. CHY-2 was explored by Govarthanan et al. (2017).

In an interesting study, Urbanek et al. (2021) showed the potential of cold-tolerant marine filamentous fungi *Geomyces* sp., *Fusarium* sp. and *Sclerotinia* sp. to degrade biodegradable plastics such as poly(e-caprolactone) (PCL), polybutylene succinate (PBS) and poly(butylene succinate-co-butylene adipate) (PBSA).

2.13.4 Pigment Production

Over the years several previously FDA-approved synthetic dyes were found to be toxic, carcinogenic and/or allergenic (Rao et al. 2017). As synthetic dyes became unfavourable with the public, natural pigments have found a surge of interest due to their sustainability, biodegradable nature and their gentler effect on the environment (Sajjad et al. 2020). Biological pigments formed by microorganisms are preferred as opposed to plant pigments as they are more stable, generally possess a greater solubility and can be easily cultivated by fermentation methods and have their pigments extracted by downstream processing (Rao et al. 2017). Between the years 2007 and 2011, international commerce of naturally sourced pigments has increased by 29% (Tuli et al. 2014). The microbiota of cold-adapted environments produces a variety of unique pigments in response to the environmental stresses already addressed in this chapter, which can be exploited for multiple applications. Several pigment-producing yeasts and filamentous fungi form the Arctic, Antarctic and Alpine regions have been isolated and identified over the years for the last decade (Sajjad et al. 2020).

Pigments produced predominantly by cold-adapted fungi include melanin, carotenoid derivatives (such as beta-carotene, lycopene, torule) and mycosporines. Pigment production is affected by external factors such as temperature and pH (Wang et al. 2017). Pigment-producing psychrophilic and psychrotrophic fungi can be utilized in the food industry, textile industry, cosmetics industry, pharmaceutical industry and other applications.

Numerous studies show the potential of pigment-producing fungi in the textile industry (Venil et al. 2020). Chadni et al. (2017) extensively reviewed the red pigment production of *Talaromyces verruculosa* and possible uses in the textile industry. Though there are several reports on cold-adapted pigment-producing fungi (Hassan et al. 2016; Sajjad et al. 2020), their potential in the textile industry is yet to be a studied in detail. Perhaps, the best-known application of natural pigments is as food colourants and additives in the food industry. *Monascus* pigments, riboflavin from *Ashbya gossypii* and beta-carotene from *Blakeslea trispora* are already being employed (Lagashetti et al. 2019).

Apart from being a colouring agent, pigments exhibit other properties, for example, antioxidant activity, antitumor activity and flavour enhancer depending on the structural conformation of the pigment. Bisht et al. (2020) identified and extracted red-coloured pigments from psychrophilic yeast *Rhodonellum psychrophilum* isolate GL8 from the high-altitude Pangong Tso Lake (Leh Ladakh, India). These red pigments were found to exhibit antimicrobial, antioxidant and cytotoxicity properties.

2.14 Agriculture

A number of psychrophilic and psychrotrophic fungi were identified to produce metabolites that are effective against phytopathogens and thus can be used as biological control agents (Pandey et al. 2018; Zucconi et al. 2020). Antarctic psychrotrophic yeasts *Leucosporidium scottii* At17 and two isolates of *Candida sake* were effective in suppressing fungal diseases caused by postharvest fungal pathogens of apple (*A. alternata*, *A. tenuissima*, *A. arborescens*, *Botrytis cinerea* and *P. expansum*), using tactics like biofilm production and synthesis of antifungal volatile compounds (Arrarte et al. 2017; Vero et al. 2012).

Phosphate-solubilizing fungi mobilize bound phosphate by converting inorganic phosphate to organic phosphate via the production of phosphatase enzymes and/or secretion organic acids, a process that is essential for plant yield and health. Cold-tolerant fungi have exhibited phosphate-solubilizing activity, which can be exploited for commercial production of biofertilizers (Hassan et al. 2016). Psychrotrophic strains of *Paecilomyces hepiali* and *Penicillium* from Himalayan soil were found to possess phosphate-solubilizing activity (Rinu and Pandey 2010). Cold-tolerant yeast strains from a crater of Xinantécatl volcano (Mexico) belonging to genera *Rhodotorula*, *Mrakia* and *Naganishia* inhibited growth of phytopathogens, with certain strains exhibiting phosphate-solubilizing activity as well (Tapia-Vázquez et al. 2020).

2.15 Conclusion

The cold-adapted fungi are a taxonomically heterogeneous group of great ecological and potential industrial importance. Historically, the cold-adapted fungi have been broadly separated into either "psychrotrophs" or "psychrophiles", though in recent years, this has largely been replaced by the more fitting terminology "eurypyscrophiles" and "stenopyscrophiles". It is an exciting era for the study of cold-adapted fungi as several fungal taxonomic investigations have been conducted throughout the world in the past decade. The diverse nature of global cold environments is reflected in varying degrees of abundance, richness and diversity of fungal communities present in various substrata such as permafrost, snow, glacial ice, lakes, rocks and marine waters. Both culture-dependent and culture-independent techniques have further contributed to the advancement of this field of study. Future directions for the field of psychrophilic and psychrotrophic fungi include identifying intrinsic and extrinsic factors influencing fungal communities, elucidating the molecular and genomic basis of adaptations to extremophilic conditions and the progress of cold-adapted fungi for industrial use. Determining the molecular basis for adaptation is essential to understanding fungi. For example, several cosmopolitan fungi have been isolated in the polar regions, but the adaptations these fungi possess are poorly understood. The function of cold-adapted fungi in their respective ecosystems still requires further investigations. Accessibility to extreme cold environments remains a problem and hinders research. The composition and ecological roles of fungal communities not only provide us with a more complete picture of a specific ecosystem but also set a working model for future comparisons. This is particularly of necessity as global climate change is drastically altering the cryosphere. Many cold-adapted fungi are known to produce an arsenal of secondary metabolites of known industrial importance such as pharmaceutical products, extracellular enzymes and pigments. However, with the exception of very few, none of the metabolites have been fully exploited and developed for commercial production. Hence, the development of large-scale production of cold-adapted fungal metabolites for commercial use remains a priority in the future.

References

Abrego N, Huotari T, Tack AJM, Lindahl BD, Tikhonov G, Somervuo P, Martin Schmidt N, Ovaskainen O, Roslin T (2020) Higher host plant specialization of root-associated endophytes than mycorrhizal fungi along an arctic elevational gradient. Ecol Evol 10:8989–9002. https://doi.org/10.1002/ece3.6604

ACIA (2005) Arctic climate impact assessment. Cambridge University Press, Cambridge

Adapa V, Ramya LN, Pulicherla KK, Rao KRSS (2014) Cold active pectinases: advancing the food industry to the next generation. Appl Biochem Biotechnol 172:2324–2337. https://doi.org/10.1007/s12010-013-0685-1

Aislabie J, Saul DJ, Foght JM (2006) Bioremediation of hydrocarbon-contaminated polar soils. Extremophiles 10:171–179. https://doi.org/10.1007/s00792-005-0498-4

Alcaíno J, Cifuentes V, Baeza M (2015) Physiological adaptations of yeasts living in cold environments and their potential applications. World J Microbiol Biotechnol 31:1467–1473. https://doi.org/10.1007/s11274-015-1900-8

Alexopoulos CJ, Mims CW (1979) Introductory mycology, 3rd edn. Wiley, New York

Almeida F, Rodrigues ML, Coelho C (2019) The still underestimated problem of fungal diseases worldwide. Front Microbiol 10:214. https://doi.org/10.3389/fmicb.2019.00214

Alshannaq A, Yu J-H (2017) Occurrence, toxicity, and analysis of major mycotoxins in food. IJERPH 14:632. https://doi.org/10.3390/ijerph14060632

AMAP (2009) Amap assessment 2009: human health in the Arctic. Arctic monitoring and assessment programme. AMAP, Oslo

Amend A, Burgaud G, Cunliffe M, Edgcomb VP, Ettinger CL, Gutiérrez MH, Heitman J, Hom EFY, Ianiri G, Jones AC, Kagami M, Picard KT, Quandt CA, Raghukumar S, Riquelme M, Stajich J, Vargas-Muñiz J, Walker AK, Yarden O, Gladfelter AS (2019) Fungi in the marine environment: open questions and unsolved problems. MBio 10:e01189. https://doi.org/10.1128/mbio.01189-18

Archer SDJ, de los Ríos A, Lee KC, Niederberger TS, Cary SC, Coyne KJ, Douglas S, Lacap-Bugler DC, Pointing SB (2016) Endolithic microbial diversity in sandstone and granite from the McMurdo dry valleys, Antarctica. Polar Biol 40:997–1006. https://doi.org/10.1007/s00300-016-2024-9

Arrarte E, Garmendia G, Rossini C, Wisniewski M, Vero S (2017) Volatile organic compounds produced by Antarctic strains of Candida sake play a role in the control of postharvest pathogens of apples. Biol Control 109:14–20. https://doi.org/10.1016/j.biocontrol.2017.03.002

Barone V, Tangherlini R, Dell'Anno D, Corinaldesi (2019) Marine fungi: biotechnological perspectives from deep-hypersaline anoxic basins. Diversity 11:113. https://doi.org/10.3390/d11070113

Batista TM, Hilario HO, de Brito GAM, Moreira RG, Furtado C, de Menezes GCA, Rosa CA, Rosa LH, Franco GR (2020) Whole-genome sequencing of the endemic Antarctic fungus Antarctomyces pellizariae reveals an ice-binding protein, a scarce set of secondary metabolites gene clusters and provides insights on Thelebolales phylogeny. Genomics 112:2915–2921. https://doi.org/10.1016/j.ygeno.2020.05.004

Birgisson H, Delgado O, García Arroyo L, Hatti-Kaul R, Mattiasson B (2003) Cold-adapted yeasts as producers of cold-active polygalacturonases. Extremophiles 7:185–193. https://doi.org/10.1007/s00792-002-0310-7

Bisht G, Srivastava S, Kulshreshtha R, Sourirajan A, Baumler DJ, Dev K (2020) Applications of red pigments from psychrophilic Rhodonellum psychrophilum GL8 in health, food and antimicrobial finishes on textiles. Process Biochem 94:15–29. https://doi.org/10.1016/j.procbio.2020.03.021

Bjorbækmo M, Carlsen T, Brysting A, Vrålstad T, Høiland K, Ugland K, Geml J, Schumacher T, Kauserud H (2010) High diversity of root associated fungi in both alpine and arctic Dryas octopetala. BMC Plant Biol 10:244. https://doi.org/10.1186/1471-2229-10-244

Blanchette RA, Held BW, Hellmann L, Millman L, Büntgen U (2016) Arctic driftwood reveals unexpectedly rich fungal diversity. Fungal Ecol 23:58–65. https://doi.org/10.1016/j.funeco.2016.06.001

Boetius A, Anesio AM, Deming JW, Mikucki JA, Rapp JZ (2015) Microbial ecology of the cryosphere: sea ice and glacial habitats. Nat Rev Microbiol 13:677–690. https://doi.org/10.1038/nrmicro3522

Boo SY, Wong CMVL, Rodrigues KF, Najimudin N, Murad AMA, Mahadi NM (2012) Thermal stress responses in Antarctic yeast, *Glaciozyma antarctica* PI12, characterized by real-time quantitative PCR. Polar Biol 36:381–389. https://doi.org/10.1007/s00300-012-1268-2

Botnen S, Thoen E, Eidesen PB, Krabberød AK, Kauserud H (2020) Community composition of arctic root-associated fungi mirrors host plant phylogeny. FEMS Microbiol Ecol 96(11):1–12. https://doi.org/10.1093/femsec/fiaa185

Botnen S, Vik U, Carlsen T, Eidesen PB, Davey ML, Kauserud H (2014) Low host specificity of root-associated fungi at an Arctic site. Mol Ecol 23:975–985. https://doi.org/10.1111/mec.12646

Bratchkova A, Ivanova V (2011) Bioactive metabolites produced by microorganisms collected in Antarctica and the Arctic. Biotechnol Biotechnol Equip 25:1–7. https://doi.org/10.5504/bbeq.2011.0116

Brunner I, Frey B, Hartmann M, Zimmermann S, Graf F, Suz LM, Niskanen T, Bidartondo MI, Senn-Irlet B (2017) Ecology of alpine macrofungi-combining historical with recent data. Front Microbiol 8:2066. https://doi.org/10.3389/fmicb.2017.02066

Butinar L, Strmole T, Gunde-Cimerman N (2011) Relative incidence of Ascomycetous yeasts in Arctic coastal environments. Microb Ecol 61:832–843. https://doi.org/10.1007/s00248-010-9794-3

Camenzuli D, Freidman BL (2015) On-site and in situ remediation technologies applicable to petroleum hydrocarbon contaminated sites in the Antarctic and Arctic. Polar Res 34:24492. https://doi.org/10.3402/polar.v34.24492

Carrasco M, Rozas JM, Alcaíno J, Cifuentes V, Baeza M (2019) Pectinase secreted by psychrotolerant fungi: identification, molecular characterization and heterologous expression of a cold-active polygalacturonase from Tetracladium sp. Microb Cell Fact 18:45. https://doi.org/10.1186/s12934-019-1092-2

Carrasco M, Villarreal P, Barahona S, Alcaíno J, Cifuentes V, Baeza M (2016) Screening and characterization of amylase and cellulase activities in psychrotolerant yeasts. BMC Microbiol 16:21. https://doi.org/10.1186/s12866-016-0640-8

Casadevall A (2018) Fungal diseases in the 21st century: the near and far horizons. Pain 3:183. https://doi.org/10.20411/pai.v3i2.249

Cavicchioli R (2015) On the concept of a psychrophile. ISME J 10:793–795. https://doi.org/10.1038/ismej.2015.160

Cavicchioli R, Charlton T, Ertan H, Omar SM, Siddiqui KS, Williams TJ (2011) Biotechnological uses of enzymes from psychrophiles. J Microbial Biotechnol 4:449–460. https://doi.org/10.1111/j.1751-7915.2011.00258.x

Chadha BS, Kaur B, Basotra N, Tsang A, Pandey A (2019) Thermostable xylanases from thermophilic fungi and bacteria: current perspective. Bioresour Technol 277:195–203. https://doi.org/10.1016/j.biortech.2019.01.044

Chadni Z, Rahaman MH, Jerin I, Hoque KMF, Reza MA (2017) Extraction and optimisation of red pigment production as secondary metabolites from Talaromyces verruculosus and its potential use in textile industries. Mycology 8:48–57. https://doi.org/10.1080/21501203.2017.1302013

Chaudhary DK, Kim J (2019) New insights into bioremediation strategies for oil-contaminated soil in cold environments. Int Biodeter Biodegr 142:58–72. https://doi.org/10.1016/j.ibiod.2019.05.001

Chuang WY, Hsieh YC, Lee T-T (2020) The effects of fungal feed additives in animals: a review. Animals 10:805. https://doi.org/10.3390/ani10050805

Coleine C, Stajich JE, Zucconi L, Onofri S, Pombubpa NE, Franks A, Buzzini P, Selbmann L (2018) Antarctic Cryptoendolithic fungal communities are highly adapted and dominated by Lecanoromycetes and Dothideomycetes. Front Microbiol 9:1392. https://doi.org/10.3389/fmicb.2018.01392

Collins T, Margesin R (2019) Psychrophilic lifestyles: mechanisms of adaptation and biotechnological tools. Appl Microbiol Biotechnol 103:2857–2871. https://doi.org/10.1007/s00253-019-09659-5

Comeau AM, Vincent WF, Bernier L, Lovejoy C (2016) Novel chytrid lineages dominate fungal sequences in diverse marine and freshwater habitats. Sci Rep 6:30120. https://doi.org/10.1038/srep30120

Convey P (2013) Antarctic ecosystems. In: Encyclopedia of biodiversity, 2nd edn. Academic Press, Amsterdam, pp 179–181

Cox F, Newsham KK, Bol R, Dungait JAJ, Robinson CH (2016) Not poles apart: Antarctic soil fungal communities show similarities to those of the distant Arctic. Ecol Lett 19:528–536. https://doi.org/10.1111/ele.12587

Cox F, Newsham KK, Robinson CH (2019) Endemic and cosmopolitan fungal taxa exhibit differential abundances in total and active communities of Antarctic soils. Environ Microbiol 21:1586–1596. https://doi.org/10.1111/1462-2920.14533

da Silva TH, Silva DAS, de Oliveira FS, Schaefer CEGR, Rosa CA, Rosa LH (2020) Diversity, distribution, and ecology of viable fungi in permafrost and active layer of maritime Antarctica. Extremophiles 24:565–576. https://doi.org/10.1007/s00792-020-01176-y

de la Torre JR, Goebel BM, Friedmann EI, Pace NR (2003) Microbial diversity of Cryptoendolithic communities from the McMurdo dry valleys, Antarctica. AEM 69:3858–3867. https://doi.org/10.1128/aem.69.7.3858-3867.2003

de Oliveira TB, Gomes E, Rodrigues A (2014) Thermophilic fungi in the new age of fungal taxonomy. Extremophiles 19:31–37. https://doi.org/10.1007/s00792-014-0707-0

Del-Cid A, Ubilla P, Ravanal M-C, Medina E, Vaca I, Levicán G, Eyzaguirre J, Chávez R (2013) Cold-active xylanase produced by fungi associated with Antarctic marine sponges. Appl Biochem Biotechnol 172:524–532. https://doi.org/10.1007/s12010-013-0551-1

Ding Z, Li L, Che Q, Li D, Gu Q, Zhu T (2016) Richness and bioactivity of culturable soil fungi from the Fildes peninsula, Antarctica. Extremophiles 20:425–435. https://doi.org/10.1007/s00792-016-0833-y

Duarte AWF, Barato MB, Nobre FS, Polezel DA, de Oliveira TB, dos Santos JA, Rodrigues A, Sette LD (2018) Production of cold-adapted enzymes by filamentous fungi from King George Island, Antarctica. Polar Biol 41:2511–2521. https://doi.org/10.1007/s00300-018-2387-1

Duarte AWF, dos Santos JA, Vianna MV, Vieira JMF, Mallagutti VH, Inforsato FJ, Wentzel LCP, Lario LD, Rodrigues A, Pagnocca FC, Pessoa Junior A, Durães Sette L (2017) Cold-adapted enzymes produced by fungi from terrestrial and marine Antarctic environments. Crit Rev Biotechnol 38:600–619. https://doi.org/10.1080/07388551.2017.1379468

Durán P, Barra PJ, Jorquera MA, Viscardi S, Fernandez C, Paz C, Mora M d l L, Bol R (2019) Occurrence of soil fungi in Antarctic pristine environments. Front Bioeng Biotechnol 7:28. https://doi.org/10.3389/fbioe.2019.00028

Edwards A, Douglas B, Anesio AM, Rassner SM, Irvine-Fynn TDL, Sattler B, Griffith GW (2013) A distinctive fungal community inhabiting cryoconite holes on glaciers in Svalbard. Fungal Ecol 6:168–176. https://doi.org/10.1016/j.funeco.2012.11.001

Ekkers DM, Cretoiu MS, Kielak AM, van Elsas JD (2011) The great screen anomaly—a new frontier in product discovery through functional metagenomics. Appl Microbiol Biotechnol 93:1005–1020. https://doi.org/10.1007/s00253-011-3804-3

Fenice M (2016) The Psychrotolerant Antarctic fungus Lecanicillium muscarium CCFEE 5003: a powerful producer of cold-tolerant Chitinolytic enzymes. Molecules 21:447. https://doi.org/10.3390/molecules21040447

Fisher MC, Gow NAR, Gurr SJ (2016) Tackling emerging fungal threats to animal health, food security and ecosystem resilience. Philos Trans R Soc B 371:20160332. https://doi.org/10.1098/rstb.2016.0332

Fisher MC, Henk Daniel A, Briggs CJ, Brownstein JS, Madoff LC, McCraw SL, Gurr SJ (2012) Emerging fungal threats to animal, plant and ecosystem health. Nature 484:186–194. https://doi.org/10.1038/nature10947

Fleischhacker A, Grube M, Kopun T, Hafellner J, Muggia L (2015) Community analyses uncover high diversity of Lichenicolous fungi in alpine habitats. Microb Ecol 70:348–360. https://doi.org/10.1007/s00248-015-0579-6

Fornbacke M, Clarsund M (2013) Cold-adapted proteases as an emerging class of therapeutics. Infect Dis Ther 2:15–26. https://doi.org/10.1007/s40121-013-0002-x

French HM (2007) The periglacial environment, 3rd edn. Wiley, West Sussex, p 458

Friedman S (2019) Emerging fungal infections: new patients, new patterns, and new pathogens. JoF 5:67. https://doi.org/10.3390/jof5030067

Fujimura KE, Egger KN (2012) Host plant and environment influence community assembly of high Arctic root-associated fungal communities. Fungal Ecol 5:409–418. https://doi.org/10.1016/j.funeco.2011.12.010

Furbino LE, Pellizzari FM, Neto PC, Rosa CA, Rosa LH (2017) Isolation of fungi associated with macroalgae from maritime Antarctica and their production of agarolytic and carrageenolytic activities. Polar Biol 41:527–535. https://doi.org/10.1007/s00300-017-2213-1

Furhan J (2020) Adaptation, production, and biotechnological potential of cold-adapted proteases from psychrophiles and psychrotrophs: recent overview. J Genet Eng Biotechnol 18:36. https://doi.org/10.1186/s43141-020-00053-7

Gerday C, Aittaleb M, Bentahir M, Chessa J-P, Claverie P, Collins T, D'Amico S, Dumont J, Garsoux G, Georlette D, Hoyoux A, Lonhienne T, Meuwis M-A, Feller G (2000) Cold-adapted enzymes: from fundamentals to biotechnology. Trends Biotechnol 18:103–107. https://doi.org/10.1016/s0167-7799(99)01413-4

Gerginova M, Manasiev J, Yemendzhiev H, Terziyska A, Peneva N, Alexieva Z (2013) Biodegradation of phenol by Antarctic strains of Aspergillus fumigatus. Z Naturforsch C 68:0384. https://doi.org/10.5560/znc.2013.68c0384

Gittel A, Bárta J, Kohoutová I, Mikutta R, Owens S, Gilbert J, Schnecker J, Wild B, Hannisdal B, Maerz J, Lashchinskiy N, Čapek P, Šantrůčková H, Gentsch N, Shibistova O, Guggenberger G, Richter A, Torsvik VL, Schleper C, Urich T (2013) Distinct microbial communities associated with buried soils in the Siberian tundra. ISME J 8:841–853. https://doi.org/10.1038/ismej.2013.219

Gleason FH, Gadd GM, Pitt JI, Larkum AWD (2017) The roles of endolithic fungi in bioerosion and disease in marine ecosystems. I General concepts. Mycology 8:205–215. https://doi.org/10.1080/21501203.2017.1352049

Godinho VM, de Paula MTR, Silva DAS, Paresque K, Martins AP, Colepicolo P, Rosa CA, Rosa LH (2019) Diversity and distribution of hidden cultivable fungi associated with marine animals of Antarctica. Fungal Biol 123:507–516. https://doi.org/10.1016/j.funbio.2019.05.001

Godinho VM, Gonçalves VN, Santiago IF, Figueredo HM, Vitoreli GA, Schaefer CEGR, Barbosa EC, Oliveira JG, Alves TMA, Zani CL, Junior PAS, Murta SMF, Romanha AJ, Kroon EG, Cantrell CL, Wedge DE, Duke SO, Ali A, Rosa CA, Rosa LH (2015) Diversity and bioprospection of fungal community present in oligotrophic soil of continental Antarctica. Extremophiles 19:585–596. https://doi.org/10.1007/s00792-015-0741-6

Gomes ECQ, Godinho VM, Silva DAS, de Paula MTR, Vitoreli GA, Zani CL, Alves TMA, Junior PAS, Murta SMF, Barbosa EC, Oliveira JG, Oliveira FS, Carvalho CR, Ferreira MC, Rosa CA, Rosa LH (2018) Cultivable fungi present in Antarctic soils: taxonomy, phylogeny, diversity, and bioprospecting of antiparasitic and herbicidal metabolites. Extremophiles 22:381–393. https://doi.org/10.1007/s00792-018-1003-1

Gonçalves VN, Carvalho CR, Johann S, Mendes G, Alves TMA, Zani CL, Junior PAS, Murta SMF, Romanha AJ, Cantrell CL, Rosa CA, Rosa LH (2015) Antibacterial, antifungal and antiprotozoal activities of fungal communities present in different substrates from Antarctica. Polar Biol 38:1143–1152. https://doi.org/10.1007/s00300-015-1672-5

Gong J-S, Lu Z-M, Li H, Zhou Z-M, Shi J-S, Xu Z-H (2013) Metagenomic technology and genome mining: emerging areas for exploring novel nitrilases. Appl Microbiol Biotechnol 97:6603–6611. https://doi.org/10.1007/s00253-013-4932-8

Goordial J, Davila A, Lacelle D, Pollard W, Marinova MM, Greer CW, DiRuggiero J, McKay CP, Whyte LG (2016) Nearing the cold-arid limits of microbial life in permafrost of an upper dry valley, Antarctica. ISME J 10:1613–1624. https://doi.org/10.1038/ismej.2015.239

Gottschalk LMF, Oliveira RA, Bon EP d S (2010) Cellulases, xylanases, β-glucosidase and ferulic acid esterase produced by Trichoderma and Aspergillus act synergistically in the hydrolysis of sugarcane bagasse. Biochem Eng J 51:72–78. https://doi.org/10.1016/j.bej.2010.05.003

Gounot A-M (1986) Psychrophilic and psychrotrophic microorganisms. Experientia 42:1192–1197. https://doi.org/10.1007/bf01946390

Govarthanan M, Fuzisawa S, Hosogai T, Chang Y-C (2017) Biodegradation of aliphatic and aromatic hydrocarbons using the filamentous fungus Penicillium sp. CHY-2 and characterization of its manganese peroxidase activity. RSC Adv 7:20716–20723. https://doi.org/10.1039/c6ra28687a

Hagestad OC, Andersen JH, Altermark B, Hansen E, Rämä T (2019) Cultivable marine fungi from the Arctic archipelago of Svalbard and their antibacterial activity. Mycology 11:230–242. https://doi.org/10.1080/21501203.2019.1708492

Hamid R, Javed S, Ahmad M, Ahmad M, Abdin M, Khan M, Musarrat J (2013) Chitinases: an update. J Pharm Bioallied Sci 5:21. https://doi.org/10.4103/0975-7406.106559

Hao J-H, Sun M (2014) Purification and characterization of a cold alkaline protease from a psychrophilic Pseudomonas aeruginosa HY 1215. Appl Biochem Biotechnol 175:715–722. https://doi.org/10.1007/s12010-014-1315-2

Hassan N, Rafiq M, Hayat M, Shah AA, Hasan F (2016) Psychrophilic and psychrotrophic fungi: a comprehensive review. Rev Environ Sci Biotechnol 15:147–172. https://doi.org/10.1007/s11157-016-9395-9

Hassett BT, Borrego EJ, Vonnahme TR, Rämä T, Kolomiets MV, Gradinger R (2019) Arctic marine fungi: biomass, functional genes, and putative ecological roles. ISME J 13:1484–1496. https://doi.org/10.1038/s41396-019-0368-1

He J, Yang Z, Hu B, Ji X, Wei Y, Lin L, Zhang Q (2015) Correlation of polyunsaturated fatty acids with the cold adaptation of Rhodotorula glutinis. Yeast 32:683–690. https://doi.org/10.1002/yea.3095

Hill PW, Broughton R, Bougoure J, Havelange W, Newsham KK, Grant H, Murphy DV, Clode P, Ramayah S, Marsden KA, Quilliam RS, Roberts P, Brown C, Read DJ, Deluca TH, Bardgett RD, Hopkins DW, Jones DL (2019) Angiosperm symbioses with non-mycorrhizal fungal partners enhance N acquisition from ancient organic matter in a warming maritime Antarctic. Ecol Lett 22:2111–2119. https://doi.org/10.1111/ele.13399

Hirose D, Hobara S, Matsuoka S, Kato K, Tanabe Y, Uchida M, Kudoh S, Osono T (2016) Diversity and community assembly of moss-associated fungi in ice-free coastal outcrops of continental Antarctica. Fungal Ecol 24:94–101. https://doi.org/10.1016/j.funeco.2016.09.005

Hutchinson MI, Powell AJ, Herrera J, Natvig DO (2019) New perspectives on the distribution and roles of thermophilic fungi. In: Fungi in extreme environments: ecological role and biotechnological significance. Springer, Cham, pp 59–80

Jones EBG, Suetrong S, Sakayaroj J, Bahkali AH, Abdel-Wahab MA, Boekhout T, Pang K-L (2015) Classification of marine Ascomycota, Basidiomycota, Blastocladiomycota and Chytridiomycota. Fungal Divers 73:1–72. https://doi.org/10.1007/s13225-015-0339-4

Juturu V, Wu JC (2012) Microbial xylanases: engineering, production and industrial applications. Biotechnol Adv 30:1219–1227. https://doi.org/10.1016/j.biotechadv.2011.11.006

Kamoun S, Furzer O, Jones JDG, Judelson HS, Ali GS, Dalio RJD, Roy SG, Schena L, Zambounis A, Panabières F, Cahill D, Ruocco M, Figueiredo A, Chen X-R, Hulvey J, Stam R, Lamour K, Gijzen M, Tyler BM, Grünwald NJ, Mukhtar MS, Tomé DFA, Tör M, Van Den Ackerveken G, McDowell J, Daayf F, Fry WE, Lindqvist-Kreuze H, Meijer HJG, Petre B, Ristaino J, Yoshida K, Birch PRJ, Govers F (2014) The top 10 oomycete pathogens in molecular plant pathology. Mol Plant Pathol 16:413–434. https://doi.org/10.1111/mpp.12190

Kilias ES, Junges L, Šupraha L, Leonard G, Metfies K, Richards TA (2020) Chytrid fungi distribution and co-occurrence with diatoms correlate with sea ice melt in the Arctic Ocean. Commun Biol 3:183. https://doi.org/10.1038/s42003-020-0891-7

Korniłłowicz-Kowalska T, Kitowski I (2012) *Aspergillus fumigatus* and other thermophilic fungi in nests of wetland birds. Mycopathologia 175:43–56. https://doi.org/10.1007/s11046-012-9582-3

Krishnan A, Convey P, Gonzalez-Rocha G, Alias SA (2014) Production of extracellular hydrolase enzymes by fungi from King George Island. Polar Biol 39:65–76. https://doi.org/10.1007/s00300-014-1606-7

Kuehn HH, Gunderson MF (1963) Psychrophilic and mesophilic fungi in frozen food products. Appl Microbiol 11(4):352–356

Kumar M, Brar A, Vivekanand V, Pareek N (2018) Bioconversion of chitin to bioactive Chitooligosaccharides: amelioration and coastal pollution reduction by microbial resources. Marine Biotechnol 20:269–281. https://doi.org/10.1007/s10126-018-9812-x

Kundys A, Białecka-Florjańczyk E, Fabiszewska A, Małajowicz J (2017) Candida antarctica lipase B as catalyst for cyclic esters synthesis, their polymerization and degradation of aliphatic polyesters. J Polym Environ 26:396–407. https://doi.org/10.1007/s10924-017-0945-1

Lagashetti AC, Dufossé L, Singh SK, Singh PN (2019) Fungal pigments and their prospects in different industries. Microorganisms 7:604. https://doi.org/10.3390/microorganisms7120604

Latijnhouwers M, de Wit PJGM, Govers F (2003) Oomycetes and fungi: similar weaponry to attack plants. Trends Microbiol 11(10):462–469. https://doi.org/10.1016/j.tim.2003.08.002

Lopez-Lopez O, Cerdan M, Siso M (2014) New extremophilic lipases and esterases from metagenomics. CPPS 15:445–455. https://doi.org/10.2174/1389203715666140228153801

Low C-Y, Rotstein C (2011) Emerging fungal infections in immunocompromised patients. F1000 Med Rep 3:14. https://doi.org/10.3410/m3-14

Mahapatra S, Banerjee D (2013) Fungal exopolysaccharide: production, composition and applications. Microbiol Insights 6:10957. https://doi.org/10.4137/mbi.s10957

Maheshwari R, Bharadwaj G, Bhat MK (2000) Thermophilic fungi: their physiology and enzymes. Microbiol Mol Biol Rev 64:461–488. https://doi.org/10.1128/mmbr.64.3.461-488.2000

Mair J, Schinner F, Margesin R (2013) A feasibility study on the bioremediation of hydrocarbon-contaminated soil from an alpine former military site: effects of temperature and biostimulation. Cold Reg Sci Technol 96:122–128. https://doi.org/10.1016/j.coldregions.2013.07.006

Margesin R, Fauster V, Fonteyne P-A (2005) Characterization of cold-active pectate lyases from psychrophilic Mrakia frigida. Lett Appl Microbiol 40:453–459. https://doi.org/10.1111/j.1472-765x.2005.01704.x

Melo IS, Santos SN, Rosa LH, Parma MM, Silva LJ, Queiroz SCN, Pellizari VH (2013) Isolation and biological activities of an endophytic Mortierella alpina strain from the Antarctic moss Schistidium antarctici. Extremophiles 18:15–23. https://doi.org/10.1007/s00792-013-0588-7

Merín MG, Mendoza LM, Farías ME, Morata de Ambrosini VI (2011) Isolation and selection of yeasts from wine grape ecosystem secreting cold-active pectinolytic activity. Int J Food Microbiol 147:144–148. https://doi.org/10.1016/j.ijfoodmicro.2011.04.004

Molnár Z, Bódai V, Szakacs G, Erdélyi B, Fogarassy Z, Sáfrán G, Varga T, Kónya Z, Tóth-Szeles E, Szűcs R, Lagzi I (2018) Green synthesis of gold nanoparticles by thermophilic filamentous fungi. Sci Rep 8:3943. https://doi.org/10.1038/s41598-018-22112-3

Muggia L, Fleischhacker A, Kopun T, Grube M (2015) Extremotolerant fungi from alpine rock lichens and their phylogenetic relationships. Fungal Divers 76:119–142. https://doi.org/10.1007/s13225-015-0343-8

Naga Padma P, Anuradha K, Reddy G (2011) Pectinolytic yeast isolates for cold-active polygalacturonase production. Innov Food Sci Emerg Technol 12:178–181. https://doi.org/10.1016/j.ifset.2011.02.001

Nagano Y, Nagahama T, Hatada Y, Nunoura T, Takami H, Miyazaki J, Takai K, Horikoshi K (2010) Fungal diversity in deep-sea sediments—the presence of novel fungal groups. Fungal Ecol 3:316–325. https://doi.org/10.1016/j.funeco.2010.01.002

Nakagawa T, Nagaoka T, Taniguchi S, Miyaji T, Tomizuka N (2004) Isolation and characterization of psychrophilic yeasts producing cold-adapted pectinolytic enzymes. Lett Appl Microbiol 38:383–387. https://doi.org/10.1111/j.1472-765x.2004.01503.x

Naranjo-Ortiz MA, Gabaldón T (2019) Fungal evolution: diversity, taxonomy and phylogeny of the fungi. Biol Rev 94:2101–2137. https://doi.org/10.1111/brv.12550

Newsham KK, Garnett MH, Robinson CH, Cox F (2018) Discrete taxa of saprotrophic fungi respire different ages of carbon from Antarctic soils. Sci Rep 8:1–11. https://doi.org/10.1038/s41598-018-25877-9

Newsham KK, Upson R, Read DJ (2009) Mycorrhizas and dark septate root endophytes in polar regions. Fungal Ecol 2:10–20. https://doi.org/10.1016/j.funeco.2008.10.005

Noffsinger C, Cripps CL, Horak E (2020) A 200-year history of arctic and alpine fungi in North America: early sailing expeditions to the molecular era. Arct Antarct Alp Res 52(1):323–340. https://doi.org/10.1080/15230430.2020.1771869

Onofri S, Selbmann L, Zucconi L, Pagano S (2004) Antarctic microfungi as models for exobiology. Planet Space Sci 52:229–237. https://doi.org/10.1016/j.pss.2003.08.019

Oses-Pedraza R, Torres-Díaz C, Lavín P, Retamales-Molina P, Atala C, Gallardo-Cerda J, Acuña-Rodríguez IS, Molina-Montenegro MA (2020) Root endophytic Penicillium promotes growth of Antarctic vascular plants by enhancing nitrogen mineralization. Extremophiles 24:721–732. https://doi.org/10.1007/s00792-020-01189-7

Oyeleye A, Normi YM (2018) Chitinase: diversity, limitations, and trends in engineering for suitable applications. Biosci Rep 38(4):1–21. https://doi.org/10.1042/bsr20180323

Pacelli C, Selbmann L, Moeller R, Zucconi L, Fujimori A, Onofri S (2017) Cryptoendolithic Antarctic black fungus Cryomyces antarcticus irradiated with accelerated helium ions: survival and metabolic activity, DNA and ultrastructural damage. Front Microbiol 8:2002. https://doi.org/10.3389/fmicb.2017.02002

Panda T, Gowrishankar BS (2005) Production and applications of esterases. Appl Microbiol Biotechnol 67:160–169. https://doi.org/10.1007/s00253-004-1840-y

Pandey A, Dhakar K, Jain R, Pandey N, Gupta VK, Kooliyottil R, Dhyani A, Malviya MK, Adhikari P (2018) Cold adapted fungi from Indian Himalaya: untapped source for bioprospecting. Proc Natl Acad Sci, India, Sect B Biol Sci 89:1125–1132. https://doi.org/10.1007/s40011-018-1002-0

Paterson R, Lima N (2017) Thermophilic fungi to dominate Aflatoxigenic/Mycotoxigenic fungi on food under global warming. IJERPH 14:199. https://doi.org/10.3390/ijerph14020199

Payne CM, Knott BC, Mayes HB, Hansson H, Himmel ME, Sandgren M, Ståhlberg J, Beckham GT (2015) Fungal cellulases. Chem Rev 115:1308–1448. https://doi.org/10.1021/cr500351c

Perini L, Gostinčar C, Anesio AM, Williamson C, Tranter M, Gunde-Cimerman N (2019a) Darkening of the Greenland ice sheet: fungal abundance and diversity are associated with algal bloom. Front Microbiol 10:557. https://doi.org/10.3389/fmicb.2019.00557

Perini L, Gostinčar C, Gunde-Cimerman N (2019b) Fungal and bacterial diversity of Svalbard subglacial ice. Sci Rep 9:20230. https://doi.org/10.1038/s41598-019-56290-5

Pernice M, Raina J-B, Rädecker N, Cárdenas A, Pogoreutz C, Voolstra CR (2019) Down to the bone: the role of overlooked endolithic microbiomes in reef coral health. ISME J 14:325–334. https://doi.org/10.1038/s41396-019-0548-z

Phitsuwan P, Laohakunjit N, Kerdchoechuen O, Kyu KL, Ratanakhanokchai K (2012) Present and potential applications of cellulases in agriculture, biotechnology, and bioenergy. Folia Microbiol 58:163–176. https://doi.org/10.1007/s12223-012-0184-8

Polizeli MLTM, Rizzatti ACS, Monti R, Terenzi HF, Jorge JA, Amorim DS (2005) Xylanases from fungi: properties and industrial applications. Appl Microbiol Biotechnol 67:577–591. https://doi.org/10.1007/s00253-005-1904-7

Poniecka EA, Bagshaw EA, Sass H, Segar A, Webster G, Williamson C, Anesio AM, Tranter M (2020) Physiological capabilities of Cryoconite hole microorganisms. Front Microbiol 11:1783. https://doi.org/10.3389/fmicb.2020.01783

Poveda G, Gil-Durán C, Vaca I, Levicán G, Chávez R (2018) Cold-active pectinolytic activity produced by filamentous fungi associated with Antarctic marine sponges. Biol Res 51:28. https://doi.org/10.1186/s40659-018-0177-4

Rahman AT, Arai T, Yamauchi A, Miura A, Kondo H, Ohyama Y, Tsuda S (2019) Ice recrystallization is strongly inhibited when antifreeze proteins bind to multiple ice planes. Sci Rep 9:2212. https://doi.org/10.1038/s41598-018-36546-2

Rämä T, Davey ML, Nordén J, Halvorsen R, Blaalid R, Mathiassen GH, Alsos IG, Kauserud H (2016) Fungi sailing the Arctic Ocean: speciose communities in North Atlantic driftwood as revealed by high-throughput amplicon sequencing. Microb Ecol 72:295–304. https://doi.org/10.1007/s00248-016-0778-9

Ramos P, Rivas N, Pollmann S, Casati P, Molina-Montenegro MA (2018) Hormonal and physiological changes driven by fungal endophytes increase Antarctic plant performance under UV-B radiation. Fungal Ecol 34:76–82. https://doi.org/10.1016/j.funeco.2018.05.006

Ranganathan A, Smith OP, Youssef NH, Struchtemeyer CG, Atiyeh HK, Elshahed MS (2017) Utilizing anaerobic fungi for two-stage sugar extraction and biofuel production from lignocellulosic biomass. Front Microbiol 8:635. https://doi.org/10.3389/fmicb.2017.00635

Rangaswami G, Venkatesan R (1961) Two fungi causing spoilage of refrigerated poultry meat. Curr Sci 30(2):64–65

Rao MPN, Xiao M, Li W-J (2017) Fungal and bacterial pigments: secondary metabolites with wide applications. Front Microbiol 8:1113. https://doi.org/10.3389/fmicb.2017.01113

Raymond-Bouchard I, Tremblay J, Altshuler I, Greer CW, Whyte LG (2018) Comparative transcriptomics of cold growth and adaptive features of a Eury- and steno-Psychrophile. Front Microbiol 9:1565. https://doi.org/10.3389/fmicb.2018.01565

Richards TA, Jones MDM, Leonard G, Bass D (2012) Marine fungi: their ecology and molecular diversity. Ann Rev Mar Sci 4:495–522. https://doi.org/10.1146/annurev-marine-120710-100802

Rinu K, Pandey A (2010) Slow and steady phosphate solubilization by a psychrotolerant strain of Paecilomyces hepiali (MTCC 9621). World J Microbiol Biotechnol 27:1055–1062. https://doi.org/10.1007/s11274-010-0550-0

Ríos A, Wierzchos J, Sancho LG, Ascaso C (2004) Exploring the physiological state of continental Antarctic endolithic microorganisms by microscopy. FEMS Microbiol Ecol 50:143–152. https://doi.org/10.1016/j.femsec.2004.06.010

Robinson CH (2001) Cold adaptation in Arctic and Antarctic fungi. New Phytol 151:341–353. https://doi.org/10.1046/j.1469-8137.2001.00177.x

Rodrigues ML (2016) Funding and innovation in diseases of neglected populations: the paradox of Cryptococcal meningitis. PLoS Negl Trop Dis 10:e0004429. https://doi.org/10.1371/journal.pntd.0004429

Rosa LH, de Sousa JRP, de Menezes GCA, da Costa CL, Carvalho-Silva M, Convey P, Câmara PEAS (2020) Opportunistic fungi found in fairy rings are present on different moss species in the Antarctic peninsula. Polar Biol 43:587–596. https://doi.org/10.1007/s00300-020-02663-w

Ruisi S, Barreca D, Selbmann L, Zucconi L, Onofri S (2006) Fungi in Antarctica. Rev Environ Sci Biotechnol 6:127–141. https://doi.org/10.1007/s11157-006-9107-y

Sajjad W, Din G, Rafiq M, Iqbal A, Khan S, Zada S, Ali B, Kang S (2020) Pigment production by cold-adapted bacteria and fungi: colorful tale of cryosphere with wide range applications. Extremophiles 24:447–473. https://doi.org/10.1007/s00792-020-01180-2

Salar RK, Aneja KR (2007) Thermophilic fungi: taxonomy and biogeography. J Agri Tech 3(1):77–107.

Santiago IF, Rosa CA, Rosa LH (2016a) Endophytic symbiont yeasts associated with the Antarctic angiosperms *Deschampsia antarctica* and *Colobanthus quitensis*. Polar Biol 40:177–183. https://doi.org/10.1007/s00300-016-1940-z

Santiago IF, Soares MA, Rosa CA, Rosa LH (2015) Lichensphere: a protected natural microhabitat of the non-lichenised fungal communities living in extreme environments of Antarctica. Extremophiles 19:1087–1097. https://doi.org/10.1007/s00792-015-0781-y

Santiago M, Ramírez-Sarmiento CA, Zamora RA, Parra LP (2016b) Discovery, molecular mechanisms, and industrial applications of cold-active enzymes. Front Microbiol 7:1–32. https://doi.org/10.3389/fmicb.2016.01408

Sarmiento F, Peralta R, Blamey JM (2015) Cold and hot Extremozymes: industrial relevance and current trends. Front Bioeng Biotechnol 3:148. https://doi.org/10.3389/fbioe.2015.00148

Satapathy S, Rout JR, Kerry RG, Thatoi H, Sahoo SL (2020) Biochemical prospects of various microbial pectinase and pectin: an approachable concept in pharmaceutical bioprocessing. Front Nutr 7:1–17. https://doi.org/10.3389/fnut.2020.00117

Selbmann L, de Hoog GS, Mazzaglia A, Friedmann EI, Onofri S (2005) Fungi at the edge of life: cryptoendolithic black fungi from Antarctic deserts. Stud Mycol 51:1–32

Selbmann L, Onofri S, Coleine C, Buzzini P, Canini F, Zucconi L (2017) Effect of environmental parameters on biodiversity of the fungal component in lithic Antarctic communities. Extremophiles 21:1069–1080. https://doi.org/10.1007/s00792-017-0967-6

Semenova TA, Morgado LN, Welker JM, Walker MD, Smets E, Geml J (2016) Compositional and functional shifts in arctic fungal communities in response to experimentally increased snow depth. Soil Biol Biochem 100:201–209. https://doi.org/10.1016/j.soilbio.2016.06.001

Sharma N, Rathore M, Sharma M (2012) Microbial pectinase: sources, characterization and applications. Rev Environ Sci Biotechnol 12:45–60. https://doi.org/10.1007/s11157-012-9276-9

Siddiqui KS (2015) Some like it hot, some like it cold: temperature dependent biotechnological applications and improvements in extremophilic enzymes. Biotechnol Adv 33:1912–1922. https://doi.org/10.1016/j.biotechadv.2015.11.001

Singh P, Roy U, Tsuji M (2016) Characterisation of yeast and filamentous fungi from Brøggerbreen glaciers, Svalbard. Polar Record 52:442–449. https://doi.org/10.1017/s0032247416000085

Singh P, Singh SM (2011) Characterization of yeast and filamentous fungi isolated from cryoconite holes of Svalbard, Arctic. Polar Biol 35:575–583. https://doi.org/10.1007/s00300-011-1103-1

Skidmore ML, Foght JM, Sharp MJ (2000) Microbial life beneath a high Arctic glacier. Appl Environ Microbiol 66:3214–3220. https://doi.org/10.1128/aem.66.8.3214-3220.2000

Sonjak S, Frisvad JC, Gunde-Cimerman N (2006) Penicillium Mycobiota in Arctic subglacial ice. Microb Ecol 52:207–216. https://doi.org/10.1007/s00248-006-9086-0

Srilakshmi J, Madhavi J, Lavanya S, Ammani K (2015) Commercial potential of fungal protease: past, present and future prospects. J Pharmaceut Chem Biol Sci 2:218–234

Tahir L, Ishtiaq Ali M, Zia M, Atiq N, Hasan F, Ahmed S (2013) Production and characterization of esterase in Lantinus tigrinus for degradation of polystyrene. Pol J Microbiol 62:101–108. https://doi.org/10.33073/pjm-2013-015

Tapia-Vázquez I, Sánchez-Cruz R, Arroyo-Domínguez M, Lira-Ruan V, Sánchez-Reyes A, del Rayo S-CM, Padilla-Chacón D, Batista-García RA, Folch-Mallol JL (2020) Isolation and characterization of psychrophilic and psychrotolerant plant-growth promoting microorganisms from a high-altitude volcano crater in Mexico. Microbiol Res 232:126394. https://doi.org/10.1016/j.micres.2019.126394

Timling I, Dahlberg A, Walker DA, Gardes M, Charcosset JY, Welker JM, Taylor DL (2012) Distribution and drivers of ectomycorrhizal fungal communities across the north American Arctic. Ecosphere 3:1–25. https://doi.org/10.1890/es12-00217.1

Timling I, Taylor DL (2012) Peeking through a frosty window: molecular insights into the ecology of Arctic soil fungi. Fungal Ecol 5:419–429. https://doi.org/10.1016/j.funeco.2012.01.009

Troncoso E, Barahona S, Carrasco M, Villarreal P, Alcaíno J, Cifuentes V, Baeza M (2016) Identification and characterization of yeasts isolated from the South Shetland Islands and the Antarctic peninsula. Polar Biol 40:649–658. https://doi.org/10.1007/s00300-016-1988-9

Tsuji M (2016) Cold-stress responses in the Antarctic basidiomycetous yeast Mrakia blollopis. R Soc Open Sci 3:160106. https://doi.org/10.1098/rsos.160106

Tsuji M, Fujiu S, Xiao N, Hanada Y, Kudoh S, Kondo H, Tsuda S, Hoshino T (2013) Cold adaptation of fungi obtained from soil and lake sediment in the Skarvsnes ice-free area, Antarctica. FEMS Microbiol Lett 346:121–130. https://doi.org/10.1111/1574-6968.12217

Tsuji M, Tanabe Y, Vincent WF, Uchida M (2018) *Mrakia arctica* sp. nov., a new psychrophilic yeast isolated from an ice island in the Canadian high Arctic. Mycoscience 59:54–58. https://doi.org/10.1016/j.myc.2017.08.006

Tsuji M, Tanabe Y, Vincent WF, Uchida M (2019) *Mrakia hoshinonis* sp. nov., a novel psychrophilic yeast isolated from a retreating glacier on Ellesmere Island in the Canadian high Arctic. Int J Syst Evol Microbiol 69:944–948. https://doi.org/10.1099/ijsem.0.003216

Tuli HS, Chaudhary P, Beniwal V, Sharma AK (2014) Microbial pigments as natural color sources: current trends and future perspectives. J Food Sci Technol 52:4669–4678. https://doi.org/10.1007/s13197-014-1601-6

Turk M, Abramoviä Z, Plemenitaä A, Gunde-Cimerman N (2007) Salt stress and plasma-membrane fluidity in selected extremophilic yeasts and yeast-like fungi. FEMS Yeast Res 7:550–557. https://doi.org/10.1111/j.1567-1364.2007.00209.x

Urbanek AK, Strzelecki MC, Mirończuk AM (2021) The potential of cold-adapted microorganisms for biodegradation of bioplastics. Waste Manag 119:72–81. https://doi.org/10.1016/j.wasman.2020.09.031

van Dorst J, Wilkins D, King CK, Spedding T, Hince G, Zhang E, Crane S, Ferrari B (2020) Applying microbial indicators of hydrocarbon toxicity to contaminated sites undergoing bioremediation on subantarctic Macquarie Island. Environ Pollut 259:113780. https://doi.org/10.1016/j.envpol.2019.113780

Vaz ABM, Rosa LH, Vieira MLA, de Garcia V, Brandão LR, Teixeira LCRS, Moliné M, Libkind D, van Broock M, Rosa CA (2011) The diversity, extracellular enzymatic activities and photoprotective compounds of yeasts isolated in Antarctica. Braz J Microbiol 42:937–947. https://doi.org/10.1590/s1517-83822011000300012

Veliz E, Martínez-Hidalgo P, Hirsch A (2017) Chitinase-producing bacteria and their role in biocontrol. AIMS Microbiol 3:689–705. https://doi.org/10.3934/microbiol.2017.3.689

Venil CK, Velmurugan P, Dufossé L, Renuka Devi P, Veera Ravi A (2020) Fungal pigments: potential coloring compounds for wide ranging applications in textile dyeing. JoF 6:68. https://doi.org/10.3390/jof6020068

Vero S, Garmendia G, González MB, Bentancur O, Wisniewski M (2012) Evaluation of yeasts obtained from Antarctic soil samples as biocontrol agents for the management of postharvest diseases of apple (malus × domestica). FEMS Yeast Res 13:189–199. https://doi.org/10.1111/1567-1364.12021

Vester JK, Glaring MA, Stougaard P (2014) Improved cultivation and metagenomics as new tools for bioprospecting in cold environments. Extremophiles 19:17–29. https://doi.org/10.1007/s00792-014-0704-3

Villarreal P, Carrasco M, Barahona S, Alcaíno J, Cifuentes V, Baeza M (2018) Antarctic yeasts: analysis of their freeze-thaw tolerance and production of antifreeze proteins, fatty acids and ergosterol. BMC Microbiol 18:66. https://doi.org/10.1186/s12866-018-1214-8

Walia A, Guleria S, Mehta P, Chauhan A, Parkash J (2017) Microbial xylanases and their industrial application in pulp and paper biobleaching: a review. 3 Biotech 7:11. https://doi.org/10.1007/s13205-016-0584-6

Walker DA, Raynolds MK, Daniëls FJA, Einarsson E, Elvebakk A, Gould WA, Katenin AE, Kholod SS, Markon CJ, Melnikov ES, Moskalenko NG, Talbot SS, Yurtsev BA, CAVM Team (2005) The circumpolar Arctic vegetation map. J Veg Sci 16:267–282. https://doi.org/10.1111/j.1654-1103.2005.tb02365.x

Wang J, Wang J, Zhang Z, Li Y, Zhang B, Zhang Z, Zhang G (2015a) Cold-adapted bacteria for bioremediation of crude oil-contaminated soil. J Chem Technol Biotechnol 91:2286–2297. https://doi.org/10.1002/jctb.4814

Wang M, Tian J, Xiang M, Liu X (2017) Living strategy of cold-adapted fungi with the reference to several representative species. Mycology 8:178–188. https://doi.org/10.1080/21501203.2017.1370429

Wang Y-T, Xue Y-R, Liu C-H (2015b) A brief review of bioactive metabolites derived from Deep-Sea fungi. Mar Drugs 13:4594–4616. https://doi.org/10.3390/md13084594

Watanabe T, Suzuki K, Sato I, Morita T, Koike H, Shinozaki Y, Ueda H, Koitabashi M, Kitamoto HK (2015) Simultaneous bioethanol distillery wastewater treatment and xylanase production by the phyllosphere yeast Pseudozyma Antarctica GB-4(0). AMB Expr 5:36. https://doi.org/10.1186/s13568-015-0121-8

Wentzel LCP, Inforsato FJ, Montoya QV, Rossin BG, Nascimento NR, Rodrigues A, Sette LD (2018) Fungi from Admiralty Bay (King George Island, Antarctica) soils and marine sediments. Microb Ecol 77:12–24. https://doi.org/10.1007/s00248-018-1217-x

Wu D-L, Li H-J, Smith D, Jaratsittisin J, Xia-Ke-Er X-F-K-T, Ma W-Z, Guo Y-W, Dong J, Shen J, Yang D-P, Lan W-J (2018) Polyketides and alkaloids from the marine-derived fungus *Dichotomomyces cejpii* F31-1 and the antiviral activity of Scequinadoline a against dengue virus. Mar Drugs 16:229. https://doi.org/10.3390/md16070229

Xiao N, Suzuki K, Nishimiya Y, Kondo H, Miura A, Tsuda S, Hoshino T (2009) Comparison of functional properties of two fungal antifreeze proteins from *Antarctomyces psychrotrophicus* and *Typhula ishikariensis*. FEBS J 277:394–403. https://doi.org/10.1111/j.1742-4658.2009.07490.x

Yu NH, Park S-Y, Kim JA, Park C-H, Jeong M-H, Oh S-O, Hong SG, Talavera M, Divakar PK, Hur J-S (2018) Endophytic and endolichenic fungal diversity in maritime Antarctica based on cultured material and their evolutionary position among Dikarya. Fungal System Evol 2:263–272. https://doi.org/10.3114/fuse.2018.02.07

Zain ul Arifeen M, Ma Y-N, Xue Y-R, Liu C-H (2019) Deep-Sea fungi could be the new arsenal for bioactive molecules. Mar Drugs 18:9. https://doi.org/10.3390/md18010009

Zhang T, Fei Wang N, Qin Zhang Y, Yu Liu H, Yan Yu L (2015) Diversity and distribution of fungal communities in the marine sediments of Kongsfjorden, Svalbard (high Arctic). Sci Rep 5:14524. https://doi.org/10.1038/srep14524

Zhang T, Wang N-F, Liu H-Y, Zhang Y-Q, Yu L-Y (2016) Soil pH is a key determinant of soil fungal community composition in the Ny-Ålesund region, Svalbard (high Arctic). Front Microbiol 7:227. https://doi.org/10.3389/fmicb.2016.00227

Zhang T, Yao Y-F (2015) Endophytic fungal communities associated with vascular plants in the high Arctic zone are highly diverse and host-plant specific. PLoS One 10:e0130051. https://doi.org/10.1371/journal.pone.0130051

Zucconi L, Canini F, Temporiti ME, Tosi S (2020) Extracellular enzymes and bioactive compounds from Antarctic terrestrial fungi for bioprospecting. Int J Environ Res Public Health 17:6459. https://doi.org/10.3390/ijerph17186459

Microbial Life in Cold Regions of the Deep Sea

3

Jupinder Kaur, A. L. Vishnu, Neha Khipla, and Jaspreet Kaur

Abstract

Deep sea ecosystem is not only the largest but also the most remote biome of the biosphere. Exploration of sea depths has resulted in the discovery of several new microbial habitats with unique nutritional composition and high microbial diversity. Microorganisms present in deep sea play a fundamental role in global biogeochemical cycles, and with their functional activities, they allow the existence of life. To survive and multiply in cold regions of deep sea, microorganisms should be able to adapt to a variety of changing conditions and stresses. Adaptations to fluctuations in temperature and pressure are possibly the most common; thus, psychropiezophilic microbes dominate in cold regions of the deep sea. These microorganisms make numerous adjustments to cope up with temperatures and pressure lower or higher than optimum. Benthic microbes exhibit both autotrophic and heterotrophic modes of nutrition in obligate oligotrophic environments of the deep sea. The rearrangement of simple metabolic strategies might help these microbes to metabolize in nutrient-poor environments. Further understanding of the genetic switches regulating the metabolism versatility at the deep sea could help us use and manipulate deep sea microbial strains for improved bioprocesses.

Keywords

Adaptions · Deep sea · Microbial diversity · Metabolism · Piezopsychrophiles

J. Kaur (✉) · A. L. Vishnu · N. Khipla · J. Kaur
Department of Microbiology, College of Basic Sciences and Humanities, Punjab Agricultural University, Ludhiana, Punjab, India
e-mail: jupinderkaur@pau.edu

R. Goel et al. (eds.), *Survival Strategies in Cold-adapted Microorganisms*, https://doi.org/10.1007/978-981-16-2625-8_3

3.1 Introduction

Ecological biodiversity is always a subject of interest for researchers, and a vast amount of information about the distribution of microorganisms around the world has also been collected. Several researchers have focused in recent years on the great potential of marine microbial treasures as a prolific producer of bioactive substances as well as a potential source of drug and antimicrobial compounds (Gimmler et al. 2016). Furthermore, the use of deep sea microbes in biogeochemical processes, biotechnology, pollution, and health has become increasingly interesting. The microbes that inhabit these unusual habitats are usually extremophiles. Extremophiles are the microbes that are capable of surviving in extreme environments. These microbes can survive in conditions like elevated (thermophilic) or low-temperature (psychrophilic), heavy ionic strength (halophilic), acid or alkaline conditions (acidophilic, alkalophilic), anaerobic environment, higher pressure (piezophilic), UV rays and polyextremophilic conditions, such as thermoacidophilic and thermohalophilic values. In cold regions of the deep sea, temperatures would be cold (2–3 °C), and the pressure can be more than 10 MPa, and thus microbes living there are called psychropiezophiles, and if the temperature is high like 400 °C (near hydrothermal vents), then microbes living there are called thermo-piezophiles, respectively (Fang et al. 2010). Extremophiles thrive in hot, cold, and high-pressure environments owing to their lipids, enzymes, and other biopolymers having specific properties/features to function in extreme conditions.

The Earth's biosphere is dominated by low-temperature ecosystems which are effectively colonized by a wide number of cold-adapted organisms. Although microorganisms, particularly, bacteria, yeasts, archaea, and protists, predominate in these cold habitats, microorganisms such as algae and microalgae have also been reported in these ecosystems. The ability of psychrophilic microorganisms to prevail in these conditions reflects their adaptability to the cold deep sea environment (Margesin and Collins 2019). This is accomplished through a set of morphological and physiological adaptations of all cellular elements, from a molecule level to whole cells and even complete ecosystems. In the deep sea, life is encountered with low temperature and high pressure. In addition to reduced thermal energy, low temperatures often contribute to more physicochemical constraints such as higher viscosity of solvents and solubility of the gases (such as oxygen and reactive oxygen species). Low temperature also lead to decreased solubility of solutes and nutrients, reduced diffusion, increased osmotic tension, desiccation, and ice formation. Various cold habitats are also marked with other extreme conditions including high salinity, oxidative stress, low nutrient levels, low water activity, and freeze-thaw cycles. Microorganisms in extreme sea interior and subglacial conditions are often subjected to the additional stress of high pressure. Thus, a multitude of synergistic adaptations are required for life in the cold biosphere, to react to not only the low-temperature threat but also the multitude of other interactive stresses imposed by particular environmental conditions. Importantly, many of these methods have multiple uses and can be used to address a variety of problems or combinations of problems. Unraveling the various interacting parameters and deciphering the precise role of a specific trait, whether it is a specific response to low temperatures or another (or other) environmental stressor (s) common to a particular habitat, is a common problem with the classification of cold-adapted microorganisms. Moreover, microorganisms do not necessarily use all resources in their "cold adaptation" toolbox. In reality, each organism will use its strategy or combination of strategies, depending on its specific requirements and the environmental parameters, and the microbial community structure.

High hydrostatic pressure (HHP) is an important parameter in the deep oceans as the average hydrostatic pressure is estimated at 38 MPa. Piezophiles are the species that survive at a pressure higher than ambient pressure (0.1 MPa) with an optimum growth rate. The effects of HHP on the physiological functioning of microbes have been studied in piezosensitive mesophilic (e.g., *Escherichia coli*) and psychrophilic bacteria (e.g., *Photobacterium profundum* SS9). In piezo-sensitive bacteria, due to HHP, compaction of lipid constituents of the cytoplasmic membrane occurs, and it turns to a rigid structure, whereas piezophiles counteract this constraint by altering the composition of the membrane lipids, particularly the ratio of monounsaturated fatty acids. For example, *P. profundum* SS9, a piezophilic bacterium, has a high ratio of unsaturated/saturated membrane lipids in the membrane which increases the membrane fluidity under HHP. The effect of HHP on piezophiles could arise through multiple pathways directly related to the composition of the membrane-like altered functioning of cellular transporters, motility, and respiratory chain components. Therefore, understanding the regulation of genes and enzymes involved in the respiratory chain of piezophiles is important to have an insight into differential mechanisms involved to counteract the effect of HHP. In this chapter, we will discuss the various concepts about psychrophiles and piezophiles, living in cold regions of the deep sea.

3.2 Deep Sea as a Microbial Habitat

3.2.1 With Low Temperature

For the past three million years, the relationship between microbial diversity and temperature is one of the most fascinating ecological phenomena (Tittensor et al. 2010). The mechanisms involved in this relationship have been explained through various hypothetical theories based on ecology and evolution. However, due to the inability to reach the deep sea region, accessible microbial diversity is still restricted to a few taxa. Also, the numerous theories suggested so far are still controversial (Brown and Thatje 2014), posing the requirement for more detailed studies with comparative analysis under natural conditions. Since current Intergovernmental Panel on Climate Change (IPCC) scenarios indicate that temperatures in most ocean regions will change rapidly in the coming decades, one of the main objectives of current ecological research is to gain a better understanding of potential responses to these changes. During the glacial/interglacial cycles of the Late Quaternary, deep sea temperatures in glacials were ~4 °C cooler than in interglacials.

Palaeoceanographic data showed 1–2 °C temperature variations in deep sea both on millennial and centennial time scales. Deep-water temperatures in the Labrador sea have demonstrated complex decadal variation at rates of change of up to 0.5 °C per decade over the last 60 years. Abrupt changes in temperature of deep-water temperature can impact physical and biological processes occurring down in the sea (Canals et al. 2006). There are significant differences in deep sea temperature, especially between oceans. There are exceptionally high deep sea temperatures in certain marginal seas, such as the Mediterranean, Red, and Sulu seas (from around 13 °C for the Mediterranean to >20 °C for the Red sea at a depth of 2000 m). Some deep waters are very cold at high latitudes, with temperatures close to −2 °C (e.g., Antarctic bottom water). Sometimes deep sea organisms are sensitive to even minor temperature changes because they encounter less seasonal variation in temperature compared to surface-sea organisms. The microorganisms that travel from shallow-sea environments to the deep sea are thought to be more temperature tolerant compared to those that originated at deep sea.

3.2.2 With High Pressure

The relationship between the rate of change in pressure and ocean depth is linear. In the shallower areas, the relative rate of pressure change is much higher with depth. For example, a microbe descending from 500 to 1000 m will experience a 101.2% pressure change, whereas an organism going from 9500 to 10,000 m will experience a pressure change of just 5.3%, while the absolute change is still ~500 dbar (decibar) per 500 m. Therefore, vertical migration does not generally account for any shifts in the regular pressure faced by benthic fauna. Likewise, the pressure difference is also negligible if an organism traverses a smooth and vast abyssal plain. If, however, an entity travels in any direction inside the trench perpendicular to the trench axis, then compression (if heading toward the axis) or decompression is encountered (if moving away from the axis). The pressure increases by 30–60 dbar per km across the abyssal plains (4000–6000 m). As the sea surface rises and falls, the atmospheric hydrostatic pressure is subject to tidal cycles regardless of changes in depth or distance traveled overground. Pressure data from the Kermadec Trench from 4329 to 8547 m, taken both in 2007 and 2009, indicate a cumulative mean tidal duration of 12.42 h 0.64 S.D. (semidiurnal), implying the presence of an internal tidal period of M2 (*lunar* semidiurnal tide). The M2 tidal cycle is one of the region's dominant semidiurnal tides and rotates around New Zealand anti-clockwise (Chiswell and Moore 1999). It is also common in conditions that are bathyal and abyssal. It was found that the mean amplitude (peak to trough) of these cycles of pressure was 1.26 dbar 0.19 S.D., approximating a swell of 1 m. In the Kermadec Trench, as well as in all other stations examined during the HADEEP project, these tidal cycles have been found as deep as 9900 m, regardless of depth (Kermadec, Tonga, Izu-Bonin, Japan, and Peru-Chile trenches). The same signature of the cycle is seen as deep as 7700 m in the North Pacific trenches. Hadal species would therefore likely be able to detect minor tidal variations in pressure.

3.3 Microbial Diversity in Deep Sea

Microbes are found everywhere on Earth. Microbial activities (nitrogen fixation, phosphate solubilization, etc.) are affected by various environmental factors and climatic changes (Kaur et al. 2014; Kaur and Gosal 2015, 2017). The deep seafloor comprises the largest ecological realm of the world. In deep sea sediments, bacteria and archaea (mostly in deep sea hyperthermal vents) and some fungi comprise the largest fraction of taxonomic richness and biomass at deep sea, playing a major role in remineralizing the organic matter as well as in nutrient cycling (Jørgensen and Boetius 2007; Wei et al. 2010). The highest sea depth reported for microbial occurrence in the deep sea is 10,898 m for bacteria *Dermacoccus abyssi* MT1.1 T (Pathom-aree et al. 2006) and *Shewanella benthica* DB21MT-2 (Kato et al. 1998; Nogi and Kato 1999). Understanding the spatial patterns of microbial diversity could pave the way toward better insight into mechanisms of diversification in the deep sea (Varliero et al. 2019) (Table 1.1).

3.4 Microbial Adaptations at Deep Sea

Microorganisms are novel living agents which can tolerate extreme environmental conditions existing on earth. Many environmental conditions on earth may be unideal for the survival of living agents. These conditions may exist normally or due to some external forces. Among them, physical extreme conditions, temperature, and pressure are important. Below are the details of various adaptation mechanisms adopted by microorganisms to thrive under deep sea extreme environmental conditions like low temperature and high pressure.

3.4.1 Low-Temperature Adaptations

Permanently cold environments existing on earth (like the deep sea and polar regions) have been successfully colonized by microbial species. Microbes surviving in deep sea or polar regions are referred to as psychrotolerants or psychrotrophs, based on their ability to grow at different temperature ranges (Morita 1975). Depending on the temperature, the abundance and composition of the microbial community vary. Variation in temperature only changes the types of microbes but not their ability to grow under such an environment. This novel property has made psychrophiles able to tolerate the effects of lower temperature like high viscosity (increases by drop-down of 2 °C) and negative effects on biochemical reactions. These effects of low temperature are successfully overcome by some of the psychrophiles (*Moritella profunda*). *Moritella profunda* has the ability to survive under a temperature range of 2–12 °C (Xu et al. 2003a, b). Other microorganisms also show adaptations to the cold environment by evolving various mechanisms which are discussed below.

Table 1.1 Microorganisms isolated from deep sea cold regions

Microbial type	Most resembles with	Isolation			Optimum growth		Reference
		Site	Source	Depth	Temp.	Pressure	
Bacteria	*Psychromonas kaikoae* JT7304T	Japan trench	Cold-seep sediment	7434 m	10 °C	50 MPa	Yayanos and Dietz (1979) and Yayanos (1986)
	Colwellia sp. strain MT41	Mariana trench	Decaying amphipod	10,476 m	8 °C	103 MPa	Yayanos et al. (1981) and Yayanos (1986)
	Colwellia hadaliensis BNL-1 T	Puerto Rico trench	Seawater	7410 m	10 °C	90 MPa	Deming et al. (1988)
	Shewanella violacea DSS12	Ryukyu trench	Sediment	5110 m	10 °C	30 MPa	Kato et al. (1995)
	Shewanella benthica DB6101	Ryukyu trench	Sediment	5110 m	10 °C	50 MPa	
	Shewanella benthica DB5501	Suruga bay	Sediment	2485 m	15 °C	60 MPa	Kato et al. (1995)
	Shewanella benthica DB6705	Japan trench	Sediment	6356 m	5 °C	60 MPa	
	Shewanella benthica DB6906	Japan trench	Sediment	6269 m	15 °C	60 MPa	
	Shewanella benthica F1A	Atlantic Ocean	Water column	4900 m	8 °C	30 MPa	Jannasch and Taylor (1984), Kato et al. (1996)
	Shewanella benthica DB172R	Izu-Bonin trench	Sediment	6499 m	10 °C	60 MPa	Kato et al. (1996)
	Shewanella benthica DB172F	Izu-Bonin trench	Sediment	6499 m	10 °C	70 MPa	
	Desulfovibrio profundus 500-1 T	Japan Sea	Sediment core 518 mbsfb	900 m	25 °C	15 MPa	Bale et al. (1997)
	Shewanella benthica DB21MT-2	Mariana trench	Sediment	10,898 m	10 °C	70 MPa	Kato et al. (1998) and Nogi and Kato (1999)
	Psychromonas profunda 2825 T	Atlantic Ocean	Sediment	2770 m	10 °C	25 MPa	Xu et al. (2003a, b)
	Colwellia piezophila Y223GT	Japan trench	Sediment	6278 m	10 °C	60 MPa	Nogi et al. (2004)
	Dermacoccus abyssi MT1.1 T	Mariana trench	Sediment	10,898 m	28 °C	40 MPa	Pathom-aree et al. (2006)

	Psychromonas sp. strain CNPT3	Central North Pacific	Decaying amphipod	5800 m	12 °C	52 MPa	Nogi et al. (2007)
	Psychromonas hadalis K41GT	Japan trench	Sediment	7542 m	6 °C	60 MPa	
	Shewanella sp. strain KT99	Kermadec trench	Amphipod homogenate	9856 m	2 °C	~98 MPa	Lauro et al. (2007)
	Carnobacterium sp. strain AT7	Aleutian trench	Water column	2500 m	20 °C	20 MPa	
	Rhodobacterales bacterium PRT1	Puerto Rico trench	Water	8350 m	2 °C	–	Eloe et al. (2011)
	Gammaproteobacteria *Alphaproteobacteria*; *Bacteroides* *Actinobacteria* and *Fermicutes*	Eastern Mediterranean deep sea	Sediment; water	2800–4400 m; 500 m–4000 m	14 °C	52 MPa	Gärtner et al. (2011)
	Rhodobacterales bacterium PRT1	Puerto Rico trench	Seawater	8350 m	10 °C	80 MPa	Zeng et al. (2009)
Yeast	*Rhodotorula rubra* and *Rhodosporidium sphaerocarpum*		Marine water	4000 m	–	40 MPa	Lorenz (1997)

3.4.1.1 Maintenance of Membrane Structure by the Generation of Unsaturated Fatty Acids

Enzymes are responsible for various conversions (Kaur et al. 2020). Due to a decrease in temperature, certain enzyme-mediated changes occur in the microbial cell membrane fatty acid profile. One such conversion is a change of saturated fatty acids to unsaturated fatty acids. This conversion is carried out by desaturase enzyme. These changes may occur to maintain optimum fluidity. Desaturase is also known to preferentially synthesize various types of fatty acids which may include short-chain fatty acids, branched-chain fatty acids, and anteiso fatty acids (Suutari and Laakso 1994). Some of the microbes involved in carrying out these functions are *Micrococcus roseus*, *Sphingobacterium antarcticus*, and *Pseudomonas syringae* (Chattopadhyay and Jagannadham 2001). Anteiso saturated fatty acid (a-C15:0) plays a major role in the survivability of psychrophiles (Annous et al. 1997). Kumar et al. (2002) described the role of hydroxy fatty acids in homeoviscous adaptation (an adaptation of lipid composition in the cell membrane) of outer membrane fluidity. It was demonstrated using *P. syringae* that when bacteria were incubated at a low temperature, there was an increased concentration of hydroxy fatty acids in lipopolysaccharides. When *Bacillus subtilis*, a mesophilic bacterium, was incubated in a psychrophilic condition, there was a transcriptional upregulation of some of the genes which were involved in coding those enzymes that degrade amino acid with branched chains (Kaan et al. 2002). Compounds like isobutyryl-CoA and α -methylbutyryl-CoA which are the intermediate product of valine and isoleucine degradation are utilized as a part of the cellular mechanism (synthesis of branched chain fatty acids) to maintain fluidity at low temperature. This indicates that not only anabolic pathways but catabolic pathways are also involved in maintaining membrane fluidity.

To thrive under cold conditions, many bacteria have evolved various other different mechanisms like the synthesis of unsaturated fatty acids in the case of *B. subtilis* under low temperature. In order to regulate glycerophospholipid, *B. subtilis* harbors a sensory system called DesK (dimeric histidine kinase). It has two domains that include five transmembrane helical domains and cytosolic kinase/phosphate domains. Under cold conditions, DesK changes from phosphatase active site to kinase active site leading to autophosphorylation. This phenomenon activates the DesR which in turn activates 5-lipid desaturase, which transforms saturated lipid acyl chains to unsaturated. Once the conditions become normal, DesK returns to the phosphatase active site, and dephosphorylation occurs, inactivates the DesR, and stalls the production of desaturase.

3.4.1.2 Cold-Shock Proteins (CSP)

A sudden downshift in temperature leads to harmful effects on the cells. Such damages are counteracted by proteins called cold-shock proteins (Phadtare 2004). These proteins are activated only under cold shock. Immediately after its synthesis, other proteins are recruited for growth synthesizes leading to the growth under cold conditions but at a slower pace (Ermolenko and Makhatadze 2002). Recent studies have revealed the role of cold-shock proteins in bacterial stress tolerance (Schmid et al. 2009).

CSPs are "small nucleic acid-binding proteins" whose length ranges between 65 and 75 amino acids (Czapski and Trun 2014). These proteins occur in psychrophiles as well as mesophiles (Jin et al. 2014). There is a total of nine CSPs (CspA to CspI) which are homologous to each other and share 46–91% similarity (Yamanaka et al. 2001). Among all, CspA plays an important role during cold shock, and its role has been described in *E. coli* (Goldstein et al. 1990).

Even though all CSPs share structural similarity, still their thermostability varies (Jin et al. 2014). CspA protein can even tolerate the mesophilic temperature of 40 °C (Lee et al. 2013). This nature of the protein helps them survive in varying temperatures (Jin et al. 2014). It was first identified in *Listeria monocytogenes* (Jin et al. 2014). At mesophilic temperature, mRNA of *csp*A is highly unstable. Usually under mesophilic range, half-life will be very less (12 s) but increases up to 20 min under cold conditions (Mitta et al. 1997). Under cold conditions, it is essential to transiently stabilize *cspA* mRNA as it plays a greater role in inducing CspA (Phadtare and Severinov 2005).

Functions of Cold-Shock Proteins

The highly conserved nucleic acid-binding domain of CSPs is called cold-shock domain (CSD) (Graumann and Marahiel 1996). Ribonucleoproteins 1 and 2 are two important nucleic acid-binding motifs of CSD (Lee et al. 2013). These help the protein to bind to its target RNA or DNA. Jiang et al. (1997) stated that the binding ability of CspA to RNA is weak and is responsible for minimal specificity for RNA. CSPs are known as molecular chaperones because they disrupt the secondary structure of RNA thereby helping transcription and translation to occur smoothly. This process is highly dependent on the mode of attachment of CSP to RNA. If CSP binds strongly to RNA, then, their role as molecular chaperon will be interrupted. CSPs are also known as anti-terminators as they terminate the formation of hairpin structures, which halt the transcription (Phadtare et al. 2002). Usually, during cold shock, CspA, CspB, CspE, CspG, and CspI are induced, but during the first temperature downshift, only CspA and CspB are synthesized (Jung et al. 2010).

3.4.1.3 Viable but Non-Culturable Cell (VBNC)

Viable but non-culturable cells (VBNC) are live bacteria that neither grow nor divide but are alive and capable of performing necessary metabolic operations for their survival. Generally, VBNC has greater physical and chemical resistance compared to culturable cells because of reduced metabolic activity and high content of peptidoglycan (Signoretto et al. 2000). Bacteria do not directly enter into VBNC state; before it, they enter into a persister cell phase (Bigger 1944). Persister cells refer to phenotypic variants in the population. Till today, persister cells are considered to be a nongrowing state of the cell. These are also known to tolerate antibiotics. Ayrapetyan et al. (2015) stated that "VBNC and persister cells are closely related states of a shared dormancy continuum." It suggests that logarithmic phase cells may enter into the persister state before entering the VBNC state.

Mechanism of VBNC Formation

The mechanism by which bacteria enter to VBNC state is not thoroughly understood. Various hypotheses have been put forward to explain the mechanism lying behind it. Among them, three important include the following: firstly, the severe conditions may lead to cells with poor quality which may result in null activity, and such cells cannot be cultured (Nystrom 2003). Secondly, it is described as a strategy of survivability in order to overcome harsh environmental conditions (Oliver 2005). Thirdly, it has been stated that genes are involved in the formation of VBNC (Ayrapetyan and Oliver 2016). The third hypothesis is widely accepted by scientists. Although the molecular mechanism of VBNC formation is not understood completely, several genes involved in its formation have been identified. One among them is the rpoS gene, which codes for the stress regulator protein RpoS which is known to enhance the efficiency of bacterial survivability under extreme conditions. If bacteria cannot produce this protein, then bacteria may enter VBNC. Research over a longer period revealed the role of ppGpp in VBNC formation. ppGpp is considered as a regulatory signaling molecule that regulates RpoS. The higher the ppGpp concentration, the greater the synthesis of RpoS, which contributes to resistance and persistence of cells under stress. Thus, ppGpp is considered as an inducer of the VBNC state.

3.4.1.4 Antifreeze Proteins

Antifreeze proteins (AFPs) refer to a class of polypeptides produced by certain animals, plants, fungi, and bacteria that enable their survivability in freezing temperature. These proteins are also known as ice structuring proteins. The main functions of AFPs are to bind to ice crystals and prevent growth and recrystallization (Collins and Margesin 2019). Unlike ethylene glycol (automotive antifreeze agent), they will not reduce freezing point but instead work in a non-colligative manner. This phenomenon enables them to be better antifreeze agents. These are known to act as an antifreeze agent at a concentration of 1/300th to 1/500th. As it is highly effective at lower concentrations, it doesn't have any side effects on the organism. A unique property of AFP is to bind to the particular ice crystals and immediately prevent their formation. AFPs mechanism is completely based on thermal hysteresis (TH) (Zhang et al. 2008). Thermal hysterisis refers to "a difference between the melting and freezing point" (busting temperature of AFP bound ice crystal). Thermodynamically favored growth of ice crystals can be inhibited by the addition of AFP between the solid ice and liquid water. Kinetically, ice growth can be inhibited by AFP that covers the water-accessible surfaces of ice.

Mechanism of AFP

The mechanism adopted by AFP is not frozen avoidance but freeze tolerance. Crystallization involves two major steps: nucleation (formation of a stable crystal nucleus) and extending the synthesis of crystals by nucleus growth. Based on the occurrence, nucleation is classified into two groups, i.e., nucleation taking place around the foreign molecule called heterogeneous nucleation and spontaneous formation of nucleus due to natural fluctuations called homogeneous nucleation.

Homogeneous nucleation occurs in the case of absolutely pure water. Crystallization is a cyclic process that may occur again and again; this is due to fluctuation in temperature within the subzero range. This fluctuation is the result of the dissolving of small crystals and the formation of larger crystals. This phenomenon may cause more damage to the cells and tissues (Hassas-Roudsari and Goff 2012).

As mentioned earlier, AFP mechanism is based on TH, which prevents the death of cells by various mechanisms like modification of ice crystal morphology (Kontogiorgos et al. 2007), recrystallization inhibition (Zhang et al. 2008), and intensifying integrity of the cell. All these properties are the result of interactions occurring between AFP, water, and ice. Freezing point depression occurs through a non-colligative mechanism (occurrence of protein between water ice interface to modify the growth of ice crystals). On to the outer world, the mechanism seems to be the adsorption-inhibition process (antifreeze agents bind to the surface of growing crystals). According to this, crystals of ice grow between adjacent antifreeze molecules with high surface curvature. High energy is required for the addition of water molecules to the convex surface. The whole process is nothing but maintenance of freezing point keeping the melting point at constant. This phenomenon is called as Kelvin effect. There are two models (mattress model and step pinning model) that justify the Kelvin effect. In the mattress model, the growth of ice crystals perpendicular to the ice surface is prevented by adsorbed molecules, whereas in the case of the step pinning model, molecules are blocked by ice growth (Bouvet and Ben 2003).

3.4.1.5 Adaptation Mechanism of Psychrophilic Enzymes

A higher degree of structural flexibility, lower thermostability, and specific activity are some of the characteristics of psychrophilic enzymes. Increased structural flexibility of psychrophilic enzymes may be restricted to a catalytic site which helps them exist in a disordered state. Increased flexibility in turn intensifies the degree of compatibility between catalytic site and substrate. This leads to an increase in substrate turnover rate and a decrease in activation energy. Multiple mechanisms have been evolved by psychrophilic enzymes to enhance their flexibility and activity and decrease thermostability. Among them, one mechanism involves reducing amino acids like arginine and proline. These amino acids are known to reduce conformational flexibility by the formation of a large number of hydrogen bonds and salt bridges. This mechanism has been observed in many psychrophilic enzymes. Some of the psychrophilic bacteria (*Shewanella* sp.) were known to have less alanine content, while others (*Psychrobacter arcticus*) lack proline/arginine content (particularly in those proteins involved in reproduction and cell division) (Zhao et al. 2010). Some other compositional variations found in the psychrophilic enzymes are increased content of methionine, asparagine, and glycine. These amino acids are found especially in the catalytic site which is known to contribute to local mobility. Increased lysine/arginine ratio is known to lower the hydrogen bond and salt bridge formation. Psychrophilic proteins with a longer external loop and reduced proline content result in less compact and highly stable proteins and also a catalytic site with more flexibility and mobility. Electron microscopic study of cold-adapted enzymes revealed that they contain a greater number of cavities with a larger size compared to that of mesophiles (Paredes et al. 2011). Larger cavities can hold a maximum number of hydrophilic groups thereby binding the large number of water molecules which enhance the flexibility by increasing internal salvation. For example, a region present near the helical lid of the psychrophilic enzyme lipase M37 found in *Photobacterium lipolyticum* consists of the surface cavity (Jung et al. 2008). The destabilizing effects of these surface cavities may provide flexibility to the helical lid thereby enhancing the lateral movement when substrate binds to it.

3.4.1.6 Piezophiles/Barophiles

Piezophiles or barophiles are organisms with the ability to survive under high pressure (depth of sea/ocean). Piezophiles are primarily found in ocean depths, with an average pressure of 10 MPa (megapascals). Some of the microbes are also found in the deepest point in the ocean (Mariana trench) where the pressure is around 110 MPa (Abe and Horikoshi 2001). *Pyrococcus yayanosii*, an extremophile, could survive in pressures ranging up to 150 MPa (Zeng et al. 2009). To counteract the effects of the elevated pressure, these organisms have evolved various mechanisms. As of yet, very little information is gathered regarding the piezophiles. Preliminary research on piezophiles indicated their potential applications in the industrial and biotechnological field (Abe and Horikoshi 2001).

3.4.2 Adaptation Mechanism of Piezophiles (High-Pressure Adaptations)

3.4.2.1 Membrane Lipid Adaptation

The effect of high pressure is similar to that of low temperature as both are involved in decreasing fluidity by increasing packing of the fatty acyl chains of phospholipids. Piezophiles found in the depths of the ocean need to acclimatize not only to high pressure but also low temperature. Intense hydrostatic pressure reduces membrane fluidity which results in the formation of the gel-like membrane that may interfere with the uptake of nutrients and cell signaling. These problems in piezophiles can be avoided by increasing the number of mono- and polyunsaturated fatty acids in their membranes. These fatty acids are difficult to be packed tightly. This nature of fatty acids makes the movement of the membrane easier (Bartlett 1999). An example is mentioned below.

Synechocystis, a phototrophic bacterium, has a two-component regulatory system to control the expression of desaturase which is involved in regulating membrane viscosity. A key regulatory element involved in inducing the expression of desB gene (codes for desaturase) is histidine kinase 33 (Hik33) (Suzuki et al. 2001). Hik33 are the key regulatory element in homeoviscous adaptation also known to regulate more than two dozens of the gene. There are several highly conserved domains in Hik33 which include HAMP domain (histidine kinases, adenyl cyclases, methyl-accepting chemotaxis proteins, and phosphatases), a leucine zipper domain which transfers signals to the 2-helix bundle in DesK, and a PAS domain (Per, Arnt, Sim sensor proteins) acting as a light-sensitive module in Hik33. Two helical regions of HAMP domain present adjacent to each other are involved in converting cold stress signal by structural modification. HAMP domain signal transmission is mediated by homo-dimeric, four-helical, parallel coiled coils. Hik33 gets activated by the enhanced molecular lipid packing (Los and Murata 2004), but the underlying sensor mechanism remains to be unknown.

3.4.2.2 Outer Membrane Porins

In response to high pressure, the membrane becomes highly rigid which has a greater influence on the movement and nutrient uptake. Many different proteins are involved in order to get acclimatized to these situations. The best example is *Photobacterium profundum* SS9, which regulates the outer membrane protein under high pressure. These bacteria consist of a specialized protein called OmpH. Its concentration increases with increasing pressure. It is usually expressed at a pressure of 28 MPa which is the minimum pressure required by the *P. profundum* SS9 for its growth (Chi and Bartlett 1993). Another protein involved is ompL which acts simultaneously along with ompH, but ompL is encoded by pressure-regulated genes that express at 0.1 MPa (decreases with increasing pressure); therefore, ompH contributes more to the pressure regulation when compared to ompL (Le Bihan et al. 2013). These both are the fourth most expressed proteins at high pressure. At elevated pressure, *P. profundum* lacking the ompH is not greatly affected. OmpH is also known to play a greater role under nutrient-deprived low-pressure conditions. There are a series of nine genes that are known to be present on the outer membrane, but the functions of these genes are not analyzed yet. Among these genes, ompC and two others, hypothetical maltoporins, are studied to some extent. These two maltoporins are known to express usually at constant pressure, while other genes, pbpra2139, express at elevated pressure. Porin-encoding genes that express at high pressure are a counter-intuitive example to show how difficult it would be to survive at high pressure. Increased porin produces many different compounds to boost survivability in the nutrient-scare ecosystem (Bartlett et al. 1993).

3.4.2.3 Membrane Transport

High hydrostatic pressure (HHP) has a greater influence on the transportation system in the bacterial cell membrane (Vezzi et al. 2005). Due to high hydrostatic pressure, fluidity is affected which may interfere with the transportation of nutrients across the membrane. HHP leads to an increase in volume, inhibits certain reactions, and makes amino acid (histidine, lysine, leucine, and tryptophan) movement difficult (Abe and Horikoshi 2001). Some of the bacteria in the sea/ocean use a higher amount of acetate and glutamate at elevated pressure. In *P. profundum* SS9, amino acid synthesis and ion transport were upregulated at 0.1 MPa. This is the best example that represents the adaptation of *P. profundum* SS9 transporters to high pressure. A variety of transporters such as ion, sugar, and phosphate transporters have isoforms known to function at varying pressure. It is essential to regulate the transporters in order to survive in the case of marine bacteria. Some hypothetical models have been explained to reveal the mechanism of transportation under high pressure, but a much detailed study has not been done yet.

3.4.2.4 Respiratory Chain

The respiration mechanism in piezophiles is quite different from other organisms to survive under extreme conditions. These consist of two kinds of electron transport systems in the inner membrane. A model organism used to study the respiratory chain in the deep sea is *Shewanella benthica* (Kato et al. 1999). A series of steps are involved in the respiratory chain of this organism: Initially, $NADH_2$ is oxidized to NAD by transferring two electrons to quinone(Q), that quinone get reduces to quinol (QH2), and it is carried out by NADH-dehydrogenase (ionic complex I). Within complex III (cytochrome c-551), electrons are exchanged between quinol and cytochrome c-551. These electrons are then passed to the active complex which covalently binds to the terminal cytochrome oxidase (a soluble protein). Later, oxygen is broken down to water by periplasmic oxidase and pumps proton to the cell. Not only periplasmic oxidase but also BC1 complex pumps proton to the cell, and this leads to the synthesis of ATP in the cytoplasm which is catalyzed by an enzyme called ATP synthase.

Under high pressure (60 MPa), the respiratory chain becomes more compact. During this situation, electrons are donated to quinol oxidase reducing the supply of oxygen to cytochrome c-551. This leads to the pumping of protons to the periplasmic vacuum. At elevated pressure, cytochrome c-552 will not be produced. The ability of a piezophile *Shewanella violacea* DSS12, to survive under high pressure depends 40% on its strain and 60% on cytochrome bc-1 complex. *Streptococcus* existing under high pressure contain two forms of soluble cytochrome. Under high hydrostatic pressure, cytochrome cA (belongs to c5 group) is constitutively expressed, whereas cytochrome cB is repressed. Three terminal oxidases exist under HHP, i.e., one terminal cytochrome c-oxidase, two bo, and bd-type quinol oxidases. At low oxygen high pressure, bd-type quinol oxidase increases, while all other terminal oxidase genes decrease. Bd-type quinol oxidase plays a greater role at high pressure and also makes a significant contribution to respiration. These kinds of variation do not occur in all types of microbes but only in piezophiles.

3.4.2.5 Motility Under High Pressure

Motility is a critical process for the survivability of bacteria as it enables bacteria to escape unfavorable conditions and helps to move toward the nutrient-rich environment. Motility occurs due to flagella. Flagella are found attached to cell envelopes and extended to extracellular space (Schuhmacher et al. 2015). The basic structure of flagella includes basal body, hook, and filament. The basal body has a C-shaped ring with a rod attached to it. All three parts of the flagellum are assembled by a type III secretion system. Approximately 25 different types of proteins are assembled in flagella. An important component of flagella is flagellin protein; 20,000–30,000 flagellin subunits are found at the distal end of the flagella. The synthesis of flagella is a highly complex process as there is the involvement of many genes. These genes are classified as early, middle, and late genes based on their involvement in the synthesis (Merino et al. 2006).

Flagellar structure and its role under high pressure are studied using two piezophiles, i.e., *Shewanella piezotolerans* WP3 and *Photobacterium profundum* SS9; these will have either polar or lateral flagella (Campanaro et al. 2005). Lateral flagella (LF) have a complex structure, encoded by 40 genes, and have higher GC content. There are two different kinds of motors for the motion of flagella, i.e., sodium-driven motors (*Shewanella piezotolerans* WP3) and proton-driven motors (*Photobacterium profundum* SS9) (Wang et al. 2008). Two genes identified to be away from the flagellar cluster were responsible for the movement of both types of flagella. Two flagellin genes (*fla*A, *fla*C) are known to regulate flagella in *S. piezotolerans* WP3 whereas three (*fla*A, *fla*C, and *fla*b) in case of *P. profundum* SS9. Recent studies have confirmed that high pressure inhibits the *fla*A and *fla*C and prevents motility. Under high physical tension and viscosity, lateral flagella enable the bacteria to swarm rather them swim. Under nutrient-deprived status, *mot*A and *fla*B are not expressed leading to non-motility. Mutants of any of the genes in the flagella inhibit either swimming or swarming. It is revealed that the developmental process is responsible for the proper functioning of flagella. Destruction of polar flagella either genetically or physically leads to activation of lateral flagella and vice versa.

3.4.2.6 Enzymes Adaptations Under High Pressure

Microbes are the richest source of enzymes; they help microbes to carry out biochemical reactions (Kaur et al. 2020). These proteins under high pressure may undergo physical damage; hence, microbes have developed a certain mechanism for their protection. Piezophilic microbes contain proteins that are homologs mesophilic proteins. When a comparison is made between them, some mesophilic proteins are found to be pressure adapted but not all. Microbes are found in the deepest region of the ocean, i.e., Mariana trench (temp. 1–4 °C, pressure 1.1 kbar) and in the hydrothermal vent (temp. ≥100 °C, pressure ~0.5 kbar), describing microbes' ability to survive by acclimatizing to varying temperature and pressure (Prieur et al. 2009). Therefore, it is confirmed that microbes' survival is based on their adapting mechanism; among them, protein protection mechanisms are explained below.

Low Stability

One of the common enzymes found in the piezophiles is DHFR (dihydrofolate reductase) which is used as a model to explain piezotolerant enzymes. Usually, enzymes found in high-pressure areas will have low stability because low stability is associated with high flexibility. So far, DFHR is less stable which is indicated by ΔG_u. Unlike fatty acid conversion in psychrophiles, low stability is not a driving force for survivability. It just prevents enzymes from being ruptured under high pressure. It is not essential to adapt a feature of low stability in the piezophilic microbe. Low stability may also exist in microbes under normal pressure. Overall, it can be inferred that low stability and greater flexibility appear to be favorable for the survival of enzymes under cold and high-pressure conditions.

High Compressibility

A common feature found in all piezophilic enzymes is high compressibility. The presence of a larger internal cavity makes the protein more susceptible to pressure unfolding. Maintaining normal protein structures under high pressure without leading them to get distorted keeps the microbes highly active (Kato et al. 1998). A study on crystal structures of piezophilic enzymes revealed that enzymes are loosely packed and highly hydrated with a large internal cavity. Piezophilic enzyme will have a greater number of small cavities instead of a single larger cavity (Fig. 3.1).

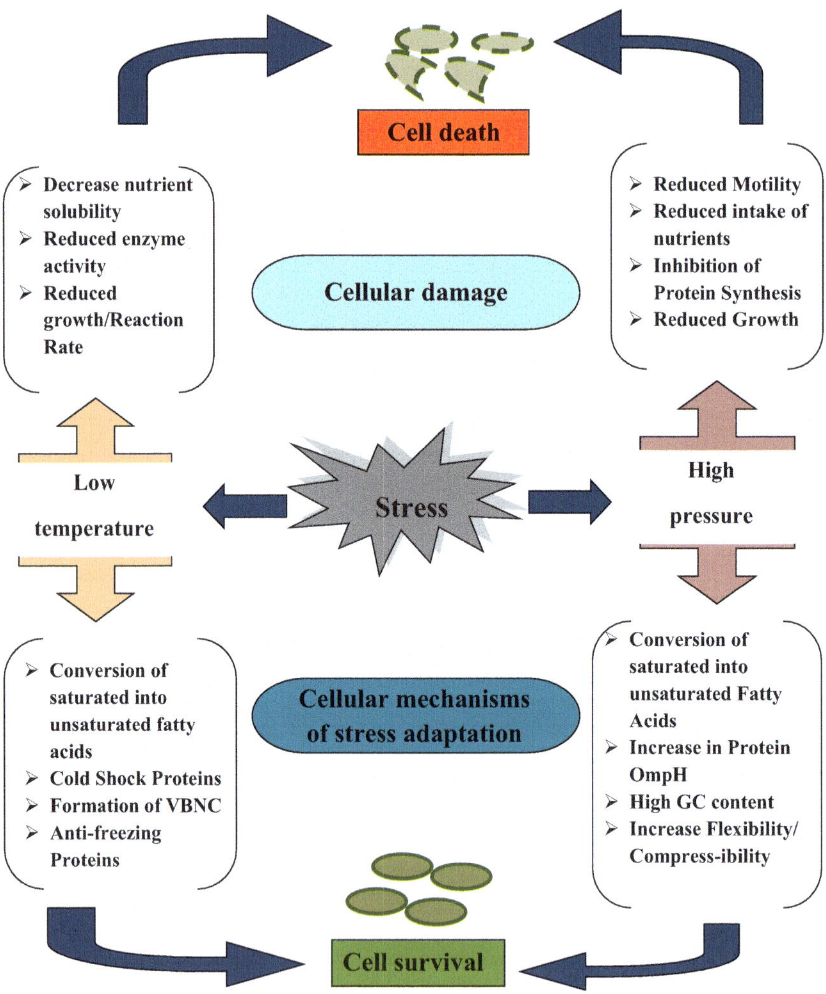

Fig. 3.1 Schematic diagram representing effects and adaptation mechanism of microbes at low temperature and high pressure

High Absolute Activity

Regardless of temperature and pressure, one mechanism adapted by microbes to retain their catalytic activity is maintaining high absolute activity. The turnover number (k_{cat}) of most of the piezophilic microbes is usually four or five times greater than other enzymes. It indicates that even activity of microbes is reduced at low temperature and high pressure, and still they can survive at their GTP. This is due to greater versatility, as they have lower stability and novel modifications to enhance the catalytic activity. But, some of the enzymes isolated from *Shewanella* normally have high catalytic activity. They do not possess any absolute activity to retain catalytic activity. Recent studies have confirmed that the catalytic activity of enzymes greatly depends on microbes rather than the condition they exist.

High Relative Activity at High Pressures

Increased relative activity helps the piezophilic microbes to maintain catalytic activity under high-pressure conditions. Many piezophilic enzymes maintain high relative activity to retain catalytic activity with available GTP. In some enzymes, increased pressure enhances both relative and absolute activity. An example is D27E, a mutant of DFHR which shows increased relative activity with a 50% increase in k_{cat} and a slight 2 kJ/mol increase in ΔG_u (Ohmae et al. 2013). An amino acid residue Asp27 in DFHR plays a central role in hydride transfer (Schnell et al. 2004). This indicates that DFHR has a slightly opened substrate-binding cleft which is the explanation for the increased activity of DFHR under high pressure. Low stability of DFHR leads them to decrease their activity above 500 bar.

3.5 Microbial Nutrition and Metabolism in Deep Sea

3.5.1 Chemistry of Deep Sea

Seawater is an open ecosystem serving as a sink of nutrients from various reservoirs. The knowledge of chemical features of seawater can provide an insight into the net functioning as well as the inflow and outflow of energy from the system. Sea surfaces have complex nutritional composition and are rich in ionic forms of nutrients especially nitrogen and sulfur. Urban storm, water runoff, irrigation drainage, agricultural runoff, creeks, rivers, and estuaries are the general portals of entry of nutrients into the marine environment (Akinde and Obire 2011). Unlike the organisms of the terrestrial ecosystem, marine organisms are dependent on dissolved forms of nutrients. Carbon, the major building block of life forms, can enter the sea interior from the atmosphere through a network of processes, where it can be stored or sequestered for millennia. Of the stored carbon on Earth, deep ocean zones constitute the largest reservoir (3150 Pmol, 1 Petamole = 1015 moles). It also corresponds to more than 50 times the amount of carbon present in the atmosphere (currently estimated as 62.5 Pmol) and more than one order of magnitude greater than all the carbon present in microbes, terrestrial vegetation, and soils combined.

Table 3.2 Major nutrient forms at deep sea levels and their source of origin [Source: Jørgensen et al. (2019), Thompson and Johnston (2017), Voss et al. (2013), Jasińska et al. (2012)]

Nutrient form	Source
Carbon	
Dissolved organic carbon (DOC)	Living plants and marine organisms (mainly phytoplankton), organic-rich detritus, or as dissolved organic carbon
Inorganic (carbonic acid, bicarbonate, and carbonate)	Atmospheric carbon dioxide
Sulfur	
Sulfate (So_4^{2-})	Sediments, weathering and leaching of rocks, biological or chemical oxidation of sulfides, sulfur partitioning, and riverine inflow
Sulfide	Sulfate (So_4^{2-}) reduction by marine microbes
Iron sulfide (FeS); pyrite (FeS_2)	Product of sulfide oxidation in the presence of C_{org} and Fe^{3+}; Pyritization of H_2S and FeS
Nitrogen	
Dissolved organic nitrogen (DON)	Rivers, atmospheric processes, wind, Ekman upwelling, biomass of surface autotrophs
No_2^-	Regeneration of particles, microbial nitrification
No_3^-	Wind, convective overturning, and Ekman upwelling, eddy activity/gyres, rivers, atmospheric deposition, shelf processes, regeneration of particles, microbial nitrification
NH_4^+	Diffusion, atmospheric deposition, regeneration of particles, decomposition of debris
N_2O	Bacterial and archaeal nitrification, intermediate of denitrification

Aside from atmospheric carbon, there are many other sources of carbon entry into the marine system (Table 3.2).

3.5.2 Microbial Metabolism in Deep Sea

Nutrient availability is a key factor for microbial existence and activity in any environment. The deep sea is an extreme oligotrophic environment that is often thought to set limits for microbial activity. However, this challenging environment is a habitat for great microbial diversity (Molari et al. 2013). The ability to survive in such contrasting extremes of temperature and pressure is assumed to have arisen from the adaptive route they followed to reprogramme their metabolism, scavenge the limiting nutrients, and bypass starvation.

The dark conditions of the deep sea enable microbes to develop photopigments; therefore, the major source of energy for benthic microbes is the downward flux of organic matter from primary producers on the sea surface (Danovaro et al. 2014). In marine environments, phytoplanktons are the primary producers (Azam and Malfatti 2007) and thereby the continuous source of organic matter for other life forms. A large fraction of their primary production is released as dissolved organic matter (DOM) into the system, either by the producers or by the degradative action of other organisms (Ducklow and Carlson 1992). Almost half of the DOM is consumed by higher-trophic-level organisms, while the remaining (1–2%) reach the benthic microbes (chemoorganoheterotrophs), which they assimilate into their biomass and re-mineralize the excess into inorganic nutrients that re-enter the nutrient cycle. Therefore, microbes play a critical role in the marine food web by organic matter turnover and establishing a balance between net energy flux (Mason et al. 2009). It is estimated that about one-third of the CO_2 produced in oceans originates from the microbial transformations at sea bottom (Arístegui et al. 2005). Despite the heterotrophic C metabolism, there is evidence for the existence of microbes with an expression of enzymes involved in autotrophic nutrition, especially the Calvin cycle and 3-hydroxypropionate pathway subsidizing autotrophic nutrition.

Benthic microbes can also metabolize several reduced inorganic compounds to accomplish heterotrophic nutrition. These compounds usually serve as an additional source of energy for support. The most prevalent of these is the oxidation of sulfur compounds via chemolithoheterotrophy (Ghosh and Dam 2009). On an average, 10% of microbes from marine environments are found with genes (*sox*) for sulfur oxidation (Venter et al. 2004). Microbes at sea sediments are also found with the ability of nitrification, the oxidation of ammonia to nitrite and nitrate. These microbes hold 21–50% of the total oxygen demand of the deep sea and mainly belong to the group of *Gammaproteobacteria* and *Archaea* (Könneke et al. 2005; Swan et al. 2011). The energy derived from nitrification is usually associated with carbon-dioxide fixation (chemolithoautotrophs) in dark regions of the deep sea. The process of carbon-dioxide fixation in these heterotrophs is not to derive biomass carbon like autotrophs but for the transformation of organic compounds into precursors of central metabolism using assimilatory carboxylases (Wuchter et al. 2006; Middelburg 2011). This is often termed as "mixotrophic nutrition," since the fixation of inorganic CO_2 without photoactivity often costs much energy to an organism, while, here, the energy is supplied from heterophony.

Despite the evidence of diverse metabolic activity at deep sea levels, it is important to understand the stress-derived metabolic alterations in microbes in their natural environment. Studies are based on quantifying the metabolic fluxes of marine microbes under conditions of varying temperature and pressure. The metabolism of model psychrophilic bacterium *Colwellia psychrerythraea* 34H at 4 °C (the temperature at natural environment) and under heat-stressed conditions was compared with cold-stressed and mesophilic *E.coli*, respectively (Jeffrey et al. 2018). Genetic analysis revealed that both bacteria had a similar metabolic network, but 34H had certain metabolic alterations that allow it to survive as an obligate psychrophile. These include the ability to suppress catabolic repression under a complex medium, activation of anaplerotic reactions to supplement TCA intermediates via CO_2 fixation, and therefore the maintenance of high cell biomass and metabolic flux. In contrast to *E. coli*, 34H favored ED pathway as the primary glycolytic route under glucose-rich medium. The potential driving force behind is the thermodynamic advantage of the ED pathway ($\Delta G = -36$ kJ/mol) to the bacterium as compared to EMP pathway (ΔG = −8 kJ/mol) at low temperature. Therefore, it can be said till further evidence that marine heterotrophs also use simplified metabolic strategies but are rearranged to overcome the thermodynamic constraints imposed by the environment.

3.6 Conclusion

The microorganisms thriving the extreme environmental conditions indeed have uniquely adapted enzymatic and metabolic systems as discussed in the chapter. These unique metabolic and enzymatic mechanisms are the most promising resources for the isolation of cold-adapted and pressure-tolerant enzyme systems. This potential appears even larger with psychrophiles than for peizophiles in terms of their diversity and potential uses in industries. Presently, huge data is available on biochemical/physicochemical reaction and protein and enzyme structure of peizizophiles and psychrophiles which open two major research avenues: (1) detailed insights in survival mechanism of these extremophiles and (2) application of these mechanisms for biotechnological applications. These extremophiles provide an immense genetic resource for manipulating industrial strain to work under extreme environmental conditions, thus delimiting the various stress factors during industrial production. Finally, combining the basic knowledge of extreme pressure and temperature effects on biochemical/physical reaction and advanced molecular biology techniques opens greater possibilities toward generating clean, efficient, and energy-saving industrial applications.

References

Abe F, Horikoshi K (2001) The biotechnological potential of piezophiles. Trends Biotechnol 19 (3):102–108

Akinde SB, Obire O (2011) In-situ physico-chemical properties of the deep Atlantic Ocean water column and their implications on heterotrophic bacterial distribution in the gulf of Guinea. Adv Appl Sci Res 2(6):470–482

Annous BA, Becker LA, Bayles DO, Labeda DP, Wilkinson BJ (1997) Critical role of anteiso-C15: 0 fatty acid in the growth of *Listeria monocytogenes* at low temperatures. Appl Environ Microbiol 63(10):3887–3894

Arístegui J, Duarte CM, Gasol JM, Alonso-Sáez L (2005) Active mesopelagic prokaryotes support high respiration in the subtropical northeast Atlantic Ocean. Geophys Res Lett 32:03608

Ayrapetyan M, Williams TC, Oliver JD (2015) Bridging the gap between viable but non-culturable and antibiotic persistent bacteria. Trends Microbiol 23(1):7–13

Ayrapetyan M, Oliver JD (2016) The viable but non-culturable state and its relevance in food safety. Curr Opin Food Sci 8:127–133

Azam F, Malfatti F (2007) Microbial structuring of marine ecosystems. Nat Rev Microbiol 5:782–791

Bale SJ, Goodman K, Rochelle PA, Marchesi JR, Fry JC, Weightman AJ, Parkes RJ (1997) *Desulfovibrio profundus* sp. nov., a novel barophilic sulfate-reducing bacterium from deep sediment layers in the Japan Sea. Int J Syst Bacteriol 47:515–521

Bartlett DH (1999) Microbial adaptations to the psychrosphere/piezosphere. J Mol Microbiol Biotechnol 1(1):93–100
Bartlett DH, Chi E, Wright ME (1993) Sequence of the ompH gene from the deep sea bacterium *Photobacterium* SS9. Gene 131(1):125–128
Bigger J (1944) Treatment of *Staphylococcal* infections with penicillin by intermittent sterilisation. Lancet 244:497–500
Bouvet V, Ben RN (2003) Antifreeze glycoproteins. Cell Biochem Biophys 39(2):133–144
Brown A, Thatje S (2014) Explaining bathymetric diversity patterns in marine benthic invertebrates and demersal fishes: physiological contributions to adaptation of life at depth. Biol Rev 89 (2):406–426
Campanaro S, Vezzi A, Vitulo N, Lauro FM, D'Angelo M, Simonato F, Bartlett DH (2005) Laterally transferred elements and high pressure adaptation in *Photobacterium profundum* strains. BMC Genomics 6(1):122
Canals M, Puig P, Durrieu d MX, Heussner S, Palan-ques A, Fabres J (2006) Flushing submarine canyons. Nature 444:354–357
Chattopadhyay M, Jagannadham M (2001) Maintenance of membrane fluidity in Antarctic bacteria. Polar Biol 24(5):386–388
Chi E, Bartlett DH (1993) Use of a reporter gene to follow high-pressure signal transduction in the deep-sea bacterium Photobacterium sp. strain SS9. J Bacteriol 175(23):7533–7540
Chiswell SM, Moore MI (1999) Internal tides in the south-West Pacific Ocean near the Kennadec ridge. J Phys Oceanogr 29:1019–1035
Collins T, Margesin R (2019) Psychrophilic lifestyles: mechanisms of adaptation and biotechnological tools. Appl Microbiol Biotechnol 103(7):2857–2871
Czapski TR, Trun N (2014) Expression of csp genes in *E. coli* K-12 in defined rich and defined minimal media during normal growth, and after cold-shock. Gene 547(1):91–97
Danovaro R, Snelgrove PV, Tyler P (2014) Challenging the paradigms of deep sea ecology. Trends Ecol Evol 29:465–475. https://doi.org/10.1016/j.tree.2014.06
Deming JW, Somers LK, StraubeWL SDG, MacDonell MT (1988) Isolation of an obligately barophilic bacterium and description of a new genus, *Colwellia* gen nov. Syst Appl Microbiol 10:152–160
Ducklow HW, Carlson CA (1992) Oceanic bacterial production. Adv Microb Ecol 12:113–181
Eloe EA, Shulse CN, Fadrosh DW, Williamson SJ, Allen EE, Bartlett DH (2011) Compositional differences in particle-associated and free-living microbial assemblages from an extreme deep-ocean environment. Environ Microbiol Rep 3:449–458
Ermolenko DN, Makhatadze GI (2002) Bacterial cold-shock proteins. Cellular Molecular Life Sci CMLS 59(11):1902–1913
Fang J, Zhang L, Bazylinski DA (2010) Deep sea piezosphere and piezophiles: geomicrobiology and biogeochemistry. Trends Microbiol 18(9):413–422
Gärtner A, Blümel M, Wiese J, Imhoff JF (2011) Isolation and characterisation of bacteria from the Eastern Mediterranean deep sea. Antonie Van Leeuwenhoek 100(3):421–435
Ghosh W, Dam B (2009) Biochemistry and molecular biology of lithotrophic sulfur oxidation by taxonomically and ecologically diverse bacteria and archaea. FEMS Microbiol Rev 33:999–1043
Gimmler A, Korn R, De Vargas C, Audic S, Stoeck T (2016) The Tara Oceans voyage reveals global diversity and distribution patterns of marine planktonic ciliates. Sci Rep 6:33555
Goldstein J, Pollitt NS, Inouye M (1990) Major cold shock protein of *Escherichia coli*. Proc Natl Acad Sci U S A 87:283–287
Graumann PL, Marahiel MA (1996) Some like it cold: response of microorganisms to cold shock. Arch Microbiol 166:293–300
Hassas-Roudsari M, Goff HD (2012) Ice structuring proteins from plants: mechanism of action and food application. Food Res Int 46:425–436
Jannasch HW, Taylor CD (1984) Deep-sea microbiology. Annu Rev Microbiol 38(1):487–487

Jasińska A, Burska D, Bolałek J (2012) Sulfur in the marine environment. Int J Oceanogr Hydrobiol 41:72–81

Jeffrey JC, Mary HA, Veronica TB, Edward EKB, Jody WD, Yinjie JT (2018) Model metabolic strategy for heterotrophic bacteria in cold ocean based on Colwellia psychrerythraea 34H. PNAS 115(49):12507–12512

Jiang W, Hou Y, Inouye M (1997) CspA, the major cold-shock protein of *Escherichia coli*, is an RNA chaperone. J Biol Chem 272:196–202

Jin B, Jeong KW, Kim Y (2014) Structure and flexibility of the thermophilic cold-shock protein of *Thermus aquaticus*. Biochem Biophys Res Commun 451:402–407

Jørgensen BB, Boetius A (2007) Feast and famine - microbial life in the deep sea bed. Nat Rev Microbiol 5:770–781. https://doi.org/10.1038/nrmicro1745

Jørgensen BB, Findlay AJ, Pellerin A (2019) The biogeochemical sulfur cycle of marine sediments. Front Microbiol 10:849

Jung S, Jeong DG, Lee MS, Lee J, Kim H, Ryu SE (2008) Structural basis for the cold adaptation of psychrophilic M37 lipase from *Photobacterium lipolyticum*. Proteins 71:476–484

Jung YH, Yi JY, Jung HJ, Lee YK, Lee HK, Naicker MC (2010) Overexpression of cold shock protein A of *Psychromonas arctica* KOPRI 22215 confers cold-resistance. Protein J 29 (2):136–142

Kaan T, Homuth G, Mader U, Bandow J, Schweder T (2002) Genome-wide transcriptional profiling of the *Bacillus subtilis* cold-shock response. Microbiology 148:3441–3455

Kato C, Li L, Nogi Y, Nakamura Y, Tamaoka J, Horikoshi K (1998) Extremely barophilic bacteria isolated from the Mariana trench, challenger deep, at a depth of 11,000 meters. Appl Environ Microbiol 64:1510–1513

Kato C, Masui N, Horikoshi K (1996) Properties of obligately barophilic bacteria isolated from a sample of deep sea sediment from the IzuBonin trench. J Mar Biotechnol 4:96–99

Kato C, Qureshi MH, Horikoshi K (1999) Pressure response in deep sea piezophilic bacteria. J Mol Microbiol Biotechnol 1(1):87–92

Kato C, Sato T, Horikoshi K (1995) Isolation and properties of barophilic and barotolerant bacteria from deep sea mud samples. Biodivers Conserv 4:1–9

Kaur J, Gosal SK (2015) Influence of weather parameters on activities of bacterial population associated with rhizospheric soil of rice crop. Ecol Environ Cons 21(3):1255–1262

Kaur J, Gosal SK (2017) Influence of environmental factors on activities of bacterial population associated with rhizospheric soil of wheat crop. Br J Environ Clim Chang 7(3):195–204

Kaur J, Gosal SK, Kaur P (2014) Effects of climate change on plant associated microbial communities and enzyme activities. Afr J Microbiol Res 8(33):3087–3093

Kaur J, Gosal SK, Khurana S (2020) Soil enzymes: an agricultural perspective. In: New perspectives in agricultural sciences. Lambert Academic Publishing, Moldova, pp 59–82

Könneke M, Bernhard AE, de la Torre JR, Walker CB, Waterbury JB, Sthal DA (2005) Isolation of an autotrophic ammonia-oxidizing marine archaeon. Nature 437:543–546

Kontogiorgos V, Regand A, Yada RY, Goff HD (2007) Isolation and characterization of ice structuring proteins from cold-acclimated winter grass extract for recrystallization inhibition in frozen foods. J Biochem 31:139–160

Kumar GS, Jagannadham MV, Ray MK (2002) Low-temperature-induced changes in composition and fluidity of lipopolysaccharides in the Antarctic psychrotrophic bacterium *Pseudomonas syringae*. J Bacteriol 184(23):6746–6749

Lauro FM, Chastain RA, Blankenship LE, Yayanos AA, Bartlett DH (2007) The unique 16S rRNA genes of piezophiles reflect both phylogeny and adaptation. Appl Environ Microbiol 73:838–845

Le Bihan T, Rayner J, Roy MM, Spagnolo L (2013) *Photobacterium profundum* under pressure: a MS-based label-free quantitative proteomics study. PLoS One 8(5):608–697

Lee J, Jeong KW, Jin B, Ryu KS, Kim EH, Ahn JH (2013) Structural and dynamic features of cold-shock proteins of *Listeria monocytogenes*, a psychrophilic bacterium. Biochemistry 52:2492–2504

Lorenz M (1997) Cultivation of fungi under stimulated deep sea conditions. Mycol Res 101 (11):1355–1395

Los DA, Murata N (2004) Membrane fluidity and its roles in the perception of environmental signals. Biochim Biophys Acta 1(2):142–157

Margesin R, Collins T (2019) Microbial ecology of the cryosphere (glacial and permafrost habitats): current knowledge. Appl Microbiol Biotechnol 103(6):2537–2549

Mason OU, Di Meo-Savoie CA, Van Nostrand JD, Zhou J, Fish MR, Giovannoni SJ (2009) Prokaryotic diversity, distribution, and insights into their role in biogeochemical cycling in marine basalts. ISME J 3:231–242

Merino S, Shaw JG, Tomás JM (2006) Bacterial lateral flagella: an inducible flagella system. FEMS Microbiol Lett 263(2):127–135

Middelburg JJ (2011) Chemoautotrophy in the ocean. Geophys Res Lett 38:L24604. https://doi.org/10.1029/2011GL049725

Mitta M, Fang L, Inouye M (1997) Deletion analysis of cspA of *Escherichia coli*: requirement of the AT-rich UP element for cspA transcription and the downstream box in the coding region for its cold shock induction. Mol Microbiol 26:321–335

Molari M, Manini E, Dell'Anno A (2013) Dark inorganic carbon fixation sustains the functioning of benthic deep sea ecosystems. Global Biogeochem Cycles 27:212–221

Morita RY (1975) Psychrophilic bacteria. Bacteriol Rev 39(2):144

Nogi Y, Hosoya S, Kato C, Horikoshi (2004) *Colwellia piezophila* sp. nov., a novel piezophilic species from deep sea sediments of the Japan trench. Int J Syst Evol Microbiol 54:1627–1631

Nogi Y, Hosoya S, Kato C, Horikoshi (2007) *Psychromonas hadalis* sp. nov., a novel piezophilic bacterium isolated from the bottom of the Japan trench. Int J Syst Evol Microbiol 57:1360–1364

Nogi Y, Kato C (1999) Taxonomic studies of extremely barophilic bacteria isolated from the Mariana trench and description of *Moritella yayanosii sp.* nov., a new barophilic bacterial isolate. Extremophiles 3:71–77

Nystrom T (2003) Nonculturable bacteria: programmed survival forms or cells at death's door? Bioessays 25:204–211

Ohmae E, Miyashita Y, Kato C (2013) Thermodynamic and functional characteristics of deep sea enzymes revealed by pressure effects. Extremophiles 17(5):701–709

Oliver JD (2005) The viable but nonculturable state in bacteria. J Microbiol 43:93–100

Paredes DI, Watters K, Pitman DJ, Bystroff C, Dordick JS (2011) Comparative void-volume analysis of psychrophilic and mesophilic enzymes: structural bioinformatics of psychrophilic enzymes reveals sources of core flexibility. BMC Struct Biol 11:42

Pathom-aree W, Nogi Y, Sutcliffe LC, Ward AL, Horokoshi K, Bull AT, Goodfellow M (2006) *Dermacoccus abyssi* sp. nov., a piezotolerant actinomycete isolated from the Mariana trench. Int J Syst Evol Microbiol 56:1233–1237

Phadtare S (2004) Recent developments in bacterial cold-shock response. Curr Issues Mol Biol 6 (2):125–136

Phadtare S, Inouye M, Severinov K (2002) The nucleic acid melting activity of *Escherichia coli* CspE is critical for transcription antitermination and cold acclimation of cells. J Biol Chem 277:7239–7245

Phadtare S, Severinov K (2005) Extended −10 motif is critical for activity of the cspA promoter but does not contribute to low-temperature transcription. J Bacteriol 187:6584–6589

Prieur D, Jebbar M, Bartlett D, Kato C, Oger P (2009) Piezophilic prokaryotes. In: Comparative High Pressure Biology. Science Publishers, Enfield, pp 285–322

Schmid B, Klumpp J, Raimann E, Loessner MJ, Stephan R, Tasara T (2009) Role of cold shock proteins in growth of *Listeria monocytogenes* under cold and osmotic stress conditions. Appl Environ Microbiol 75:1621–1627

Schnell JR, Dyson HJ, Wright PE (2004) Structure, dynamics, and catalytic function of dihydrofolate reductase. Annu Rev Biophys Biomol Struct 33:119–140

Schuhmacher JS, Thormann KM, Bange G (2015) How bacteria maintain location and number of flagella? FEMS Microbiol Rev 39(6):812–822

Signoretto C, del Mar LM, Tafi MC, Canepari P (2000) Cell wall chemical composition of *Enterococcus faecalis* in the viable but nonculturable state. Appl Environ Microbiol 66 (5):1953–1959

Suutari M, Laakso S (1994) Microbial fatty acids and thermal adaptation. Crit Rev Microbiol 20 (4):285–328

Suzuki I, Kanesaki Y, Mikami K, Kanehisa M, Murata N (2001) Cold-regulated genes under control of the cold sensor Hik33 in *Synechocystis*. Mol Microbiol 40(1):235–244

Swan BK, Martinez-Garcia M, Preston CM, Sczyrba A, Woyke T, Lamy D (2011) Potential for chemolithoautotrophy among ubiquitous bacteria lineages in the dark ocean. Science 333:1296–1300

Thompson M, Johnston S (2017) Storage of carbon by marine ecosystems and their contribution to climate change mitigation. In: Marine ecosystems and blue carbon. Greenpeace Research Laboratories, Exeter

Tittensor DP, Mora C, Jetz W, Lotze HK, Ricard D, Berghe EV, Worm B (2010) Global patterns and predictors of marine biodiversity across taxa. Nature 466(7310):1098–1101

Varliero G, Bienhold C, Schmid F, Boetius A, Molari M (2019) Microbial diversity and connectivity in deep sea sediments of the South Atlantic Polar Front. Front Microbiol 10:665

Venter JC, Remington K, Heidelberg JF, Halpern AL, Rusch DB, Eisen JA (2004) Environmental genome shotgun sequencing of the Sargasso Sea. Science 304:66–74

Vezzi A, Campanaro S, D'angelo M, Simonato F, Vitulo N, Lauro FM, Romualdi C (2005) Life at depth: *Photobacterium profundum* genome sequence and expression analysis. Science 307 (5714):1459–1461

Voss M, Bange HW, Dippner JW, Middelburg JJ, Montoya JP, Ward B (2013) The marine nitrogen cycle: recent discoveries, uncertainties and the potential relevance of climate change. Philos Trans R Soc B 368(1621):20130121

Wang F, Wang J, Jian H, Zhang B, Li S, Wang F, Hu S (2008) Environmental adaptation: genomic analysis of the piezotolerant and psychrotolerant deep-sea iron reducing bacterium *Shewanella piezotolerans* WP3. PLoS One 3(4):e1937

Wei CL, Rowe GT, Briones EE, Boetius A, Soltwedel T, Caley MJ (2010) Global patterns and predictions of seafloor biomass using random forests. PLoS One 5:e15323. https://doi.org/10.1371/journal.pone.0015323

Wuchter C, Abbas B, Coolen MJL, Herfort L, van Bleijswijk J, Timmers P (2006) Archaeal nitrification in the ocean. PNAS USA 103:12317–12322

Xu Y, Nogi Y, Kato C, Liag Z, Ruger HJ, Kegel DD, Glansdroff N (2003a) *Psychromonas profunda* sp. nov., a psychropiezophilic bacterium from deep Atlantic sediments. Int J Syst Evol Microbiol 53:527–532

Xu Y, Nogi Y, Kato C, Liang Z, Ruger HJ, De Kegel D, Glansdorff N (2003b) *Moritella profunda* sp. nov. and *Moritella abyssi* sp. nov., two psychropiezophilic organisms isolated from deep Atlantic sediments. Int J Syst Evol Microbiol 53:533–538

Yamanaka K, Zheng W, Crooke E, Wang YH, Inouye M (2001) CspD, a novel DNA replication inhibitor induced during the stationary phase in *Escherichia coli*. Mol Microbiol 39:1572–1584

Yayanos AA (1986) Evolutional and ecological implications of the properties of deep sea barophilic bacteria. PNAS USA 83:9542–9546

Yayanos AA, Dietz AS, Vanboxtel R (1979) Isolation of a deep sea barophilic bacterium and some of its growth characteristics. Science 205:808–810

Yayanos AA, Dietz AS, Boxtel RV (1981) Obligately barophilic bacterium from the Mariana trench. PNAS USA 78:5212–5215

Zeng X, Birren JL, Fouquet Y, Cherkashov G, Jebbar M, Querellou J, Oger P, Banovita MAC, Xiao X, Prieur D (2009) Pyrococcus CH1, an obligate piezophilic hyperthermophile: extending the upper pressure-temperature limits for life. ISME J 3:873–876

Zhang C, Zhang H, Wang L, Gao H, Guo XN, Yao HY (2008) Effect of carrot (*Daucus carota*) antifreeze proteins on texture properties and frozen dough and volatile compounds of crumb. LWT- Food Sci Technol 41:1029–1036

Zhao JS, Deng Y, Manno D, Hawari J (2010) *Shewanella spp.* genomic evolution for a cold marine lifestyle and in-situ explosive biodegradation. PLoS One 5:9109

Adaptation to Cold Environment: The Survival Strategy of Psychrophiles

4

Shruti Pathania, Preeti Solanki, Chayanika Putatunda, Ravi Kant Bhatia, and Abhishek Walia

Abstract

Earth, the wonderland inhabited by huge diversity of microbial life, has more than three-quarters of its habitat occupied by ubiquitous microorganisms. These habitats include deep dark ocean, the subzero polar and extreme cold alpine regions. A class of extremophilic microorganisms known as psychrophiles have successfully colonized these cold conditions. This unique class of extremophiles can easily thrive at very low temperatures, i.e. below 15 °C or lower than this where otherwise no any other life is able to survive. In order to survive under such harsh environmental conditions, these microorganisms require much specialized adaptation features to maintain the best homeostasis to perform various metabolic activities to sustain their growth and development. These features of psychrophiles may also be exploited by various industries to develop novel catalysts and therapeutics to establish successful biochemical processes for the betterment of mankind. This chapter will provide updated information on

S. Pathania
Department of Microbiology, Dr. YSP University of Horticulture and Forestry, Solan, Himachal Pradesh, India

P. Solanki
Multi-Disciplinary Research Unit, Pt. B. D. Sharma Institute of Medical Sciences, Rohtak, Haryana, India

C. Putatunda
Department of Microbiology, Om Sterling Global University, Hisar, Haryana, India

R. K. Bhatia
Department of Biotechnology, Himachal Pradesh University, Shimla, Himachal Pradesh, India

A. Walia (✉)
Department of Microbiology, CSK HPKV, Palampur, Himachal Pradesh, India
e-mail: abhishek@hillagric.ac.in

R. Goel et al. (eds.), *Survival Strategies in Cold-adapted Microorganisms*,
https://doi.org/10.1007/978-981-16-2625-8_4

psychrophilic microbes, various specialized adaptations to extreme conditions like morphological features and molecular adaptations along with some other survival strategies adopted by these extremophiles to thrive in the coldest environment of the earth.

Keywords

Psychrophiles · Antifreeze proteins · Microbial adaptations · Cold-adapted enzymes

4.1 Introduction

Cold environments are predominant over the Earth since its origin. During its evolution and transformation to present form, various habitats originated which are occupied by different microorganisms, whereas some dark, deep habitats inside the ocean and colder regions of earth are still occupied by cold-loving microorganisms referred to as psychrophiles. These extremophilic cold-loving bacteria or archaea generally prefer lower temperature of about 15–0 °C or lower for their growth (Barria et al. 2013). These microorganisms need a broad variety of adaptations at all levels to survive in extreme conditions, beginning from the cell structure to the metabolites to the specific biocatalysts to catalyse various anabolic and catabolic functions. Certain specialized stabilizing factors have been synthesized by these extremophiles which interfere with each other as well as with other structural proteins to aid in their compactness, kinetic stability and thermodynamic stability making them immune to chemical denaturation and harsh environments (Bhatia et al. 2020; Cavicchioli 2016). These different factors must be avoided in order to deal with low temperatures in order to conduct biochemical processes effectively in order to make life simpler under these extreme conditions. Cold-adapted proteins/factors must be structurally modified to improve their cold tolerance role since these factors are often heat sensitive, whereas the majority of these related proteins from their mesophilic counterparts are otherwise stable (De-Maayer et al. 2014). As a consequence, different molecular modifications and evolutions influence their properties and characteristics. Thus, it has been proposed that strong evolutionary pressure can modify nucleotides through substitution, addition and deletion to alter the respective amino acid in the polypeptide chain that ultimately alters their activity towards low temperatures (Maccario et al. 2015). Psychrophiles rising in the cold world have evolved special mechanisms at their molecular and protein levels to regulate their various intrinsic and extrinsic functions to sustain in the freezing conditions. This chapter offers new perspectives and comprehensive details about the different traits and characteristics that psychrophiles have embraced in order to live under intense circumstances in the varied ecosystems of the world, where otherwise life cannot thrive.

4.2 Ecological Adaptability of Psychrophiles

Almost 80% of our planet's area is permanently cold, with temperature below 5 °C (Margesin and Miteva 2011). The major fraction of this low-temperature environment is represented by the deep sea (nearly 71% of the Earth is covered by oceans, and 90% of the ocean volume is below 5 °C), followed by permafrost (24% of land surface), snow (35% of land surface), glaciers (10% land surface) and sea ice (13% of coverage space). Other cold environments are cold water lakes, cold soils (especially subsoils), cold deserts and caves (Rodrigues and Tiedje 2008). The diversity of cold-adapted microbes associated with various aquatic and terrestrial and aquatic cold environments has recently been comprehensively reviewed (Piette et al. 2011). Several physiochemical constraints were put on cellular function of microbes by lower temperature negatively influencing cell integrity, membrane fluidity, macromolecular interactions, solute diffusion rates and enzyme kinetics (Tehei and Zaccai 2005). Therefore, in order to grow, survive and reproduce, they must adopt themselves by changing their chemical composition of cell membrane and small molecules within the cells (De-Maayer et al. 2014). Microbes have remarkable potential to successfully colonize or adapt to harsh cold environmental conditions, where the temperature seems to be around −20 °C and survival seems to be almost impossible (Morgan-Kiss et al. 2006). Despite all these challenges to survive in these extreme environments, a remarkable diversity of microbes of bacteria, fungi (yeast) and microalgae exists. Although cold-adapted microbes are found in diverse environment conditions, their culture-dependent and culture-independent studies were mostly done in sea ice and deep sea water permafrost environments (Ganzert et al. 2007; Cavicchioli 2006). Psychrophilic bacteria represent Gram-negative bacteria (e.g. *Psychrobacter*, *Pseudoalteromonas*, *Moraxella*, *Polaromonas*, *Psychroflexus*, *Moritella*, *Pseudomonas* and *Vibrio* species), Gram-positive bacteria (e.g. *Micrococcus*, *Arthrobacter* and *Bacillus* species), *Archaea* (such as *Methanococcoides*, *Methanogenium* and *Halorubrum* species), yeast (*Candida* and *Cryptococcus* species), fungi (such as *Cladosporium* and *Penicillium*) and microalgae (*Chloromonas*) displaying striking cold-adaptive characteristics. Ecological limiting factors such as water availability and nutrient, pressure, salinity, UV irradiation and temperature are all characteristic properties of cold environments (Cary et al. 2010).

Water plays a crucial role in the survival of cold-adaptive microbes, as its role is more important and great in cold environments (Franks et al. 1990). To perform its critical function under harsh environments, cold-adaptive microbes need to improve catalytic efficiency of protein, improve mechanism to stabilize macromolecules their needed function and improve cell envelope (Mazur 1977; Nedwell 1999; Herbert and Bell 1977). Freezing, thawing, nutrient availability and drought are common processes in cold regions on earth including polar regions and high altitude (Bolter et al. 2005). Low temperature decreases the thermal notion of atoms and molecules decreasing or slowing down all the metabolic processes effecting its survival and reproduction (De-Maayer et al. 2014; Mackelprang et al. 2017). Below the freezing point, water availability of solvent is reduced affecting metabolic processes. Rigidity of protein increases with decrease in temperature that indirectly decreases the reaction rate. Continued freezing and thawing influences the membrane structure and function of microbial membrane lipids of cold-adaptive microbes (Hazel 1995). Reduction in activity of membrane-bound enzyme slows the rate of lateral protein diffusion within plane of membrane bilayer and induces cluster formation of integral membrane protein challenges faced during transition from fluid to gel phase (Jagannadham et al. 2000). Insufficient kinetic energy to overcome the enzyme activation barrier resulted in the slow rate of chemical reaction, main factor preventing growth at low temperature.

There are some psychrophiles that encounter cold and high pressure both at the same time, called as piezopsychrophile. High pressure has similar effect on cell membrane and protein as high temperature, i.e. they reduce the fluidity of membrane by increasing packing of fatty acyl chains of phospholipid, nutrient uptake and cell signalling mechanism. The conformation of cellular protein changes hydrostatic pressure (Bartlett 1999). The active site of catalytic domain reduces/inhibits the enzymatic reaction under overpressure. Selection of highly stable protein models and production of osmolyte, chaperone proteins are adaptive mechanisms adopted by cold-adaptive microbe for survival (Richardson 1981). Presence of high hydrostatic pressure leads to upregulation of chaperone protein which helps in proper folding of proteins (Oger and Jebbar 2010). In cold environments, primary producer of food webs are photoautotrophic phytoplanktons, and many algal classes (*Bacillariophyceae*, *Chlorophyceae*, *Chrysophyceae*, *Dinophyceae*, *Cryptophyceae*, *Prymnesiophyceae*, *Cyanobacteria* and *Prasinophyceae*) are identified in the polar ice communities (Dieser et al. 2010). UV-A and UV-B have damaging effects to phototrophic organisms, i.e. UV-B affects nucleotides by damaging cells and at the protein level by inhibiting photosynthesis that leads to cell death, later causing formation of reactive oxygen species (ROS) that attack protein, lipid and nucleotide (Correa-Llantén et al. 2012). The "coloured snow" generally found in alpine snowfield at high altitudes (>2500 m) and polar regions persists throughout the summer. At this altitude, algal cells are exposed to more light with PAR levels as high as 5000μmol m^{-2} s^{-1} during summer and 30% more UV compared to phototrophic organism (Smith et al. 1992). To inhibit the effect *C. nivalis* synthesize secondary carotenoid astxanthin causing red coloration of cells and accumulate around chloroplast as lipid droplets and also accumulate phenolic compounds and mycosporine-like amino acids to maximize photoprotection (Remias et al. 2005). Microbial extremophiles' nature and distribution inhabiting earth's polar regions provides valuable information for development of instruments and operational techniques needed to recognize evidence of extinct life elsewhere in the cosmo. *Chlamydomonas nivalis*, *Ankistrodesmus*, *Chloromonas* and *Raphidonema* are among the most common species of snow algae (Duval et al. 2000).

4.3 Environmental Adaptability of Psychrophiles

A larger area of the Earth's biosphere is covered by cold blanket with temperature below 5 °C. These cold habitats include oceans (envelop 70% of the Earth's surface), high peaks of Alps polar regions (arctic circle) and rocky mountains, Himalayan regions (Hamdan 2018). Despite harsh and unfavourable conditions, theses habitats are potentially colonized by huge amounts of cold-adaptive microbes, i.e. bacteria, fungi (yeast) and microalgae broadly subdivided as psychrophiles and psychrotrophs/psychrotolerants (Margesin and Collins 2019). These microbes have accumulated a multiplicity of approaches and mechanisms that enable them to survive inhabiting permanent cold environments (Kawahara 2017). Psychrophiles are true extremophiles because they are exposed to not only the cold condition but also high pressure (piezopsychrophile), high salt concentration (halopsychrolphiles) and absence of light (troglo-psychrophiles). Researchers have reported cyanobacteria, i.e. *Umbilicaria aprine* from Antarctica that carries out photosynthesis at −17 °C (Schroeter 1994) and yeast (*Rhodotorula glutinis*) causing food spoilage at −18 °C (Collins and Buick 1989).

The main challenges encountered by psychrophiles are low temperature that directly affects the rate of biochemical reaction, and another one is viscosity of aqueous environment. A vast array of unique adaptive features was required to be adapted by psychrophiles to survive in harsh and extremely cold environments.

4.3.1 Membrane Fluidity

The most adaptive strategies adopted by the microbes in freezing environment are the ability to regulate and modulate the fluidity of membrane and provide interface between external and internal environment to overcome the deleterious effects of harsh environments (Morita 1975). An increase in membrane rigidity, activation of membrane-associated sensor and upregulation of membrane fluidity are characteristics adopted by microbes under freezing environment (Shivaji and Prakash 2010). Physical properties and function of membranes are affected at low temperature, as it leads to decrease in membrane fluidity, the onset of gel-phase transitions and loss of cell function. Modification of lipid fatty acyl chains associated with cell membrane helps maintain optimum membrane fluidity. Desaturases, a cold-shock activated enzyme, converts saturated acyl fatty acids to unsaturated acyl chains; an increase in methyl branched fatty acid helps cold microbial cells survive under harsh environmental conditions. Change in composition plays a crucial role in enhancing membrane fluidity by introducing steric constraints that reduce the number of interaction and change the packing order in the membrane. Kerekes and Nagy (1980) reported an increase in degree of unsaturated fatty acid in *Candida*, *Leucosporidium* and *Torulopsis*. Similarly, in *Psychrobacter urativorans*, increase in 14% cell lipid content and enhanced synthesis of wax ester synthase in *Psychrobacter arcticus* when exposed to cold temperature (Ayala-del-Río et al. 2010).

4.3.2 Cold-Shock and Heat-Shock Responses

Bacteria encounter varied changing environmental conditions to cope with stress and adaptation to changing environmental conditions. A shift in temperature downshift or upward decreases cell membrane fluidity, affecting protein secretion and active transport (Phadtare and Severinov 2010). Low temperature affects fluidity of the membrane, efficiency of translocation and translation due to formation of secondary structure of DNA and RNA and protein folding (Phadtare 2004). A rapid drop in temperature leads to synthesis of cold-induced proteins (Cips) in microbes and increases severity of protein. Csps (cold-shock proteins) are small nucleic acid-binding proteins (65–75 amino acids in length), highly conserved but vary in thermostability (Graumann and Marahiel 1996). In *E.coli*, numerous Cips have been recognized, e.g. cold-shock proteins (csps), RNA helicase csdA, exoribonuclease PNPase and RNase R (Jones et al. 1987). After the downshift in temperature, a cold-shock response is initiated, whose synthesis declines with time and other protein synthesis increases (Ermolenko and Makhatadze 2002). Cold-shock domain (CSD) having two nucleic acid-binding motifs 1 and 2 that facilitate binding of DNA and RNA is a highly conserved nucleic acid domain synthesized by Csps. Csps also work as chaperons preventing misfolding of RNA to maintain functional conformation and suppressing mutation that affects RNA structure; they are also linked with biofilms and persister cell formation, thus helping microbes deal with environmental stress. Jung et al. (2010) reported a CspA of psychrophilic *Psychromonas arctica* that helps in cell survival at low temperature. Cpn60 and Cpn 10 are cold-adaptive chaperons named as artic express commercialized by Agilent Technologies (Vancuren and Hill 2019).

4.3.3 Antifreeze Proteins (AFPs)

AFPs are a survival strategy for psychrophiles in cold environments. They synthesize some unique proteins that prevent growth and recrystallization of ice. A diverse group of AFP was isolated and characterized from varied sources (i.e. fishes, insects and plants). AFP has been recovered from different sources including snow mould fungi (Hoshino et al. 2003), sea ice diatoms (Janech et al. 2006), snow algae (Raymond 2011) and bacteria (Garnham et al. 2008). Bacteria that reported AFP activity are *Rhodococcus* sp., *Enterobacter agglomerans*, *Stenotrophomonas maltophilia*, *Psychrobacter* sp., *P. fluorescens* and *Marinomonas protea* (Santiago et al. 2016). AFP leads to depression of freezing points of the solution thereby preventing ice growth (Celik et al. 2010). The difference in freezing and melting point is referred as thermal hysteresis (TH) resulted in adsorption of AFP on the crystal surface of ice. That resulted in ice growth on the convex surface between adjacent AFPs thus decreasing the freezing point (Kristiansen et al. 2011). Muñoz et al. (2017) isolated antifreeze protein (AFN) from Antarctic bacterial culture GU3.1.1 that helps maintain frozen food cell structure, having great potential value for frozen food industry. Similarly, a novel antifreeze protein Afp1 was isolated from psychrophilic yeast *Glaciozyma antarctica* by Hashim et al. (2012) exhibiting both thermal hysteresis (TH) and ice recrystallization inhibition (RI) properties (Hashim et al. 2012).

4.3.4 Cryoprotectants

Exposure of microbes to cold environments leads to osmotic drainage and dehydration of cell interior that leads to deactivation of enzymes and negative effect on function of cell and its survival. But psychrophiles have evolved to prevent this cold aggregation of protein by maintaining optimum membrane fluidity through secretion of exopolymeric substances, i.e. extracellular polymeric substances (EPS), ice-nucleating proteins, compatible solutes and bio-surfactants (Bar Dolev et al. 2005). To prevent cell from shrinkage, these organic osmolytes accumulate inside the cell and suppress the freezing of solutes inside the cell membrane, thereby protecting the cell from cryo-injuries. Mechanical disruption to cell membranes caused by ice provides protection by secreting extracellular polymeric substances (EPS). *Colwellia psychrerythraea*, a sea ice bacteria, produce EPS providing protection to the cell membrane (Thomas and Dieckmann 2002; Junge et al. 2019). Psychrophilic microbial cell forms biofilms that provide protection against invasive ice crystal damage as well as acquire the nutrient within the channels (Bowman et al. 2003). Scavenging of free radicals and proper folding and stabilizing of proteins are an essential role played at low temperature (Mykytczuk et al. 2016). Psychrophilic diatom (Melosira arctica) and cold-tolerant bacterium (*Colwellia psychrerythraea*) secrete exopolymeric substances that cause alterations in the desalination and microstructure of growing ice, and also enhance ice crystal disorder and increase pore density. An increase in accumulation of threonine, sarcosine and valine was observed in *Mesorhizobium* sp. strain N33 when grown at 4 °C that probably acts as a cryoprotectant for microbes (Ghobakhlou et al. 2015).

4.3.5 Cold-Adapted Enzymes

The demand for enzymes with novel and potential characteristics has been rising continuously in multiple industrial processes. Microbes living in harsh environmental conditions like pH, temperature, high pressure and salt are potential sources of extremozymes having potential biotechnological application. A vast majority of metabolic adaptations are required by microbes to adapt and survive under harsh environmental conditions. Temperature lower or below the melting point enhances the compactness of protein (enzyme) thereby decreasing the chemical reaction rate and alters the conformational movement required for catalysis.

By decreasing temperature dependence of the reaction, enzymes maintain high activity at low temperature and improve the flexibility of active site required by substrate binding (K_m). Secondly, to maintain specificity of catalytic centre and to improve catalysis at low temperature by better accessibility of active site and favourable electrostatic interactions with the substrate. The stability curve of psychrophiles shows that they are stable at room temperature, indicating effect of hydrophobic forces in protein folding. The higher stability of cold-adaptive enzymes is driven at low temperature by change in entropy (Feller and Gerday 2004). Cold-adaptive enzymes can be exploited to replace with mesophilic enzymes as they have economic benefits by decreasing the industrial energy expenditure. Researchers are already exploiting their benefits by production of biogas, cost-effective lignocellulosic conversion, their utilization in treatment of petroleum-contaminated sites and food additive such as dietary supplement in human diets, aquaculture and livestock; enzymes like subtilisin, protease and lipase have potential application in detergent industries. *Bacillus subtilis* subsp. *subtilis* A-53, isolated from the seashore of Kyungsang (Korea), produce potential carboxymethyl cellulose able to hydrolyse cellulosic material for the industrial conversion of biomass of fermentable sugars (Kim et al. 2009). Neelamegam et al. (2012) isolated a potential bacterium, i.e. *Bacillus licheniformis* AU01 from marine sediments in India capable of synthesizing cellulose which is thermostable and resists to high pH and is thus found applicable for hydrolysis of lignocellulosic materials for ethanol production (Neelamegam et al. 2012).

4.3.6 Carotenoid Pigments

Extreme environment causes several structural damages to non-adaptive organisms; to survive in this harsh environment, organism must possess mechanisms to protect themselves. Microbes isolated from icy environments have thick cell walls and polysaccharide capsules (Priscu et al. 2006) to overcome stress, i.e. intracellular cellular concentration, decreased cell size and physical cell ruptures caused by thawing and freezing. Bacterial groups, i.e. *Bacillus* and *Actinomycetes*, from icy environments synthesize a high concentration of pigment, i.e. glaciers (Foght et al. 2004) or marine surface or ice cores. It was observed that carotenoid pigment may play an essential function in modulation of fluidity within cell membrane of bacteria. Solar radiations are highly damaging to biological system of microbes; to overcome this, microbes have employed UV screening or absorbing components, i.e. carotenoid pigments, scytonemin or amino acid (mycosporine); these compounds provide protection to photosynthetic and non-photosynthetic bacteria (Sinha and Hader 2008).

4.3.7 Protein Folding in Psychrophiles

Enzymes are the catalyst performing most of the enzymatic reaction in microbes, and their efficiency depends upon the temperature. Psychrophiles living at low temperature have metabolic flux close to mesophiles. That means their enzymes have been modified to overcome the cooling effect on reaction rates. Structural analysis has revealed that the change in thermostability of psychrophilic enzyme contributed to a discrete change in amino acid composition. Increase in flexibility of molecular structure enables accommodation of substrate at temperatures lower than the freezing point of water. Kahlke and Thorvaldsen (2012) performed a conformative bioinformatical analysis of *Vibrionaceae* genomes to characterize membrane proteins, signal peptides and transmembrane proteins and observed that amino acids, i.e. Lys (K), Asp (D) and Thr (T), were statistically increased under cold-adaptive environment that provides conformational stability to proteins synthesized. Psychrophile can optimally grow at temperature 15 °C. As compared to mesophiles, heat stress occurs at a very low temperature in psychrophiles. They possess some gene encoding a complete set of heat-shock protein (Hsp) by specific system that enables the expression and function of Hsps at a relatively low temperature. The focus on microbial biodiversity and abundance and also the functional activity, biogeography and adaptation of psychrophilic microorganisms varied in different cold environments.

4.3.7.1 Marine Environment

Almost 71% of the earth's surface is occupied by the ocean, and majority of the area is over 3000 m deep. This extreme environment is characterized by low temperature with hydrothermal vents (temperature 370 °C), hadal zone (>6000 m; the deepest part of the marine environment) and abyssal zone (3000–6000 m) characterized by enhanced hydrostatic pressure (up to 110 MPa) and the absence of solar radiation (Lauro and Bartlett 2008; Nogi 2008). Bacteria, archaea, protists and yeasts communities are the major biomass in oceans and also responsible for 98% of primary production (Whitman et al. 1998). *Gammaproteobacteria* are the predominates among varied psychrophilic and piezophilic microbes in the deep sea. Majority of culturable cold-adaptive microbes are affiliated to genera *Colwellia*, *Moritella*, *Photobacterium*, *Psychromonas*, *Marinomonas* and *Shewanella* species (Dang et al. 2009; Lauro and Bartlett 2008; Nogi 2008). A higher amount of unsaturated fatty acids in the cell membranes provides flexibility to the cell membrane, ease of solute movement and nutrient availability to the cell. Low availability of nutrients in the deep ocean leads to widespread production of extracellular hydrolytic enzymes (amylase, protease, lipases and DNases), which explains the ecological role of cold-adaptive enzymes helping in biocycling of elements in deep sea and providing food to the other simplified microbes (Dang et al. 2009). Chemoautotrophs are the dominant group in deep seas and sediments of ocean, e.g. ammonia oxidizing *Crenarchaeota* (2000–3000 m depth) (Francis et al. 2005; Nakagawa et al. 2007), playing a role in global nitrogen cycle (Francis et al. 2005). Sea ice is another most extensive and extreme cold habitat covering almost an area of 30 million km^2 in polar ocean (Collins et al. 2008) with varied ice matrix brine channels providing some extreme and harsh conditions, i.e. high salinity (35–200 psu) and pH, low temperature (0–35 °C) and solar radiation for the microbial communities (Collins et al. 2008; Junge and Swanson 2008). The diverse communities of diatoms persisting in deep oceans contribute as the primary producer in polar oceans (Collins et al. 2008). Deep sea extremophiles have attracted the attention of researchers searching for novel bioactive substances synthesized by these potential microbes.

4.3.7.2 Non-marine Environment

Almost 70% of the earth's surface is covered by the seas and oceans that harbour a variety of life forms The great majority of marine environment belongs to two domains, i.e. *Bacteria* and *Archaea*, capable of survival in extreme chemical and physical environments. The largest fraction of the global biosphere has temperature permanently below 5 °C (Feller and Gerday 2004; Cavicchioli 2006; Siddiqui and Cavicchioli 2006; Margesin and Miteva 2011). Geographically dispersed alpine regions, vast tracts of the deep sea, geologically specific subterranean caverns and climatically challenged regions of permafrost are diverse types of environment, having plethora of cold-adaptive microbes, i.e. archaea, bacteria, eucar*ya* and viruses. Metagenomics, small subunit ribosomal RNA (SSU rRNA) sequencing and fluorescent in situ hybridization (FISH) are among the molecular genetic approaches adopted to study the microbial diversity of these icy environments. Antarctic lakes contain varied molecular signatures of archaea aerobic haloarchaea and strictly anaerobic methanogens (Bowman et al. 2003). *Arthrobacter*, *Gelidibacter*, *Halobacillus*, *Marinobacter*, *Pseudoalteromonas*, *Psychrobacter*, *Shewanella* and *Sphingomonas* are among the most prominent Antarctic isolates. They have physiological and genetic traits that enable them to effectively compete and proliferate under the specific physiochemical regimes. The methanogenic archaea and homoacetogenic bacteria are chemolithoautrophic producers of organic matter capable of utilizing hydrogen and CO_2/CO for growth.

Psychrophiles employ small RNA-binding proteins (Rbps) that generally accumulate in cold stress having an important role in termination of transcription. During cold and osmotic stress, RNA-binding motif gets upregulated by Rbps (Maruyama et al. 1999). Ehira et al. (2003) studied that expression of rbp gene increases at low temperature in psychrophilic cyanobacteria *Oscillatoria* sp. SU1 during cold stress. *Psychrobacter cryohalolentis*, a eurypsychrophilic bacterium isolated from Siberian permafrost, showed an enhanced upregulation of specific energy conservation and biosynthetic pathways and an increase in cytoplasmic pool of ATP and specific carbon substrate utilization pathways (Amato and Christner 2009). Ice damage, oxidative insult and osmotic balance are some of the stress microbe encounter at temperature low enough to form ice (Williams et al. 2010). Microbes synthesize EPS that protect cell against mechanical disruption. *Colwellia psychrerythraea* produce EPS that resulted in biofilm formation providing protection against invasive ice crystal in glaciers (Thomas and Dieckmann 2002). Similarly, many psychrophilic microbes (*Gammaproteobacteria*, i.e. *Colwellia*, *Moritella*, *Photobacterium*, *Psychromonas*, *Marinomonas* and *Shewanella*) have cell membrane composition with high proportion of unsaturated fatty acids (Margesin and Miteva 2011). Diverse classes of enzyme (e.g. oxidoreductases, glucanases, hydrolases, hydrogenases, isomerases, nucleic acid-modifying enzymes) have evolved to effectively function ranges from 0 to 100 °C to maintain its structure and function (Siddiqui and Cavicchioli 2006). *Bacillus psychrosaccharolyticus*, a cold-adaptive microbe, synthesize alanine racemase capable of thriving at temperature optimum of 0 °C (Okubo et al. 1999).

4.3.7.3 Glacier Environment

Glacial environment is referred to as a life-preserving environment. Polar regions account for 15% of the area (glaciers are added as well as permafrost). These cold temperature biotropes are colonized by cold-adapted organisms, i.e. psychrophiles representing three domains of life, i.e. bacteria, archaea and eukarya. A vast array of adaptive features are required by psychrophiles to survive and function under low temperature. Priscu and Christtner (2004) initially describe the concept of ice as a habitat for cold-adapted organisms. Viable microbial population reported beneath Canadian high arctic regions and polythermal glaciers in Alaskan (Bhatia et al. 2006), occurrence of yeasts in glacial meltwater river originated from glaciers of Argentinean Patagonia, bacterial population beneath glacier in Swiss Alps, yeast in alpine glacier cryoconites (Margesin et al. 2002). Psychrophiles generally grow in three habitats in the glacial environment: (1) liquid veins between ice crystals—solutes from ice crystals are concentrated in intestinal vein allowing these veins to retain liquid even at temperature –35 °C; (2) unfrozen water film which is nanometers thick allowing cells to get immobilized forming a film of water; and (3) polycrystalline ice trap individual microbes within the crystals having mineral inclusions. Microbes entrapped in ice carry out survival metabolic without cell division, which only requires a specific quantity energy available, utilized only for macromolecular damage repair. The reduced metabolic activity in ancient bacteria removed from ice can easily recover from spontaneous macromolecular damages, i.e. nucleic acid depurination and amino acid racemization. The reduced rate of diffusion of ion nutrients and waste products is a survival strategy for the microbes to survive on ice conditions. High-fidelity RNA replication and reduced activity of ribozyme enable the synthesis of replication products for extended periods of time at –7 °C indicating that mutations in RNA sequences are able to confer adaptive traits to the microbe (Montiel 2000). At subzero temperature, reduced enzyme activity and slow chemical reaction rates, denaturation of protein, increased water viscosity, decreased cell membrane fluidity and limited availability of water for performing biochemical reactions are some special challenges that microbes cope with to survive (Hassan et al. 2016; Wynn-Williams 1990).

4.4 Adaptation to Cold Habitat

4.4.1 Morphological Features

Optimum growth temperatures: Generally, psychrophiles multiply at temperatures >15 °C, up to 20 °C. They are named as (Atlas and Bartha 1998; Cavicchioli and Siddiqui 2004):

1. Eurypsychrophile—having wide environmental tolerance.
2. Stenopsychrophile—having narrow environmental tolerance.

Factors governing the morphological and physiological adaptations in psychrophiles may be summarized under "Genetic elements and regulation of their expression" explained further under "Molecular aspects" and Environmental Factors (Cavicchioli 2016). The latter may further be classified as:

1. Biotic factors: predatory microorganisms, viruses, inter-cellular interactions, natural habit and habitat.
2. Abiotic factors: pH, buffers, salinity, dissolved gasses like oxygen and CO_2, nutrients, toxins and antibiotics.

Adaptations in the envelopes and cell membrane: the lipid composition of the cell membrane in psychrophiles dictates the membrane fluidity, through shorter-chain fatty acids and decreased lipid saturation, at low temperatures, preventing the freezing or crystallization of the membrane (Table 4.1).

4.4.2 Molecular Aspects

The era of metagenomics has unveiled the genetic characterization of non-culturable novel microorganisms, suggesting their role and basis of adaptations to their natural habitats. Genomics coupled with proteomics helps in interpreting not only the functional potential of the gene but also the role of proteins in establishing the favourable adaptations in the microorganisms. Analysis of transcriptome and metatranscriptome has elucidated the mechanistic and metabolic adaptations that enable psychrophiles to survive in cold temperatures.

1. Regulation of genome and proteome: These molecular adaptations include changes at transcriptional levels and regulation of genes involved in metabolism, biosynthetic pathways and stress responses (Table 4.2).
2. Role of enzymes in psychrophiles: Structural and functional alterations in enzymes enable psychrophiles to adapt and survive under cold environment. Cold adaptation results from preferred selection of high enzymatic activity at low temperatures and lowered preference for thermostability (Feller and Gerday 2004). Enzymes in psychrophiles are adapted to have higher K_{cat}, i.e. faster turnover, and lower K_m, i.e. higher affinity for their substrate, thus increasing the overall enzymatic efficiency (Baraúna et al. 2017). Characteristics features of enzymes present in psychrophiles in comparison with meso−/thermophilic homologs are summarized in Table 4.3 (Feller and Gerday 1997; Smalås et al. 2000; Fields and Somero 1998; Russell 2000; Gianese et al. 2002; D'Amico et al. 2003; Georlette et al. 2004).
3. Cold-shock proteins (CSPs) and cold acclimatization/adaptation proteins (CAPs): CSPs are RNA chaperone proteins that regulate processes of transcription and

Table 4.1 Cold adaptations in cell envelopes and membranes of bacteria

Psychrophile	Adaptations in the envelopes and cell membrane	References
Gram-negative bacteria		
Pseudoalteromonas haloplanktis	↑ cell envelope genes	Médigue et al. (2005)
Psychrobacter arcticus	Genes of membrane lipids and cell wall synthesis	Bergholz et al. (2009)
Sphingopyxis alaskensis	↑ expression of proteins for biosynthesis of envelope, cell wall, cell membrane, and exopolysaccharides at 10 °C	Ting et al. (2010)
*Pseudomonas syringae n*Lz4w	Alterations in the configuration and fluidity of lipopolysaccharides ↑ polymyxin B sensitivity ↑ hydroxy fatty acids	Kumar et al. (2002)
Pseudomonas extremaustralis	Core lipopolysaccharide glycosyltransferase regulates cell envelope's flexibility and turgor pressure, surface area/volume ratio, and cell permeability, depending on temperature	Benforte et al. (2018)
Pseudoalteromonas sp.	Exopolysaccharides with mannose as primary constituent	Caruso et al. (2018b)
Winogradskyella, *Colwellia*, and *Shewanella* genera	Freeze-thaw survival through extracellular polymeric substances	Caruso et al. (2018a)
Colwellia psychrerytharea	Proteins for synthesis, ramification, and cis-isomerization of polyunsaturated fatty acids	Nunn et al. (2015)
Gram-positive bacteria		
Exiguobacterium sibiricum	↑ biosynthesis of cell wall at −2.5 °C	Rodrigues et al. (2008)
Planococcus halocryophilus Or1	Cell envelope having encrustations containing calcium carbonate (20%), peptidoglycan (50%), and choline (29%) at −15 °C ↑ peptidoglycan synthetase and precursors ↑ carbonic anhydrase ↑ hydrophobicity and ↑ fatty acid saturation Inactive fatty acid desaturases	Mykytczuk et al. (2013, 2016), Ronholm et al. (2015)

translation in prokaryotes through destabilizing the secondary RNA structures formed due to exposure to cold environment (Phadtare 2004; Baraúna et al. 2017). CSP studies in *E. coli* can be broadly classified into CSPs I and II (Table 4.4).

Table 4.2 Molecular targets and resultant changes for cold adaptations in psychrophiles

Targeted genes/pathways	Resultant changes
• Cold-shock proteins (CSPs A–E and G) and cold adaptation/acclimatization proteins • Antifreeze proteins or ice-nucleating proteins • RNA/DNA helicases • Protein chaperones • Osmoprotectants • Proteins of oxidative stress response • Exopolysaccharide synthesis • Fatty acid desaturases • Trehalose synthase • Membrane modifications • pH homeostasis pathway • Production of compatible solutes	• ↑ Membrane fluidity • Cell wall thickening • Survival strategies and stress responses for cryoprotection • Enzyme response to decompose organic matter upon thawing • Acetogenesis • Heterotrophic methanogenesis • ↓ Freezing point for the cytoplasmic aqueous phase and molecular stabilization • Resist turgor pressure and prevent water loss

Jia et al. (1997), Russell (1997), Hébraud and Potier (1999), Jagannadham et al. (2000), Varin et al. (2012), Coolen and Orsi (2015), Liljeqvist et al. (2015), Koo et al. (2016), Mackelprang et al. (2017).

Table 4.3 Characteristic features and molecular adaptations in enzymes of psychrophiles

Molecular modification in the enzymes	Characteristics developed in enzymes due to modifications	Cold adaptations in psychrophilic enzymes
↓ Stabilizing interactions in 3° protein structure ↑ Number and length of hydrophobic loops ↑ Size of hydrophobic core ↑ Glycine content ↓ Proline and arginine content ↓ Intramolecular bonds: H-bonds and salt bridges	↓ Thermostability (often), or ↑ Thermolability ↑ Structural flexibility for better enzyme-substrate interaction	↓ Optimum temperature: 0–20 °C ↓ Activation energy requirement ↑ Catalytic efficiency at low temperatures

Table 4.4 Cold-shock protein characteristics in *E. coli* (Phadtare 2004; Pierce et al. 2011; Baraúna et al. 2017)

CSP Group I	Expression induced due to cold shock but dramatically decreased afterwards
	CspA, CspB, CspG, CspI: RNA chaperones
	CsdA: DEAD-box RNA helicase, induces coccobacilli shape, RNA chaperone
	RbfA: Ribosome binding factor, causes ribosome maturation
	PNPase: Enzyme, catalyzes phosphorolysis of sspolyribonucleotides; responsible for decrease in CSPs after cold-shock response
CSP group II	Expression induced during acclimatization phase for maintenance of cellular functions. RecA, IF-2, H-NS, GyrA, Hsc66, and HscB

4.4.3 Other Special Features

Microorganisms living in extremely cold environments have also been reported to possess altered metabolic lifestyles for survival in such harsh conditions. Microbes living in lower-temperature conditions also have to face the additional problem of higher probability of encountering reactive oxygen species owing to higher solubility of oxygen at lower temperatures. In order to deal with this problem, common oxidative pathways like EMP, TCA cycle, electron transport chain, etc. are generally depressed in such conditions (Piette et al. 2012). However, there is still a lot of gap in our understanding regarding the mechanisms involved in the depression as well as alternate mechanisms of metabolism (Tribelli and Lopez 2018).

Ayala-Del-Río et al. (2010) have reported that *Psychrobacter arcticus* 273-4 possessed several gluconeogenesis enzymes but lacked glycolytic genes and phosphotransferase system genes. Induction of glyoxylate cycle genes was reported by Aliyu et al. (2016) in polyextremophilic *Nesterenkonia* sp. AN1 found in Antarctic soil. Similarly, the glyoxylate pathway has also been observed to play a significant role in *Planococcus halocryophilus* Or1, isolated from high Arctic permafrost (Raymond-Bouchard et al. 2017). Another alternative metabolic approach has been observed in the case of *Pseudomonas extremaustralis*, isolated from Antarctica, where ethanol oxidation genes were observed to be indispensable for low-temperature growth by Tribelli et al. (2015). Another approach used by microorganisms to survive in extreme conditions includes fatty acid metabolism involving intermediates like acetyl CoA and propionyl-CoA (Ting et al. 2010). Evidence also suggests that amino acid metabolism can also serve as a source of propionyl-CoA (Ting et al. 2010). So, it appears that for growth in cold conditions, the conventional pathways are repressed, and alternative pathways or modified pathways become significant for generation of energy and biomolecules. Carotenoids also offer another line of defence in several microorganisms residing in low-temperature environments (Dieser et al. 2010). Presence of several genes associated with carotenoid biosynthesis was observed in *Arthrobacter* sp. isolated from Antarctic soil (Dsouza et al. 2015). The primary action of carotenoids is thought to be by alleviation of stress associated with UV radiation damage as well as oxidative stress (Dsouza et al. 2015).

Another metabolic adaptation which seems to be important in cold conditions is related to compatible solutes. Apart from the conventional role of osmoprotectant, they may also act as a cryoprotectant as well as carbon, nitrogen and energy sources. Cryoprotective action may be mediated via mechanisms like prevention of aggregation of macromolecules, prevention or delaying the freezing of cytoplasm by depressing its freezing point and stabilization of membranes (Collins and Deming 2013). Additionally, they may also act as scavenger of free radicals, thus helping in tolerance of oxidative stress. In the case of the Antarctic bacterium *Pseudoalteromonas haloplanktis* TAC125, cryotolerance seemed to be associated with spermine glutathione and ornithine, mediated via glutathione metabolism (Mocali et al. 2017). Similarly, in the case of *Mesorhizobium* sp. strain N33, isolated from the Arctic region, enhancement in levels of sarcosine, threonine and valine was observed on growing the microbe at lower temperature indicating possible role of this compound in tolerating cold conditions (Ghobakhlou et al. 2015). Kumar et al. (2020) analysed the proteomic response of psychrophilic *Pseudomonas helmanticensis*, from the Himalayan mountains, and observed higher expression of enzymes involving biosynthesis of compatible solutes like proline, polyamines, etc.

Many bacteria also make use of another strategy of survival under extremely cold conditions. This strategy involves antifreeze proteins. The primary activity of these proteins is to protect the cells from cryodamage arising from ice crystals by preventing or controlling their formation within the cytoplasm or in the immediate vicinity of the cell (Cheung et al. 2017). These proteins exercise the cryoprotective action via lessening of recrystallization of ice and thermal hysteresis (Bowman 2017). Muñoz et al. (2017) carried out the structural and functional analysis of antifreeze proteins from Antarctic bacterial isolates and reported their potential application in protection of frozen foods.

Apart from this, many bacteria rely on polyhydroxyalkanoates (PHA) for survival in cold conditions. Though the exact mechanism of protection is not clear, in many microbes like *Pseudomonas extremoautralis*, there are evidences of survival benefits provided by the PHAs (Catone et al. 2014). The protective action of the PHAs may be attributed to their association with global stress response (Ruiz et al. 2001). It has also been suggested that a cryoprotective role of polyhydroxybutyrate monomer may be due to protection from intracellular and extracellular ice crystal injuries (Obruca et al. 2016). The potential role of PHAs in cold tolerance has also been indicated by the presence of genes encoding PHA synthases in the bacterial community culture in samples from Antarctic freshwater environments (Ciesielski et al. 2014).

4.5 R&D Effort Innovation Technologies to Find Specific Adaptations

Psychrophilic microbes offer a unique opportunity to better understand life under extremely harsh and dynamic environmental conditions replete with frigid temperatures, freezing-thawing cycles, low water activity, high oxidative stress, etc. These environments which appear barren to the naked eyes for most parts of the years are teeming with microorganisms. Traditionally, earlier efforts to study these microorganisms have relied on the conventional cultivation-based techniques; however, nowadays, newer techniques are gaining importance in unravelling the mysterious lives of the psychrophilic microorganisms. "Omics"-based techniques are especially playing a pivotal role in exploring the adaptations of these microorganisms (Junge et al. 2019). Allen et al. (2009) carried out the genome sequencing of the psychrophilic *Methanococcoides burtonii* and concluded that cold adaptations of this archaeon could be attributed to evolution encompassing thousands of years through genome plasticity (mediated by horizontal gene transfers, transposases as well as nucleotide skews). Comparison of psychrophilic microbes with mesophilic microbes can also help in deciphering the mechanisms of cold adaptations. Metpally and Reddy (2009) carried out comparative proteomic analyses of several microorganisms and observed elevated levels of amino acids which add to flexibility in the coil regions of proteins of psychrophiles as compared to the proteins of mesophilic origin.

The role of "omics" in analysing the adaptations in cold environments may be depicted with the help of the model organism *Exiguobacterium antarcticum*, isolated from biofilm in Ginger Lake situated in the Antarctic Peninsula (Carneiro et al. 2012). The genome sequencing of this microbe was carried out by Carneiro et al. (2012), and the gene expression of this bacterium in low-temperature conditions was analysed by the transcriptomic and proteomic approach by Dall'Agnol et al. (2014). The genomic analysis led to identification of 2772 coding sequences, 9 rRNA operons as well as 76 pseudogenes (Carneiro et al. 2012). Transcriptomic data revealed differential expression of 564 genes at lower temperatures, and proteomic data showed differential expression of 73 proteins (Dall'Agnol et al. 2014). In-depth analysis of these data led to the conclusion that CSP1 was the major cold-shock protein in this bacterium. Further, evidence of post-translational modification of this gene was also observed by them. These kinds of large amounts of data were generated by omic studies and were used for carrying out bibliomic analysis for reconstruction of fatty acid biosynthesis pathway by Kawasaki et al. (2016).

Another unique approach put forth by Mocali et al. (2017) included the integrated genomic and phenomic approach for assessing the bacterial adaptations to low-temperature environments. The integrated high-throughput phenotypic data and genomic data of *Pseudoalteromonas haloplanktis* TAC125, grown at low temperatures, was compared with another bacterium *Pseudoalteromonas* sp. TB41. Apart from the expected mechanisms like alteration of membrane fatty acids, synthesis of cold-shock proteins, etc., a novel possible cold adaptation mechanism involving protein S-thiolation regulated by glutathione and glutathionylspermidine was observed by Mocali et al. (2017). Thus, by bringing together the data from different "omic" approaches, we may achieve higher levels of understanding about the psychrophilic lifestyles.

The applicability of metagenomics in providing information about the diversity of the microbial populations as well as the activities occurring within the microbial communities of the extremely cold environments is well established (Koo et al. 2016, 2018). These techniques also allow discovery of novel cold-adapted or cold-active enzymes as well as identification of novel metabolites (Aliyu et al. 2017). In a pioneering effort, Mackelprang et al. (2017) analysed the microbial survival strategies in the case of microbes in permafrost with the help of metagenomics and reported enhanced involvement of functions like stress response, environmental sensing, scavenging of detrital biomass, horizontal gene transfer, chemotaxis, dormancy, etc. For obtaining an ever clearer insight into the lives of microbial community in cryoenvironments, transcriptomics and especially metatranscriptomics offer a better option (Raymond-Bouchard and Whyte 2017). Hultman et al. (2015) carried out a metatranscriptomic study of permafrost from Fairbanks, Alaska, and observed lower metabolic activity in the permafrost as compared to the active layer. However, they also reported elevated levels of transcripts involved in methane oxidation in the permafrost sample.

In various "omic"-based studies, the significance of computers is well established. However, computer-based simulations also play a crucial role in analysing the structure and functions of various enzymes. For example, Sočan et al. (2020) elucidated the anomalous temperature optimum in cold-adapted alpha amylases with the help of computer-based simulations.

4.6 Conclusion

Life in cryoenvironments present a plethora of challenges to the microorganisms living there. Not only they have to face extremely low temperatures and freezing-thawing damages, but also they are exposed to radiation damage, low water activity, lower nutritional concentrations, oxidative stress, etc. Lower temperatures also have a detrimental effect on the membrane fluidity as well as functioning of enzymes. Microorganisms have developed various means of combating these difficulties with the help of various morphological, physiological and genetic adaptations. With the advances in research techniques, especially the "omics"-based approaches, many novel metabolites as well as many aspects of the cold-adapted microbial lifestyles are being unravelled. Although many of the strategies used by the psychrophilic microorganisms have been deciphered, there are still many glaring gaps in our understanding. So, more research efforts need to be directed towards comprehending the intricacies of psychrophilic metabolisms as well as their interactions among each other as well as with their environment.

References

Aliyu H, De Maayer P, Cowan D (2016) The genome of the Antarctic polyextremophile *Nesterenkonia* sp. AN1reveals adaptive strategies for survival under multiple stress conditions. FEMS Microbiol Ecol 92:fiw032. https://doi.org/10.1093/femsec/fiw032

Aliyu H, De Maayer P, Sjöling S, Cowan DA (2017) Metagenomic analysis of low-temperature environments. In: Margesin R (ed) Psychrophiles: from biodiversity to biotechnology. Springer, Cham, pp 389–421. https://doi.org/10.1007/978-3-319-57057-0_16

Allen MA, Lauro FM, Williams TJ, Burg D, Siddiqui KS, Francisci DD et al (2009) The genome sequence of the psychrophilic archaeon, *Methanococcoides burtonii*: the role of genome evolution in cold adaptation. ISME J 3:1012–1035. https://doi.org/10.1038/ismej.2009.45

Amato P, Christner BC (2009) Energy metabolism response to low-temperature and frozen conditions in *Psychrobacter cryohalolentis*. Appl Environ Microbiol 75:711–718

Atlas RM, Bartha R (1998) Microbial ecology: fundamentals and applications, 4th edn. Benjamin/ Cummings Science Publishing, Menlo Park

Ayala-del-Río HL, Chain PS, Grzymski JJ, Ponder MA, Ivanova N, Bergholz PW et al (2010) The genome sequence of *Psychrobacter arcticus* 273-4, a psychroactive Siberian permafrost bacterium, reveals mechanisms for adaptation to low-temperature growth. Appl Environ Microbiol 76(7):2304–2312. https://doi.org/10.1128/AEM.02101-09

Bar Dolev M, Bernheim R, Shuaiqi Guo S (2005) Putting life on ice: Bacteria that bind to frozen water. J Royal Soc Interface 13(121):20160210. https://doi.org/10.1098/rsif.2016.0210

Baraúna RA, Freitas DY, Pinheiro JC, Folador ARC, Silva A (2017) A proteomic perspective on the bacterial adaptation to cold: integrating OMICs data of the Psychrotrophic bacterium *Exiguobacterium antarcticum* B7. Proteomes 5:9. https://doi.org/10.3390/proteomes5010009

Barria C, Malecki M, Arraiano CM (2013) Bacterial adaptation to cold. Microbiology 159:2437–2443. https://doi.org/10.1099/mic.0.052209-0

Bartlett DH (1999) Microbial adaptations to the psychrosphere/piezosphere. J Mol Microbiol Biotechnol 1(1):93–100

Benforte FC, Colonnella MA, Ricardi MM, Solar Venero EC, Lizarraga L, López NI, Tribelli PM (2018) Novel role of the LPS core glycosyltransferaseWapH for cold adaptation in the Antarctic bacterium *Pseudomonas extremaustralis*. PLoS One 13(2):e0192559. https://doi.org/10.1371/journal.pone.0192559

Bergholz PW, Bakermans C, Tiedje JM (2009) *Psychrobacter arcticus* 273-4 uses resource efficiency and molecular motion adaptations for subzero temperature growth. J Bacteriol 191 (7):2340–2352. https://doi.org/10.1128/JB.01377-08

Bhatia M, Sharp M, Foght J (2006) Distinct bacterial communities exist beneath a high Arctic polythermal glacier. Appl Environ Microbiol 72(9):5838–5845. https://doi.org/10.1128/AEM.00595-06

Bhatia RK, Ullah S, Hoque MZ, Ahmad I, Yang YH, Bhatt AK, Bhatia SK (2020) Psychrophiles: a source of cold-adapted enzymes for energy efficient biotechnological industrial processes. J Environ Chem Eng 9:1. https://doi.org/10.1016/j.jece.2020.104607

Bolter M, Soethe N, Horn R, Uhlig C (2005) Seasonal development of microbial activity in soils of northern Norway. Pedosphere 15:716–727

Bowman JP (2017) Genomics of psychrophilic bacteria and archaea. In: Margesin R (ed) Psychrophiles: from biodiversity to biotechnology. Springer, Cham, pp 345–387. https://doi.org/10.1007/978-3-319-57057-0_15345

Bowman JP, Mancuso Nichols C, Gibson JAE (2003) *Algoriphagus ratkowskyi* gen. nov., sp nov., *Brumimicrobium glaciale* gen. Nov., sp nov., *Cryomorpha ignava* gen. Nov., sp nov and *Crocinitomix catalasitica* gen. Nov., sp nov., novel flavobacteria isolated from various polar habitats. Int J Syst Evol Microbiol 53:1343–1355

Carneiro AR, Ramos RTJ, Dall'Agnol H, Pinto AC, Soares SC, Santos AR, Guimarães LC, Almeida SS, Baraúna RA, Graças DA et al (2012) Genome sequence of *Exiguobacterium antarcticum* B7, isolated from a biofilm in ginger Lake, King George Island, Antarctica. J Bacteriol 194:6689–6690. https://doi.org/10.1128/JB.01791-12

Caruso C, Rizzo C, Mangano S, Poli A, Di Donato P, Finore I, Nicolaus B, Di Marco G, Michaud L, Lo Giudice A (2018a) Production and biotechnological potential of extracellular polymeric substances from sponge-associated Antarctic Bacteria. Appl Environ Microbiol 84(4):e01624–e01617. https://doi.org/10.1128/AEM.01624-17

Caruso C, Rizzo C, Mangano S, Poli A, Di Donato P, Nicolaus B, Di Marco G, Michaud L, Lo Giudice A (2018b) Extracellular polymeric substances with metal adsorption capacity produced by *Pseudoalteromonas* sp. MER144 from Antarctic seawater. Environ Sci Pollut Res Int 25 (5):4667–4677. https://doi.org/10.1007/s11356-017-0851-z

Cary SC, McDonald IR, Barrett JE, Cowan DA (2010) On the rocks: the microbiology of Antarctic dry valley soils. Nat Rev Microbiol 8(2):129–138. https://doi.org/10.1038/nrmicro2281

Catone MV, Ruiz JA, Castellanos M, Segura D, Espin G, López NI (2014) High polyhydroxybutyrate production in *Pseudomonas extremaustralis* is associated with differential expression of horizontally acquired and core genome polyhydroxyalkanoate synthase genes. PLoS One 9:e98873. https://doi.org/10.1371/journal.pone.0098873

Cavicchioli R (2006) Cold-adapted archaea. Nat Rev Microbiol 4:331–343. https://doi.org/10.1038/nrmicro1390

Cavicchioli R (2016) On the concept of a psychrophile. ISME J 10:793–795. https://doi.org/10.1038/ismej.2015.160

Cavicchioli R, Siddiqui KS (2004) Cold adapted enzymes. In: Pandey A, Webb C, Soccol CR, Larroche C (eds) Enzyme technology. Asiatech Publishers, New Delhi, pp 615–638

Celik Y, Drori R, Graham L, Yee-Foong M, Davies PL, Braslavsky I (2010) In: Furukawa Y, Sazaki G, Uchida T, Watanabe N (eds) Freezing and melting hysteresis measurements in solutions of hyperactive antifreeze protein from an antarctic bacteria physics and chemistry of ice. Hokkaido University Press, Sapporo

Cheung RC, Ng TB, Wong JH (2017) Antifreeze proteins from diverse organisms and their applications: an overview. Curr Protein Pept Sci 18:262–283. https://doi.org/10.2174/1389203717666161013095027

Ciesielski S, Górniak D, Mozejko J, Swiatecki A, Grzesiak J, Zdanowski M (2014) The diversity of bacteria isolated from Antarctic freshwater reservoirs possessing the ability to produce polyhydroxyalkanoates. Curr Microbiol 69:594–603. https://doi.org/10.1007/s00284-014-0629-1

Collins M, Buick R (1989) Effect of temperature on the spoilage of stored peas by *Rhodotorula glutinis*. Food Microbiol 6:135–141. https://doi.org/10.1016/S0740-0020(89)80021-8

Collins RE, Carpenter SD, Deming JW (2008) Spatial heterogeneity and temporal dynamics of particles, bacteria, and pEPS in Arctic winter sea ice. J Mar Syst 74:902–917. https://doi.org/10.1016/j.jmarsys.2007.09.005

Collins RE, Deming JW (2013) An inter-order horizontal gene transfer event enables the catabolism of compatible solutes by *Colwellia psychrerythraea* 34H. Extremophiles 17:601–610. https://doi.org/10.1007/s00792-013-0543-7

Coolen MJ, Orsi WD (2015) The transcriptional response of microbial communities in thawing Alaskan permafrost soils. Front Microbiol 6:197. https://doi.org/10.3389/fmicb.2015.00197

Correa-Llantén DN, Amenábar MJ, Blamey JM (2012) Antioxidant capacity of novel pigments from an Antarctic bacterium. J Microbiol 50:374–379. https://doi.org/10.1007/s12275-012-2029-1

Dall'Agnol HPMB, Baraúna RA, de Sá PHCG, Ramos RTJ, Nóbrega F, Nunes CIP, Graças DA, Carneiro AR, Santos DM, Pimenta AMC et al (2014) Omics profiles used to evaluate the gene expression of *Exiguobacterium antarcticum* B7 during cold adaptation. BMC Genomics 15:986. https://doi.org/10.1186/1471-2164-15-986

D'Amico S, Marx JC, Gerday C, Feller G (2003) Activity-stability relationships in extremophilic enzymes. J Biol Chem 278(10):7891–7896. https://doi.org/10.1074/jbc.M212508200

Dang HY, Zhu H, Wang J, Li TG (2009) Extracellular hydrolytic enzyme screening of culturable heterotrophic bacteria from deep-sea sediments of the southern Okinawa trough. World J Microbiol Biotechnol 25:71–79. https://doi.org/10.1007/s11274-008-9865-5

De-Maayer P, Anderson D, Cary C (2014) Some like it cold: understanding the survival strategies of psychrophiles. EMBO Rep 15(5):508–517. https://doi.org/10.1002/embr.201338170

Dieser M, Greenwood M, Foreman CM (2010) Carotenoid pigmentation in Antarctic heterotrophic bacteria as a strategy to withstand environmental stresses. Arct Antarct Alp Res 42:396–405. https://doi.org/10.1657/1938-4246-42.4.396

Dsouza M, Taylor MW, Turner SJ, Aislabie J (2015) Genomic and phenotypic insights into the ecology of *Arthrobacter* from Antarctic soils. BMC Genomics 16:36. https://doi.org/10.1186/s12864-015-1220-2

Duval B, Shetty K, Thomas WH (2000) Phenolic compounds and antioxidant properties in the snow alga *Chlamydomonas nivalis* after exposure to UV light. J Appl Phycol 11:559–566. https://doi.org/10.1023/A:1008178208949

Ehira S, Hamano T, Hayashida T, Kojima K, Nakamoto H, Hiyama T, Ohmori M, Shivaji S, Sato N (2003) Conserved temperature-dependent expression of RNA-binding proteins in cyanobacteria with different temperature optima. FEMS Microbiol Lett 225:137–142. https://doi.org/10.1016/S0378-1097(03)00503-2

Ermolenko DN, Makhatadze GI (2002) Bacterial cold-shock proteins. Cell Mol Life Sci 59:1902–1913. https://doi.org/10.1007/PL00012513

Feller G, Gerday C (1997) Psychrophilic enzymes: molecular basis of cold adaptation. Cell Mol Life Sci 53(10):830–841. https://doi.org/10.1007/s000180050103

Feller G, Gerday C (2004) Psychrophilic enzymes: hot topics in cold adaptation. Nat Rev Microbiol 1(3):200–208. https://doi.org/10.1038/nrmicro773

Fields PA, Somero GN (1998) Hot spots in cold adaptation: localized increases in conformational flexibility in lactate dehydrogenase A4 orthologs of Antarctic notothenioid fishes. Proc Natl Acad Sci U S A 95(19):11476–11481. https://doi.org/10.1073/pnas.95.19.11476

Foght JM, Aislabie J, Turner S, Brown CE (2004) Culturable bacteria in subglacial sediments and ice from two southern hemisphere glaciers. Microb Ecol 4:329–340. https://doi.org/10.1007/s00248-003-1036-5

Francis CA, Roberts KJ, Beman JM, Santoro AE, Oakley BB (2005) Ubiquity and diversity of ammonia-oxidizing archaea in water columns and sediments of the ocean. Proc Natl Acad Sci U S A 102:14683–14688. https://doi.org/10.1073/pnas.0506625102

Franks F, Mathias SF, Hatley RH (1990) Water, temperature and life. Philos Trans R Soc Lond A B326:517–533. https://doi.org/10.1098/rstb.1990.0029

Ganzert L, Jurgens G, Munster U, Wagner U (2007) Methanogenic communities in permafrost-affected soils of the Laptev Sea coast, Siberian Arctic, characterized by 16S rRNA gene fingerprints. FEMS Microbiol Ecol 59:476–488. https://doi.org/10.1111/j.1574-6941.2006.00205.x

Garnham CP, Gilbert JA, Hartman CP, Campbell RL, Laybourn-Parry J, Davies PL (2008) A Ca^{2+}-dependent bacterial antifreeze protein domain has a novel beta-helical ice-binding fold. Biochem J 411:171–180. https://doi.org/10.1042/BJ20071372

Georlette D, Blaise V, Collins T, D'Amico S, Gratia E, Hoyoux A, Marx JC, Sonan G, Feller G, Gerday C (2004) Some like it cold: biocatalysis at low temperatures. FEMS Microbiol Rev 28 (1):25–42. https://doi.org/10.1016/j.femsre.2003.07.003

Ghobakhlou AF, Johnston A, Harris L, Antoun H, Laberge S (2015) Microarray transcriptional profiling of Arctic *Mesorhizobium* strain N33 at low temperature provides insights into cold adaption strategies. BMC Genomics 16:383. https://doi.org/10.1186/s12864-015-1611-4

Gianese G, Bossa F, Pascarella S (2002) Comparative structural analysis of psychrophilic and meso- and thermophilic enzymes. Proteins 47(2):236–249. https://doi.org/10.1002/prot.10084

Graumann PL, Marahiel MA (1996) Some like it cold: response of microorganisms to cold shock. Arch Microbiol 166:293–300. https://doi.org/10.1007/s002030050386

Hamdan A (2018) Psychrophiles: ecological significance and potential industrial application. S Afr J Sci 114:1–6. https://doi.org/10.17159/sajs.2018/20170254

Hashim NHF, Bharudin I, Nguong DLS, Higa S (2012) Characterization of Afp1, an antifreeze protein from the psychrophilic yeast *Glaciozyma antarctica* PI12. Extremophiles 17(1):12–23. https://doi.org/10.1007/s00792-012-0494-4

Hassan N, Rafiq M, Hayat M, Shah AA, Hasan F (2016) Psychrophilic and psychrotrophic fungi: a comprehensive review. Rev Environ Sci Biotechnol 15:147–172

Hazel JR (1995) Thermal adaptation in biological membranes: is homeoviscous adaptation the explanation. A Rev Physiol 57:19–42. https://doi.org/10.1146/annurev.ph.57.030195.000315

Hébraud M, Potier P (1999) Cold shock response and low temperature adaptation in psychrotrophic bacteria. J Mol Microbiol Biotechnol 1(2):211–219

Herbert RA, Bell CR (1977) Growth characteristics of an obligately psychrophilic *Vibrio* sp. Arch Microbiol 113:215–220. https://doi.org/10.1007/BF00492028

Hoshino T, Kiriaki M, Nakajima T (2003) Novel thermal hysteresis proteins from low temperature basidiomycete, *Coprinus psychromorbidus*. CryoLetters 24:135–142

Hultman J, Waldrop MP, Mackelprang R et al (2015) Multi-omics of permafrost, active layer and thermokarst bog soil microbiomes. Nature 521:208–212. https://doi.org/10.1038/nature14238

Jagannadham MV, Chattopadhyay MK, Subbalakshmi C, Vairamani M, Narayanan K, Rao CM, Shivaji S (2000) Carotenoids of an Antarctic psychrotolerant bacterium, *Sphingobacterium antarcticus*, and a mesophilic bacterium, *Sphingobacterium multivorum*. Arch Microbiol 173 (5-6):418–424. https://doi.org/10.1007/s002030000163

Janech M, Krell A, Mock T, Kang J-S, Raymond J (2006) Ice-binding proteins from sea ice diatoms (*Bacillariophyceae*). J Phycol 42:410–416. https://doi.org/10.1111/j.1529-8817.2006.00208.x

Jia Z, DeLuca CI, Chao H, Davies PL (1997) Structural basis for the binding of a globular antifreeze protein to ice. Nature 384(6606):285–288. https://doi.org/10.1038/384285a0. Erratum in: nature 385 (6616): 555

Jones PG, Van Bogelen RA, Neidhardt FC (1987) Induction of proteins in response to low temperature in *Escherichia coli*. J Bacteriol 169:2092–2095. https://doi.org/10.1128/jb.169.5.2092-2095.1987

Jung YH, Yi JY, Jung HJ, Lee YK, Lee HK (2010) Overexpression of cold shock protein a of *Psychromonas arctica* KOPRI 22215 confers cold resistance. Protein J 29:136–142. https://doi.org/10.1007/s10930-010-9233-9

Junge K, Cameron K, Nunn B (2019) Diversity of psychrophilic Bacteria in sea and glacier ice environments—insights through genomics, metagenomics, and proteomics approaches. In: Sas S, Dash HR (eds) Microbial diversity in the genomic era. Academic Press, Amsterdam, pp 197–216. https://doi.org/10.1016/B978-0-12-814849-5.00012-5

Junge K, Swanson BD (2008) High-resolution ice nucleation spectra of sea-ice bacteria: implications for cloud formation and life in frozen environments. Biogeosciences 5:865–873

Kahlke T, Thorvaldsen S (2012) Molecular characterization of cold adaptation of membrane proteins in the *Vibrionaceae* core-genome. PLoS One 12:51761. https://doi.org/10.1371/journal.pone.0051761

Kawahara H (2017) Cryoprotectants and ice-binding proteins. In: Psychrophiles: from biodiversity to biotechnology. Springer, Cham, pp 237–257

Kawasaki R, Baraúna RA, Silva A, Carepo MSP, Oliveira R, Marques R, Ramos RTJ, Schneider MPC (2016) Reconstruction of the fatty acid biosynthetic pathway of *Exiguobacterium antarcticum* B7 based on genomic and bibliomic data. Biomed Res Int 2016:7863706. https://doi.org/10.1155/2016/7863706

Kerekes R, Nagy G (1980) Membrane lipid composition of a mesophilic and psychrophilic yeast. Acta Aliment 9:93–98. https://doi.org/10.1016/S0007-1536(86)80056-0

Kim BK, Lee B-H, Lee Y-J, Jin H (2009) Purification and characterization of carboxymethyl cellulose isolated from a marine bacterium, *Bacillus subtilis* subsp. subtilis A-53. Enzyme Microb Technol 44(6):411–416. https://doi.org/10.1016/j.enzmictec.2009.02.005

Koo H, Hakim JA, Fisher PR, Grueneberg A, Andersen DT, Bej AK (2016) Distribution of cold adaptation proteins in microbial mats in Lake Joyce, Antarctica: analysis of metagenomic data by using two bioinformatics tools. J Microbiol Methods 120:23–28. https://doi.org/10.1016/j.mimet.2015.11.008

Koo H, Hakim JA, Morrow CD, Crowley MR, Andersen DT, Bej AK (2018) Metagenomic analysis of microbial community compositions and cold-responsive stress genes in selected Antarctic lacustrine and soil ecosystems. Life 8:29. https://doi.org/10.3390/life8030029

Kristiansen E, Ramløv H, Højrup P, Pedersen SA, Hagen L, Zachariassen KE (2011) Structural characteristics of a novel antifreeze protein from the longhorn beetle *Rhagium inquisitor*. Insect Biochem Mol Biol 41:109–117

Kumar GS, Jagannadham MV, Ray MK (2002) Low-temperature-induced changes in composition and fluidity of lipopolysaccharides in the Antarctic psychrotrophic bacterium *Pseudomonas syringae*. J Bacteriol 184(23):6746–6749. https://doi.org/10.1128/jb.184.23.6746-6749.2002

Kumar S, Suyal DC, Yadav A, Shouche Y, Goel R (2020) Psychrophilic *Pseudomonas helmanticensis* proteome under simulated cold stress. Cell Stress Chaperones 25:1025–1032. https://doi.org/10.1007/s12192-020-01139-4

Lauro FM, Bartlett DH (2008) Prokaryotic lifestyles in deep sea habitats. Extremophiles 12:15–25. https://doi.org/10.1007/s00792-006-0059-5

Liljeqvist M, Ossandon FJ, González C, Rajan S, Stell A, Valdes J, Holmes DS, Dopson M (2015) Metagenomic analysis reveals adaptations to a cold-adapted lifestyle in a low-temperature acid mine drainage stream. FEMS Microbiol Ecol 91(4):011. https://doi.org/10.1093/femsec/fiv011

Maccario L, Sanguino L, Vogel TM, Larose C (2015) Snow and ice ecosystems: not so extreme. Res Microbiol 166:782–795. https://doi.org/10.1016/j.resmic.2015.09.002

Mackelprang R, Burkert A, Haw M, Mahendrarajah T, Conaway CH, Douglas TA, Waldrop MP (2017) Microbial survival strategies in ancient permafrost: insights from metagenomics. ISME J 11(10):2305–2318. https://doi.org/10.1038/ismej.2017.93

Margesin R, Collins T (2019) Microbial ecology of the cryosphere (glacial and permafrost habitats): current knowledge. Appl Microbiol Biotechnol 103:2537–2549. https://doi.org/10.1007/s00253-019-09631-3
Margesin R, Miteva V (2011) Diversity and ecology of psychrophilic microorganisms. Res Microbiol 162:346–361. https://doi.org/10.1016/j.resmic.2010.12.004
Margesin R, Zacker G, Schinner F (2002) Characterization of heterotrophic microorganisms in alpine glacier cryoconite. Arct Antarct Alp Res 34:88–93
Maruyama K, Sato N, Ohta N (1999) Conservation of structure and cold regulation of RNA-binding proteins in cyanobacteria: probable convergent evolution with eukaryotic glycine-rich RNAbinding proteins. Nucleic Acids Res 27:2029–2036
Mazur P (1977) The role of intracellular freezing in the death of cells cooled at supraoptimal rates. Cryobiology 14:251–272. https://doi.org/10.1016/0011-2240(77)90175-4
Médigue C, Krin E, Pascal G, Barbe V, Bernsel A, Bertin PN, Cheung F, Cruveiller S, D'Amico S, Duilio A, Fang G, Feller G, Ho C, Mangenot S, Marino G, Nilsson J, Parrilli E, Rocha EP, Rouy Z, Sekowska A, Tutino ML, Vallenet D, von Heijne G, Danchin A (2005) Coping with cold: the genome of the versatile marine Antarctica bacterium *Pseudoalteromonas haloplanktis* TAC125. Genome Res 15(10):1325–1335. https://doi.org/10.1101/gr.4126905
Metpally RPR, Reddy BVB (2009) Comparative proteome analysis of psychrophilic versus mesophilic bacterial species: insights into the molecular basis of cold adaptation of proteins. BMC Genomics 10:11. https://doi.org/10.1186/1471-2164-10-11
Mocali S, Chiellini C, Fabiani A, Decuzzi S, de Pascale D, Parrilli E, Tutino ML, Perrin E, Bosi E, Fondi M, Giudice AL, Fani R (2017) Ecology of cold environments: new insights of bacterial metabolic adaptation through an integrated genomic-phenomic approach. Sci Rep 7:839. https://doi.org/10.1038/s41598-017-00876-4
Montiel PO (2000) Soluble carbohydrates (trehalose in particular) and cryoprotection in polar biota. Cryo Letters 21:83–90
Morgan-Kiss RM, Priscu JC, Pocock T, Gudynaite-Savitch L, Huner NP (2006) Adaptation and acclimation of photosynthetic microorganisms to permanently cold environments. Microbiol Mol Biol Rev 70(1):222–252. https://doi.org/10.1128/MMBR.70.1.222-252.2006
Morita RY (1975) Psychrophilic bacteria. Bacteriol Rev 39(2):144–167
Muñoz PA, Márquez SL, González-Nilo FD, Márquez-Miranda V, Blamey JM (2017) Structure and application of antifreeze proteins from Antarctic bacteria. Microb Cell Fact 16:138. https://doi.org/10.1186/s12934-017-0737-2
Mykytczuk NC, Foote SJ, Omelon CR, Southam G, Greer CW, Whyte LG (2013) Bacterial growth at -15 °C; molecular insights from the permafrost bacterium *Planococcus halocryophilus* Or1. ISME J7(6):1211–1226. https://doi.org/10.1038/ismej.2013.8
Mykytczuk NCS, Lawrence JR, Omelon CR, Southam G, Whyte LG (2016) Microscopic characterization of the bacterial cell envelope of Planococcus halocryophilus Or1 during subzero growth at -15°C. Polar Biol 39:701–712. https://doi.org/10.1007/s00300-015-1826-5
Nakagawa T, Mori K, Kato C, Takahashi R, Tokuyama T (2007) Distribution of cold-adapted ammonia-oxidizing microorganisms in the deep ocean of the northeastern Japan Sea. Microb Environ 22:365–372. https://doi.org/10.1264/jsme2.22.365
Nedwell DB (1999) Effect of low temperature on microbial growth: lowered affinity for substrates limits growth at low temperature. FEMS Microbiol Ecol 30:101–111. https://doi.org/10.1111/j.1574-6941.1999.tb00639.x
Neelamegam A, Rajeswari MV, Sivaramasamy E, Renga T (2012) Marine bacterium *Bacillus licheniformis* AU01 by utilizing cellulosic wastes. Waste Biomass Valoriz 3:305–310. https://doi.org/10.1007/s12649-012-9113-y
Nogi Y (2008) Bacteria in the deep sea: psychropiezophiles. In: Margesin R, Schinner F, Marx JC, Gerday C (eds) Psychrophiles: from biodiversity to biotechnology. Springer, Berlin Heidelberg, p 73e82
Nunn BL, Slattery KV, Cameron KA, Timmins-Schiffman E, Junge K (2015) Proteomics of Colwellia psychrerythraea at subzero temperatures - a life with limited movement, flexible

membranes and vital DNA repair. Environ Microbiol 17(7):2319–2335. https://doi.org/10.1111/1462-2920.12691

Obruca S, Sedlacek P, Krzyzanek V, Mravec F, Hrubanova K, Samek O, Kucera D, Benesova P, Marova I (2016) Accumulation of poly(3-hydroxybutyrate) helps bacterial cells to survive freezing. PLoS One 11:e0157778. https://doi.org/10.1371/journal.pone.0157778

Oger PM, Jebbar M (2010) The many ways of coping with pressure. Res Microbiol 161 (10):700–809. https://doi.org/10.1016/j.resmic.2010.09.017

Okubo Y, Yokoigawa K, Esaki N, Soda K, Kawai H (1999) Characterization of psychrophilic alanine racemase from *Bacillus psychrosaccharolyticus*. Biochem Biophys Res Commun 256:333–340

Phadtare S (2004) Recent developments in bacterial cold-shock response. Curr Issues Mol Biol 6 (2):125–136

Phadtare S, Severinov K (2010) RNA remodeling and gene regulation by cold shock proteins. RNA Biol 7:788–795. https://doi.org/10.4161/rna.7.6.13482

Pierce A, Gillette D, Jones PG (2011) *Escherichia coli* cold shock protein CsdA effects an increase in septation and the resultant formation of coccobacilli at low temperature. Arch Microbiol 193 (5):373–384. https://doi.org/10.1007/s00203-011-0682-0

Piette F, D'Amico S, Mazzucchelli G, Danchin A, Leprince P, Feller G (2011) Life in the cold: a proteomic study of cold-repressed proteins in the Antarctic bacterium *Pseudoalteromonas haloplanktis* TAC125. Appl Environ Microbiol 77(11):3881–3883. https://doi.org/10.1128/AEM.02757-10

Piette F, Leprince P, Feller G (2012) Is there a cold shock response in the Antarctic psychrophile *Pseudoalteromonas haloplanktis*? Extremophiles 16:681–683. https://doi.org/10.1007/s00792-012-0456-x

Priscu JC, Christner BC, Foreman CM, Royston Bishop G (2006) Biological material in ice cores. In: Elias SA (ed) Encyclopedia of quaternary science, vol 2. Elsevier, London, pp 1156–1166

Priscu JC, Christtner BC (2004) Earth's icy biosphere. In: Bull AT (ed) Microbial diversity and bioprospecting. ASM Press, Washington, pp 130–145

Raymond JA (2011) Algal ice-binding proteins change the structure of sea ice. Proc Natl Acad Sci 108:E198. https://doi.org/10.1073/pnas.1106288108

Raymond-Bouchard I, Chourey K, Altshuler I, Iyer R, Hettich RL, Whyte LG (2017) Mechanisms of subzero growth in the cryophile *Planococcus halocryophilus* determined through proteomic analysis. Environ Microbiol 19:4460–4479. https://doi.org/10.1111/1462-2920.13893

Raymond-Bouchard I, Whyte LG (2017) From transcriptomes to Metatranscriptomes: cold adaptation and active metabolisms of Psychrophiles from cold environments. In: Margesin R (ed) Psychrophiles: from biodiversity to biotechnology. Springer, Cham, pp 437–457. https://doi.org/10.1007/978-3-319-57057-0_18

Remias D, Lütz-Meindl U, Lütz C (2005) Photosynthesis, pigments and ultrastructure of the alpine snow alga *Chlamydomonas nivalis*. Eur J Phycol 40(3):259–268. https://doi.org/10.1080/09670260500202148

Richardson JS (1981) The anatomy and taxonomy of protein structure. Adv Protein Chem 34:167–339. https://doi.org/10.1016/S0065-3233(08)60520-3

Rodrigues DF, Ivanova N, He Z, Huebner M, Zhou J, Tiedje JM (2008) Architecture of thermal adaptation in an *Exiguobacterium sibiricum* strain isolated from 3 million year old permafrost: a genome and transcriptome approach. BMC Genomics 9:547. https://doi.org/10.1186/1471-2164-9-547

Rodrigues DF, Tiedje JM (2008) Coping with our cold planet. Appl Environ Microbiol 74 (6):1677–1686. https://doi.org/10.1128/AEM.02000-07

Ronholm J, Raymond-Bouchard I, Creskey M, Cyr T, Cloutis EA, Whyte LG (2015) Characterizing the surface-exposed proteome of *Planococcus halocryophilus* during cryophilic growth. Extremophiles 19(3):619–629. https://doi.org/10.1007/s00792-015-0743-4

Ruiz JA, López NI, Fernández RO, Méndez BS (2001) Polyhydroxyalkanoate degradation is associated with nucleotide accumulation and enhances stress resistance and survival of

Pseudomonas oleovorans in natural water microcosms. Appl Environ Microbiol 67:225–230. https://doi.org/10.1128/AEM.67.1.225-230.2001

Russell NJ (1997) Psychrophilic bacteria--molecular adaptations of membrane lipids. Comp Biochem Physiol A Physiol 118(3):489–493. https://doi.org/10.1016/s0300-9629(97)87354-9

Russell NJ (2000) Toward a molecular understanding of cold activity of enzymes from psychrophiles. Extremophiles 4(2):83–90. https://doi.org/10.1007/s007920050141

Santiago M, Ramirez-Sarmiento CA, Zamora RA, Parra LP, Discovery (2016) Molecular mechanisms, and industrial applications of cold-active enzymes. Front Microbiol 7:1408–1410. https://doi.org/10.3389/fmicb.2016.01408

Schroeter B (1994) In situ photosynthetic differentiation of the green algal and the cyanobacterial photobiont in the crustose lichen *Placopsis contortuplicata*. Oecologia 98:212–220

Shivaji S, Prakash J (2010) How do bacteria sense and respond to low temperature. Arch Microbiol 192:85–95. https://doi.org/10.1007/s00203-009-0539-y

Siddiqui KS, Cavicchioli R (2006) Cold-adapted enzymes. Annu Rev Biochem 75:403–433

Sinha RP, Hader DP (2008) UV-protectants in cyanobacteria. Plant Sci 174:278–289. https://doi.org/10.1016/j.plantsci.2007.12.004

Smalås AO, Leiros HK, Os V, Willassen NP (2000) Cold adapted enzymes. Biotechnol Annu Rev 6:1–57. https://doi.org/10.1016/s1387-2656(00)06018-x

Smith RC, Prézelin BB, Baker KS, Bidigare RR, Boucher N, Coley TR, Karentz D, MacIntyre S, Matlick HA, Menzies D (1992) Ozone depletion: ultraviolet radiation and phytoplankton biology in Antarctic waters. Science 255:952–959. https://doi.org/10.1126/science.1546292

Sočan J, Purg M, Åqvist J (2020) Computer simulations explain the anomalous temperature optimum in a cold-adapted enzyme. Nature Comm 11:2644. https://doi.org/10.1038/s41467-020-16341-2

Tehei M, Zaccai G (2005) Adaptation to extreme environments: macromolecular dynamics in complex systems. Biochem Biophys Acta 1724(3):404–410. https://doi.org/10.1016/j.bbagen.2005.05.007

Thomas DN, Dieckmann GS (2002) Ocean science-Antarctic Sea ice-a habitat for extremophiles. Science 295:641–644

Ting L, Williams TJ, Cowley MJ, Lauro FM, Guilhaus M, Raftery MJ, Cavicchioli R (2010) Cold adaptation in the marine bacterium, *Sphingopyxis alaskensis*, assessed using quantitative proteomics. Environ Microbiol 12(10):2658–2676. https://doi.org/10.1111/j.1462-2920.2010.02235.x

Tribelli PM, Lopez NI (2018) Reporting key features in cold-adapted bacteria. Life 8:8. https://doi.org/10.3390/life8010008

Tribelli PM, Venero ECS, Ricardi MM, Gómez-Lozano M, Iustman LJR, Molin S, López NI (2015) Novel essential role of ethanol oxidation genes at low temperature revealed by transcriptome analysis in the Antarctic bacterium *Pseudomonas extremaustralis*. PLoS One 10:e0145353. https://doi.org/10.1371/journal.pone.0145353

Vancuren SJ, Hill JE (2019) Update on cpnDB: a reference database of chaperonin sequences. Database 2019:baz033. https://doi.org/10.1093/database/baz033

Varin T, Lovejoy C, Jungblut AD, Vincent WF, Corbeil J (2012) Metagenomic analysis of stress genes in microbial mat communities from Antarctica and the high Arctic. Appl Environ Microbiol 78(2):549–559. https://doi.org/10.1128/AEM.06354-11

Whitman WB, Coleman DC, Wiebe WJ (1998) Prokaryotes e the unseen majority. Proc Natl Acad Sci U S A 95:6578–6583

Williams TJ, Burg DW, Raftery MJ, Poljak A, Guilhaus M et al (2010) Global proteomic analysis of the insoluble, soluble, and supernatant fractions of the psychrophilic archaeon *Methanococcoides burtonii*. Part I: the effect of growth temperature. J Proteome Res 9:640–652

Wynn-Williams DD (1990) Ecological aspects of Antarctic microbiology. In: Marshall KC (ed) Advances in microbial ecology. Springer, Boston, pp 71–146

Enzymatic Behaviour of Cold Adapted Microbes

5

Niharika Rishi, Rachna Yadav, Shweena Krishnani, and Rajni Singh

Abstract

Cold adapted microbes inhabiting at near-zero temperatures produce psychrophilic enzymes to withstand such harsh environments and sustain their life cycles. They are the requisites for the adaption of psychrophilic organisms in the cold. A remarkable characteristic demonstrated by cold active enzymes is their high catalytic activity (k_{cat}) at low temperatures which is obtained at the expense of substrate affinity (K_m) and by minimizing their reaction dependence on the temperature. This is achieved by increasing their molecular flexibility and destabilizing the active site by decreasing the strength of weak interactions along with removal of factors responsible for stabilization, subsequently refining the active site dynamics. These activity-stability-flexibility relationships are understood with the help of folding funnel model. In addition, to overcome the free energy (ΔG) barrier of the active site, they exhibit distinct partition of energy into its enthalpic and entropic constituents in contrast to their mesophilic orthologues. These characteristics of cold active enzymes make them commercially useful for several biotechnological applications such as wastewater treatment, molecular biology, food industry, etc.

Keywords

Cold active enzymes · Enzyme activity · Substrate affinity · Protein stability · Psychrophiles

N. Rishi · R. Yadav · S. Krishnani · R. Singh (✉)
Amity Institute of Microbial Technology, Amity University Uttar Pradesh, Noida, Uttar Pradesh, India
e-mail: rsingh3@amity.edu

R. Goel et al. (eds.), *Survival Strategies in Cold-adapted Microorganisms*,
https://doi.org/10.1007/978-981-16-2625-8_5

5.1 Introduction

The earth's biosphere is abundantly occupied by cold territories exhibiting temperatures lesser than 5 °C. This is mainly due to the fact that 71% of our planet's surface is submerged in oceans which have temperature ranging from 2 to 4 °C below 1000 m depth. These inhospitable and extreme environments are colonized by cold adapted microbes like bacteria, yeast, unicellular algae and fungi.

These extremophiles have adapted to such harsh environments and show such high biocatalytic activity due to the presence of the psychrophilic enzymes which are the key determinants of the cold adaptation at the molecular level. It may appear contradicting to the biochemical Q_{10} guide which states that with each 10 °C reduction in the temperature, enzyme activity gets halved (Åqvist et al. 2017). Psychrophilic enzymes don't disobey this concept; instead, they have shifted their peak activities to lower temperatures (Brenchley 1996). There is a vast array of functional and structural features which have been acquired by these enzymes to do so. The high specific activity of these enzymes enables them to compensate for the exponential decrease in the biochemical reactions with the decrease in temperature, and also, they exhibit high biocatalytic activity due to the absence of several non-covalent stabilizing interactions. Destabilization also results in improved flexibility and conformation.

One of the most important properties of psychrophilic enzymes is high specific activity at low temperature which is due to low requirement of activation energy for accommodation of substrates, particularly macromolecules. This is a result of a more flexible framework which subsequently decreases the cost of conformational changes desired for binding to the substrate. These enzymes also have to overcome another problem that at low temperatures, ice formation takes place which increases the risk of crystal formation inside the cell which may result in cell death. This is resolved by the expression of antifreeze protein (Gounot and Russell 1999).

Many psychrophilic organisms are irreversibly adapted to cold environments and are heat labile (Singh and Singh 2015). Heat liability, high specificity and biocatalytic activity make them important for several biotechnological uses. In this chapter, we are going to study about the main structural and functional properties of cold adapted enzymes important for the survival at low temperature and the kinetic and thermodynamic challenge along with several industrial applications which make them important for future studies.

5.2 Psychrophillic Enzymes

Psychrophilic enzymes show biocatalytic activity at low temperature approximately below 20 °C (Gerday et al. 2000). These biocatalysts attain this high activity at such extreme environments by increasing their flexibility and decreasing their stability. Most of these enzymes are thermolabile and cannot function at higher temperatures. Crystal structure of the cold active enzyme is quite important in order to study several characteristics of cold active enzymes and for comparison with the mesophilic and thermophilic enzymes.

5.2.1 Cold Adapted Activity

5.2.1.1 Inactivation and Unfolding

The most common characteristic which is observed in the psychrophiles is heat-labile activity regardless of the structural stability. Moreover, the most unstable and heat-labile region of the enzyme is found to be the active site which is majorly involved in the catalytic cycle (Fig. 5.1). This property results in easier inactivation which makes them valuable for industrial utilization.

It has been observed that the cold active enzymes are a lot more flexible than the mesophilic and thermophilic enzymes as they have localized flexibility in specific regions of their molecular structure (Feller 2013). This has been seen in cold active carbonic anhydrase and isocitrate dehydrogenase, and they have a stable overall structure except for the catalytic regions (Fedøy et al. 2007).

In low-temperature conditions, thermal energy provided by the environment is low, so in order to compensate that, psychrophilic enzymes need to be flexible. Hence, flexibility is considered accountable for the poor chemical and thermal stability of psychrophilic enzymes. Mesophilic and thermophilic proteins are more rigid and stable when compared to the psychrophilic counterparts; therefore, when temperature rises, the psychrophiles lost their activity quite soon, even before the protein unfolding occurs, while the rest are inactivated by the heat after protein unfolding (Fig. 5.2). The multidomain cold active enzymes such as α-amylase derived from *Pseudoalteromonas haloplanktis* have unstable catalytic regions, but the non-catalytic parts are as stable as the mesophilic proteins (Siddiqui et al. 2005).

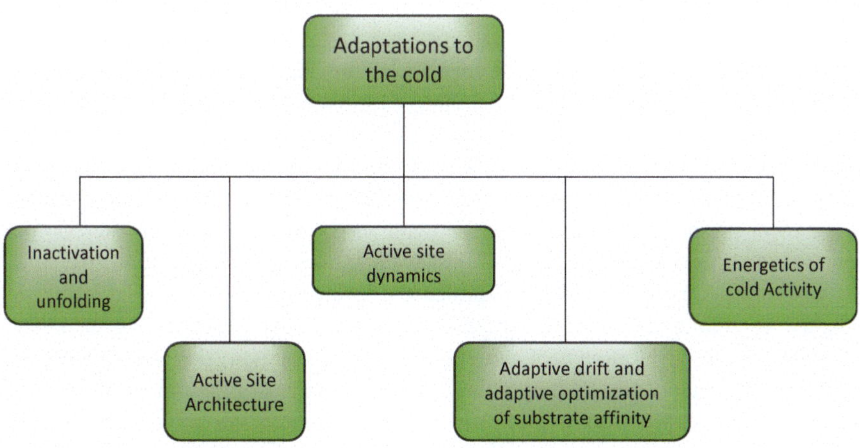

Fig. 5.1 Adaptations to cold

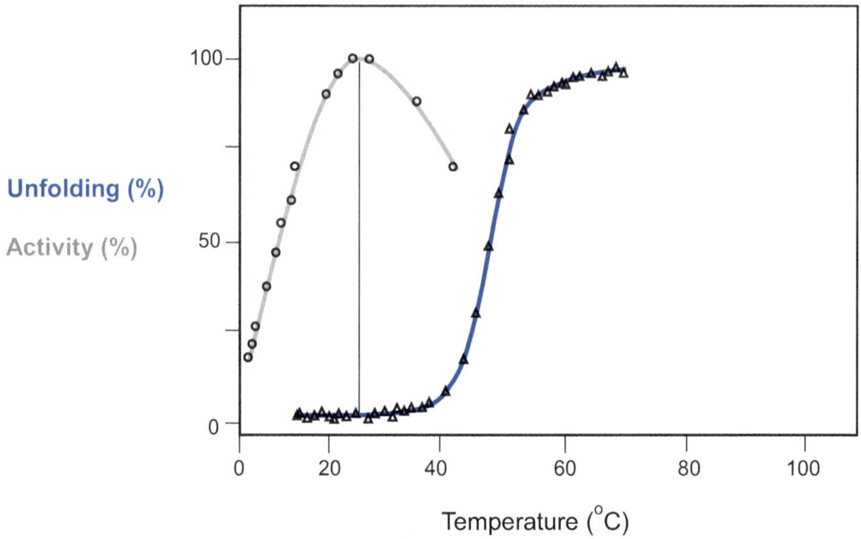

Fig. 5.2 Inactivation and unfolding of psychrophilic enzymes (D'Amico et al. 2003)

5.2.1.2 Active Site Architecture

Crystal structures of cold active enzymes have a significant role in investigation of the properties of their thermolabile and cold-adapted active sites. The active site is the most flexible and unstable domain of the cold active enzymes, due to its involvement in the catalytic activity. If we get in detail of these catalytic sites, the side chains of the active site which take part in the biocatalytic process are strongly conserved along with all the residues making the catalytic cleft. From this, it can be demonstrated that no change in the catalytic centre is accountable for the distinct characteristics of the cold active enzymes, and they can be approached without any amino acid substitution in the catalytic region. As a result, some other changes localized in a different region of the active site are accountable for enhancing the catalytic efficiency and improved dynamics of the active site (Saxena and Singh 2015; Sočan et al. 2019) However, certain conformational adjustments are observed at the active site of cold active enzymes such as formation of bigger opening of the catalytic cleft which can be accomplished by removal of the bulky side chains and substituting it with smaller groups. Bigger active site facilitates convenient release and exit of products and therefore lessens the effect of a rate-limiting step on the reaction rate (Struvay and Feller 2012). The adaptations are reported in the psychrophilic citrate synthase. In Ca^{2+}, Zn^{2+} protease (Gerday et al. 2000) derived from cold adapted pseudomonas species, supplementary Ca^{2+} ion, stretches the backbone involved in the development of the entrance of the catalytic site and enhance its accessibility, which subsequently cuts the high energy cost required for binding of substrates and hence decrease the amount of activation energy needed for the generation of the enzyme substrate complex. Another important parameter for catalytic activity at cold temperature is the production of electrostatic surface potential by the charged and polar groups. The potential generated attracts the substrate before any linkage is observed between the enzyme and the substrate. In comparison to thermophiles and mesophiles, a significant difference has been observed in the amount of electrostatic potentials generated in psychrophiles such as psychrophilic citrate synthase, trypsin, malate dehydrogenase, etc.

5.2.1.3 Active Site Dynamics

Most of the cold active enzymes are characterized as heat labile. This feature suggests that the catalytic side chains of the active site of the enzyme play a significant role in the cold adapted activity of these enzymes. It also improves the approachability of the enzyme to the substrate and liberation of product. Non-specific psychrophilic enzymes have larger range of specificity in comparison to the thermophilic and mesophilic counterparts. This is the result of broader and deeper binding pocket and highly flexible active site. Due to high flexibility, the cold active alcohol dehydrogenase is able to oxidize huge bulky alcohols, and α-amylase is able to accustom macromolecular polysaccharides (Hiteshi and Gupta 2014). It is also observed that the flexible sites of cold active enzymes are not able to hold on to short oligosaccharides and similar substrates efficiently. Moreover, in α-amylase, the inhibition pattern suggests that it can form ternary enzyme, substrate and inhibitor complexes, while the mesophilic counterparts can only form binary substrate, enzyme inhibitor complexes (Sočan et al. 2020).

5.2.1.4 Adaptive Drift and Adaptive Optimization of Substrate Affinity

Psychrophilic enzymes have low affinity toward the substrate due to adaptations in the active site dynamics which results in weak binding between the active site of the enzyme and substrate. The enzyme and substrate affinity is measured by the K_m value (refer to Table 1.1). High K_m is observed in cold active enzymes which shows its low substrate affinity (Chiuri et al. 2009). As observed in α-amylase which is found to be active on macromolecular substrates, it has 30-fold high K_m values which means it has 30-fold low substrate affinity. It reflects that psychrophilic enzymes achieve the high catalytic activity at the expense of K_m. K_m value needs to be high to maximize the reaction rate. This adaptive drift is demonstrated by cold active α- amylase and lactate dehydrogenase.

However, some of the psychrophilic enzymes counteract the adaptive drift of K_m to enhance their substrate affinity. K_m value is linked with the regulatory activity of the intracellular enzymes, e.g. intracellular chitobiase enzyme, which catalyses the chitin hydrolysis derived from psychrophilic *Arthrobacter* species (Table 5.1).

5.2.1.5 Comparative Structural Analysis of Extremophiles

Cold active enzymes are mostly illustrated by increased molecular flexibility, which enables good interaction with the substrates, and by reduced activation energy constraints in contrast to their homologues. Therefore, high flexibility could rationalize these features: heat-labile activity and increased catalytic efficiency in cold environments. The high molecular flexibility of cold active enzymes, in comparison to their mesophilic and thermophilic homologues, is the consequence of

Table 5.1 Kinetic parameters of different psychrophilic enzymes and their orthologues

		k_{cat} (min^{-1})	K_m (mM)	Ref.
Imidase	Mesophile	1500	1.0	Siddiqui et al. (2013)
	Psychrophile	25,700	1.6	
Cellulase	Mesophile	0.6	1.5	Siddiqui and Cavicchioli (2006)
	Psychrophile	11	6.0	
α-Amylase	Mesophile	700	0.10	Feller et al. (1996)
	Psychrophile	2148	0.50	

amalgamation of many distinct characteristics: deteriorating intramolecular bonds, reduced hydrophobic core compactness, increased number of hydrophobic side chains which are revealed to the solvent, elongated and increased number of hydrophilic loops, lesser proline and arginine residues and increased glycine residues. However, every protein family has adopted its own distinct approach to enhance complete or local molecular flexibility by utilizing one or more than one of these conformational adaptations (Tronelli et al. 2007).

5.2.1.6 Composition of Amino Acids

Amino acids such as alanine, aspartic acid, serine and threonine are majorly preferred in psychrophiles in comparison to their mesophilic counterparts. However, glutamic acid and leucine residues are less preferred in psychrophilic proteomes. Neutral AA residues are majorly favoured in psychrophiles, while the charged, basic, aromatic and hydrophilic residues are quite less preferred (Metpally and Reddy 2009).

5.2.1.7 Secondary Structural Elements

Constitution of amino acids of cold active and mesophilic proteomes in three main secondary structure components: α-helices, β-sheets and coils. Psychrophiles have substantially reduced amount of residues in the α-helix and significantly increased amount of residues in the coil regions. AA such as glutamic acid, phenylalanine leucine, asparagine and tyrosine are in substantially less number in α-helices of psychrophilic proteomes, while AA such as alanine, aspartic acid, glycine, serine, threonine and valine are substantially increased in the coil region of psychrophilic proteomes. The AA glutamic acid is substantially reduced in the coil region of cold active proteomes. β-sheets of psychrophiles don't have any major differences than mesophiles (Aghajari et al. 1998).

5.2.1.8 Comparative Proteome Analysis

Approximately, 10–25% of sequences from each proteome show best hit homologues from constituents of other thermal groups. This percentage relies upon the size of the proteome under consideration. If a high number of proteins are present in a query proteome, then increased percentage of hits are observed from the subject proteomes searched. On average 16.7% and 17.1% orthologous proteins in psychrophilic and mesophilic proteomes are observed in a study.

5.2.1.9 Amino Acid Substitution Pattern

Psychrophilic proteins preclude having the AA glutamic acid, phenylalanine leucine, asparagine and tyrosine, while residues alanine, aspartic acid, glycine, serine and threonine are preferred as compared to mesophilic counterparts (Sindhu et al. 2017).

5.3 Kinetics and Energetics of Cold Activity

Enzyme activity is highly dependent on the temperature, and k_{cat} is the catalytic constant which refers to the maximum number of substrate molecules which are transformed to product per active site per unit of time. As per the Michaelis-Menten mechanism, catalytic constant is the first-order rate constant utilized for converting the enzyme-substrate complex to enzyme and product. According to the transition state theory, the stable and activated enzyme substrate complex is in equilibrium with the inactivated ground-state enzyme substrate complex:

$$E + S \leftrightarrow ES \leftrightarrow ES^* \overset{k_{cat}}{\rightarrow} E + P$$

and dependence of temperature on the K_{cat} is given as follows which is equivalent to the Arrhenius law (Feller et al. 1996):

$$k_{cat} = K\frac{k_B T}{h} e^{-\Delta G^{\#}/RT} \tag{5.1}$$

where K = transmission coefficient (usually close to one), Boltzmann constant (k_B) = $1.38 \times 10^{-23} JK^{-1}$, Plank constant, h = 6.63×10^{-34}Js, Universal constant, R = 8.31 JK^{-1} mol^{-1}, ΔG = Gibbs energy.

Here, ΔG represents the free energy of activation or variation of the Gibbs energy between the activated enzyme substrate complex ES^*, and the ground state is represented as ES (D'Amico et al. 2002).

As a point of reference, the catalytic activity of a mesophilic enzyme is reduced by 20 to 80 times when the temperature drops from 37 °C to 0 °C in a biochemical reaction. This is one of the major factors due to which the growth of the organism could not occur in low-temperature ranges.

The above equation is applicable for the exponential increase in the catalytic activity of the enzyme with the temperature. Several models have been anticipated in order to stimulate the effect of heat inactivation. The elementary characteristics of psychrophilic enzymes acquired for the adaptation in the low temperatures are the following: (1) the major physiological adaptation is that the cold active organisms produce large amount of enzymes which have almost tenfold enhanced specific catalytic activity; this adaptation is to counterbalance the impact of low biochemical reaction rates at low temperatures. (2) These organisms have modified their optimum maximum activity temperatures to the lower range; as a consequence, they show poor stability, and their unfolding and inactivation occur at higher temperatures. Several studies have revealed certain relationships between the stability, activity and flexibility of these enzymes (Fig. 5.3). Definitely, increased low temperature activity

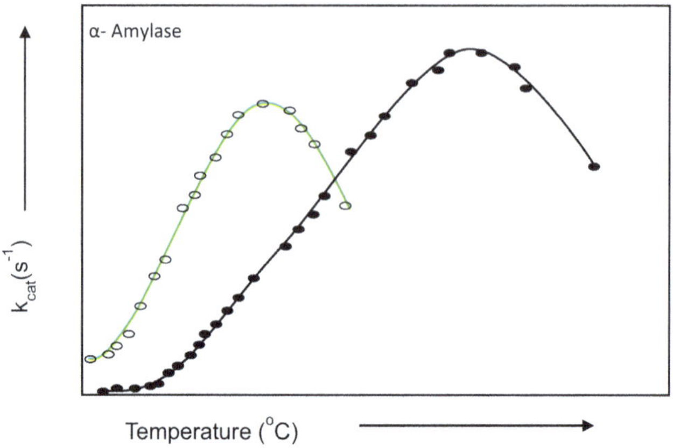

Fig. 5.3 Enzyme activity dependence on temperature. Activity of cold active α-amylase (open symbols, green line) from *P. haloplanktis AHA* and its mesophilic orthologue (symbols closed, black line) from *B. amyloliquefaciens BAA* depicts properties of cold active enzymes: cold activity and heat-labile behaviour (Marx et al. 2004)

has emerged from high flexibility, but it results in enhanced mobility which reflects the decreased stability.

Referring to (5.1), enhanced catalytic activity of psychrophilic enzymes is linked to the reduced free energy of activation denoted as $\Delta G^{\#}$. In order to decrease the height of the activation energy barrier, two approaches are introduced. In the first one, the evolutionary pressure increases the K_m to maximize the reaction rate. In accordance to the transition state theory, as the enzyme and substrate come across each other, the enzyme-substrate complex formed falls into an energy pit. To continue the reaction, a transition state ES^* needs to be achieved, which further breaks down into enzyme and the product. The length of activation energy barrier midst the ground (ES) and transition/active state ES^* is the free activation energy. The lower is this barrier, the more the activity there is. Psychrophilic enzymes have low affinity for the substrate due to which the energy pit is less deep for the enzyme-substrate complex. It illustrates that the height of the energy barrier is decreased, increasing the activity.

The second strategy takes into account the temperature dependency of the reaction which is catalysed by the psychrophilic enzyme. The classic Gibbs Helmholtz reaction shows that the free energy of activation, i.e. $\Delta G^{\#}$ comprises both enthalpic and entropic terms.

$$\Delta G^{\#} = \Delta H^{\#} - T\Delta S^{\#} \tag{5.2}$$

According to (5.1) and (5.2), k_{cat} can be rewritten as

$$k_{cat} = K\frac{k_B T}{h} e^{-\left(\Delta H^{\#}/RT - \Delta S^{\#}/R\right)} \tag{5.3}$$

This equation shows that k_{cat} is dependent upon the $\Delta H^{\#}$ and $\Delta S^{\#}$ values. $\Delta H^{\#}$ (enthalpy of activation) shows the dependence of the enzyme activity on temperature. They are related as follows: the lower the value of $\Delta H^{\#}$,the less the variation would be in activity with temperature (Feller and Gerday 2003). The psychrophiles tend to have low enthalpy value which exhibits that their reaction rate is less effected as the temperature is reduced. Therefore, this decrease in activation enthalpy is considered as one of the main reasons for the low-temperature adaptation in psychrophiles. This is a result of the reduced number of enthalpy-driven interactions occurring in psychrophiles. These interactions are also related to the structural stability of the protein folded. And as a consequence, the domain having the active site structure should be more flexible. The entropic contribution $T\Delta S^{\#}$ is found to be larger and negative in the psychrophilic enzymes. This has been inferred as the huge decrease in the disorder among the ground state with its comparatively loose conformation and the well-organized and compact transition state. The thermolability of psychrophilic enzymes shows a macroscopic explanation for this thermodynamic parameter (Feller 2003). As a result of catalytic site flexibility, the enzyme-substrate complex occupies a wider distribution of conformational states interpreted into high entropy of this state, in comparison to the mesophilic and thermophilic counterparts. Moreover, a broader distribution of ground-state ES is not accompanied by a weaker substrate binding strength, as witnessed in several cold active enzymes. In conclusion, it can be interpreted that the typical activation parameters of cold active enzymes are produced by kinetic stimulations.

5.4 Conformational Stability

Cold active enzymes tend to have weak conformational stability and are fragile in contrast to their thermophilic and mesophilic homologues. They have very few stable interactions and heat-tolerant proteins which can reduce the unstable behaviour. The thermodynamics of the low stability can be demonstrated by various techniques, and one of them is calorimetry. Calorimetric analysis of psychrophilic enzymes in comparison to the mesophilic and thermophilic counterparts revealed that they display discrete stability patterns from one another (Cipolla et al. 2012). These patterns have progressed from simple profile, unstable psychrophilic enzymes to stable but complicated thermophiles. These characteristics have been adapted using a set of parameters: (1) psychrophilic enzymes unfold as the temperature decreases, and it is referred to as the melting point or temperature of half denaturation T_m; (2) the region under the transition corresponds to the total amount of heat absorbed in the process of unfolding, that is, calorimetric enthalpy ΔH_{cal}, which is related to the enthalpic contribution concerned with maintaining the stable native form and is remarkably low in cold active enzymes (Fig. 5.4). As we move from psychrophiles to mesophiles to thermophiles, increase in the value of the calorimetric enthalpy is observed. This indicates that cold active enzymes undergo cooperative unfolding; the molecular edifice is stabilised by some weak interactions, and the commotion in these weak interaction affects the overall structure of the enzyme

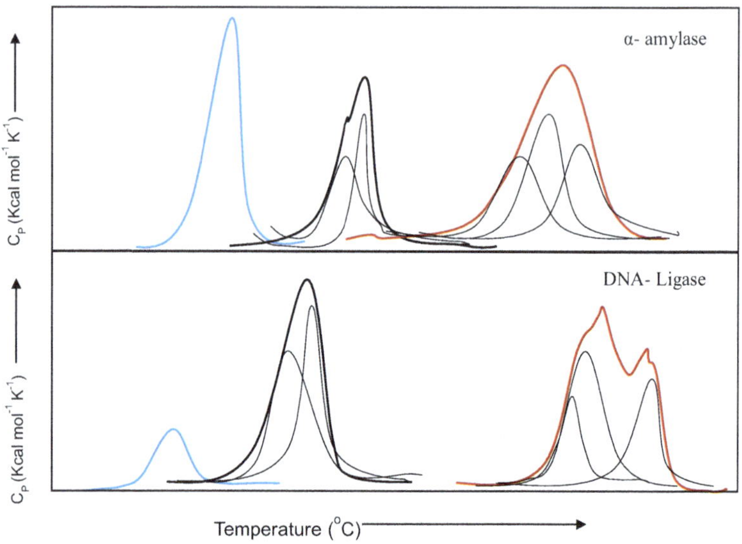

Fig. 5.4 Thermal unfolding of psychrophilic (blue), mesophilic (black) and thermophilic (red) proteins. Thermograms of α-amylases and DNA ligases recorded by microcalorimetry. The psychrophilic proteins illustrated by low T_m (top of transition) and calorimetric enthalpy ΔHcal (region under transition), by cooperative transition and absence of stability domains (depicted by thin black lines in stable proteins) (D'Amico et al. 2001)

which subsequently causes unfolding (Georlette et al. 2003). The unfolding of these enzymes is a all-or-none process which reflects the uniformly low stability of its structure. Unlike psychrophilic enzymes, the thermophilic and mesophilic enzymes have configurational domains of distinct stabilities which unfold independently. However, the unfolding process in the psychrophilic enzymes is quite reversible in comparison to the other homologues, due to the low temperature of unfolding and presence of hydrophobic core clusters. These characteristics inhibit the process of aggregation and are responsible for the reversible unfolding of psychrophilic enzymes (Feller and Gerday 1997).

5.4.1 Structural Origin of Low Stability

Low stability of cold adapted enzymes is a well-known characteristic along with role of flexibility in cold adapted activity. Several structural factors are responsible for this behaviour. Generally, it is known that the structural factors responsible for stabilizing the conformation of these enzymes are attenuated in strength and are low in number. The major elements of stability are conformational factors, hydrophobicity and most importantly the weak interactions between structural molecules. This includes clustered glycine residues which are responsible for the movement, absence of proline in loops which increases the flexibility of the chain

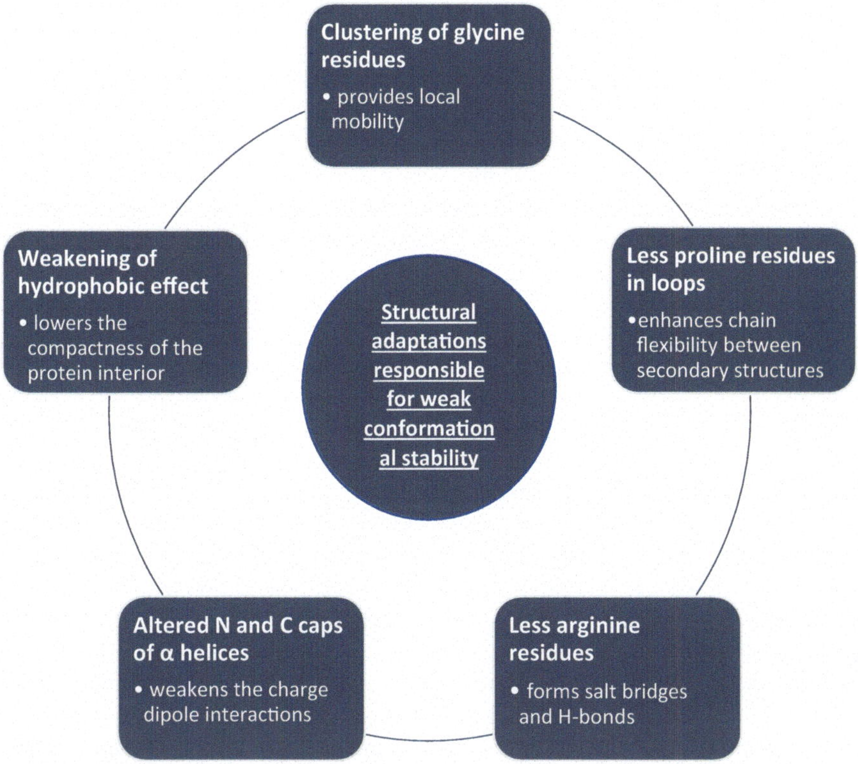

Fig. 5.5 Structural adaptations responsible for weak conformational stability

(Georlette et al. 2004) and less arginine content which forms several salt bridges and hydrogen bonds. Most of the weak interactions such as aromatic interactions, hydrogen bonds, ion pairs, etc. are found to be limited, and weak hydrophobic effect is observed in the core clusters. This collectively results in less compact protein interior. Cofactors responsible for stability are commonly found to be bound in a loose fashion and consequently favouring unzipping. Various solvent exposed ion pairs are absent on the surface of the protein due to which nonpolar ions get exposed to the outer environment or medium (entropy-driven destabilizing factor). Negative charge is also present abundantly which enhances the interaction with the solvent. The molecular plasticity of the outer shell is increased due to these factors. In multimeric proteins and enzymes, the coherence in the monomers is decreased due to the reduction in the number and strength of interaction. However, different families of proteins have adapted distinct strategies in order to reduce stability with the help of combined or individual structural alterations (Chao et al. 2020) (Fig. 5.5).

5.5 Folding Funnel Model of Cold Active Enzymes

Wolynes, Onuchic and Thirumalia introduced the folding funnel model which is a pictorial representation of protein folding (Onuchic et al. 1997). Various kinetic properties of cold active enzymes are merged in a folding funnel-based model in order to express the activity-stability association in the different extremophilic enzymes on the basis of free energy of folding (E) and conformational diversity. Energy landscapes are depicted in Fig. 5.6. Every part of the model portrays different features.

The upper most part of the model depicts the high-energy, unstable state, but as we move down the funnel, there is increase in the stability of the protein and decrease in the free energy. Therefore, the lowermost section of the funnel is occupied by the stable and folded state having minimum free energy. The length of the funnel represents the free energy of folding which correlates to the conformational stability (Ma et al. 1999). The top edge represents the unstable and unfolded state in random coil conformation; however, in cold active enzymes, it is found that a limited number of stabilizing interactions and high hydrophobic loops are available along with more glycine composition and less arginine and proline composition than that of thermophilic or mesophilic homologs. Also, an increased number of glycine clusters are found and few disulphide bonds due to which the funnel of psychrophilic enzymes is broader and placed at a higher energy level in comparison to the mesophilic and thermophilic counterparts during protein folding; the structure starts achieving conformational stability and reduction in free energy level is observed. Yet, ruggedness in the funnel slopes of the thermophilic proteins is observed due to

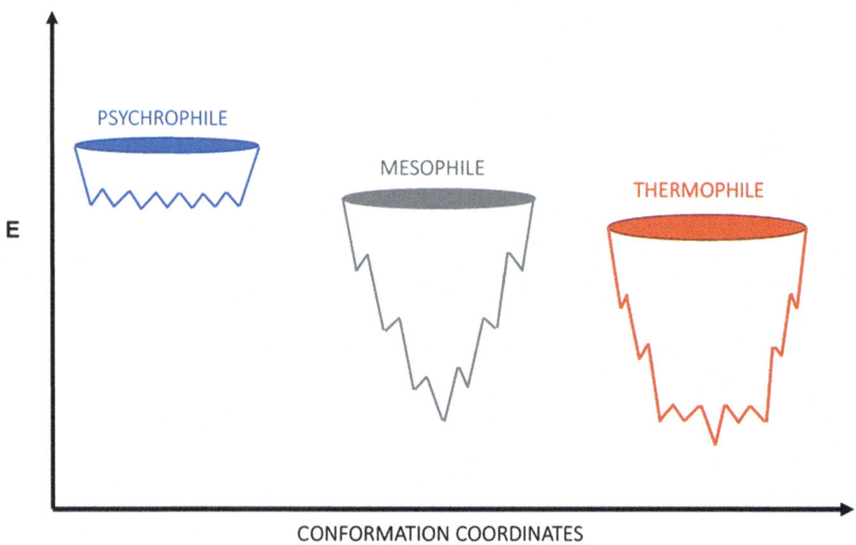

Fig. 5.6 Folding funnel model of psychrophilic, mesophilic and thermophilic enzymes (Papaleo et al. 2011; Gerday 2013)

the formation of intermediate states which result in lower cooperativity in the folding-unfolding process (Mereghetti et al. 2010). Contrary to it, in the cold active enzymes, the funnel slopes are smooth as intermediate states are absent and less stable domains and interactions are found. As a consequence, unfolding occurs cooperatively.

5.6 Psychrophillic Enzymes in Biotechnology

Psychrophilic organisms have gained comparatively less attention than the thermophiles, despite their various potential applications in applied fields. But in recent years, an increased interest of biotechnologists has been observed for the psychrophiles. Cold active enzymes have several properties which make them important for commercial and industrial use: (1) as most of the cold active enzymes are heat labile, they can be easily inactivated by moderate-heat treatment (Margesin and Schinner 1994). Selective inactivation of their enzymatic activity can also be done in a complex medium (2) as a result of high enzymatic activity at low temperature, amount of enzyme required is quite less, and it also reduces the process time thereby decreasing the energy use (Fig. 5.7).

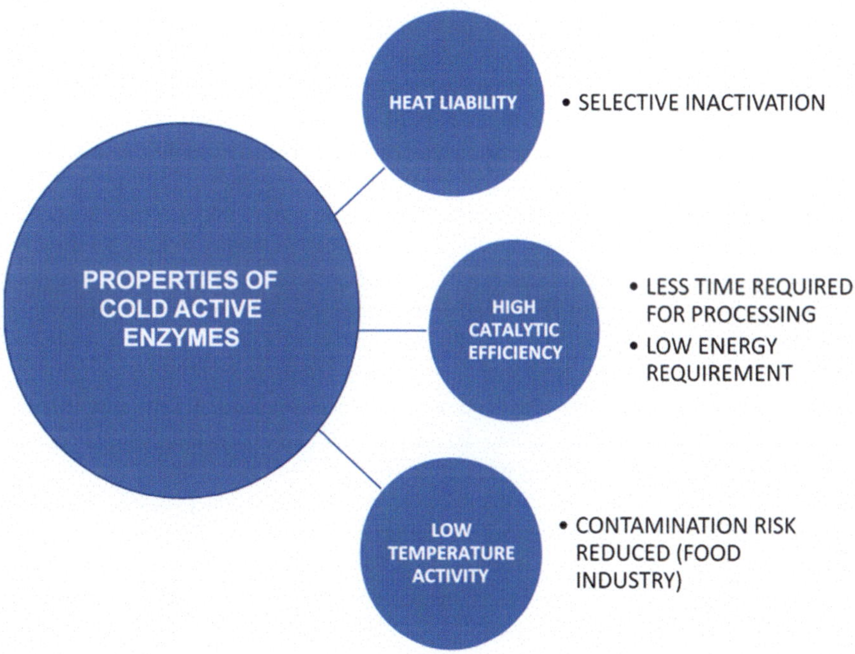

Fig. 5.7 Properties of cold active enzymes advantageous for industrial use

5.6.1 Heat Lability in Molecular Biology

Psychrophilic enzymes maintain high activity at low temperatures mainly by decreasing the temperature dependence of the reaction that is catalysed. Ribonuclease H (RNase H) explicitly cuts the RNA strand of the hybrid of RNA and DNA. This enzyme is found to be produced by several organisms and shows its involvement in DNA replication and repair. On the basis of AA sequence composition and related data, we can classify RNase H enzymes into two major categories, Type 1 and Type 2 RNases H (Ohtani et al. 1999), which is subsequently divided into three subcategories which are viral RNase H, bacterial RNase HI and eukaryotic RNase H1. Type 2 RNase H family is further divided into four subcategories, bacterial RNases HII and HIII, archaeal RNase HII and eukaryotic RNase H2. Out of the different RNase H enzymes, *E. coli* RNase HI, which constitutes bacterial Type 1 RNases H, has been extensively researched upon for structural and functional analysis. *E. coli* RNase HI is characterized by these characteristics: monomeric and small in size (155 AA residues).

5.6.2 Application of Cold Active Enzymes for Manufacturing Chemicals and Wastewater Treatment

Chemicals can be enzymatically synthesized, and one of the best examples of it is the production of acrylamide by *Rhodococcus* sp.-derived nitrile hydratase. Immobilized strain of *Rhodococcus* sp. and *Pseudomonas chlororaphis* can be utilized for the continuous production of acrylamide at low temperatures. Along with that, nitrile hydratase is used for the synthesis of R. Mandelic and nicotinamide as well, which are used in the pharmaceutical industry. Lipases are also in high demand by the pharmaceutical industries as the cold active lipases can be used for the production of optically active esters and volatile compounds. A psychrophilic *Acinetobacter* sp. produced lipase which catalyses the ester synthesis reaction of n-hexane and is highly active at –25 °C, while scallop hepatopancreas lipase is highly active at 10 °C. Furthermore, lipases are absolute for the production of nitrogenated compounds and show high stereospecificity during chemical synthesis similar to esterase (Maiangwa et al. 2015; Cabrera and Blamey 2018).

Cold active enzymes can be used as bioremediation agents for wastewater treatment at low temperature. Lipases derived from *Bacillus cereus* HSS are efficient in degrading the oily wastes from wastewater by forming grease traps (Hassan et al. 2018). Cold active catalases can degrade the excess hydrogen peroxide from the wastewater, and due to this ability of catalases, they can be also be used in the cold pasteurization of milk as permitted by the Food and Agriculture Organization. Concentration of 0.05–0.25% of hydrogen peroxide can be added to milk as a preservative if all the hydrogen peroxide can be degraded by using immobilized catalase after processing (Table 5.2).

Table 5.2 Industrial application of psychrophilic enzymes at low temperature (Sarmiento et al. 2015; Margesin and Schinner 1994)

Applications	Enzymes	Advantages
Biotechnology		
Molecular biology	Alkaline phosphatases	Dephosphorylation of 5′ end of a linearized fragment of DNA
	Uracil-DNA N-glycosylases (UNGs)	Release of free uracil from uracil-containing DNA
	Nucleases	Digestion of all types of DNA and RNA
Protoplast formation	Cell wall digesting enzyme phosphatase	High viability
Enzymatic synthesis		
	Lipase nitrile hydratase, etc	For volatile and heat sensitive materials
Food industry		
Modification of constituents	Galactosidase, lipase	Keeping freshness
Improvement of taste and flavour	Protease, lipase, etc.	Keeping freshness
Removal of fish skin	Protease	Maintaining product quality
Clarification of fruit juice	Pectinase, cellulase	Keeping fragrance
Detergent		
	Lipases	Breaking down of lipid stains
	Proteases	Breaking down of protein stains
	Amylases	Breakdown starch-based stains
	Cellulases	Wash of cotton fabrics
	Mannanases	Degradation of mannan or gum
	Pectate lyases	Pectin-stain removal activity
Wastewater treatment		
	Catalase	Less energy consumption

5.6.3 Cold Active Enzymes Used in the Food Industry

Microbes are utilized in food fermentation processes for a long time and are still useful within the preparation of several food items. Microbic enzymes play a significant role in food industries as they are more stable than plant and animal enzymes.

Alpha-amylases. Cold active amylases could be of interest for baking to enhance bread softness and forestall obstruction, since they will be simply inactivated throughout cooking. A patent developed with Novozymes involves a *Bacillus licheniformis* enzyme whose specific activity was increased at temperatures from 10 to 60 °C by protein engineering (Borchert et al. 2004). A second patent developed with the commercial partner Coldzymes ApS, Greenland describes a system for the heterologous expression of a *Clostridium* α-amylase retaining activity at temperatures lower than 10 °C (Mangiagalli et al. 2020).

β-D-Galactosidases hydrolyse milk sugar (lactose) into glucose and galactose and catalyse the transgalactosylation of lactose, which is used in the synthesis of galacto-oligosaccharides. β-galactosidases are used in the dairy industry to produce lactose-free products (Ohgiya et al. 1999). Lactose hydrolysis is of benefit in lactose intolerance and increases milk sugar. Furthermore, during the production of ice cream, treatment with β-D-galactosidases prevents the milk from developing lactose crystals which make the texture uneven. Many industrial β-D-galactosidases are secreted by mesophilic microbes having optimum temperature of the range 30–60 °C. To avoid spoilage, lactose hydrolysis is generally done at 30–40 °C for 4 h or at 5–10 °C for 24 h.

Proteases find wide application in food processing, including brewing, bakery, dairy, meat tenderization and the production of hydrolysates from meat, fish, gelatin and soy. Presently, the commonest enzymatic meat tenderizers are cysteine proteases, such as bromelain, papain, actinidin and ficin from fruits. They are thermostable, and most remain active upon heating at 70 °C. The inherent thermolability of CAEs is a desirable feature of meat tenderizers, since it would allow enzyme inactivation at cooking temperatures.

5.7 Future Prospects

Even though substantial advancements have been attained in understanding the enzymatic adaptations to such harsh environmental conditions, certain questions remain as an area of concern, regarding their structure and function relationships in cold active proteins.

Folding at low temperature. As we know, the folding process must be majorly dependent on the temperature at which the individual synthesizes the polypeptides. This develops the question "What is the process by which these polypeptides are produced and regulated at such low temperatures, and what is the role of chaperons and the rate-limiting step of the folding process?" (Siddiqui and Cavicchioli 2006) *Engineering cold activity.* This entails introducing the properties of cold active enzymes to an already commercially used enzyme in order to gain more industrial benefits. But engineering the psychrophilic activity in a mesophilic enzyme has not been reported yet, and the main concern is the large complexity of amino acid substitution and interactions leading to cold activity that have been acquired in the course of evolution. *Flexibility.* As we know, to catalyse a biochemical reaction at such low temperature, the cold active enzymes have acquired a flexible configuration of the active site through evolution and natural selection. This peculiarity makes them fascinating for understanding models in protein research: for protein evolution, folding and related fields. But of course, the following challenge is to explain the molecular flexibility in context of type, amplitude and timescale of molecular motions.

5.8 Conclusion

It can be concluded that psychrophilic enzymes are the significant factors required for the low-temperature adaptation of cold adapted microbes, and the correlation between the activity-flexibility-stability is one of the features accountable for the dominant adaptive characteristics of psychrophilic enzymes. The molecular flexibility of cold active enzymes improves the capability of the protein to undergo rapid conformational changes in cold environments. However, flexibility is achieved at the expense of stability of the native space and substrate affinity. The low stability of the native state is the result of a few numbers of weak interactions and makes the cold active enzymes heat labile. Another important characteristic of the cold active enzyme is the large opening of the catalytic cleft caused by the small deletions in the loops surrounding the catalytic region and by the substitution of the bulky chains with the light ones. Lower enthalpic and entropic values are accountable for the decrease in the activation energy barrier of the cold active enzymes. The distinct characteristics of the cold active enzymes make them useful for the industrial use in food industry, molecular biology, detergents and wastewater treatment. Yet, a need for better understanding of enzyme evolution and dynamics is required for more extensive application of psychrophilic enzymes.

References

Aghajari N, Feller G, Gerday C, Haser R (1998) Crystal structures of the psychrophilic alpha-amylase from Alteromonas haloplanctis in its native form and complexed with an inhibitor. Protein Sci 7:564–572. https://doi.org/10.1002/pro.5560070304

Åqvist J, Isaksen G, Brandsdal B (2017) Computation of enzyme cold adaptation. Nat Rev Chem 1:0051. https://doi.org/10.1038/s41570-017-0051

Borchert TV, Svendsen A, Andersen C, Nielsen B, Nissem TL, Kjaerulff S (2004) Exhibit alterations in at least one of the following properties relative to parent alpha-amylase: improved pH stability at pH 8–10.5, improved calcium ion stability at pH 8–10.5, increased specific activity at 10 to 60 °C. US Patent No 6673589 B2

Brenchley J (1996) Psychrophilic microorganisms and their cold-active enzymes. J Ind Microbiol Biotechnol 17:432–437. https://doi.org/10.1007/BF01574774

Cabrera MÁ, Blamey JM (2018) Biotechnological applications of archaeal enzymes from extreme environments. Biol Res 51:37. https://doi.org/10.1186/s40659-018-0186-3

Chao YC, Merritt M, Schaefferkoetter D, Evans TG (2020) High-throughput quantification of protein structural change reveals potential mechanisms of temperature adaptation in Mytilus mussels. BMC Evol Biol 20:28. https://doi.org/10.1186/s12862-020-1593-y

Chiuri R, Maiorano G, Rizzello A, del Mercato LL, Cingolani R, Rinaldi R, Maffia M, Pompa PP (2009) Exploring local flexibility/rigidity in psychrophilic and mesophilic carbonic anhydrases. Biophys J 96:1586–1596. https://doi.org/10.1016/j.bpj.2008.11.017

Cipolla A, Delbrassine F, Da Lage JL, Feller G (2012) Temperature adaptations in psychrophilic, mesophilic and thermophilic chloride-dependent alpha-amylases. Biochimie 94:1943–1950. https://doi.org/10.1016/j.biochi.2012.05.013

D'Amico S, Claverie P, Collins T, Georlette D, Gratia E, Hoyoux A, Meuwis MA, Feller G, Gerday C (2002) Molecular basis of cold adaptation. Philos Trans R Soc Lond B Biol Sci 357:917–925. https://doi.org/10.1098/rstb.2002.1105

D'Amico S, Gerday C, Feller G (2001) Structural determinants of cold adaptation and stability in a large protein. J Biol Chem 276:25791–25796. https://doi.org/10.1074/jbc.m102741200
D'Amico S, Marx JC, Gerday C, Feller G (2003) Activity-stability relationships in extremophilic enzymes. J Biol Chem 278:7891–7896. https://doi.org/10.1074/jbc.M212508200
Fedøy AE, Yang N, Martinez A, Leiros HK, Steen IH (2007) Structural and functional properties of isocitrate dehydrogenase from the psychrophilic bacterium Desulfotalea psychrophila reveal a cold-active enzyme with an unusual high thermal stability. J Mol Biol 372:130–149. https://doi.org/10.1016/j.jmb.2007.06.040
Feller G (2003) Molecular adaptations to cold in psychrophilic enzymes. Cell Mol Life Sci 60:648–662. https://doi.org/10.1007/s00018-003-2155-3
Feller G (2013) Psychrophilic enzymes: from folding to function and biotechnology. Scientifica (Cairo) 2013:512840. https://doi.org/10.1155/2013/512840
Feller G, Gerday C (1997) Psychrophilic enzymes: molecular basis of cold adaptation. Cell Mol Life Sci 53:830–841. https://doi.org/10.1007/s000180050103
Feller G, Gerday C (2003) Psychrophilic enzymes: hot topics in cold adaptation. Nat Rev Microbiol 1:200–208. https://doi.org/10.1038/nrmicro773
Feller G, Narinx E, Arpigny JL, Aittaleb M, Baise E, Genicot S, Gerday C (1996) Enzymes from psychrophilic organisms. FEMS Microbiol Rev 18:189–202. https://doi.org/10.1016/0168-6445(96)00011-3
Georlette D, Blaise V, Collins T, D'Amico S, Gratia E, Hoyoux A, Marx JC, Sonan G, Feller G, Gerday C (2004) Some like it cold: biocatalysis at low temperatures. FEMS Microbiol Rev 28:25–42. https://doi.org/10.1016/j.femsre.2003.07.003
Georlette D, Damien B, Blaise V, Depiereux E, Uversky VN, Gerday C, Feller G (2003) Structural and functional adaptations to extreme temperatures in psychrophilic, mesophilic, and thermophilic DNA ligases. J Biol Chem 278:37015–37023. https://doi.org/10.1074/jbc.m305142200
Gerday C (2013) Psychrophily and catalysis. Biology (Basel) 2:719–741. https://doi.org/10.3390/biology2020719
Gerday C, Aittaleb M, Bentahir M, Chessa JP, Claverie P, Collins T, D'Amico S, Dumont J, Garsoux G, Georlette D, Hoyoux A, Lonhienne T, Meuwis MA, Feller G (2000) Cold-adapted enzymes: from fundamentals to biotechnology. Trends Biotechnol 18:103–107. https://doi.org/10.1016/s0167-7799(99)01413-4
Gounot AM, Russell NJ (1999) Physiology of cold-adapted microorganisms. In: Margesin R, Schinner F (eds) Cold-adapted organisms. Springer, Berlin, Heidelberg
Hassan SWM, Abd El Latif HH, Ali SM (2018) Production of cold-active lipase by free and immobilized marine Bacillus cereus HSS: application in wastewater treatment. Front Microbiol 9:2377. https://doi.org/10.3389/fmicb.2018.02377
Hiteshi K, Gupta R (2014) Thermal adaptation of α-amylases: a review. Extremophiles 18:937–944. https://doi.org/10.1007/s00792-014-0674-5
Ma B, Kumar S, Tsai CJ, Nussinov R (1999) Folding funnels and binding mechanisms. Protein Eng 12:713–720. https://doi.org/10.1093/protein/12.9.713
Maiangwa J, Ali MS, Salleh AB, Rahman RN, Shariff FM, Leow TC (2015) Adaptational properties and applications of cold-active lipases from psychrophilic bacteria. Extremophiles 19:235–247. https://doi.org/10.1007/s00792-014-0710-5
Mangiagalli M, Brocca S, Orlando M, Lotti M (2020) The "cold revolution" Present and future applications of cold-active enzymes and ice-binding proteins. J Biotechnol 55:5–11. https://doi.org/10.1016/j.nbt.2019.09.003
Margesin R, Schinner F (1994) Properties of cold-adapted microorganisms and their potential role in biotechnology. J Biotechnol 33:1–14. https://doi.org/10.1016/0168-1656(94)90093-0
Marx JC, Blaise V, Collins T, D'Amico S, Delille D, Gratia E, Hoyoux A, Huston AL, Sonan G, Feller G, Gerday C (2004) A perspective on cold enzymes: current knowledge and frequently asked questions. Cell Mol Biol (Noisy-le-Grand) 50:643–655

Mereghetti P, Riccardi L, Brandsdal BO, Fantucci P, De Gioia L, Papaleo E (2010) Near native-state conformational landscape of psychrophilic and mesophilic enzymes: probing the folding funnel model. J Phys Chem B 114:7609–7619. https://doi.org/10.1021/jp911523h

Metpally RP, Reddy BV (2009) Comparative proteome analysis of psychrophilic versus mesophilic bacterial species: Insights into the molecular basis of cold adaptation of proteins. BMC Genomics 10:11. https://doi.org/10.1186/1471-2164-10-11

Ohgiya S, Hoshino T, Okuyama H, Tanaka S, Ishizaki K (1999) Biotechnology of enzymes from cold-adapted microorganisms. In: Margesin R, Schinner F (eds) Biotechnological applications of cold-adapted organisms. Springer, Berlin, Heidelberg

Ohtani N, Haruki M, Morikawa M, Crouch RJ, Itaya M, Kanaya S (1999) Identification of the genes encoding Mn2+-dependent RNase HII and Mg2+-dependent RNase HIII from Bacillus subtilis: classification of RNases H into three families. Biochemistry 38:605–618. https://doi.org/10.1021/bi982207z

Onuchic JN, Luthey-Schulten Z, Wolynes PG (1997) Theory of protein folding: the energy landscape perspective. Annu Rev Phys Chem 48:545–600. https://doi.org/10.1146/annurev.physchem.48.1.545

Papaleo E, Tiberti M, Invernizzi G, Pasi M, Ranzani V (2011) Molecular determinants of enzyme cold adaptation: comparative structural and computational studies of cold- and warm-adapted enzymes. Curr Protein Pept Sci 7:657–683. https://doi.org/10.2174/1389203711109070657

Sarmiento F, Peralta R, Blamey JM (2015) Cold and hot extremozymes: industrial relevance and current trends. Front Bioeng Biotechnol 3:148. https://doi.org/10.3389/fbioe.2015.00148

Saxena R, Singh R (2015) MALDI-TOF MS and CD spectral analysis for identification and structure prediction of a purified, novel, organic solvent stable, fibrinolytic metalloprotease from Bacillus cereus B80. Biomed Res Int 2015:527015. https://doi.org/10.1155/2015/527015

Siddiqui KS, Cavicchioli R (2006) Cold-adapted enzymes. Annu Rev Biochem 75:403–433. https://doi.org/10.1146/annurev.biochem.75.103004.142723

Siddiqui KS, Feller G, D'Amico S, Gerday C, Giaquinto L, Cavicchioli R, Bacteriol J (2005) The active site is the least stable structure in the unfolding pathway of a multidomain cold-adapted alpha-amylase. J Bacteriol 187:6197–6205. https://doi.org/10.1128/jb.187.17.6197-6205.2005

Siddiqui KS, Williams TJ, Wilkins D, Yau S, Allen MA, Brown MV, Cavicchioli R (2013) Psychrophiles. Annu Rev Earth Planet Sci 41:87–115. https://doi.org/10.1146/annurev-earth-040610-133514

Sindhu R, Binod P, Madhavan A, Beevi US, Mathew AK, Abraham A, Pandey A, Kumar V (2017) Molecular improvements in microbial α- amylases for enhanced stability and catalytic efficiency. Bioresour Technol 245:1740–1748. https://doi.org/10.1016/j.biortech.2017.04.098

Singh AK, Singh M (2015) Molecular mechanism of cold adaptation in bacteria. Aperito J Cell Mol Biol 1:104. https://doi.org/10.14437/AJCMB-1-104

Sočan J, Isaksen GV, Brandsdal BO, Åqvist J (2019) Towards rational computational engineering of psychrophilic enzymes. Sci Rep 9:19147. https://doi.org/10.1038/s41598-019-55697-4

Sočan J, Purg M, Åqvist J (2020) Computer simulations explain the anomalous temperature optimum in a cold-adapted enzyme. Nat Commun 11:2644. https://doi.org/10.1038/s41467-020-16341-2

Struvay C, Feller G (2012) Optimization to low temperature activity in psychrophilic enzymes. Int J Mol Sci 13:11643–11665. https://doi.org/10.3390/ijms130911643

Tronelli D, Maugini E, Bossa F, Pascarella S (2007) Structural adaptation to low temperatures--analysis of the subunit interface of oligomeric psychrophilic enzymes. FEBS J 274:4595–4608. https://doi.org/10.1111/j.1742-4658.2007.05988.x

An Overview of Survival Strategies of Psychrophiles and Their Applications

6

Sonal Sharma, Urvija Chaturvedi, Krishna Sharma, Anukool Vaishnav, and Harikesh Bahadur Singh

Abstract

Psychrophiles are capable of surviving under extreme cold conditions, subzero temperatures. They have adapted various mechanisms like altered membrane fluidity, antifreeze proteins, cold shock proteins, chaperones, trehalose, exopolysaccharides, synthesis of carotenoid pigments, production of ice nucleating proteins, decreased flagellar motility, etc. Psychrophiles mainly find their application in environmental bioremediation, in preventing food spoilage, as cell factories for production of various enzymes, and also in degradation of oil spills in oceans. They have proved to be a boon for the agriculture due to their plant growth-promoting properties at low temperatures. Development of microbial consortium and genetic engineering may be fruitful in the coming future in plant biotechnology. This chapter describes the cold tolerance mechanisms in psychrophilic microorganisms and the application of such microbes in different industrial sectors and agriculture. We also included the gaps and overcome strategies in the agriculture application of cold-tolerant microorganisms.

Keywords

Extremophiles · Psychrophiles · Antifreeze proteins · Psychrophilic enzymes · Agriculture

S. Sharma · U. Chaturvedi · K. Sharma · A. Vaishnav · H. B. Singh (✉)
Department of Biotechnology, Institute of Applied Sciences & Humanities, GLA University, Mathura, Uttar Pradesh, India

R. Goel et al. (eds.), *Survival Strategies in Cold-adapted Microorganisms*,
https://doi.org/10.1007/978-981-16-2625-8_6

6.1 Introduction

In the last few decades, the pioneer conditions in which existence will thrive are regularly fluctuating with higher ranges of temperature, pH, pressure, radiation, salinity, energy, and restriction on supplements. Under such a wide range of parameters, not only can microorganisms thrive on earth, but they can also survive extreme space conditions (Horneck et al. 2010; Yamagishi et al. 2018). When taking extremophilic (instead of extremotolerant) organisms into account, it is imperative to remember that these living beings are exceptionally adjusted for extreme conditions establishing the standard under which the life form can metabolically and biochemically operate. In the course of the recent years, researchers have been captivated by the interesting life forms that occupy extraordinary conditions. All such organisms, regarded as extremophiles, live in such intolerably hazardous or even lethal environment in which other life forms cannot survive such as high acidic or alkaline pH, high and low pressure, high salt condition, and cold and hot springs (Rampelotto 2013).

The majority of extremophiles are microorganisms in which *Archaea* accounts a major portion. In *Archaea*, most of the organisms are hyperthermophilic, acidophilic, alkaliphilic, and halophilic microorganisms. The archaeal strain *Methanopyrus kandleri* 116 has been reported to show growth at 122 °C temperature, while the genus *Picrophilus* has potential to grow at 0.0 pH to 0.06. In *Eubacteria*, cyanobacteria are the prominent species adapted to different extreme environments. Cyanobacteria can survive to hypersaline, high-metal, and xerophilic conditions. On the other hand, fungi are also reported to survive in mining areas, high pH conditions, hot and cold deserts and metal-contaminated water. In eukaryotic invertebrates, *Tardigrade* is known as polyextremophiles for surviving under extreme temperatures ranging from −272 °C to 151 °C, 6000 atm pressure, and radiation environment (Rampelotto 2013).

6.2 Types of Extremophiles

6.2.1 Thermophiles

Some microbial life can survive at moderately high temperatures, between 45 °C and 80 °C, and are known as thermophiles. In fact, hyperthermophiles are especially outrageous thermophiles with ideal temperatures of over 80 °C (Madern et al. 2006). In numerous geothermally warmed districts on Earth, such microorganisms include volcanic deposits infiltrated by hot gases and deep-sea hydrothermal vents (Nakagawa and Takai 2006). Usually, such extreme regions are abundant in decreased synthetics from the inside of the Earth, and subsequently, numerous thermophilic microorganisms are chemoautotrophs (Amend et al. 2003). During chemoautotrophic nature, acidic condition is created in the surrounding environment due to production of acid and product. These reactions produce sulfuric acid as a result of removing energy by oxidizing sulfur compounds, thereby also rendering geothermal waters extremely acidic. Thus, many heat-loving microorganisms are additionally adjusted to highly acidic environments (Satyanarayana et al. 2005). For instance, *Picrophilus* spp. are reported to survive at pH 0.7 and at temperature of 60 °C (Schleper et al. 1995). In Japan, they were secluded from volcanically induced dry land. Hydrothermal deep ocean vent populations are located near subsurface volcanoes and at the border between seawater and magma, typically kilometers underneath the sea surface (Desbruyères et al. 2000). As no light is visible and the oxygen content is extremely limited, chemoautotrophic anaerobes are the large proportion of thermophilic isolates found in these areas.

For high temperatures, the subatomic explanation for changes to outrageous conditions has been more intensively researched than for any other parameter. Biomolecules, for example, catalysts, denature, lose their potential at high temperatures, and then subsequently halt metabolism at high temperatures. In addition, membrane's fluidity rises exponentially, disrupting the cell. A number of mobile diversifications are presented by thermophiles to protect them. The membrane lipids comprise greater straight and saturated fatty acids than mesophiles (Ulrih et al. 2009). By supplying the exact degree of fluidity needed for membrane operation, this allocates thermophiles to expand at elevated temperatures. Thermophilic proteins seem to be smaller and mainly greater in a few instances, which can also contribute to prolonged stability (Kumar and Nussinov 2001). An additional mechanism for protection of proteins is the action of chaperones, which facilitate the refolding of denatured proteins (Jaenicke 1996). In addition, monovalent and divalent salts upgrade the stability of nucleic acids (Hickey and Singer 2004). Another approach to stabilize out DNA is more compaction of genome into chromatin (Marguet and Forterre 1998).

6.2.2 Psychrophiles

Psychrophiles have an ideal growth temperature of 15 °C and an upper limit of 20 °C that expand at or below 0 °C (Rothschild 2007). Such microbes establish in an assortment of cold conditions, from the stratosphere to the deep ocean. A significant part of the deep sea is at a stable temperature of 2 °C, while liquid sea water can also be cooled to below 0 °C across the polar ice caps, when the usual salt content of ocean water (3.4%) takes the freezing edge down to −1.8 °C (Atkins and Locke 2004). At the point when the seawater freezes up, salt turns out to be progressively gathered in little compartments. Under these conditions, the edge of water freezing may be discouraged to −20 °C (Margesin et al. 2008). Table 6.1 represents the diversity of psychrophiles in different areas.

Psychrophiles such as *Psychrobacter cryopegellai* have shown regular digestion and metabolic activity at frozen temperature such as −10 °C (Rodrigues et al. 2009). For this cause, numerous psychrophiles are halophiles as well (microorganisms that develop in elevated salt concentrations). To endure and thrive at low temperatures, psychrophiles need to beat a few issues identified with perpetual cold conditions. Enzymes also become unbending at minimal temperatures, and solute

Table 6.1 List of psychrophiles present in different habitats

Soil	River	Lake water	Stream water
Acidobacteria sp., *Ascomycota* sp., *Zygomycota* sp., *Methylobacter* sp., *Methylosinus* sp., *Eurotium* sp., *Aspergillus* sp., *Deinococcus* sp., *Chytridiomycetes* sp., *Streptococcus* sp., *Crenarchaeota* sp., *Cryptococcus* sp., *Mrakia* sp., *Arthrobacter* sp., *Actinobacteria* sp.	*Actinobacteria* sp. *Firmicutes* sp. *Proteobacteria* sp. *Pseudomonas fluorescens Trematomus* sp.	*Actinobacteria* sp., *Hydrogenophilus thermoluteolus*, *Methanococcoides frigidum*, *Methanococcoides burtonii*	*Cyanobacteria* sp.

concentrations are at elevated, possibly destructive levels (Cavicchioli 2006). In addition, ice crystals can pierce the cell membranes until the water is frozen, compromising cellular integrity (D'Amico et al. 2006). Membranes of psychrophiles comprise of expanded degrees of unsaturated fatty acids that further growth with the reduction in temperature so that it will modulate membrane fluidity (Nichols et al. 2004). At low temperatures, psychrophiles generate cold-adapted catalysts that have highly explicit activities (Feller and Gerday 2003). These enzymes are able to maintain transcription and translation at low temperatures. The existence of some genes active at low temperatures has also been seen in studies (Goodchild et al. 2004). In addition, antifreeze proteins are also involved to resist such bacteria under cold environment by reducing formation of ice (Gilbert et al. 2004).

6.2.3 Acidophiles

Acidophiles are microorganisms that can grow at pH 2.0 optimally (Morozkina et al. 2010). Sulfur and its minerals are oxidized by acidophiles to gain energy that creates intense acidic conditions (Rohwerder and Sand 2007). In reality, the greater part of the firm acidophilic microorganisms characterized has been detached from volcanic regions or corrosive mine seepage. Some *Archaea* sp. like *Picrophilus oshimae* and *P. torridus* were isolated from volcanically heated soils of 60 °C (Schleper et al. 1995). The intracellular pH value of these microorganisms is held at 4.6, while the other acidophiles retain their pH at 6.0. In Iron Mountain, California, *Ferroplasma acidarmanus* was secluded from acid mine discharge and is capable of increasing at a pH of 0 (Dopson et al. 2004). The Tinto River is regarded as a fascinating model for acidic conditions due to its scale and convenient accessibility. Low-pH conditions can cause protein denaturation in a cell. Many organisms survive in such conditions by inducing more neutral amino acid production (Baker-Austin and Dopson 2007).

6.2.4 Alkaliphiles

Alkaliphiles are a group of microorganisms that can grow at pH above 9.0 (Horikoshi 1999) as shown by the hyperalkaline spring waters in lake and deserts and semi-arid environments, including the broad desert in the western United States and other regions of the world with elevated Ca^{2+} concentrations induced by silicate serpentinization (Grant 2003).

There can be a range of microorganisms living at a pH of 10.5 (Martins et al. 2001). In the soda lakes of Maqarin, Jordan, microbial populations exist at 12.9 pH (Pedersen et al. 2004). Alkaliphiles are often excluded from natural habitats, which often appear to contain elevated levels of NaCl, and are therefore referred to as haloalkalophiles (Gareeb and Setati 2009). In alkaline conditions, the convergences of hydrogen particles are extremely short, and cells experience difficulty utilizing ATP synthase to deliver energy and other fundamental particles (Krulwich et al. 1998). By continuously pumping in certain ions and exporting others to preserve their interior at near neutrality, base-loving microbes overcome these concerns. In addition, the alkalophile cell wall functions as a defense system from harsh environmental conditions (Horikoshi 2006).

6.2.5 Halophiles

Halophiles are microorganisms that thrive from around 10% sodium chloride to saturation at elevated salt concentrations, and a few of them can also flourish in salt crystals (Das Sarma 2002). Aquatic ecosystems with variable salinity, salt marshes, surface salt lakes, subterranean salt lakes, and several other areas are the habitats where halophilic microorganisms are located (Litchfield and Gillevet 2002). The Great Salt Lake in Utah and the Dead Sea in the Middle East are two of the largest and most examined recent hypersaline conditions. The hypersaline conditions are created in Antarctica regions, where high salt content can preserve water in liquid state under −20 °C temperature (Madern et al. 2006; Das Sarma 2006).

All these adapted microorganisms produce more number of solutes inside the cell to maintain osmotic balance. In their cells, halophilic *Archaea* retain extraordinarily high amounts of potassium chloride (Oren 2004). Under saturated salt conditions, all the halophiles follow the same machinery as thermophiles for survival (Michael et al. 1999). Therefore, in conjunction with thermophilic and mesophilic proteins, researchers explored the sequences of amino acids, arrangements, and functional properties of halophilic proteins to acquire intuition into the evolutionary techniques (Michael et al. 1999).

6.2.6 Piezophiles

The microorganisms which, under high hydrostatic pressure conditions, can protect themselves under elevated atmospheric pressure are referred to as piezophiles (Abe and Horikoshi 2001). Piezophiles are more prevalent in the depths of the ocean and Earth's crust. These microorganisms have been isolated at a depth of 10.5 km from the lowest portion of the ocean and have the potential to survive at pressures of up to 110 Mpa at 2 °C and 40 Mpa at temperatures above 100 °C (Yayanos 1995).

These microbes are accustomed to high temperatures and limited resources in subsurface environments embedded within the Earth's crust. Iron-reducing bacterial species were recovered from the Siijan (Sweden) granite at a depth of 6.7 km (Kotelnikova 2002). Complex microbial habitats have been confirmed in extreme conditions like in South African gold mines, inside freshly mined rocks about 3 km down the earth (Takai et al. 2001). Similarly, in the Columbia River basin, methanogenic microbes were obtained from several hundred meters (Washington State, USA) (Thomas-Keprta et al. 1997).

Some researchers have calculated that "deep biota" approaches the aggregate total of all surface living systems (Dartnell 2007). These subterranean fissures are suitable ecosystems in several aspects, since they have a stable atmosphere with steady chemical energy flow. These ecosystems often shield microbes from harmful radiations. The challenge in collecting samples from deep-sea environments and various complexities involved in performing biochemical experiments in the laboratory under high pressure conditions are the major reasons responsible for making it one of the least studied areas in the field of science. There are recent reports, though, in progress. Interactions between protein and protein are very susceptible to changes in strain, which may be the cause for the dissociation of enzymes and inhibition of gene expression (Sharma et al. 2002; Nakasone et al. 1998). Under extreme conditions, lipid membrane molecules stack closer, resulting in diminished fluidity of the membrane (Bartlett 2002). Increment in the level of unsaturated fatty acids in their membranes can also circumvent this issue (Aertsen et al. 2009). Table 6.2 represents categorized microbes for different extreme conditions with their ecological importance.

6.3 Survival Strategies Adapted by Psychrophiles

Around 10% of the area is surrounded by ice and glaciers that are not ideal for the reproduction of normal living beings in this vast planet consisting of living creatures. Prokaryotic life has occupied much of our planet's evolutionary history, expanding to fill nearly all available environmental niches, and it is a current reality that cold conditions are prevalent on earth. Psychrophilic microorganisms have effectively colonized all types of extreme situations. A portion of these life forms, contingent upon their ideal development temperature, are likewise known by the terms psychrotolerant or psychrotroph (Morita 1975). The potential of psychrophiles to flourish and propagate at low temperatures suggests that core challenges intrinsic to

Table 6.2 Types of microbes with their ecological importance

Types of microorganisms	Examples	Ecological importance
Thermophiles and hyperthermophiles	*Methanopyrus kandleri*, *Geobacillus stearothermophilus*, *Caldicellulosiruptor*, *Thermococcus*, *Sulfolobus*, *Thermotoga*, *P. furiosus*, *T. kodakarensis*	Thermophilic microorganisms have demonstrated many metabolic capacities and may have biotechnological use in anaerobic processes either as a source of thermostable enzymes or as an inoculum. In addition, by forming a syntrophy with hydrogenotrophic *Archaea*, they can accelerate protein degradation
Psychrophiles	Bacteria: *Arthrobacter* sp., *Psychrobacter* sp., *Chryseobacterium greenlandense*, *Halomonas*, *Pseudomonas* Lichens: *Umbilicaria antarctica*, *Xanthoria elegans*	In cold climates, psychrophiles play an essential part in bioremediating fat-contaminated waste water and eliminating harmful substances such as hydrocarbons, heavy metals, and fuel oils
Acidophiles	Archaea: *Sulfolobus solfataricus*, *Halobacteriaceae* *Bacteria*: *Acidobacteria*, *Alicyclobacilli*, *Acetobacter* *Eukarya*: *Urotricha*, *Dunaliella acidophila*, *Mucor racemosus*	Acid-stable enzymes have great applications in food and beverages industries
Alkaliphiles	*Natronomonas*, *Halorhodospira halochloris*, *Thiohalospira alkaliphila*	In industrial applications, alkaliphiles have had a great influence. Biological detergents produce alkaline enzymes that have been formed from alkaliphiles, such as alkaline cellulases and/or alkaline proteases. The commercial development of cyclodextrin by alkaline cyclomaltodextrin glucanotransferase is another significant use. This enzyme lowered the cost of processing and opened the way for vast amounts of cyclodextrin in foodstuffs, pesticides, and pharmaceuticals. Alkali-treated wood pulp has also been documented to be biologically bleached by xylanase produced by alkaliphiles
Halophiles	*Wallemia chthyophaga*, *Chromohalobacter beijerinckii*, *Tetragenococcus halophilus*	Halophiles have great potential in saline soil remediation
Piezophiles	*Shewanella*, *Colwellia*, *Photobacterium*, *Moritella*, *and Psychromonas*, *Pyrococcus yayanosii*, *Pyrococcus abyssi*	In the food industry; have great potential in industrial and biotechnological perspective. The high efficiency of the detergent has been demonstrated by piezophilic proteins

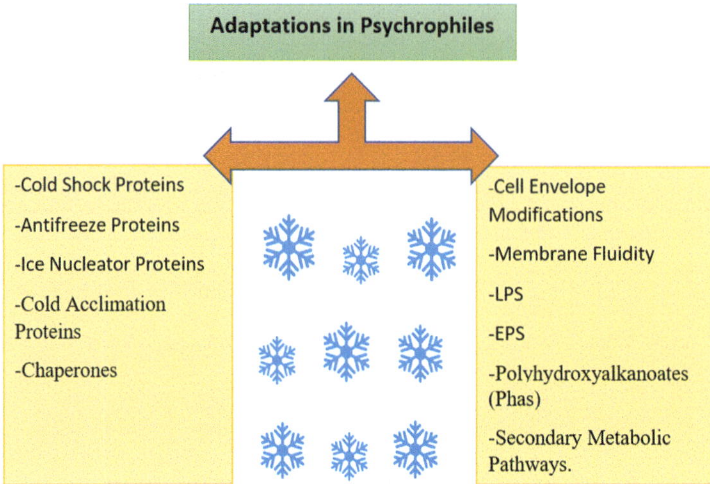

Fig. 6.1 Representation of different cellular and molecular adaptation strategies of psychrophiles under cold conditions

chronically cold conditions have been solved. These challenges include membrane fluidity, enzyme inactivation, ceasing of transcription and translation machinery, etc. Cold-adapted species such as *Moritella profunda* and many more have developed several changes at physical and genetic level within them to protect against low temperatures (Xu et al. 2003). In addition, most experiments have discussed the heat susceptibility of cold-adapted microorganisms, which is no longer the source of cold tolerance, although it is of concern. A lot of testing has been conducted in recent decades to assess the ability of psychrophiles. Psychrophiles can retain temperatures ranging from 10 °C to –20 °C, and the optimum temperature for most of them is 5 °C. However, in a permafrost bacterium, the low-temperature cutoff of psychrophiles was not addressed, and proliferation was accounted for at −12 °C and metabolic potential at −20 °C.

The strategies adapted by different organisms to survive under cold conditions result in different prospects regarding fundamental attributes of various biological processes like genetic sequence responsible for construction of macromolecules which are stable even in extreme conditions and biochemical limitations which may alter stability of macromolecules (Fig. 6.1). Such microorganisms adapted to extreme environments possess a broad range of metabolic diversity along with uncommon physiological potential to support their survival. Though the molecular, biochemical, or physiological strategies adapted by such microorganisms are not fully clarified, still, a detailed study of the involved pathways is of great importance for biotechnological sectors. Their adaptability and tolerance to such extreme conditions make them a better alternative for mesophilic enzymes which may have potential to remain active in extreme environmental conditions. Industrial biotechnology harbors wide application of such extraordinary microbes, as few enzymes may express polyextremophilicity (i.e., tolerance to more than one extreme condition) which are beneficial.

Psychrophiles inhabit many unique attributes like presence of unsaturated fatty acids in cell membranes, as unsaturated fatty acid has the capability to remain in liquid state under low temperatures and enable transport of solutes, antifreeze proteins, cold shock proteins, and cryoprotectants. Survival in low temperature is a combined result of various changes in fluidity of membrane, decreased levels RNA and protein synthesis, and alteration in structures of ribosomes, which ultimately modify the functioning of cellular machinery. Accumulation of glycine, betaine, sucrose, and mannitol for synthesis of antifreeze proteins which prevent ice crystals is also the type of technique used by psychrophiles to confirm their survival. Production of exopolysaccharides among psychrophiles is regarded as a mechanism against cryoprotection (Salwan and Sharma 2020).

Few compounds which possess properties of an osmoprotectant and cryoprotectant and also can act as source of carbon, nitrogen, and energy include sucrose, glycerol, glycine, sorbitol, and mannitol, as they reduce the freezing point of cytoplasm and prevent accumulation of macromolecules and increase stability of cell membrane. Collins and Deming (2013) conducted a study on *P. haloplanktis* and found that the uptake of compounds which confirm tolerance to low temperatures such as spermine, glutathione, and ornithine was enhanced due to thiolation of protein-S, which is regulated by glutathionyl spermidine, and glutathione seemed to show a possible cold adaptation mechanism (Mocali et al. 2017). Genes coding for proteins which are responsible for synthesis and breakdown of nitrogen reserves polymers and polyamides were found during genome analysis of *Colwellia psychrerythraea* (Nunn et al. 2015). In a study, *Mesorhizobium* sp. strain N33 was found to grow at 4 °C and accumulate a number of different cryoprotective compounds like sarcosine and threonine (Ghobakhlou et al. 2015).

Synthesis of polyhydroxyalkanoates (PHAs) is common in psychrophiles as it has an important physiological role. PHAs enable the survival and resistance of bacteria to various environmental stresses along with ability to produce and degrade fatty acids. Phasins are among the important PHAs that could help in stress protection and fitness enhancement (Mezzina and Pettinari 2016).

6.3.1 Cell Membrane Fluidity

Membrane fluidity is an important aspect of cellular functioning. Temperature at the either ends of biotic thermal range alters membrane fluidity. Cell membrane structure, composition, and response to varied temperature ranges differ among *Eubacteria* and *Archaea* (Deming 2002). In psychrophiles, membrane lipid composition is different from other organisms with increased ratio of polyunsaturated and saturated fatty acids that confers their survival in extreme climates (Guan et al. 2013).

Metabolic imbalance and growth cessation can be observed as a result of low temperature. Cell membranes and envelopes are considered as a vital link between the cell and its surrounding environment; thus, modification in its structure is an important aspect of cold adaptation. Commonly involved mechanisms at low temperatures to manage fluidity of membrane and avoid rigidness include alteration in lipid composition of cell membrane, preferring smaller chains, and lowering saturation of lipids (De Maayer et al. 2014). According to a study, widening of cell wall provides protection from cell damage caused due to formation of ice and osmotic pressure as observed in *E. sibiricum* at −2.5 °C. Growth of *Planococcus halocryophilus* Or1 in cold conditions was supported by synthesis of cell membrane, cell wall, and components of envelop along with uncommon cell envelop attributed by encrustations around the cell during subzero growth at −15 °C (Mykytczuk et al. 2013).

Many studies prove that few genes are responsible for maintenance of lipid membrane and cell wall synthesis at low temperatures as observed in *P. arcticus* (Bergholz et al. 2009). Cell wall, membrane related proteins and envelope synthesis was 289 enhanced gradually. In the Antarctic strains of *Pseudomonas syringae* and *P. extremaustralis rise*, high amount of hydroxy fatty acids was observed along with modification in fluidity and constitution of LPS (Benforte et al. 2018). Increasing hydrophobicity and amount of calcium carbonate, peptidoglycan, and choline along the cell membrane is closely related to cold adaptation. Copies of genes responsible for biosynthesis of carbonic anhydrase and peptidoglycan increased in low-temperature conditions indicating precipitation of calcium carbonate by microbes (Mykytczuk et al. 2016). *P. halocryophilus* at low temperature is known to possess increased fatty acid saturation as at low temperatures, fatty acid desaturases are in inactive state (Ronholm et al. 2015). Alteration in lipid composition was noted when the microbial cells were incubated at low temperatures. The level of unsaturated fatty acids increased in bacteria and yeasts with decrease in growth temperature which is beneficial for proper membrane functioning (Russell et al. 1995; Berry and Foegeding 1997). Decreases in temperature result in transformation of cellular fluid components to gel which inhibits the proper functioning of proteins causing leakage in microbial cell membrane. Exopolysaccharides, which constitute the polymers around the microbial cells, are a major aspect supporting tolerance to cold conditions (Tribelli and López 2018). However, change in constituents of membrane enables the membrane to maintain its fluidity, thus preventing gel formation and promoting survival of microbes at low temperatures. Carotenoid pigments have been recognized as fluidity modulators in various Antarctic bacteria.

6.3.2 Antifreeze Proteins (AFPs)

The antifreeze proteins have the ability to reduce the size and shape of ice crystals, which was first time observed in *Micrococcus cryophilus* and *Rhodococcus erythropolis* soil bacterium (Duman and Olsen 1993). The AFPs are present between the solid ice and liquid water and thus prevent the growth of ice crystals. At low concentrations of antifreeze proteins, ice recrystallization inhibition occurs.

Complementary structures present on ice crystals act as binding sites for antifreeze proteins and result in formation of thermal hysteresis thereby increasing the capacity of organism to survive in low temperature (Jia and Davies 2002).

Freeze avoidance and freeze tolerance are the major categories of antifreeze proteins. Freeze avoidance refers to avoiding low-temperature conditions generally by mobile organisms and mainly depends on high thermal hysteresis activity (D'Amico et al. 2006). Freeze tolerance involves minimization of damage caused to immobile organisms especially by ice recrystallization inhibition (Middleton et al. 2012). The antifreeze proteins studied in Antarctica Lake bacteria *Marinomonas primoryensis* were found to be Ca^{2+} dependent and hyperactive, while AFP showing antifreeze and ice nucleating activities was identified in Arctic plant growth-promoting rhizobacterium *Pseudomonas putida* GR12-2 (Muryoi et al. 2004; Gilbert et al. 2004). Ice structuring and shaping is also a type of adaptation in psychrophiles. According to Wilson et al. (2006), AFPs from *Pseudomonas putida* GR12-2 alter the morphology of ice in its supercooled state into either a hexagonal or hexagonal bipyramid.

6.3.3 Cold Shock Proteins

Cold shock proteins (CSPs) are involved in nucleic acid protection through binding with binding motifs RNP1 and RNP2. In bacteria, CSPs are activated during downshift of temperature to tolerate cold stress and are also present under general circumstances to manage other biological functions like enhancement of normal growth and stress adaptation responses. CSPs counteract the effects of cold shock by mimicking nucleic acid chaperons. Enzymes involved in transcription and translation like elongation factor, RNA polymerase, and peptidyl prolyl *cis-trans* isomerase are adapted to be optimally active at low temperatures. While cold shock proteins (CSPs) and RNA helicases are overexpressed at low temperatures (Lim et al. 2000). Thus, CSPs efficiently cope with deleterious effect of downshift in temperature by reestablishment of membrane fluidity by incorporation of unsaturated fatty acids into their membranes along with restoration of ribosome function and inducing proper protein folding. CSPs also help the cell handle various other conditions like osmosis, starvation, pH, and ethanol stress tolerance. Goldstein et al. (1990) first described CspA in *E. coli* and suggested that its homologous proteins are cold inducible and involved in chaperone role for RNA protection. In a study, CspA expression was found to be induced in *E. coli under repeated* freezing and thawing (Jung et al. 2010). The production of trehalose and other exopolysaccharides also has an important role in the survival of psychrophiles (Phadtare 2004). It was observed that, under extreme cold environments like the Antarctica, microbes produced more numbers of exopolysaccharide to protect cells and adhere on the surface and for water retention (Krembs et al. 2002; Nichols et al. 2005a, b).

6.4 Applications of Cold Adapted Microbes

6.4.1 Psychrophilic Enzymes in Different Industries

Psychrophiles are the cellular factories for production of various enzymes that remain active in the presence of detergents, oxidants, and alkaline environments, thus, expressing their potential in several industries. Various psychrophiles have proved themselves as better sources of cellulase which can be used in food industry, wastewater, and soil bioremediation and molecular biology. Psychrophilic cellulases having potential at lower washing temperatures and reduced water consumption are preferred. Cellulases, lipases, and proteases from psychrophiles also find application in environmental bioremediation, food industry, and molecular biology, and a lot is still unexploited (Souza et al. 2015). The cryophilic enzymes limit the undesirable reactions which may occur at higher temperatures and thus can be used for enzymatic reactions which require low temperatures like in the food industry to prevent spoilage and alter nutritional value and taste of heat-labile substrates. The cryophilic enzymes are economically beneficial due to low-energy requirements and higher catalytic efficiency at low temperatures. As they are functional at low temperatures, they can be used to precede a reaction along with preventing spoilage at temperatures where microbial contamination is less prevalent. Heat inactivation as an alternative to chemical inactivation is facilitated by psychrophilic enzymes, evidencing their role in vaccine and other large-scale industries.

Psychrophilic enzymes which are heat labile find their application in molecular biosciences where sequential reactions take place and each enzyme needs to be inactivated after each step. Thus, heat inactivation of enzymes may be attained by just a mild increase in temperature upholding the double-stranded DNA in stable state (Cavicchioli et al. 2011).

In the food industries, psychrophilic microorganisms and their metabolites have a wide variety of utilization. Coagulating enzymes have a great benefit in regulating case coagulation in order to preserve the consistency of whey from the cheese industry. For example, Marzyme®, Rennilase 50TL, and Moelilase are used in the market of developed countries (Divya and Naga 2015). Applications of β-galactosidase obtained from psychrophilic bacteria will produce 70–80% of product yields, which is far greater than the processes derived from mesophilic microorganisms utilizing enzymes. The commercial cold activated neutral protease is derived predominantly from *Bacillus subtilis* and is brought to market under the trade name Eutrase.

6.4.2 Use of Psychrophilic Microorganisms in Bioremediation

Psychrophilic microorganisms in the ecosystem have the potential to biodegrade different substances. They can be effective at low temperatures in bioremediation or multiple contaminants. Dodecane, hexadecane, naphthalene, and toluene were identified to be mineralized at cold temperatures by strains of bacteria (Watson et al. 1978). It has been manifested in laboratory and field studies, by means of particular bacterial cultures. A few of the decaying genes like naphthalene (ndoB) and toluene (xylE, todC I) were also recorded in psychrotrophic bacteria. Likewise, *Rhodococcus* sp. strain Q15 has been found to degrade n-alkanes and diesel fuel at low temperature (Mahdieh et al. 2014).

6.4.3 Role of Psychrophiles in Medicine and Pharmaceuticals

In order to develop antifungal, anticancer, and antitumor agents, multiple strains of bacteria, *Streptomyces*, and fungi have been identified. A species of *Alteromonas* that synthesizes 2,3-indolinedione (isatin) has been identified. This formulation preserves the shrimp *Palaemon macrodactylus* from *Lagenidium callinectes*, which is a pathogenic fungus from *Moraxella* sp. and *Flavobacterium* sp. Antiviral and antitumor medicines can be manufactured. Polysaccharide, an antitumor, has been known as Narinactin. The *Bacillus subtilis* protease and amylase combination is effective in removing dental plaque(s) (Ramana et al. 2000). Poly-unsaturated fatty acids (PUFA) isolated from psychrophilic archaea have also conferred its use in development of neutraceuticals and other dietary supplements.

6.4.4 Role of Psychrophiles in Domestic Purposes

In domestic operations, the enzymes of psychrophilic species can be utilized. Low-temperature cleaning of clothes can preserve fabric colors and reduce power consumption. Few enzymes that are applied to detergents for the hydrolysis of macromolecular stains, such as subtilisin, lipase, and glycosidases, are poorly effective at tap water levels; psychrophilic enzymes may replace them (Feller and Gerday 2003).

6.4.5 Application of Psychrophiles in Textile-Based Industries

The extremozymes of few psychrophilic microbes are an extremely good source of enzymes for the textile industries. Since the twentieth century, amylases are often used for desizing. The most recent commercial developments are the use of cellulases for denim finishing and lacquers for clothing effluent decolorization and textile bleaching. It also encourages the production of environmentally sustainable fiber manufacturing systems and techniques to increase the efficiency of the finished product (Araújo et al. 2009).

6.5 Psychrophiles Used in Fine Chemical Synthesis

Colwellia psychrerythraea are reported to produce polyhydroxyalkanoate (PHA), a polyester that has thermoplastic and elastomeric properties, and are of great economic interest. During chemical process, cold adapted esterase(s) and lipases are mainly used for industrial purposes (Methé et al. 2005).

6.6 Role of Psychrophiles in Agriculture

The microbiomes found in low-temperature conditions are of great importance to the agriculture sector as they are permanently adapted to such extreme conditions. Diverse genera of bacteria like *Sphingobacterium* sp., *Octadecabacter* sp., *Hymenobacter roseosalivarius*, *Flavobacterium* sp., *Oleispira* sp., *Glaciimonas frigoris*, and *Psychrobacter pocilloporae* have been found in cold environments. Various species of psychrotrophs have been determined among different domains, i.e., bacteria, archaea, and fungi. Few microbes isolated from the cold deserts of Northwestern Himalayas reported plant growth-promoting (PGP) properties which included *Arthrobacter nicotianae*, *Brevundimonas* sp., *Paenibacillus tylopili*, and *Pseudomonas* sp. (Yadav et al. 2015a, b). In the northern regions of India, some psychrophiles like *Arthrobacter methylotrophus* and *Pseudomonas rhodesiae* have been recovered from wheat plant (Verma et al. 2016). Several economically beneficial *Bacillus* sp. possessing efficient plant growth-promoting potential have been determined by Yadav et al. (2015a, b). *Pseudomonas* and *Exiguobacterium* are considered best PGP at low temperatures.

Among the wide diverse psychrotrophic microbiome, various bio-inoculants enriched with PGP potential can be used to enhance crop production by assisting nitrogen fixation, stimulating phytohormones and release of siderophores, and facilitating solubilization and uptake of minerals like phosphorus, potassium, and zinc, by expressing antagonistic activity against plant pathogens as a biocontrol agent, or by inducing resistance against the pathogen. Psychrophiles can also be used for biodegradation of agricultural residues and wastes. Other studies developed microbial consortium comprised of three bacterial species, namely, *Eupenicillium crustaceum*, *Paceliomyces* sp., and *Bacillus atrophaeus*, for degradation of agricultural residues and the use of generated compost for enhancing soil fertility (Ajar et al. 2017; Shukla et al. 2016). Antifreeze proteins are also used for cryopreservation and improving quality of frozen food. Moreover, a soil actinomycete has been identified as *Streptomyces* sp. which produced antibiotics showing toxicity against many gram-positive bacteria at low temperatures (Ogata et al. 1971).

6.7 Conclusion and Future Prospectives

Psychrophiles are of great importance in agriculture and industrial biotechnology. Although several researches have been successful in exploring cold-tolerant microbes and their derivatives, still, much is to be explored with diversity in cold-tolerant microbes. In many hill agriculture regions, cold temperatures and less fertile soils are two major challenges. Under cold temperature, soils are acidic and phosphorus-deficient, limiting crop productivity. In such areas, green biotechnology has a great impact on agricultural productivity of small farm holders and their economies. Several efforts have been made by the scientific community to increase crop productivity through application of bio-inoculants. However, current biofertilizers used in cold climates have been found ineffective. The results obtained so far indicate that cold-tolerant or cold-loving microbes are more effective as compared to general biofertilizers. But still, much work is needed to explore more diversity of psychrophiles and finally achieve the desired bio-inoculant formulations which could perform efficiently under cold climate conditions.

Likewise, microbial symbionts associated with plants growing at low temperature are also required to explore the management of cold stresses under natural environment. In addition, we suggest further research to explain symbiotic mechanism for applying microbial species in general condition. A promising approach in this research area is the use of metagenomics sequencing to identify potential symbionts and their metabolic characteristics. We recommend exploring yeast endophytes or unicellular algae in cold environment, as they are majorly reported in symbiotic mechanism. We also recommend the study on secondary metabolites of symbiotic microbes associated with cold-tolerant plants for producing antimicrobial compounds and ameliorating chilling and freezing events in plants.

References

Abe F, Horikoshi K (2001) The biotechnological potential of piezophiles. Trends Biotechnol 19: 102–108

Aertsen A, Meersman F, Hendrick MEG, Vogel RF, Michiels CW (2009) Biotechnology under high pressure: applications and implications. Trends Biotechnol 27:434–441

Ajar NY, Priyanka V, Vinod K, Shashwati GS, Anil KS (2017) Extreme cold environments: a suitable niche for selection of novel Psychrotrophic microbes for biotechnological applications. Adv Biotech Micro 2(2):555584

Amend JP, Rogers KL, Shock EL, Gurrieri S, Inguaggiato S (2003) Energetics of chemolithoautotrophy in the hydrothermal system of Vulcano Island, southern Italy. Geobiology 1:37–58

Araújo R, Silva C, Machado R, Casal M, Cunha AM, Rodriguez-Cabello JC, Cavaco-Paulo A (2009) Proteolytic enzyme engineering: a tool for wool. Biomacromolecules 10(6):1655–1661

Atkins PW, Locke JW (2004) Physical chemistry, 7th edn. Oxford University Press, Oxford

Baker-Austin C, Dopson M (2007) Life in acid: pH homeostasis in acidophiles. Trends Microbiol 15:165–171

Bartlett DH (2002) Pressure effects on in vivo microbial processes. Biochim Biophys Acta 1595: 367–381

Benforte FC, Colonnella MA, Ricardi MM, Solar Venero EC, Lizarraga L, López NI, Tribelli PM (2018) Novel role of the LPS core glycosyltransferase WapH for cold adaptation in the Antarctic bacterium *Pseudomonas extremaustralis*. PLoS One 13(2):e0192559

Bergholz PW, Bakermans C, Tiedje JM (2009) *Psychrobacter arcticus* 273-4 uses resource efficiency and molecular motion adaptations for subzero temperature growth. J Bacteriol 191: 2340–2352

Berry ED, Foegeding PM (1997) Cold temperature adaptation and growth of microorganisms. J Food Prot 60(12):1583–1594
Cavicchioli R (2006) Cold-adapted archaea. Nat Rev Microbiol 4:331–343
Cavicchioli R, Charlton T, Ertan H, Mohd Omar S, Siddiqui KS, Williams TJ (2011) Biotechnological uses of enzymes from psychrophiles. J Microbial Biotechnol 4(4):449–460
Collins RE, Deming JW (2013) An inter-order horizontal gene transfer event enables the catabolism of compatible solutes by *Colwellia psychrerythraea* 34H. Extremophiles 17:601–610
D'Amico S, Collins T, Marx JC, Feller G, Gerday C (2006) Psychrophilic microorganisms: challenges for life. EMBO Rep 7(4):385–389
Dartnell L (2007) Extremophiles. In: Dartnell L (ed) Life in the universe: a beginner's guide. Oneworld Publications, Cambridge
Das Sarma, S. (2002). Arora P. Halophiles, encyclopedia of life sciences; Nature Publishing Group: London
Das Sarma S (2006) Extreme halophiles are models for astrobiology. Microbe 1:120–126
De Maayer P, Anderson D, Cary C, Cowan DA (2014) Some like it cold: understanding the survival strategies of psychrophiles. EMBO Rep 15(5):508–517
Deming JW (2002) Psychrophiles and polar regions. Curr Opin Microbiol 5:301–309
Desbruyères FD, Almeida A, Biscoito M, Comtet T, Khripounoff A, Le Bris N, Sarradin PM, Segonzac M (2000) A review of the distribution of hydrothermal vent communities along the northern mid-Atlantic ridge: dispersal vs. environmental controls. Hydrobiologia 440:201–216
Divya K, Naga PP (2015) Psychrophilic yeast isolates for cold-active lipase production. Int J Sci Progr Res 10(2):93–97
Dopson M, Baker-Austin C, Hind A, Bowman JP, Bond PL (2004) Characterization of Ferroplasma isolates and Ferroplasma acidarmanus sp. nov., extreme acidophiles from acid mine drainage and industrial bioleaching environments. Appl Environ Microbiol 70:2079–2088
Duman JG, Olsen TM (1993) Thermal hysteresis protein activity in bacteria, fungi, and phylogenetically diverse plants. Cryobiology 30(3):322–328
Feller G, Gerday C (2003) Psychrophilic enzymes: hot topics in cold adaptation. Nat Rev Microbiol 1(3):200–208
Gareeb AP, Setati ME (2009) Assessment of alkaliphilic haloarchaeal diversity in Sua pan evaporator ponds in Botswana. Afr J Biotechnol 8:259–267
Ghobakhlou AF, Johnston A, Harris L, Antoun H, Laberge S (2015) Microarray transcriptional profiling of Arctic *Mesorhizobium* strain N33 at low temperature provides insights into cold adaption strategies. BMC Genomics 16:383
Gilbert JA, Hill PJ, Dodd C, Laybourn-Parry J (2004) Demonstration of antifreeze protein activity in Antarctic lake bacteria. Microbiology 150(Pt 1):171–180
Goldstein J, Pollitt NS, Inouye M (1990) Major cold shock protein of Escherichia coli. Proc Natl Acad Sci U S A 87(1):283–287. https://doi.org/10.1073/pnas.87.1.283
Goodchild A, Saunders NFW, Ertan H, Raftery M, Guilhaus M, Curmi PMG, Cavicchioli R (2004) A proteomic determination of cold adaptation in the Antarctic archaeon, *Methanococcoi desburtonii*. Mol Microbiol 53:309–321
Grant WD (2003) Alkaline environments and biodiversity. In: Gerdsy C, Glansdorff N (eds) Extremophiles: basic concepts. Encyclopedia of Life Support Systems, Paris
Guan Z, Tian B, Perfumo A, Goldfine H (2013) The polar lipids of *Clostridium psychrophilum*, an aenaerobic psychrophile. Biochem Biophys Acta 1831:1108–1112
Hickey DA, Singer GA (2004) Genomic and proteomic adaptations to growth at high temperature. Genome Biol 5:117.1–117.7
Horikoshi K (1999) Alkaliphiles: some applications of their products for biotechnology. Microbiol Mol Biol Rev 1999(63):735–750
Horikoshi K (2006) Alkaliphiles: genetic properties and applications of enzymes. Springer, Berlin
Horneck G, Klaus DM, Mancinelli RL (2010) Space microbiology. Microbiol Mol Biol Rev 74: 121–156. https://doi.org/10.1128/MMBR.00016-09

Jaenicke R (1996) Stability and folding of ultrastable proteins: eye lens crystallins and enzymes from thermophiles. FASEB J 10:84–92

Jia Z, Davies PL (2002) Antifreeze proteins: an unusual receptor-ligand interaction. Trends Biochem Sci 27(2):101–106

Jung YH, Yi JY, Jung HJ, Lee YK, Lee HK, Naicker MC, Uh JH, Jo IS, Jung EJ, Im H (2010) Overexpression of cold shock protein a of *Psychromonas arctica* KOPRI 22215 confers cold-resistance. Protein J 29(2):136–142

Kotelnikova S (2002) Microbial production and oxidation of methane in deep subsurface. Earth Sci Rev 58:367–395

Krembs C, Eicken H, Junge K, Deming JW (2002) High concentrations of exopolymeric substances in Arctic winter sea ice: implications for the polar ocean carbon cycle and cryoprotection of diatoms. Deep-Sea Res I 49:2163–2181

Krulwich TA, Ito M, Hicks DB, Gilmour R, Guffanti AA (1998) pH homeostasis and ATP synthesis: studies of two processes that necessitate inward proton translocation in extremely alkaliphilic *Bacillus* species. Extremophiles 2:217–222

Kumar S, Nussinov R (2001) How do thermophilic proteins deal with heat? Cell Mol Life Sci 58: 1216–1233

Lim J, Thomas T, Cavicchioli R (2000) Low temperature regulated DEAD-box RNA helicase from the Antarctic archaeon, *Methanococcoides burtonii*. J Mol Biol 297:553–556

Litchfield CD, Gillevet PM (2002) Microbial diversity and complexity in hypersaline environments: a preliminary assessment. J Ind Microbiol Biotechnol 28:48–55

Madern D, Ebel C, Zaccai G (2006) Halophilic adaptation of enzymes. Extremophiles 2000(4):91–98

Mahdieh H, Moghaddam S, Soltani J (2014) Psychrophilic endophytic fungi with biological activity inhabit Cupressaceae plant family. Symbiosis 63:79–86

Margesin R, Schinner F, Marx JC, Gerday C (2008) Psychrophiles: from biodiversity to biotechnology. Springer, New York

Marguet E, Forterre P (1998) Protection of DNA by salts against thermodegradation at temperatures typical for hyperthermophiles. Extremophiles 2:115–122

Martins RF, Davids W, Al-Sond WA, Levander F, Radström P, Hatti-Kaul R (2001) Starch-hydrolyzing bacteria from Ethiopian soda lakes. Extremophiles 5:135–144

Methé BA, Nelson KE, Deming JW, Momen B, Melamud E, Zhang X, Moult J, Madupu R, Nelson WC, Dodson RJ, Brinkac LM, Daugherty SC, Durkin AS, DeBoy RT, Kolonay JF, Sullivan SA, Zhou L, Davidsen TM, Wu M, Huston AL, Lewis M, Weaver B, Weidman JF, Khouri H, Utterback TR, Feldblyum TV, Fraser CM (2005) The psychrophilic lifestyle as revealed by the genome sequence of Colwellia psychrerythraea 34H through genomic and proteomic analyses. Proc Natl Acad Sci U S A 102(31):10913–10918. https://doi.org/10.1073/pnas.0504766102. Epub 2005 Jul 25

Mezzina MP, Pettinari MJ (2016) Phasins multifaceted polyhydroxyalkanoate granule-associated proteins. Appl Environ Microbiol 82:5060–5067

Michael T, Madigan M, Orent A (1999) Thermophilic and halophilic extremophiles. Curr Opin Microbiol 2:265–269

Middleton AJ, Marshall CB, Faucher F, Bar-Dolev M, Braslavsky I, Campbel, l R. L., et al. (2012) Antifreeze protein from freeze-tolerant grass has a beta-roll fold with an irregularly structured ice-binding site. J Mol Biol 416:713–724. https://doi.org/10.1016/j.jmb.2012.01.032

Mocali S, Chiellini C, Fabiani A, Decuzzi S, Pascale D, Parrilli E, Tutino ML, Perrin E, Bosi E, Fondi M et al (2017) Ecology of cold environments: new insights of bacterial metabolic adaptation through an integrated genomic-phenomic approach. Sci Rep 7:839

Morita RY (1975) Psychrophilic bacteria. Bacteriol Rev 39:144–167

Morozkina EV, Slutskaya ES, Fedorova TV, Tugay TI, Golubeva LI, Koroleva OV (2010) Extremophilic microorganisms: biochemical adaptation and biotechnological application. Appl Biochem Microbiol 46:1–14

Muryoi N, Sato M, Kaneko S, Kawahara H, Obata H, Yaish MW, Griffith M, Glick BR (2004) Cloning and expression of afpA, a gene encoding an antifreeze protein from the arctic plant growth-promoting rhizobacterium pseudomonas putida GR12-2. J Bacteriol 186(17):5661–5671

Mykytczuk NCS, Foote SJ, Omelon CR, Southam G, Greer CW, Whyte LG (2013) Bacterial growth at −15 °C; molecular insights from the permafrost bacterium *Planococcus halocryophilus* Or1. ISME J 7:1211–1226

Mykytczuk NCS, Lawrence JR, Omelon CR, Southam G, Whyte LG (2016) Microscopic characterization of the bacterial cell envelope of *Planococcus halocryophilus* Or1 during subzero growth at −15 °C. Polar Biol 39:701–712

Nakagawa S, Takai K (2006) The isolation of thermophiles from deep-sea hydrothermal environments. In: Rainey FA, Oren A (eds) Methods in microbiology: extremophiles. Elsevier, New York

Nakasone K, Ikegami A, Kato C, Usami R, Horikoshi K (1998) Mechanisms of gene expression controlled by pressure in deep-sea microorganisms. Extremophiles 2:149–154

Nichols, C. A., Guezennec, J., & Bowman, J. P. (2005a). Bacterial exopolysaccharides from extreme marine environments with special consideration of the southern ocean, sea ice, and deep-sea hydrothermal vents: a review. Marine biotechnology Springer, New York, 7(4), 253–271

Nichols CM, Lardiere SG, Bowman JP, Nichols PD, Gibson JAE, Guezennec J (2005b) Chemical characterization of exopolysaccharides from Antarctic marine bacteria. Microb Ecol 49:578–589

Nichols D, Miller MR, Davies NW, Goodchild A, Raftery M, Cavicchioli R (2004) Cold adaptation in the Antarctic archaeon, Methanococcoides burtonii, involves membrane lipid unsaturation. J Bacteriol 186:8508–8515

Nunn BL, Slattery KV, Cameron KA, Timmins-Schiffman E, Junge K (2015) Proteomics of *Colwellia psychrerythraea* at subzero temperatures—a life with limited movement, flexible membranes and vital DNA repair. Environ Microbiol 17:2319–2335

Ogata K, Yoshida N, Ohsugi M, Tani Y (1971) Studies on antibiotics produced by psychrophilic microorganisms. Agric Biol Chem 35(1):79–85

Oren A (2004) Adaptation of halophilic archaea to life at high salt concentrations. In: Lauchli A, Luttge U (eds) Salinity: environment—plants—molecules. Springer, Dordrecht

Pedersen K, Nilsson E, Arlinger J, Hallbeck L, O'Neill A (2004) Distribution, diversity and activity of microorganisms in the hyper-alkaline spring waters of Maqarin in Jordan. Extremophiles 8: 151–164

Phadtare S (2004) Recent developments in bacterial cold-shock response. Curr Issues Mol Biol 6: 125–136

Ramana KV, Singh L, Dhaked RK (2000) Biotechnological application of psychrophiles and their habitat to low temperature. J Sci Ind Res 59:87–101

Rampelotto PH (2013) Extremophiles and extreme environments. Life (Basel, Switzerland) 3(3):482–485. https://doi.org/10.3390/life3030482

Rodrigues DF, Jesus EC, Ayala-del-Río HL, Pellizari VH, Gilichinsky D, Sepulveda-Torres L, Tiedje JM (2009) Biogeography of two cold-adapted genera: psychrobacter and exiguobacterium. ISME J 3:658–665

Rohwerder T, Sand W (2007) Oxidation of inorganic sulfur compounds in acidophilic prokaryotes. Eng Life Sci 7:301–309

Ronholm J, Raymond-Bouchard I, Creskey M, Cyr T, Cloutis EA, Whyte LG (2015) Characterizing the surface-exposed proteome of *Planococcus halocryophilus* during cryophilic growth. Extremophiles 19:619–629

Rothschild L (2007) Extremophiles: defining the envelope for the search for life in the universe. In: Pudritz R, Higgs P, Stone J (eds) Planetary systems and the origins of life. Cambridge University Press, Cambridge

Russell NJ, Evans RI, terSteeg PF, Hellemons J, Verheul A, Abee T (1995) Membranes as a target for stress adaption. Int J Food Microbiol 28:255–261

Salwan R, Sharma V (2020) Physiological and biotechnological aspects of extremophiles. Academic Press, London, pp 13–22

Satyanarayana T, Raghukumar C, Shivaji S (2005) Extremophilic microbes: diversity and perspectives. Curr Sci 89:78–90

Schleper C, Puehler G, Holz I, Gambacorta A, Janekovic D, Santarius U, Klenk HP, Zillig W (1995) *Picrophilus* gen. nov. fam. nov.: a novel aerobic, heterotrophic, thermoacidophilic genus and family comprising archaea capable of growth around pH 0. J Bacteriol 177:7050–7059

Sharma A, Scott JH, Cody GD, Fogel ML, Hazen RM, Hemley RJ, Huntress WT (2002) Microbial activity at gigapascal pressures. Science 295:1514–1516

Shukla L, Suman A, Yadav AN, Verma P, Saxena AK (2016) Syntrophic microbial system for ex-situ degradation of paddy straw at low temperature under controlled and natural environment. J Appl Biol Biotechnol 4(2):30–37

Souza TV, Araujo JN, da Silva VM, Liberato MV, Pimentel AC, Alvarez TM, Squina FM, Garcia W (2015) Chemical stability of a cold-active cellulase with high tolerance toward surfactants and chaotropic agent. Biotechnol Rep 9:1–8

Takai K, Moser DP, DeFlaun M, Onstott TC, Fredrickson JK (2001) Archaeal diversity in waters from deep south African gold mines. Appl Environ Microbiol 67:5750–5760

Thomas-Keprta KL, Wentworth SJ, McKay DS, Taunton AE, Allen CC, Romanek CS, Gibson EK (1997) Subsurface terrestrial microfossils from Columbia River basalt samples: analogs of features in Martian meteorite Allan Hills 84001. Meteorit Planet Sci 32:128–129

Tribelli PM, López NI (2018) Reporting key features in cold-adapted bacteria. Life (Basel, Switzerland) 8(1):8

Ulrih NP, Gmajner D, Raspor P (2009) Structural and physicochemical properties of polar lipids from thermophilic archaea. Appl Microbiol Biotechnol 84:249–260

Watson K, Arthur H, Morton H (1978) Thermal adaptation in yeast: obligate psychrophiles are obligate aerobes, and obligate thermophiles are facultative anaerobes. J Bacteriol 136(2):815–817

Wilson SL, Kelley DL, Walker VK (2006) Ice-active characteristics of soil bacteria selected by ice-affinity. Environ Microbiol 8(10):1816–1824

Xu Y, Nogi Y, Kato C, Liang Z, Ruger HJ, De Kegel D, Glansdorff N (2003) *Moritella profunda* sp. nov. and *Moritella abyssi* sp. nov., two psychropiezophilic organisms isolated from deep Atlantic sediments. Int J Syst Evol Microbiol 53:533–538

Yadav AN, Sachan SG, Verma P, Saxena AK (2015a) Prospecting cold deserts of north western Himalayas for microbial diversity and plant growth promoting attributes. J Biosci Bioeng 119(6):683–693

Yadav AN, Verma P, Sachan S, Kaushik R, Saxena AK (2015b) Mitigation of cold stress for growth and yield of wheat (Triticum aestivum L.) by psychrotrophic pseudomonads from cold deserts of Indian Himalayas. In: Proceeding of 56th AMI and international symposium on emerging discoveries in microbiology

Yamagishi A, Kawaguchi Y, Hashimoto H, Yano H, Imai E, Kodaira S et al (2018) Environmental data and survival data of *Deinococcus aetherius* from the exposure facility of the Japan experimental module of the international space station obtained by the Tanpopo mission. Astrobiology 18:1369–1374. https://doi.org/10.1089/ast.2017.1751

Yayanos AA (1995) Microbiology to 10,500 meters in the deep-sea. Annu Rev Microbiol 49:777–805

Microbial Genes Responsible for Cold Adaptation

7

Vandana Singh

Abstract

Cold environments constitute about 85% of earth's biosphere and is primarily dominated by microorganisms particularly bacteria, fungi, yeasts, archaea, and viruses. These microbes inhabit the low-temperature environments successfully as they have several physiological adaptive strategies that help them to overcome extreme low temperatures. The microbial cells respond to low temperature shift by induction of cold shock proteins (Csps) that help facilitate the cells' adaptation to low temperature. Due to recent surge in "omic" technologies, the amount of omics data has increased considerably and has helped further the understanding of cold adaptation. In this chapter, we aim to describe and discuss main physiological adaptation strategies of the cold-adapted microorganisms and the genes involved in those adaptations, referencing genomic, and transcriptomic studies that have significantly contributed to our current knowledge in the area.

Keywords

Cold-adapted microorganisms · Cold shock response · Psychrophiles · CspA

7.1 Introduction

Earth's biosphere is predominantly cold, at temperatures below 5 °C, largely due to oceans that cover about 70% of the Earth's surface. Additionally, ice in the polar arctic regions, including Antarctica and North American and European land portions within the Arctic circle, contributes to the cold environment on the land surface and

V. Singh (✉)
Accenture Federal Services, Arlington, VA, USA

R. Goel et al. (eds.), *Survival Strategies in Cold-adapted Microorganisms*,
https://doi.org/10.1007/978-981-16-2625-8_7

makes up about 85% of the Earth's biosphere to permanently experience temperatures below 5 °C (Cowan et al. 2007; Hamdan 2018).

In these cold environments, advanced forms of life are absent, yet, cold-adapted microorganisms particularly bacteria, fungi, yeasts, archaea, and viruses grow harmoniously and inhabit the cold environments successfully. As an important environmental factor, temperature has profound effects on the cellular functions in all organisms. Owing to the high surface to volume ratio in prokaryotes, they must cope up extremely with changes in temperature and adapt to a given temperature regime. Based on their ability to grow at high, intermediate, and low temperature, microorganisms have been grouped into three categories and classified as thermophilic, mesophilic, and psychrophilic microorganisms (Smirnova et al. 2001). The term cold-tolerant organisms was proposed by Morita (1975) for the microorganisms with the capability to grow at 0 °C.

7.2 Cold-Adapted Microorganisms

The cold-adapted microorganisms are distinguished from mesophiles by their ability to grow at low temperature. Cold-adapted microbes include both prokaryotic and eukaryotic organisms and thus represent a significant portion of the living world and, therefore, widely spread in nature. The cold-adapted microorganisms are generally subdivided into psychrophiles and psychrotrophs (or psychrotolerants) (Margesin and Collins 2019) (Fig. 7.1). The ability to grow near 0 °C is a characteristic of both psychrophiles and psychrotrophs. However, psychrotrophs survive at temperature below 0 °C but grow optimally at 20–25 °C, and the upper limit temperature is as high as 40 °C. On the contrary, psychrophiles are more adapted to a lower temperature and grow optimally at less than 15 °C with an upper limit of 20 °C (Morita 1975). At low temperatures, psychrophiles and psychrotrophs have slower metabolic rates and higher catalytic efficiencies than mesophiles. Psychrophiles are more often

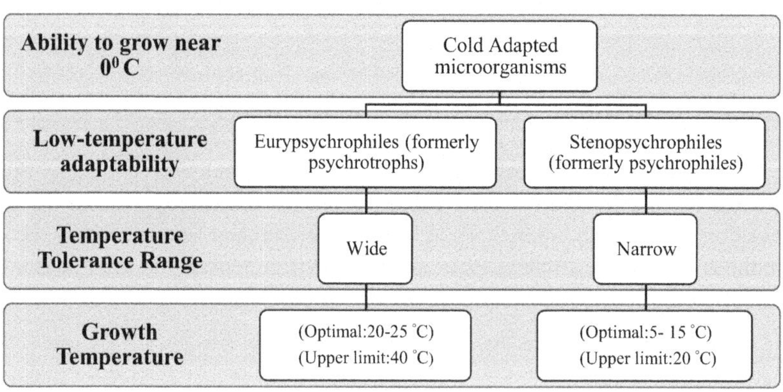

Fig. 7.1 Classification of cold-adapted microorganisms based on their low-temperature growth adaptability

isolated from permanently cold habitats, whereas psychrotrophs are more dominant in environments that undergo thermal fluctuations.

7.2.1 Diversity of Cold-Adapted Microorganisms

There are studies that reported the presence of bacteria living in permafrost soil and in sea ice where the temperature is around $-20\ ^{\circ}C$. Psychrophiles are autotrophic or heterotrophic, aerobic or anaerobic, spore formers and non-spore formers, and phototrophs and nonphototrophs. Among the bacteria that have been identified, it appears that psychrophiles are widespread in the domain. Bacteria with majority isolates come from Gram-negative divisions *Bacteroidetes* and *Proteobacteria* (Moyer and Morita 2007). The most commonly reported cold-adapted microorganisms include:

1. *Bacteria*:
   (a) Gram-negative *Alphaproteobacteria*, *Betaproteobacteria*, and *Gammaproteobacteria* (*Pseudomonas* sp. and *Vibrio* sp.) and the *Cytophaga-Flavobacterium-Bacteriodes* phylum.
   (b) Gram-positive bacteria *Coryneforms*, *Arthrobacter* sp., and *Micrococcus* sp. are the most commonly found Gram-positive bacteria.
2. *Archaea*: In deep sea waters, archaea and bacteria are found in equivalent numbers. *Methanogenium* and *Methanococcus* are widespread genera.
3. *Cyanobacteria*: *Oscillatoria*, *Phormidium*, and *Nostoc commune* are dominant in most of the Antarctic habitats (Pandey et al. 2004).
4. *Psychrophilic yeasts*: *Cryptococcus* sp. have been isolated from Antarctic soils (Deming 2002).

7.2.2 Strategies for Cold Adaptation

Extreme habitat such as cold temperatures poses acute physicochemical restrictions on the cellular functions. There are several recognized challenges impacting the microorganism at temperature downshift. These include:

1. Decline in fluidity of the membrane.
2. Diffusion and reaction rates are diminished.
3. Enzyme activities decrease profoundly but to different extents.
4. RNA secondary structures become stabilized that impacts initiation of translation.
5. Superhelical density of the DNA is too high for opening of the double helix.
6. Ribosome functionality is impacted at low temperatures, and the ribosomal adaptation can restore the proper functionality.
7. Protein folding may be too slow or inefficient or even misfolding of the proteins (Bakermans et al. 2012; De Maayer et al. 2014).

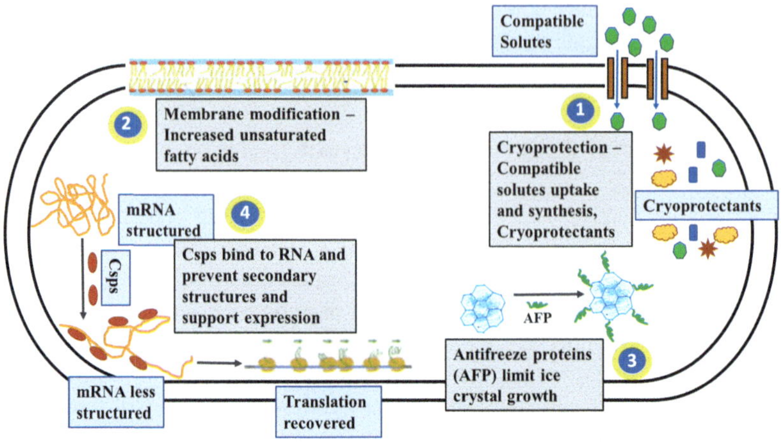

Fig. 7.2 Cold-adapted microorganisms possess different strategies and features to maintain cellular functionality and integrity at low temperatures: (1) Compatible solutes and cryoprotectants help in cryoprotection. (2) Membrane composition modulation with an increase in unsaturated fatty acids helps maintain the cell membrane fluidity. (3) Inhibition of formation of ice crystal protects the cell from damage. (4) Cold shock proteins (Csps) serve as molecular chaperones and bind to RNA/DNA to support expression by retaining the RNA primary structure and prevent secondary structure formation

Cold-adapted microorganisms have several adaptive strategies that enable them to overcome the negative influence of cold temperatures and to survive at low temperatures (Rodrigues and Tiedje 2008; Piette et al. 2011). The different strategies of cold adaptation in bacteria occur at biochemical and molecular levels to overcome extreme low temperatures. Firstly, the cold-adapted microorganisms accumulate cryoprotectants and increase the compatible solute uptake and synthesis for maintaining homeostasis in the cell (Robinson 2001). Secondly, cold-adapted organisms modify their lipid composition by increasing unsaturated fatty acids to maintain cell membrane integrity under low-temperature environments. Additionally, the presence of antifreeze proteins/ice-binding proteins is an important adaptation in the cold-adapted microbes. These ice-binding proteins inhibit the formation of ice crystals. Lastly, cold shock proteins (Csps) are another most important adaptation at low temperature which helps microbial cells by acting as molecular chaperones and bind to the RNA and prevent supercoiling and secondary structure formation of RNA. This inhibition of RNA secondary structure formation eventually helps in resuming the translation and expression of proteins. The adaptations of cold-adapted microorganisms are shown in Fig. 7.2.

7.3 Cold Adaptation Genes

Extremophiles have attracted much attention because of their wide range of biotechnological applications and potential (Margesin and Feller 2010) and, also, to understand their biochemical mechanisms of adaptation to extreme temperatures, pH, and salinity. These microorganisms and their biomolecules have a huge biotechnological potential because of their unique ability to maintain enzymatic reactions. Further, they are being utilized in various bioprocesses to be carried out at low temperature, in addition to their role in natural decomposition of organic matter and nutrient recycling at low-temperature habitats. These characteristics of the extremophiles have led to considerable research on the mechanism of adaption to cold stress in the last few decades. Furthermore, the development of the "omic" technology has also advanced research in this field considerably.

Microorganisms encounter changing environmental conditions and have evolved different strategies to survive and adapt to changing environments. The cold-adapted microorganisms harbor several cold adaptation genes such as cold shock proteins, RNA helicases, and oxidative stress. Acclimatization to rapid drop in temperature (cold shock) is achieved by transient overexpression of cold-induced proteins (Cips) in bacteria (Phadtare 2004). Many Cips have been identified in *E. coli*. One of the most important and crucial Cips produced in cold conditions is cold shock protein (Csp) family (Yamanaka et al. 1998). Other examples of Cips are RNA helicase *CsdA* (Charollais et al. 2004), exoribonucleases, PNPase and RNaseR (Phadtare 2012), initiation factors 2α and 2β, NusA, and RecA (Jones et al. 1987). The Csps have been extensively studied, and their involvement in various cellular processes such as the regulation of membrane fluidity, transcription, translation, and protein folding is well-known (Phadtare 2004).

In the mesophilic bacterium such as *E. coli*, these cold shock proteins are transiently induced upon exposure to a sudden downshift in temperature as a mechanism to cope with the cold-induced stress. On the other hand, in the cold-adapted bacteria or psychrophiles, even though lower temperature is optimal for growth, there are still genetic and physiological strategies employed for coping with the low temperature. This adaptation in part comes either from novel genes involved in cold adaptation or from constitutively expressing the genes that are, however, only induced upon a cold shock in mesophilic or thermophilic bacteria. This suggests that there is a differential regulation of gene expression of similar set of genes in mesophilic and psychrophilic bacteria at low temperatures. Due to this differential gene expression, wherein some genes are transiently induced during cold shock while others are consistently expressed at low temperatures, the terms cold-induced proteins (CIPs) and cold acclimation proteins (CAPs) are used to distinguish between two classes. In general, cold-induced genes in mesophiles may be cold acclimation genes in cold-adapted microorganisms (e.g., *cspA* in *E. coli* compared with *Arthrobacter globiformis* (Cavicchioli et al. 2009).

7.3.1 Cold Shock Response

The shift of the microbial cells, which are in log phase and increasing in number tremendously, from the optimal growth temperature to low temperature results in the cold shock response. There is momentary halt in cell growth upon cold shock in most of the bacteria. During this period, general protein synthesis is severely inhibited. However, cold shock proteins are significantly induced under low-temperature stress conditions and are a group of proteins acting primarily as RNA chaperones to prevent the misfolding of mRNA (Phadtare et al. 1999). Following the cold shock response, the cells adapt to low temperature and normal cell growth resumes, and at this time, the synthesis of these proteins decreases (Jones et al. 1987). The role of Csps during cold shock response is well established; however, recent studies suggest a wider role and their involvement in stress tolerance of bacteria (Schmid et al. 2009; Duval et al. 2010; Loepfe et al. 2010; Michaux et al. 2012; Schärer et al. 2013; Wang et al. 2014; Derman et al. 2015). The cold shock proteins (CSPs) are a conserved family of small, acidic proteins (molecular weight of approximately 7 kDa) containing the highly conserved nucleic acid-binding cold shock domain (CSD). CSD contains two nucleic acid-binding motifs, ribonucleoproteins 1 and 2 (Lee et al. 2013), that facilitate binding to target DNA and RNA (Chaikam and Karlson 2010). The length of proteins encoded by genes *cspA* to *cspI* varies from 65 to 75 amino acids (Jin et al. 2014). Many members of CSPs can be classified as either CIPs and/or CAPs.

7.3.1.1 Cold Shock Response in *E. coli*

The Csps in *E. coli* have been studied in detail. Abrupt lowering of temperature (from 37 °C to 10 °C) induces CspA expression and increases up to 200-fold within minutes (Lindquist and Mertens 2018; Gottesman 2018). CspA family consists of nine homologous proteins (CspA to CspI), out of which only CspA, CspB, CspG, and CspI are cold shock inducible. Based on the level of protein expression after cold shock induction, Csps are classified into two distinct classes: class I proteins are found at very low levels at optimal growth temperatures and are rapidly and significantly induced to very high levels when cells are shifted to low temperatures, whereas class II proteins are expressed at 37 °C and their post-cold shock induction is not as remarkable (<ten-fold) (Phadtare 2004). CspA, CspB, CspG, CspI, CsdA, RbfA, NusA, and PNP are class I proteins, whereas IF-2, H-NS, and GyrA are class II proteins (Table 7.1).

In a recent study aimed at reexamining the prior findings, cold shock response-induced overall changes in bacteria were assessed using genomic methods. CspA unfolds double-stranded RNAs to aid in translation, whereas RNase R degrades misfolded RNAs: both CspA and RNase R were identified as major players (Zhang et al. 2018).

7.3.1.2 Cold Shock Response in *B. subtilis*

The CSP family from *B. subtilis* includes three members, viz., CspB, CspC, and CspD. These are small homologous proteins and show sequence identities of

Table 7.1 *E. coli* cold shock proteins

Protein class	Member proteins	Presumed function
Class I (>ten-fold induction)	Csp family: CspA, CspB, CspG, CspI	RNA chaperones
	CsdA (DEAD-box RNA helicase family protein)	Ribosomal-associated protein with RNA unwinding activity (Charollais et al. 2004; Iost and Dreyfus 2006)
	RbfA	Ribosomal binding factor (Xia et al. 2003)
	NusA	Termination and antitermination of transcription
	Polynucleotide phosphorylase (PNPase)	Exoribonuclease, CspA homologues repression by selective *cspA* mRNA degradation at 15 °C (Phadtare 2012)
Class II (<ten-fold induction)	RecA	Recombination factor
	GyrA	Topoisomerase DNA gyrase subunit
	IF-2	Translation initiation factor
	Histone-like protein H-NS	Nucleoid-associated DNA-binding protein
	Hsc66, HscB	Hsc66 is encoded by *hscA* gene, molecular chaperone
	Dihydrolipoamide transferase	
	Pyruvate dehydrogenase	
	Trigger factor	Molecular chaperone (Phadtare and Severinov 2010)
	Trehalose-synthesizing proteins	Up to eightfold increased synthesis of trehalose whose accumulation enhances cell viability at 4 °C (Kandror et al. 2002)

72–80%. The structure of CspB from *B. subtilis* and CspA from *E. coli* reveals a similar β-barrel structure made of five compact antiparallel β-strands (β1–β5) (Schindelin et al. 1994). Two RNA-binding motifs, RNP1 and RNP2, are located on the β2 and β3 strands, respectively. CspB is important at both temperatures, low and optimal temperature. On the other hand, CspC is primarily functional at low temperature and CspD at optimal temperature (Schindler et al. 1999). In *B. subtilis*, a low-temperature-inducible *des* gene is induced upon a temperature downshift from 37 to 20 °C and resulted in a 10- to 15-fold increase in transcription. This gene encodes the membrane phospholipid desaturase which acts as a two-component system DesR-DesK. This system senses an increase in membrane rigidity at low temperatures (Beranová et al. 2010). Low-temperature-induced conformational change in DesK results in activated autokinase activity. Upon activation, DesR is phosphorylated by DesK. Phosphorylated DesR binds to DNA and leads to desaturase gene expression induction (De Mendoza 2014).

7.3.1.3 Cold Shock Response in Psychrotrophs and Psychrophiles

In psychrotrophic bacteria, a basal set of Csps that already exists is dramatically induced and expressed at very high levels upon even a mild shock. At more acute cold shock, additional Csps are synthesized. In *Psychrobacter cryohalolentis* K5, five CIPs, viz., two ribosomal proteins (S2 and the Ctc form of L25), two elongation factors (EF-Ts and the EF-Tu-related TypA), and the cold shock protein CspA, were participating in translation (Fedorov et al. 2001). The analyses of psychrophilic *Psychrobacter arcticus* genome, transcriptome, and proteome revealed that genes for energy metabolism and carbon substrate incorporation were negatively regulated, whereas membrane fluidity maintenance genes and a DEAD-box RNA helicase protein A were positively regulated upon cold shock (Kuhn 2012).

7.3.2 Cold Acclimation Proteins

Cold acclimation proteins (Caps) are a group of proteins which are expressed consistently and continuously at a greater level in the cells adapted to constant growth at low temperatures. In psychrotrophic bacterium *Arthrobacter globiformis* SI55, concentration of 18 peptides was shown to increase in the cells growing at 4 °C as opposed to the cells growing at 25 °C. Furthermore, the group suggested the presence of early and late Caps, where early Caps were believed to be the proteins that showed significantly induced expression levels in cells growing at 4 °C, whereas late Caps were thought to be a group of proteins present at high concentrations only in 4 °C steady-state cells (Berger et al. 1996). Caps have also been observed in two other psychrotrophic bacteria, *Bacillus psychrophilus* (Whyte and Inniss 1992) and *Pseudomonas fragi* (Hebraud et al. 1994), when grown at 0 and 4 °C, respectively. These proteins might be responsible for some metabolic function(s) at low temperature. In *A. globiformis*, a proteolytic system that is specific to low temperature has been identified and indicates that these Caps could function either as cold-specific proteases to remove denatured proteins, otherwise deleterious for the cell, or as antifreeze proteins or as enzymes involved in the presumed large antifreeze macromolecule synthesis (Potier et al. 1987).

7.4 Genomic Studies

There is a recent surge in "omic" studies that has widened our understanding of cold adaptation. In the last few years, genome-level studies to assess low-temperature survival approaches of psychrophiles have been well explored (Saunders et al. 2003; Methé et al. 2005; Raymond-Bouchard et al. 2018a, b). The exponential increase in psychrophilic omics data, viz., genomics, proteomics, transcriptomics, and metagenomics, has initiated several sophisticated assessments of global protein composition changes and microbial adaptation mechanisms (Fig. 7.3). Due to the availability of genomes of psychrophilic organism and their comparative mesophilic sister species, significant insights into the ecological fitness traits and metabolic versatilities of psychrophiles across different temperature ranges have been derived (Casanueva et al. 2010). Additionally, a large number of transcriptome, proteome, metabolome, and metagenome data have been used to study the differential

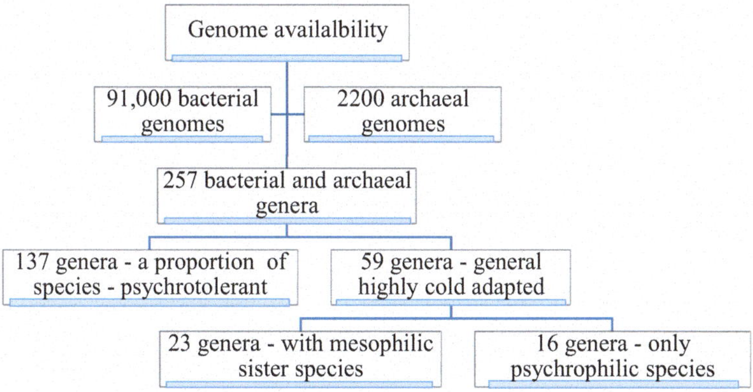

Fig. 7.3 Genome availability in cold-adapted bacteria

expression of genes or proteins in microorganisms that are grown at different low temperatures (De Maayer et al. 2014).

Out of the 137 genomes of psychrotolerant microorganisms, 134 were psychrophilic bacteria, and the rest three were archaeal genomes. Several cold-adapted bacterial and archaeal strains, from diverse environments, have complete or draft genomes (Bowman 2017). Some phyla and class with most common genus are listed below:

1. *Bacterial genomes*
   (a) Actinobacteria: *Cryobacterium*
   (b) Firmicutes: *Bacillus*, *Carnobacterium*, and *Planococcus*
   (c) Bacteroidetes: *Bizionia*, *Flavobacterium*, and *Lacinutrix*
   (d) Proteobacteria:
      - *Alphaproteobacteria*: *Octadecabacter*
      - *Betaproteobacteria*: *Polaromonas*
      - *Gammaproteobacteria*: *Colwellia*, *Moritella*, *Psychromonas*, *Psychrobacter*, and *Shewanella*
      - *Deltaproteobacteria*: *Desulfovibrio*
2. *Archaeal genomes*
   (a) *Methanococcus*

7.4.1 Psychrotrophic Microorganisms

The genomic sequencing study of *Pseudomonas* sp. strain BGI-2, a psychrotrophic bacterium isolated from Batura Glacier, identified the presence of 11 exopolysaccharide (EPS)-producing genes in the BGI-2 genome, while 7 other mesophilic *Pseudomonas* sp. had none of these genes (Pervaiz et al. 2019). Another psychrotrophic enteropathogenic bacterium *Yersinia pseudotuberculosis*, with optimal growth of 28 °C and which can also proliferate temperatures as low as 0 °C, was subjected to transcriptomic study at both temperatures to observe differential gene expression. The study revealed genes involved in glycolysis, transcription, and translation were positively regulated at 3 °C. Additionally, protein synthesis was maintained by cold shock proteins encoded by genes *yptb3585*, *yptb3586*, and *yptb2414* and transcription factors, such as Rho, IF-1, and RbfA. RNA helicases CsdA, RhlE, and DbpA, responsible for unfolding secondary structures of nucleic acids, were also remarkably overexpressed at 3 °C (Virtanen et al. 2018).

Recently, the transcriptome study of cold shock response in three species of psychrotrophic lactic acid bacteria (LAB) showed that *cspA* gene was primary gene and DEAD-box RNA helicase genes (*cshA*, *cshB*) played an important role. Gene network analysis and clustering result suggest that ribosomal proteins, tRNA and rRNA modification, and ABC and efflux MFS transporter genes were part of the cold shock response machinery, since they were clustered with known cold shock response genes, in all three species (Duru et al. 2021).

A comparative study of transcriptomes of a eurypsychrophile *Rhodococcus* sp. JG3 and a stenopsychrophile *Polaromonas* sp. Eur3 1.2.1, both isolated from permafrost, showed that both had many common cold-adaptive strategies. Some of the shared approaches include translation induction, EPS synthesis, upregulation of nutrient transport, and modulation cell membrane features. Another common cold-adaptive response was recombination and genomic redundancy (Raymond-Bouchard et al. 2018a, b).

In case of *Colwellia psychrerythraea* 34H, a number of cold adaptation traits such as EPS synthesis and polyunsaturated fatty acids for membrane fluidity, oxidative stress management, and compatible solutes were identified (Methé et al. 2005). Proteomic study on the same at different temperatures confirmed the same observations specifically for EPS and osmolyte management. At -10^0 C, DNA repair and chemotaxis were relatively more emphasized. Cell envelope modification and iron and nitrogen uptake also occurred at this temperature (Nunn et al. 2015).

7.4.2 Microbial Physiological Adaptations

Various physiological adaptations of cold-adapted microorganisms are shown in Table 7.2.

7.4.3 Cell Membrane Modulation

The first genome of a psychrophile was obtained for *Colwellia psychrerythraea* 34H, isolated from Arctic marine sediments, and the study described the proteins involved in the synthesis, ramification, and cis-isomerization of polyunsaturated fatty acids (Methé et al. 2005). Since then, there have been many more sequencing studies, and proteomic and genomic analyses show cold-specific adaptations (Lauro et al. 2011; Collins 2015). A study using the transcriptome data of *E. antarcticum* B7

Table 7.2 Omic studies revealing physiological adaptations of cold-adapted microorganisms

Cold-induced challenge	Adaptation mechanism	Biological macromolecules involved	References
Cell membrane fluidity regulation	Increased membrane permeability by composition modulation	Polyunsaturated fatty acids Fatty acid synthase Carotenoid biosynthesis genes	Sinetova and Los (2016) Ting et al. (2010) Dsouza et al. (2015)
Osmoprotection and cryoprotection	Cytoplasmic macromolecule stabilization	Compatible solutes—glucose, trehalose, glycogen, alanine, and glycerol	Ghobakhlou et al. (2015) Liljeqvist et al. (2015)
Freeze protection	Reduced cytoplasmic freezing point, prevent water molecules from ice crystal formation	Small antifreeze proteins Ice nucleation proteins	Celik et al. (2013) Virtanen et al. (2018)
Cryoprotection/ freeze protection	Cryoprotectants	Production of extracellular compounds	Methé et al. (2005)
Transport and diffusion	Transport system proteins upregulated	Membrane transporters and energy production protein modifications	Welsh (2000) Bakermans et al. (2012)
RNA/DNA secondary structure	Differential gene expression—CSP, CAP	RNA-binding proteins and helicases, signal transduction proteins	Liljeqvist et al. (2015)
Substrate oxidation	Facilitate oxidation	Different components of oxidation system	Liljeqvist et al. (2015)

showed a potential change in the metabolic pathway of fatty acids in response to cold (Kawasaki et al. 2016). The quantitative proteomic approach was applied to describe the enzymes involved in de novo synthesis of fatty acids in *Sphingopyxis alaskensis* (Ting et al. 2010).

At low temperatures, fatty acid biosynthesis gene is downregulated, whereas genes associated with desaturation are upregulated in *two psychrotrophic species—Exiguobacterium sibiricum* 255-15 and *Psychrobacter arcticus* 273-4 (Rodrigues et al. 2008; Bergholz et al. 2009).

Modulation in the cell membrane occurs at various levels such as variation in the ratio of length of acyl chain and proportion of unsaturated and branched chain fatty acid types. Taken together, these comprise of a homeostatic mechanism that involves transient activation of enzymes by a thermosensory two-component system DesR-DesK, which senses an increase in membrane inflexibility at cold temperatures. Similar study elucidating the differences between the psychrotolerant and mesophilic bacteria indicated that psychrotolerant strains can better modulate and fluidize their membranes by either desaturation of fatty acids or changing branched chain fatty acids (BCFA) levels. In *Listeria monocytogenes*, this is achieved by increasing anteiso-C15:0 level (Annous et al. 1997).

7.4.4 Osmoprotection and Cryoprotection: Compatible Solutes

Compatible solutes are small water-soluble organic compounds that are important for osmoprotection and cryoprotection. These compounds can decrease the freezing point of the cytoplasm, prevent macromolecule formation by hindering molecule aggregation, and increase membrane and proteins stability at low temperatures (Collins and Deming 2013). Cold-adapted microorganisms accumulate compatible solutes, such as trehalose, glycine betaine, and choline, from the outside environment through transporters (Hoffmann and Bremer 2017) and via de novo synthesis (Roberts 2005). Most psychrophilic phyla contain betaine-choline-carnitine transporter (BCCT)-type uptake systems that transport a combination of glycine betaine, choline, carnitine, or (hydroxy) ectoine.

Genome analyses of *Colwellia psychrerythraea* 34H, isolated from Arctic marine sediments, show an increase in transporter families involved in uptake of compatible solutes. There are five presumable transporters of the betaine/carnitine/choline transporter family, homologs to the ATP-binding cassette (ABC) transport family involved in direct uptake of glycine betaine and genes encoding a BetI family regulator that functions in production and regulation of glycine betaine in the genome of *C. psychrerythraea* 34H (Methé et al. 2005).

Metagenomic sequencing of the planktonic fractions from a low-temperature acidic environment identified psychrotolerant acidophile, *Acidithiobacillus ferrivorans*, as the most abundant microorganism. Metagenomic analysis of *At. ferrivorans*-like species showed the presence of complete trehalose synthase (TS), TreYZ and TreT pathways, and the gene for trehalose-6-phosphate synthase (TPS). This study revealed the occurrence of multiple trehalose synthesis pathways which suggests the significance of trehalose as a potential adaptive strategy to low temperature in this environment (Liljeqvist et al. 2015).

Cryoprotectants are chemical substances that after low-temperature exposure are accumulated in the cell and prevent cold-induced protein denaturation and aggregation and maintain an optimum membrane fluidity and homoeostasis in the cell. In the same study, the genes associated with cryoprotection were detected. This included the presence of glycine betaine synthesis gene along with betaine transcriptional regulator (*betI*) and gene-encoding choline dehydrogenase (*betA*). The sucrose synthase and sucrose-phosphate synthase genes that are associated with sucrose synthesis were also predicted in biofilm and the planktonic metagenome (Liljeqvist et al. 2015).

Transcriptome analysis of *Rhodococcus* sp. JG3 showed a threefold increase of choline transporter *betT* and an increase of fivefold in members of the *opuABCD* system, which is responsible for uptake of the glycine betaine, at −5 °C in *Rhodococcus* sp. JG3 when compared to 25 °C (Raymond-Bouchard et al. 2018a, b). Another metabolomic analysis study revealed an increase in the accumulation of sarcosine, threonine, and valine when Arctic isolate *Mesorhizobium* sp. strain N33 was grown at 4 °C, suggesting these compounds to be acting as cryoprotectants (Ghobakhlou et al. 2015).

7.4.5 Freeze Protection

Antifreeze proteins (AFPs) are capable of modification of the ice crystal structure and inhibit recrystallization/growth of ice, in supercooled conditions. AFPs adsorb to ice crystals and thereby create thermal hysteresis, which is the process of separation of melting and freezing temperature, and thus arrest the process of freezing and lowering the growth temperature (Jia and Davies 2002). The planktonic fraction from a low-temperature acidic environment, which is inhabited predominantly by psychrotolerant acidophile *Acidithiobacillus ferrivorans*, included a gene presumed to encode a type II antifreeze protein (Liljeqvist et al. 2015) previously identified in *Te. saanensis* (Rawat et al. 2012).

7.4.6 Extracellular Compounds

Extracellular polysaccharides (EPS) are thought to provide cryoprotection and antifreeze properties at some level. Genome sequence of *Colwellia psychrerythraea*, a marine psychrophilic bacterium, encodes presumed members of the extracellular factor subfamily of σ-70 transcription factors. These transcription factors have several roles such as EPS biosynthesis regulation and of paralogous families of glycosyltransferases, which also are involved in extracellular polysaccharide synthesis (Methé et al. 2005).

In *Polaromonas* sp. Eur3 1.2.1 and *Rhodococcus* sp. JG3, comparative transcriptome analysis revealed an increase in the number of transcripts, which are thought to be associated with EPS biosynthesis, at respective low temperatures when compared to the high temperatures for both microorganisms. These transcripts included cellulose synthase, EPS biosynthesis proteins, and mannose-6-phosphate isomerase and some others (Raymond-Bouchard et al. 2018a, b).

7.4.7 Transport and Diffusion

A metaproteomic study of bacterioplankton from Antarctic Peninsula coastal surface waters showed that the most prevalent bacterial and archaeal proteins in the metaproteome were components of ATP-binding cassette (ABC) transporters (13.7% of the total metaproteome). Most of the ABC transport proteins were periplasmic-binding proteins (PBPs) (Williams et al. 2012). Ferric iron transport is upregulated in *P. cryohalolentis* K5 at low temperature (Bakermans et al. 2012).

In the transcriptome analysis of *Rhodococcus* sp. JG3 and *Polaromonas* sp. Eur3 1.2.1, numerous transcripts for transporters were increased at respective low temperatures when compared to high temperatures for both microorganisms. However, some of the transporters that were increased were different in both species, but some others were common. In *Rhodococcus* sp. JG3, mainly branched chain amino acid and peptide transporters as well as dicarboxylate symporter were strongly induced at −5 °C compared to 25 °C, whereas in *Polaromonas* sp. Eur3 1.2.1 polar amino acid transport system and two tripartite tricarboxylate transporter receptor components were upregulated. Additionally, there was a positive regulation and an increase of sulfate transport and MFS family efflux pumps in both microorganisms. The study suggested that low-temperature-induced challenges such as decreased diffusion rates and diminished nutrient uptake are tackled by an increase in transporters at low temperatures that enables effective nutrient uptake and higher number of transporters helps subdue the impacts of diminished diffusion rates and therefore this is a crucial strategy for psychrophiles to overcome low-temperature-induced transport-related hindrance (Raymond-Bouchard et al. 2018a, b).

7.4.8 RNA/DNA Secondary Structure

A comparative study of transcriptomes of a eurypsychrophile *Rhodococcus* sp. JG3 and a stenopsychrophile *Polaromonas* sp. Eur3 1.2.1 indicated positive regulation and increase in genes involved in translation. *Rhodococcus* sp. JG3 exhibits positive regulation of 24 genes such as translation elongation factor EF-Ts at −5 °C when compared to 25 °C. A queuine tRNA-ribosyltransferase and an *O*-acetyl-ADP-ribose deacetylase (involved in suppression of RNase activity), *rbfA* (class I protein of Csp family with >tenfold induction), and a superfamily II RNA helicase (including DEAD/DEAH-box family proteins) were also positively regulated and showed an increase in expression. In *Polaromonas* sp. Eur3 1.2.1, transcripts were expressed at higher levels for several ribosome-associated proteins, viz., RNA helicase (for RNA unwinding) and IF-1 (for initiation of translation), at 0 °C compared to 20 °C (Raymond-Bouchard et al. 2018a, b).

7.4.9 Substrate Oxidation

All components of Fe^{2+} oxidation system were detected in the metagenomes from the low-temperature acidic environment biofilm and plankton. However, the expression of the genes has not been confirmed (Liljeqvist et al. 2015).

Different components of cytochrome oxidase genes were detected in the *At. ferrivorans*-like species. The detected cytochrome oxidase genes are cytochrome *bo* terminal oxidase (*cyoABCD*), an aa_3-type cytochrome *c* oxidase (*coxABCD*), and its other subunits. Additionally, two cbb_3 genes (*ccoN* and *ctaG*) and a cytochrome *bd* ubiquinol oxidase gene (*cydA*) were also discovered (Liljeqvist et al. 2015).

7.5 Conclusion

Psychrophilic microorganisms have developed the ability to grow at low temperature by developing various strategies, such as antifreeze proteins, cryoprotectants, and modulation of membrane lipid composition, to overcome low-temperature stress. Cold shock proteins are a family of proteins that are induced and highly expressed at low temperatures and are responsible for low-growth adaptation by controlling nucleic-acid secondary structures, protein folding, and transcription and translations. Recent surge in genomics and proteomics studies leads to a surge in omic data and enabled comparative studies that help our understanding of cold adaption in microbes. There are several omic studies revealing different physiological adaptations of cold-adapted microorganisms enabling them to survive in cold environments.

References

Annous BA, Becker LA, Bayles DO, Labeda DP, Wilkinson BJ (1997) Critical role of anteiso-C15:0 fatty acid in the growth of Listeria monocytogenes at low temperatures. Appl Environ Microbiol 63:3887–3894. https://doi.org/10.1128/AEM.63.10.3887-3894

Bakermans C, Bergholz PW, Rodrigues DF, Vishnivetskaya TA, Ayala-Del-Rio H, Tiedje JM (2012) Genomic and expression analyses of cold-adapted microorganisms. In: Miller RV, Whyte LG (eds) Polar microbiology: life in a deep freeze. ASM Press, New York, 126–155. https://doi.org/10.1128/9781555817183.ch6

Beranová J, Mansilla MC, de Mendoza D, Elhottová D, Konopásek I (2010) Differences in cold adaptation of *Bacillus subtilis* under anaerobic and aerobic conditions. J Bacteriol 192:4164–4171

Berger F, Morellet N, Menu F, Potier P (1996) Cold shock and cold acclimation proteins in the psychrotrophic bacterium *Arthrobacter globiformis* SI55. J Bacteriol 178:2999–3007. https://doi.org/10.1128/jb.178.11.2999-3007.1996

Bergholz PW, Bakermans C, Tiedje JM (2009) *Psychrobacter arcticus* 273-4 uses resource efficiency and molecular motion adaptations for subzero temperature growth. J Bacteriol 191:2340–2352. https://doi.org/10.1128/JB.01377-08

Bowman JP (2017) Genomics of psychrophilic bacteria and archaea. In: Margesin R (ed) Psychrophiles: from diversity to biotechnology. Springer, Cham, pp 345–387. https://doi.org/10.1007/978-3-319-57057-0_15

Casanueva A, Tuffin M, Cary S, Cowan D (2010) Molecular adaptations to psychrophily: the impact of 'omic' technologies. Trends Microbiol 18(8):374–381. https://doi.org/10.1016/j.tim.2010.05.002

Cavicchioli R, Saunders N, Thomas T (2009) Cold-shock response in microorganisms. In: Gerday C, Glansdorff N (eds) Extremophiles. Encyclopedia of Life Support Systems Publishers, Oxford, pp 1–26

Celik Y, Drori R, Pertaya-Braun N, Altan A, Barton T, Bar-Dolev M, Groisman A, Davies PL, Braslavsky I (2013) Microfluidic experiments reveal that antifreeze proteins bound to ice crystals suffice to prevent their growth. Proc Natl Acad Sci USA 110:1309–1314

Chaikam V, Karlson DT (2010) Comparison of structure, function and regulation of plant cold shock domain proteins to bacterial and animal cold shock domain proteins. BMB Rep 43:1–8. https://doi.org/10.5483/BMBRep.2010.43.1.001

Charollais J, Dreyfus M, Iost I (2004) CsdA, a cold-shock RNA helicase from Escherichia coli, is involved in the biogenesis of 50S ribosomal subunit. Nucleic Acids Res 32:2751–2759. https://doi.org/10.1093/nar/gkh603

Collins RE (2015) Microbial evolution in the cryosphere. In: Bakermans C (ed) Microbial evolution under extreme conditions. De Gruyter, Berlin, pp 31–55

Collins RE, Deming JW (2013) An inter-order horizontal gene transfer event enables the catabolism of compatible solutes by *Colwellia psychrerythraea* 34H. Extremophiles 17:601–610

Cowan DA, Casanueva A, Stafford W (2007) Ecology and biodiversity of cold-adapted microorganisms. In: Gerday C, Glansdorff N (eds) Physiology and biochemistry of extremophiles, 1st edn. ASM Press, New York, pp 119–132

De Maayer P, Anderson D, Cary C, Cowan DA (2014) Some like it cold: understanding the survival strategies of psychrophiles. EMBO Rep 15:508–517. https://doi.org/10.1002/embr.201338170

De Mendoza D (2014) Temperature sensing by membranes. Annu Rev Microbiol 68:101–116. https://doi.org/10.1146/annurev-micro-091313-103612

Deming JW (2002) Psychrophiles and polar regions. Curr Opin Microbiol 5:301–309

Derman Y, Söderholm H, Lindström M, Korkeala H (2015) Role of csp genes in NaCl, pH, and ethanol stress response and motility in Clostridium botulinum ATCC3502. Food Microbiol 46:463–470. https://doi.org/10.1016/j.fm.2014.09.004

Dsouza M, Taylor MW, Turner SJ, Aislabie J (2015) Genomic and phenotypic insights into the ecology of Arthrobacter from Antarctic soils. BMC Genom 16:36. https://doi.org/10.1186/s12864-015-1220-2

Duru IC, Ylinen A, Belanov S, Pulido AA, Paulin L, Auvinen P (2021) Transcriptomic time-series analysis of cold and heat-shock response in psychrotrophic lactic acid bacteria. BMC Genomics 22:28. https://doi.org/10.1186/s12864-020-07338-8

Duval BD, Mathew A, Satola SW, Shafer WM (2010) Altered growth, pigmentation, and antimicrobial susceptibility properties of Staphylococcus aureus due to loss of the major cold shock gene *cspB*. Antimicrob Agents Chemother 54:2283–2290. https://doi.org/10.1128/AAC.01786-09

Fedorov R, Meshcheryakov V, Gongadze G, Fomenkova N, Nevskaya N, Selmer M (2001) Structure of ribosomal protein TL5 complexed with RNA provides new insights into the CTC family of stress proteins. Acta Crystallogr D Biol Crystallogr 57:968–976

Ghobakhlou AF, Johnston A, Harris L, Antoun H, Laberge S (2015) Microarray transcriptional profiling of Arctic Mesorhizobium strain N33 at low temperature provides insights into cold adaption strategies. BMC Genomics 16:383. https://doi.org/10.1186/s12864-015-1611-4

Gottesman S (2018) Chilled in translation: adapting to bacterial climate change. Mol Cell 70:193–194

Hamdan A (2018) Psychrophiles: ecological significance and potential industrial application. S Afr J Sci 114:1–6

Hebraud M, Dubois E, Potier P, Labadie J (1994) Effect of growth temperatures on the protein levels in a psychrotrophic bacterium, *Pseudomonas fragi*. J Bacteriol 176:4017–4024

Hoffmann T, Bremer E (2017) Guardians in a stressful world: the Opu family of compatible solute transporters from Bacillus subtilis. Biol Chem 398:193–214

Iost I, Dreyfus M (2006) DEAD-box RNA helicases in Escherichia coli. Nucleic Acids Res 34:4189–4197. https://doi.org/10.1093/nar/gkl500

Jia Z, Davies PL (2002) Antifreeze proteins: an unusual receptor–ligand interaction. Trends Biochem Sci 27:101–106

Jin B, Jeong KW, Kim Y (2014) Structure and flexibility of the thermophilic cold-shock protein of *Thermus aquaticus*. Biochem Biophys Res Commun 451:402–407. https://doi.org/10.1016/j.bbrc.2014.07.127

Jones PG, VanBogelen RA, Neidhardt FC (1987) Induction of proteins in response to low temperature in *Escherichia coli*. J Bacteriol 169:2092–2095

Kandror O, DeLeon A, Goldberg AL (2002) Trehalose synthesis is induced upon exposure of *Escherichia coli* to cold and is essential for viability at low temperatures. Proc Natl Acad Sci USA 99:9727–9732

Kawasaki R, Baraúna RA, Silva A, Carepo MSP, Oliveira R, Marques R, Ramos RTJ, Schneider MPC (2016) Reconstruction of the fatty acid biosynthetic pathway of *Exiguobacterium antarcticum* B7 based on genomic and bibliomic data. Biomed Res Int 2016:7863706. https://doi.org/10.1155/2016/7863706

Kuhn E (2012) Toward understanding life under subzero conditions: the significance of exploring psychrophilic "cold-shock" proteins. Astrobiology 12:1078–1086. https://doi.org/10.1089/ast.2012.0858

Lauro FM, Allen MA, Wilkins D, Williams TJ, Cavicchioli R (2011) Psychrophiles: genetics, genomics, evolution. In: Horikoshi K (ed) Extremophiles handbook. Springer, Tokyo, pp 865–890

Lee J, Jeong KW, Jin B, Ryu KS, Kim EH, Ahn JH, Kim Y (2013) Structural and dynamic features of cold-shock proteins of *Listeria monocytogenes*, a psychrophilic bacterium. Biochemistry 52:2492–2504. https://doi.org/10.1021/bi301641b

Liljeqvist M, Ossandon FJ, González C, Rajan S, Stell A, Valdes J, Holmes DS, Dopson M (2015) Metagenomic analysis reveals adaptations to a cold-adapted lifestyle in a low-temperature acid mine drainage stream. FEMS Microbiol Ecol 91:fiv011. https://doi.org/10.1093/femsec/fiv011

Lindquist JA, Mertens PR (2018) Cold shock proteins: from cellular mechanisms to pathophysiology and disease. Cell Commun Sig 16:63. https://doi.org/10.1186/s12964-018-0274-6

Loepfe C, Raimann E, Stephan R, Tasara T (2010) Reduced host cell invasiveness and oxidative stress tolerance in double and triple *csp* gene family deletion mutants of *Listeria monocytogenes*. Foodborne Pathog Dis 7:775–783. https://doi.org/10.1089/fpd.2009.0458

Margesin R, Collins T (2019) Microbial ecology of the cryosphere (glacial and permafrost habitats): current knowledge. Appl Microbiol Biotechnol 103:1–13

Margesin R, Feller G (2010) Biotechnological applications of psychrophiles. Environ Technol 31:835–844

Methé BA, Nelson KE, Deming JW, Momen B, Melamud E, Zhang X, Moult J, Madupu R, Nelson WC, Dodson RJ, Brinkac LM, Daugherty SC, Durkin AS, DeBoy RT, Kolonay JF, Sullivan SA, Zhou L, Davidsen TM, Wu M, Huston AL, Lewis M, Weaver B, Weidman JF, Khouri H, Utterback TR, Feldblyum TV, Fraser CM (2005) The psychrophilic lifestyle as revealed by the genome sequence of *Colwellia psychrerythraea* 34H through genomic and proteomic analyses. Proc Natl Acad Sc USA 102:10913–10918

Michaux C, Martini C, Shioya K, Lecheheb SA, Budin-Verneuil A, Cosette P, Sanguinetti M, Hartke A, Verneuil N, Giard J-C (2012) CspR, a cold shock RNA-binding protein involved in the long-term survival and the virulence of *Enterococcus faecalis*. J Bacteriol 194:6900–6908. https://doi.org/10.1128/JB.01673-12

Morita RY (1975) Psychrophilic bacteria. Bacteriol Rev 39:144–167

Moyer CL, Morita RY (2007) Psychrophiles and psychrotrophs. In: Encyclopedia of life science. https://doi.org/10.1002/9780470015902.a0000402.pub2

Nunn BL, Slattery KV, Cameron KA, Timmins-Schiffman E, Junge K (2015) Proteomics of Colwellia psychrerythraea at subzero temperatures – a life with limited movement, flexible membranes and vital DNA repair. Environ Microbiol 17(7):2319–2335. https://doi.org/10.1111/1462-2920.12691

Pandey KD, Shukla SP, Shukla PN, Giri DD, Singh JS, Singh P, Kashyap AK (2004) Cyanobacteria in Antarctica: ecology, physiology and cold adaptation. Cell Mol Biol 50:575–584

Pervaiz A, Aamer AS, Fariha H, Haiyuan C, Ana S, Feng C (2019) Draft genome sequence of a cold-adapted Pseudomonas sp. Strain, BGI-2, isolated from the Ice of Batura glacier, Pakistan. https://doi.org/10.1128/MRA.00320-19

Phadtare S (2004) Recent developments in bacterial cold-shock response. Curr Issues Mol Biol 6:125–136

Phadtare S (2012) *Escherichia coli* cold-shock gene profiles in response to over-expression/deletion of CsdA, RNase R and PNPase and relevance to low-temperature RNA metabolism. Genes Cells 17:850–874. https://doi.org/10.1111/gtc.12002

Phadtare S, Severinov K (2010) RNA remodeling and gene regulation by cold shock proteins. RNA Biol 7:788–795. https://doi.org/10.4161/rna.7.6.13482

Phadtare S, Alsina J, Inouye M (1999) Cold-shock response and cold-shock proteins. Curr Opin Microbiol 2:175–180. https://doi.org/10.1016/s1369-5274(99)80031-9

Piette F, D'Amico S, Mazzucchelli G, Danchin A, Leprince P, Feller G (2011) Life in the cold: a proteomic study of cold-repressed proteins in the Antarctic bacterium *Pseudoalteromonas haloplanktis* TAC125. Appl Environ Microbiol 77:3881–3883

Potier P, Drevet P, Gounot AM, Hipkiss AR (1987) Protein turnover in a psychrotrophic bacterium: proteolytic activity in extracts of cells at different temperatures. FEMS Microbiol Lett 44:267–271

Rawat SR, Männistö MK, Starovoytov V, Goodwin L, Nolan M, Hauser L, Land M, Davenport KW, Woyke T, Häggblom MM (2012) Complete genome sequence of *Terriglobus saanensis* type strain SP1PR4(T), an Acidobacteria from tundra soil. Stand Genomic Sci 7:59–69. https://doi.org/10.4056/sigs.3036810

Raymond-Bouchard I, Goordial J, Zolotarov Y, Ronholm J, Stromvik M, Bakermans C, Whyte LG (2018a) Conserved genomic and amino acid traits of cold adaptation in subzero-growing Arctic permafrost bacteria. FEMS Microbiol Ecol 94:fiy023

Raymond-Bouchard I, Tremblay J, Altshuler I, Greer CW, Whyte LG (2018b) Comparative transcriptomics of cold growth and adaptive features of a eury- and steno-psychrophile. Front Microbiol 9:1565. https://doi.org/10.3389/fmicb.2018.01565

Roberts MF (2005) Organic compatible solutes of halotolerant and halophilic microorganisms. Saline Syst 1:5. https://doi.org/10.1186/1746-1448-1-5

Robinson CH (2001) Cold adaptation in Arctic and Antarctic fungi. New Phytol 151:341–353

Rodrigues DF, Ivanova N, He Z, Huebner M, Zhou J, Tiedje JM (2008) Architecture of thermal adaptation in an *Exiguobacterium sibiricum* strain isolated from 3 million year old permafrost: a genome and transcriptome approach. BMC Genom 9:547. https://doi.org/10.1186/1471-2164-9-547

Rodrigues DF, Tiedje JM (2008) Coping with our cold planet. Appl Environ Microbiol 74:1677–1686

Saunders NF, Thomas T, Curmi PM, Mattick JS, Kuczek E, Slade R, Davis J, Franzmann PD, Boone D, Rusterholtz K, Feldman R, Gates C, Bench S, Sowers K, Kadner K, Aerts A, Dehal P, Detter C, Glavina T, Lucas S, Richardson P, Larimer F, Hauser L, Land M, Cavicchioli R (2003) Mechanisms of thermal adaptation revealed from the genomes of the Antarctic archaea *Methanogenium frigidum* and *Methanococcoides burtonii*. Genome Res 13:1580–1588. https://doi.org/10.1101/gr.1180903

Schärer K, Stephan R, Tasara T (2013) Cold shock proteins contribute to the regulation of listeriolysin O production in *Listeria monocytogenes*. Foodborne Pathog Dis 10:1023–1029. https://doi.org/10.1089/fpd.2013.1562

Schindelin H, Jiang W, Inouye M, Heinemann U (1994) Crystal structure of CspA, the major cold shock protein of *Escherichia coli*. Proc Natl Acad Sci USA 91:5119–5123

Schindler T, Graumann PL, Perl D, Ma S, Schmid FX, Marahiel MA (1999) The family of cold shock proteins of *Bacillus subtilis* stability and dynamics *in vitro* and *in vivo*. J Biol Chem 274:3407–3413

Schmid B, Klumpp J, Raimann E, Loessner MJ, Stephan R, Tasara T (2009) Role of cold shock proteins in growth of *Listeria monocytogenes* under cold and osmotic stress conditions. Appl Environ Microbiol 75:1621–1627. https://doi.org/10.1128/AEM.02154-08

Sinetova MA, Los DA (2016) New insights in cyanobacterial cold stress responses: genes, sensors, and molecular triggers. Biochim Biophys Acta 1860(11 Pt A):2391–2403. https://doi.org/10.1016/j.bbagen.2016.07.006

Smirnova A, Li H, Weingart H, Aufhammer S, Burse A, Finis K, Schenk A, Ullrich MS (2001) Thermoregulated expression of virulence factors in plant-associated bacteria. Arch Microbiol 176:393–399

Ting L, Williams TJ, Cowley MJ, Lauro FM, Guilhaus M, Raftery MJ, Cavicchioli R (2010) Cold adaptation in the marine bacterium, *Sphingopyxis alaskensis*, assessed using quantitative proteomics. Environ Microbiol 12:2658–2676. https://doi.org/10.1111/j.1462-2920.2010.02235.x

Virtanen JP, Timonen RK, Jaakkola K, Salin N, Korkeala H (2018) Changes in transcriptome of *Yersinia pseudotuberculosis* IP32953 grown at 3 and 28°C detected by RNA sequencing shed light on cold adaptation. Front Cell Infect Microbiol 8:16

Wang Z, Wang S, Wu Q (2014) Cold shock protein A plays an important role in the stress adaptation and virulence of *Brucella melitensis*. FEMS Microbiol Lett 354:27–36. https://doi.org/10.1111/1574-6968.12430

Welsh DT (2000) Ecological significance of compatible solute accumulation by micro-organisms: from single cells to global climate. FEMS Microbiol Rev 24:263–290

Whyte LG, Inniss WE (1992) Cold shock proteins and cold acclimation proteins in a psychrotrophic bacterium. Can J Microbiol 38:1281–1285

Williams TJ, Long E, Evans F, Demaere MZ, Lauro FM, Raftery MJ, Ducklow H, Grzymski JJ, Murray AE, Cavicchioli R (2012) A metaproteomic assessment of winter and summer bacterioplankton from Antarctic peninsula coastal surface waters. ISME J 6:1883–1900. https://doi.org/10.1038/ismej.2012.28

Xia B, Ke H, Shinde U, Inouye M (2003) The role of RbfA in 16S rRNA processing and cell growth at low temperature in *Escherichia coli*. J Mol Biol 332:575–584

Yamanaka K, Fang L, Inouye M (1998) The CspA family in *Escherichia coli*: multiple gene duplication for stress adaptation. Mol Microbiol 27:247–255

Zhang Y, Burkhardt DH, Rouskin S, Li GW, Weissman JS, Gross CA (2018) A stress response that monitors and regulates mRNA structure is central to cold shock adaptation. Mol Cell 70:274–286

Survival Strategies in Cold-Adapted Microorganisms

8

Deepika Goyal, Shiv Swaroop, Om Prakash, and Janmejay Pandey

Abstract

The vast majority of the entire solar system, including Earth's environment, is characterized by extreme temperatures. Noticeably, approximately 80% of Earth's biosphere is either periodically or permanently cold and has temperatures below 5 °C. Thus, it could be concluded that low-temperature regions are the predominant "extreme" environment on Earth. The low temperatures are exceedingly hostile for the growth and survival of life. However, a significant number of scientific studies have reported the identification of cold-adapted psychrophilic microorganisms that have managed to not only survive but also thrive under hostile environments. The ability of psychrophilic microorganisms to thrive under cold environments is accredited to innate capabilities, e.g., structural adjustment of enzymes, maintenance of membrane fluidity, expression of cold-shock proteins, and presence of compatible solutes. The underlying mechanisms for survival strategies among cold-adapted microorganisms could be broadly classified under two broad categories: (1) environmental and physiological adaptations and (2) molecular adaptations. This book chapter aims to provide a detailed account of various strategies that have a significant role in the growth and survival of cold-adapted psychrophilic microorganisms under low-temperature environments. It provides crucial insight into fundamental concepts (e.g., metabolic processes, energy generation, stress resistance, cell envelop characteristics,

D. Goyal · J. Pandey (✉)
Department of Biotechnology, Central University of Rajasthan, Kishangarh, Rajasthan, India
e-mail: janmejay@curaj.ac.in

S. Swaroop
Department of Biochemistry, Central University of Rajasthan, Kishangarh, Rajasthan, India

O. Prakash
National Centre for Microbial Resource (DBT-NCMR), Pune, Maharashtra, India

R. Goel et al. (eds.), *Survival Strategies in Cold-adapted Microorganisms*,
https://doi.org/10.1007/978-981-16-2625-8_8

etc.) pertaining to microbial life under cold environments. This chapter also highlights some of the significant findings from recent studies performed with "omics" approaches (including metagenomics, metatranscriptomics, and metaproteomics) to determine the critical adaptation functions expressed by microbial communities inhabiting the cold regions.

Keywords

Earth's environment · Cold region · Psychrophilic microorganisms · Cold adaptation · Survival strategies · Omics studies

8.1 Introductions

The growth and survival of life on Earth depend on several discrete yet interdependent life-sustaining abiotic parameters such as temperature, pH, nutrient, humidity, etc. Among them, temperature is one of the most critical parameters. Although life has been reported from diverse environmental niches having a wide range of temperatures (i.e., from –20 °C to 110 °C), survival at extremes of temperature is exceptionally challenging. The cold ecological niches characterized by average annual temperatures of <5 °C are the most widespread extremes of temperature on Earth. The most prominent cold regions include vast tracts of the deep seas and oceans, regions of permafrost, parts of Northern America and Europe within the Arctic circle, and areas of great mountains (the Alps, the Himalayas, and the Rocky Mountains) and diverse polar reaches of Arctic and Antarctic regions (Siddiqui et al. 2013). In comparison, the lesser prominent cold regions include the mesosphere and the stratosphere. Besides, there are artificial cold regions such as cold storages, deep freezers, fridges, etc. Collectively, these regions make for approximately 80% of the total Earth's biosphere (De Maayer et al. 2014).

The hostile nature of cold regions is exceptionally hazardous for any life-form; still, a plethora of psychrophilic microorganisms (eubacteria and archaea) bestowed with unique cold adaptations not only survive but also exhibit robust growth under cold environments (Siddiqui et al. 2013). Such microorganisms are broadly classified into two categories: (1) psychrophiles and (2) psychrotolerants. Psychrophiles like cold temperatures exhibit optimum growth at temperatures lower than 15 °C and do not grow at temperatures greater than 20 °C; the psychrotolerant microorganisms do not like cold temperature; they grow optimally at 20–25 °C, but they can survive temperatures as low as 0 °C. One of the common observations with cold-adapted microorganisms is that psychrophilic microorganisms are predominantly found within the cold marine ecosystems that are permanently cold, e.g., oceanic waters. Psychrotolerant microorganisms are more frequently observed in terrestrial ecosystems that experience fluctuation of cold temperature from moderately to extremely cold. This generalization is somewhat oversimplified as there is a gross lack of data on the temperature profile of various ecosystems across the Earth's biosphere. Another method for classifying cold-adapted microorganisms is based on the range of cold temperature that the microorganisms can adapt to. According to this approach, cold-adapted microorganisms are classified into two groups: (1) eurypsychrophile, which can tolerate a wide range of cold temperatures, and (2) stenopsychrophile, which can tolerate only a narrow range of cold temperatures.

Interestingly, the limits of lower temperature that psychrophiles can sustain have never been clearly defined; however, a few reports have suggested −12 °C as the limit for reproductive functions and −20 °C as the limit for metabolic processes (Clarke et al. 2013). The lack of comprehensive data on temperature profiles of different cold regions across the Earth's aquatic and terrestrial niches has proven to be the major bottleneck in defining the precise limits of low temperatures wherein psychrophilic microbial communities survive (De Maayer et al. 2014). According to a few discrete reports, important metabolic and cellular functions have been reported to occur at temperatures lower than −20 °C (Clarke et al. 2013). The subzero temperature causes a severe negative effect on all metabolic and cellular functions. The negative effects are executed by altering the vital forces that determine growth and survival of life, e.g., water viscosity, solute diffusion rates, membrane fluidity, enzyme kinetics, and macromolecular interactions (Rodrigues and Tiedje 2008).

Cold-adapted microorganisms and their survival strategy under extreme cold have provoked scientific interest for a substantial period now. It is argued that an understanding of how life functions in low-temperature environments cold would be vital for understanding the life of microorganisms and the evolution of microbial life under extreme environments. Results from such studies would be extremely helpful in the exploration of the existence of life in unfathomed ecological niches. They may also provide important clues regarding many of the phenomena related to adaptive microbial physiology, including microbial dormancy, microbial persistence, microbial biofilm formation, etc., that have an extremely important role in microbial survival and microbial pathogenesis. It is also expected to provide pertinent insight into some of the poorly understood yet very important phenomenon, e.g., the viable but noncultivable (VBNC) state. Most importantly, understanding the survival mechanisms of cold-adapted microorganisms may also help develop technologies for controlling microbial life, including the pathogenic bacteria, archaea, and fungi, which thrive in cold-stored food materials. Consequently, studies on unraveling the survival mechanisms and microbial adaptation to cold stress have received considerable attention in the recent past; several studies have been carried out to understand and explore the cold-adapted biotechnological potential microorganisms and their useful biomolecules. Despite the increased interest in the applied aspects of psychrophiles and psychrotolerant microorganisms, the major focus of most of the studies related to cold-adapted microorganisms continues to address the very fundamental questions pertinent to their survival. An example of such questions is how do cold-adapted microorganisms sequester growth rate-limiting nutrient from the environment at low temperature?

8.2 Survival Strategies of Cold-Adapted Microorganisms: Initial Studies

Initial studies on the characterization of survival strategies of cold-adapted microorganisms focused primarily on addressing the unique capability of psychrophilic microorganisms to grow at low temperatures but not at mesophilic temperatures. Results obtained with initial studies indicated toward the alteration of lipid components and their relative abundance in the biological members as the most prominent adaptation shown by cold-adapted microorganisms (Russell 1990). Other studies indicated that a higher proportion of unsaturated fatty acids in the cell membrane, which increases its fluidity and alteration of protein conformation at low temperature, is the most commonly observed adaptation among psychrophilic and psychrotolerant microorganisms (Gounot 1986). Subsequent studies highlighted that the changes observed in lipid structure and composition are because of both genotypic and phenotypic adaptations. In contrast, changes in the protein structure and function are primarily genotypic. A few studies also highlighted that the major alterations observed in protein structure and function are related to protein translation machinery (Russell 1990).

Further, progress about survival strategies of cold-adapted microorganisms remained scanty for nearly two to three decades due to the nonavailability of cultivable representative cold-adapted microorganisms. Microbial communities within the extreme ecological niches are highly complex and consist of hundreds to thousands of microbial species that have never been cultivated in laboratory conditions. In the absence of pure microbial cultures of cold-adapted microorganisms, it was perceivably difficult to obtain and comprehend the survival strategies implemented by cold-adapted microorganisms for their survival. The adaptations exhibited by cold-adapted microorganisms as characterized through initial studies were classified as physiological adaptations.

8.2.1 Physiological Adaptations Exhibited by Cold-Adapted Microorganisms

Physiological adaptations shown by cold-adapted microorganisms during their growth on cold temperature have been largely identified by comparing the microorganisms grown naturally at different temperatures. Physiological adaptations exhibited by psychrotolerant and psychrophilic microorganisms are usually quite complex to study because they encompass many factors. Several extrinsic factors, including abiotic (e.g., temperature, pH, salinity, nutrient flux, redox conditions), biotic (e.g., cell-cell interactions, viruses), and ecological factors (e.g., sea ice versus seawater, particle attached versus free living), can greatly influence the growth properties of individual microorganisms. The most prominently described physiological adaptation exhibited by cold-adapted microorganisms includes membrane adaptations by altering the ratio of saturated fatty acid in membrane phospholipids, changes in lipid class composition, and reduced size and charge of lipid head groups.

These adaptations affect the phospholipid packing and conversion of trans- to cis-isomeric fatty acids.

Besides several intrinsic factors, such as genomic component and gene expression regulation, also defines the physiological state of different cells. Due to such complex control mechanisms, only a few microorganisms have evolved to successfully colonize the extremes of low and high temperature (Saunders et al. 2003; Reid et al. 2006). Therefore, it is extremely difficult to carry out studies to compare the adaptive traits and physiological adaptations of psychrophiles, mesophiles, and hyperthermophiles, which belong to the same taxonomic lineage. This limitation has been circumvented to a reasonable extent by examining microorganisms that grow at different optimal growth temperatures even though they belong to different taxonomic groups. These studies have provided significant insight, yet they are relatively minuscule. Therefore, studies based on genome composition and expression (e.g., genomics, transcriptomics, and proteomics) of psychrophiles have proven particularly valuable for determining the cold-adapted microorganisms' adaptive mechanisms. Results from these studies must be linked to the knowledge obtained from direct measurements of physiological adaptation (e.g., alteration of morphology, growth rate, macromolecular synthesis, solute composition, membrane lipid composition, and nutrient perturbation).

8.2.2 Structural Alterations of Proteins/Enzymes in Cold-Adapted Microorganisms

Proteins in general and enzymes that catalyze hundreds of biochemical reactions within the cell are the cornerstones of the life and adaptation of life to various challenges faced by any cell. Therefore, it could be easily presumed that microbial life adaptation to the low temperature must involve proteins that sustain their functions at cold temperatures. This presumption was aptly confirmed with isolation and characterization of a few proteins, e.g., DNA polymerase, RNA polymerase, ribonuclease, and alkaline phosphatase, from a cold-adapted microorganism. These proteins were observed to have structural flexibility, enabling them to remain functional even at very low temperatures. These observations present an acceptable explanation for the high levels of some cold-adapted enzymes' specific activity (Feller and Gerday 1997). Comparing the amino acid composition of proteins purified from the cold-adapted microorganisms and the similar kind of proteins purified from their mesophilic counterparts has also provided imperative clues regarding proteins' structural basis of cold-adapted microorganisms. The major observations from studies on amino acid sequence comparison indicate the presence of significantly fewer proline, arginine, cysteine, and hydrophobic amino acid residues in proteins of cold-adapted microorganisms. The lesser number of cysteine residues also results in a decrease in the number of disulfide bonds. Proteins with lesser disulfide bonds exhibit the characteristics of low rigidity. (D'Amico et al. 2001, 2002). Besides, many proteins from the cold-adapted microorganisms also show an increased abundance of polar residues (Siddiqui and Cavicchioli 2006).

Recent studies have addressed the abovementioned generalized observation related to proteins from the cold-adapted microorganisms. In one such study, few proteins were purified from a psychrotolerant strain of *Arthrobacter* that was isolated from dry valley soil from the Antarctic. β-galactosidase purified from this strain was found to be ~5 times more active than a β-galactosidase obtained purified from a mesophilic *Escherichia* coli. The amino acid sequence comparison between these two proteins revealed the excepted decrease in the number of proline residues in β-galactosidase purified from the cold-adapted *Arthrobacter*.

Interestingly, a few earlier studies focusing on comparing amino acid sequence and composition of the proteins from thermophilic and mesophilic microorganisms had suggested that the distribution of hydrogen bonds and amino acid compositions does not differ significantly between the mesophilic and thermophilic proteins. The pairwise comparisons also did not define any significant structural difference between the thermostable and mesophilic proteins (Panasik et al. 2000). Therefore, the generalization regarding the amino acid composition, distribution, and structural characteristics of the proteins from cold-adapted microorganisms must be made with utmost caution. Recent findings suggest that instead of overgeneralizing, it should be mentioned that a few subtle differences at the amino acid sequence level and their implications in synergistic molecular interactions are more important for the activity of proteins from cold-adapted microorganisms.

8.2.3 Alterations Ensuring Biomembrane Fluidity in Cold-Adapted Microorganism

Potentially the most obvious adverse effects of decreasing temperature are the changes in the biomembrane fluidity. With decreasing temperatures, the membranes become more and more rigid. It is critical to sustaining membrane fluidity for sustaining life. Therefore, microorganisms adapted to cold temperatures ensure their biomembrane's fluidity via the conversion of saturated fatty acids into unsaturated fatty acids. It is achieved by the activity of a few key enzymes, including desaturases, which get induced upon exposure of the microbial cell to low temperatures. This process is complemented by a preferential biosynthesis of short-chain fatty acids, branched chain fatty acids, and anteiso-fatty acids (Russell and Fukunaga 1990; Hassan et al. 2020). Besides, an increase in the amount of hydroxy fatty acids has also been reported to enhance the biomembrane's fluidity in the case of cold-adapted microorganisms (Nichols et al. 1993; Tribelli and López 2018). The significance of alteration in fatty acid composition as a critical feature of survival strategy implemented by cold-adapted microorganisms has also been experimentally validated in a few recent studies. For example, a study carried out with DNA microarray on transcripts of *Bacillus subtilis* upon its growth at cold temperatures showed significant downregulation of genes that encode for enzymes involved in the degradation of branched chain amino. Another important finding from this study was the observation that genes encoding for enzymes involved in the degradation of isoleucine and valine. This observation was explained with the rationale that the degradation intermediates of both isoleucine and valine are utilized for biosynthesis of branched chain fatty acids. In other words, as a survival strategy implemented by cold-adapted microorganisms when growing at low temperature, they increase the degradation of certain amino acids and utilize the degradation intermediates for biosynthesis of fatty acids required for maintaining the fluidity of biomembrane fluidity through both anabolic and catabolic pathways.

8.2.4 Other Subtle Adaptations Exhibited by Cold-Adapted Microorganisms

Apart from those mentioned above and widely acknowledged common adaptations exhibited by cold-adapted microorganisms, discrete reports have shown various subtle adaptions for surviving cold environments. These subtle adaptions include the following:

1. *Lowering of metabolic activities*: For a long time, there was a notion that cold-adapted bacteria survive the subzero temperatures by acquiring a metabolically inert state wherein they don't divide. However, in recent past, this notion has been broken. Several studies have demonstrated metabolic activity and growth among bacteria at subzero temperatures (Chattopadhyay 2006). However, the metabolic activities of cold-adapted microorganisms when grown at subzero temperatures reduce significantly. Reduced metabolic activity of a native bacterial population, obtained from Siberian permafrost, has been observed in terms of increased generation time of ~20 days and ~160 days when grown at temperatures of −10 °C and −20 °C, respectively. In contrast, the same bacterial community grows with generation time of only 1 day when grown at 5 °C (Rivkina et al. 2000).
2. *Upregulation of cold-shock proteins*: A set of transiently induced proteins during exposure to cold environments are typically referred to as cold-shock proteins. Characteristically, these kinds of proteins have been reported from both mesophilic and psychrophilic microorganisms. Yet another class of proteins that are found exclusively in psychrophilic microorganisms is referred to as cold acclimation proteins (Caps). Noticeably, both cold-shock proteins and cold acclimation proteins are upregulated during the growth of cold-adapted microorganisms at low temperatures. At function level, the cold-shock proteins are involved in either DNA modifications or controlling the replication, transcription, or protein stability. Some of the prominent examples include RNA-binding proteins, transcription factor, Hsc 66, trigger factor, several acyl lipid desaturases, and γ-glutamyl transpeptidase (Chattopadhyay 2006).
3. *Presence of alternative RNA degradosome*: RNA degradosomes are protein complexes that ensure the stability of cellular RNA. Many of the cold-adapted microorganisms, e.g., *Pseudomonas syringae* strain Lz4W, have been shown to have the presence of more than one kind of RNA degradosome (Purusharth et al. 2005). The exact significance of such alternative RNA degradosomes is not yet determined; however, it is proposed that some of the RNA degradosomes can

degrade RNAs without involvement of ATP; thereby, they may help the cell in the conservation of energy at low temperature.
4. *Cytosolic accumulation of cryoprotectants*: Cryoprotectants are chemical substances known to protect against cellular damage when the cells are subjected to extremes of cold temperatures. They are reported to prevent cold-induced aggregation of proteins and maintain their optimum activity at low temperatures.

Despite these observations, the survival strategies used by cold-adapted microorganisms were rather poorly understood till the end of the twentieth century. The lack of cultivable cold-adapted microorganism was a major bottleneck in the characterization of the elaborate biochemical and molecular mechanism that contributes to the overall survival of microorganisms within the cold environment. Fortunately, during the late 1990s to early 2000s, the explorations of cold niches and cold-adapted microorganisms like other ecological niches could escalate with the advent of two complementary scientific approaches, i.e., (1) culture independent molecular microbiology, metagenomics, and (2) high-throughput genome sequencing.

8.2.5 Metagenomics- and Genomics-Based Studies of Cold-Adapted Microorganisms

The implementation of metagenomics and high-throughput genome sequencing approaches during the early 2000s successfully defined the hidden microbial diversity of various ecological niches, including the cold regions. It revolutionized the ability to analyze microbial communities of extreme ecological niches, including those of the cold regions (Sjöling and Cowan 2008). It has also led to a rapid increase in functional diversity comprehension by discovering genes pertinent to various metabolic pathways identified either by sequencing or activity-based screening strategies. A noticeable observation from metagenomics and high-throughput genome sequencing-based studies on cold-adapted microorganisms was that their survival strategies encompass many mechanisms beyond altering the composition of fatty acid within the biological membrane or structure of certain critical proteins. It was also noted that the survival strategy implemented by cold-adapted microorganisms involves molecular evolution to counteract the harmful effects of the freezing of water in a low-energy cold environment. The molecular evolution is observed in cold-adapted microorganisms and cold-adapted microbial communities. These include (1) the presence of novel genes unique to cold-adapted microorganisms and (2) the occurrence of a consistent change in the amino acid composition of proteins that destabilize protein structures (Casanueva et al. 2010).

One of the first metagenomic studies pursued for determining the phylogenetic diversity and metabolic potential of the microbial community associated with the cold region was carried out on glacial ice of the Northern Schneeferner, Germany (Simon et al. 2009). The microbial community composition assessment revealed that the glacial ice is predominantly occupied by the *Betaproteobacteria*, *Bacteroidetes*, and *Actinobacteria.* A high level of genetic diversity was observed concerning the metabolic gene clusters that would encode for proteins required for metabolic degradation of organic substrates. Expectedly, the analyses of sequences obtained from this study also revealed the presence of genes involved in long-established adaptations associated with cold-adapted microorganisms, e.g., genes regulating the synthesis of cryoprotectants and unsaturated fatty acids. This study provided several important insights into the adaptations exhibited by microbial life thriving within the glacial ice, a worthy representative ecological niche of frozen habitats on Earth. Another metagenomic study revealed the presence of several environmental stress response genes, particularly genes for exopolysaccharide biosynthesis, membrane adaptations, and osmotic stress-associated genes in the metagenomic sequence of microbial mats from ice ponds in Antarctica and the high Arctic (Varin et al. 2012). These observations provided valuable novel insight into the survival strategies that are potentially implemented by cold-adapted microorganisms. Noticeably, most of these strategies were never identified with studies on physiological adaptation approaches.

As an offshoot, the implementation of metagenomics and high-throughput genomics on cold-adapted microorganisms also led to the successful discovery of cold-active high-value biomolecules, e.g., proteins, enzymes, peptides, secondary metabolites, etc. In one such example, a novel cold-adapted lipase LipEH166 was isolated from an intertidal flat's metagenome (Kim et al. 2009). Characteristically, the amino acid sequence of LipEH166 did not show significant amino acid similarity to a previously characterized lipolytic enzyme. The amino acid sequence similarity was limited to only the consensus region that defines lipolytic enzyme activity. These observations suggested that LipEH166 is a novel cold-adapted alkaline lipase (Kim et al. 2009).

With accumulating observations from metagenomics and high-throughput genome sequencing studies, as well as biochemistry, molecular biology, protein biochemistry, and structure biology studies, the scientific community started appreciating the diversity of mechanisms implemented by cold-adapted microorganisms as the survival strategy to withstand the adverse conditions of the cold environmental niches. Consequently, studies related to unrevealing the survival strategy used by cold-adapted microorganisms have progressed to attaining a systems biology status. Some of the relevant questions asked in this approach include how do cold-adapted microorganisms (1) sense the cold environment; (2) uptake nutrients at temperatures wherein water is frozen; (3) ensure fluidity of the biological membranes; (4) carry out essential metabolic, reproductive, and cellular functions; and (5) overcome thermodynamic interference at low temperatures.

8.3 Systems Biology Studies of Cold-Adapted Microorganisms

8.3.1 Comparative Genomic Studies of Cold-Adapted Microorganisms

The understanding of cold-adapted microorganisms and their survival has greatly enhanced over the last decade with the availability of draft and completed genome sequences of psychrophilic bacteria and archaea. As of March 2021, there are

417,394 eubacterial and 2235 archaeal genome assemblies available. Out of these, ~300 genome sequences belong to bacterial and archaeal strains obtained from low-temperature ecosystems (De Maayer et al. 2014). Further, many of these genome sequences have been subjected to comparative genomic analyses with taxonomically related mesophilic microorganisms' genomes. Such comparative genome analyses have been successfully utilized to identify the molecular determinants of cold adaptations by defining the presence/absence of genes in microorganisms isolated from ecological niches characterized by cold vs. moderate temperatures. In an example study, the genome sequence of a cold-adapted *Alteromonas* sp. SN2 was compared to the genome sequences of two mesophilic strains of *Alteromonas macleodii*; it was observed that the cold-adapted strain has the presence of 15 genomic islands that might contribute to the ecological fitness of this strain in the cold marine environment (Math et al. 2012). In another study, the comparative genomic analyses of *Halobacterium* sp. tADL, a psychrophilic strain isolated from a cold water lake in Antarctica, revealed the presence of novel genes that would encode for bacteriorhodopsin and polyhydroxyalkanoate biosynthesis (DeMaere et al. 2013). This study also revealed that the haloarchaea obtained from the Antarctic lake have genomic characteristics consistent with a high level of gene exchange for enabling the selection of ecotypes that are compatible for the maintenance of sympatric speciation within the polar lake system (DeMaere et al. 2013).

The capabilities of genomics, metagenomics, and comparative genomics have been appropriately complemented by the advent of other "omics" technologies, viz., transcriptomics and proteomics. These technologies have enabled a series of very sophisticated analyses for addressing adaptive changes shown by cold-adapted microorganisms, specifically at the level of their transcript and protein composition (Casanueva et al. 2010).

8.3.2 Transcriptomics Studies of Cold-Adapted Microorganisms

Transcriptomics is defined as the study of differential expression of genes. It is most often carried out comparatively, wherein the samples for the study are subject to different physiological conditions. Subsequently, their transcripts are extracted, sequenced, and analyzed to define which genes are expressed or what transcripts are generated when the sample was subject to different conditions. This approach has been implemented for studying the survival strategy and unraveled the molecular mechanism of the adaption in psychrophilic and psychrotolerant microorganisms. Studies carried out with transcriptome profiling have provided important insight into the adaptation and survival mechanisms that were rarely revealed by any other approach (Raymond-Bouchard et al. 2018).

In an example study, cold-adapted archaeon *Methanolobus psychrophilus* strain R15 was analyzed with comparative transcriptomics of the cultures grown at 18 °C and 4 °C, respectively. The results showed that the genes for methanogenesis were downregulated, whereas those encoding for the RNA polymerase complex and a putative exosome complex were strongly upregulated during its growth at cold temperature. This observation suggests that growth under cold temperature may require exosome-mediated RNA decay (Chen et al. 2012). Interestingly, genes encoding for chaperonin protein complexes (Thermosome and GroES/E.L.) were also upregulated during the growth of strain R15 at 4 °C. Strain R15 also showed exhibited cold-enhanced expression of genes that encode for proteins involved in oxygen detoxification, e.g., superoxide reductase (SOR), superoxide dismutase (SOD), and catalase oxidant-removing system. During growth on cold environments, strain R15 also exhibited upregulation of 71 single-component systems and 50 two-component systems involved in signal transduction. These observations indicate an enhanced oxidative tolerance and signal transduction in R15 (a cold-adapted microorganism) during its adaptation to the cold environment (Chen et al. 2012). In another example, transcriptome profiling of a psychrophilic bacteria, viz., *Exiguobacterium sibiricum* strain 255-15, during its growth at 2.5 °C and 39 °C revealed several genes involved in more diverse functions, e.g., DNA replication, transcription and translation, carbohydrate and amino acid metabolism, and cell membrane biosynthesis were differentially expressed (Rodrigues et al. 2008).

Psychrobacter arcticus strain 273-4 is an important model organism for studying life within freezing environments because it can grow at subzero temperatures. It can also grow very efficiently under limited nutrient availability and low-water activity conditions. The genome and transcriptome of strain 273-4 have been analyzed thoroughly for revealing specific molecular adaptations that allow it to survive under freezing environments. Such studies have shown that during the growth of *P. arcticus* 273-4 at subzero temperatures, many of its genes related to carbon substrate assimilation and energy metabolism are downregulated, whereas genes involved in the maintenance of biomembranes, cell walls, and nucleic acid motion are upregulated (Kuhn 2012). Interestingly, during the growth at freezing temperature, the expression of either protein or RNA chaperones does not get upregulated; however, its cold-shock protein DEAD-box RNA helicase protein A (CsdA-Psyc_1082) gets significantly upregulated. This observation proves CsdA-Psyc_1082 as one of the key proteins for sustaining life under freezing temperatures (Kuhn 2012).

These examples clearly showed how transcriptomics has led to significant value addition toward the characterization of new perspectives on the survival strategies implemented by cold-adapted microorganisms.

8.3.3 Proteomics Studies of Cold-Adapted Microorganisms

In comparison to genomics, metagenomics, and transcriptomics, proteomics-based approaches have been widely believed to be more suitable and capable of defining the survival strategies used by cold-adapted microorganisms. Such a belief is based on the direct relationship of the proteome with the phenotype of the cell/organisms. In other words, the existence of a novel gene, unique mutation, or even the presence of a gene transcript cannot lead to any phenotypic change if the corresponding protein(s) is not expressed in a functional state. In view of this, a few studies have been carried out in the recent past to investigate the global proteome expression of cold-adapted microorganisms when they are grown in cold environments.

Many proteomics-based studies have validated the earlier understanding regarding the protein-level adaptations exhibited by cold-adapted microorganisms. For example, it is reported that during their growth in a cold environment, psychrophiles express proteins that comprise of a significantly higher fraction of amino acids with small or neutral side chains. These amino acids contribute to higher protein flexibility. On the other hand, amino acids with aliphatic, basic, aromatic, and hydrophilic side chains are underrepresented in psychrophiles' proteins. Additionally, it has been recorded that some amino acids, viz., serine, aspartic acid, threonine, and alanine, are overrepresented in the coil regions of secondary structures, whereas glutamic acid and leucine are underrepresented in the helical regions of the cold-induced proteins expressed by psychrophilic bacteria (Metpally and Reddy 2009).

A differential proteomic analysis study of Himalayan low-temperature diazotroph *Pseudomonas migulae* strain S10724 was carried out using conventional two-dimensional electrophoresis and MALDI-TOF-MS-based proteomic approach. Results from this analysis showed a total of 66 differentially expressed proteins and revealed several mechanisms that might be involved in low-temperature adaptation, nitrogen fixation, general stress adaptation, protein and nucleic acid synthesis, energy metabolism, cell growth/maintenance, etc. (Suyal et al. 2014). This study also identified two proteins (viz., NifU family SUF system FeS assembly protein and membrane protein, a suppressor for copper sensitivity B precursor) as unexpected exclusively expressed during nitrogen fixation and growth at low temperature.

8.4 Conclusion and Future Prospects

With nearly three decades of studies on cold-adapted microorganisms and more specifically on the survival strategies/molecular adaptation exhibited by them, a lot of vital information has cumulatively emerged. During the past 10–15 years, the expansion of knowledge in this field has gained major impetus with the advent of "omics"-based technologies; consequently, the nature of many cold-adapted proteins and the metabolic pathways that get upregulated upon exposure to low-temperature is now relatively well understood. The understanding of these proteins and metabolic pathways has collectively helped reveal the large picture of microbial physiology and survival strategy at low temperatures. An important concept that has emerged out of systems biology or omics-based studies of cold-adapted microorganisms is that the underlying mechanisms for microbial tolerance to a variety of abiotic stress are interlinked in a manner that was not perceived earlier. Some of other more recent observations, e.g., a mesophilic strain of *E. coli* when transformed for expressing two chaperonin genes that were obtained from an Antarctic bacterium, are groundbreaking in the field, and they provide the prospect of not only basic research studies but also applied research work for critical issues of the cold regions. One such issue is the spillage of petroleum/petroleum products in Antarctic regions during the past two decades due to the increased anthropogenic activities. It has taken the shape of a growing global environmental concern. However, treatment of the spillage sites in cold regions is considered to be extremely challenging. Bioremediation of petroleum spillage using recombinant microbial strains expressing cold adaptation-related genes and capable of petroleum biodegradation can provide the possible solution.

References

Casanueva A, Tuffin M, Cary C, Cowan DA (2010) Molecular adaptations to psychrophily: the impact of 'omic' technologies. Trends Microbiol 18:374–381

Chattopadhyay M (2006) Mechanism of bacterial adaptation to low temperature. J Biosci 31:157–165

Chen Z, Yu H, Li L, Hu S, Dong X (2012) The genome and transcriptome of a newly described psychrophilic archaeon, M ethanolobus psychrophilus R 15, reveal its cold adaptive characteristics. Environ Microbiol Rep 4:633–641

Clarke A, Morris GJ, Fonseca F, Murray BJ, Acton E, Price HC (2013) A low temperature limit for life on earth. PLoS One 8:e66207

D'Amico S, Claverie P, Collins T, Georlette D, Gratia E, Hoyoux A, Meuwis M-A, Feller G, Gerday C (2002) Molecular basis of cold adaptation. Philos Trans R Soc Lond B Biol Sci 357:917–925

D'Amico S, Gerday C, Feller G (2001) Structural determinants of cold adaptation and stability in a large protein. J Biol Chem 276:25791–25796

De Maayer P, Anderson D, Cary C, Cowan DA (2014) Some like it cold: understanding the survival strategies of psychrophiles. EMBO Rep 15:508–517

DeMaere MZ, Williams TJ, Allen MA, Brown MV, Gibson JA, Rich J, Lauro FM, Dyall-Smith M, Davenport KW, Woyke T (2013) High level of intergenera gene exchange shapes the evolution of haloarchaea in an isolated Antarctic lake. Proc Natl Acad Sci 110:16939–16944

Feller G, Gerday C (1997) Psychrophilic enzymes: molecular basis of cold adaptation. Cell Mol Life Sci 53:830–841

Gounot A-M (1986) Psychrophilic and psychrotrophic microorganisms. Experientia 42:1192–1197

Hassan N, Anesio AM, Rafiq M, Holtvoeth J, Bull I, Haleem A, Shah AA, Hasan F (2020) Temperature driven membrane lipid adaptation in glacial psychrophilic bacteria. Front Microbiol 11:824

Kim E-Y, Oh K-H, Lee M-H, Kang C-H, Oh T-K, Yoon J-H (2009) Novel cold-adapted alkaline lipase from an intertidal flat metagenome and proposal for a new family of bacterial lipases. Appl Environ Microbiol 75:257–260

Kuhn E (2012) Toward understanding life under subzero conditions: the significance of exploring psychrophilic "cold-shock" proteins. Astrobiology 12:1078–1086

Math RK, Jin HM, Kim JM, Hahn Y, Park W, Madsen EL, Jeon CO (2012) Comparative genomics reveals adaptation by Alteromonas sp. SN2 to marine tidal-flat conditions: cold tolerance and aromatic hydrocarbon metabolism. PLoS One 7:e35784

Metpally RPR, Reddy BVB (2009) Comparative proteome analysis of psychrophilic versus mesophilic bacterial species: insights into the molecular basis of cold adaptation of proteins. BMC Genomics 10:1–10

Nichols DS, Nichols PD, McMeekin TA (1993) Polyunsaturated fatty acids in Antarctic bacteria. Antarct Sci 5:149–149

Panasik N Jr, Brenchley JE, Farber GK (2000) Distributions of structural features contributing to thermostability in mesophilic and thermophilic α/β barrel glycosyl hydrolases. Biochim Biophys Acta (BBA) 1543:189–201

Purusharth RI, Klein F, Sulthana S, Jäger S, Jagannadham MV, Evguenieva-Hackenberg E, Ray MK, Klug G (2005) Exoribonuclease R interacts with endoribonuclease E and an RNA helicase in the psychrotrophic bacterium Pseudomonas syringae Lz4W. J Biol Chem 280:14572–14578

Raymond-Bouchard I, Tremblay J, Altshuler I, Greer CW, Whyte LG (2018) Comparative transcriptomics of cold growth and adaptive features of a eury-and steno-psychrophile. Front Microbiol 9:1565

Reid I, Sparks W, Lubow S, McGrath M, Livio M, Valenti J, Sowers K, Shukla H, MacAuley S, Miller T (2006) Terrestrial models for extraterrestrial life: methanogens and halophiles at Martian temperatures. Int J Astrobiol 5:89

Rivkina E, Friedmann E, McKay C, Gilichinsky D (2000) Metabolic activity of permafrost bacteria below the freezing point. Appl Environ Microbiol 66:3230–3233

Rodrigues DF, Ivanova N, He Z, Huebner M, Zhou J, Tiedje JM (2008) Architecture of thermal adaptation in an Exiguobacterium sibiricum strain isolated from 3 million year old permafrost: a genome and transcriptome approach. BMC Genomics 9:1–17

Rodrigues DF, Tiedje JM (2008) Coping with our cold planet. Appl Environ Microbiol 74:1677–1686

Russell N (1990) Cold adaptation of microorganisms. Philos Trans Roy Soc Lond 326:595–611

Russell N, Fukunaga N (1990) A comparison of thermal adaptation of membrane lipids in psychrophilic and thermophilic bacteria. FEMS Microbiol Lett 75:171–182

Saunders NF, Thomas T, Curmi PM, Mattick JS, Kuczek E, Slade R, Davis J, Franzmann PD, Boone D, Rusterholtz K (2003) Mechanisms of thermal adaptation revealed from the genomes of the Antarctic archaea Methanogenium frigidum and Methanococcoides burtonii. Genome Res 13:1580–1588

Siddiqui KS, Cavicchioli R (2006) Cold-adapted enzymes. Annu Rev Biochem 75:403–433

Siddiqui KS, Williams TJ, Wilkins D, Yau S, Allen MA, Brown MV, Lauro FM, Cavicchioli R (2013) Psychrophiles. Annu Rev Earth Planet Sci 41:87–115

Simon C, Wiezer A, Strittmatter AW, Daniel R (2009) Phylogenetic diversity and metabolic potential revealed in a glacier ice metagenome. Appl Environ Microbiol 75:7519–7526

Sjöling S, Cowan DA (2008) Metagenomics: microbial community genomes revealed. In: Psychrophiles: from biodiversity to biotechnology. Springer, Cham, pp 313–332

Suyal DC, Yadav A, Shouche Y, Goel R (2014) Differential proteomics in response to low temperature diazotrophy of Himalayan psychrophilic nitrogen fixing Pseudomonas migulae S10724 strain. Curr Microbiol 68:543–550

Tribelli PM, López NI (2018) Reporting key features in cold-adapted bacteria. Life 8:8

Varin T, Lovejoy C, Jungblut AD, Vincent WF, Corbeil J (2012) Metagenomic analysis of stress genes in microbial mat communities from Antarctica and the high Arctic. Appl Environ Microbiol 78:549–559

Microbial Adaptations Under Low Temperature

9

Dikchha Singh, Shobit Thapa, K. Pandiyan, Malay Kumar Das, Avijit Pramanik, and Hillol Chakdar

Abstract

Microorganisms are ubiquitous and diverse microbes inhabit low-temperature niches. More than three quarters of the Earth's surface is either occasionally cold or permanently frozen, making it a predominant habitat in the world. Despite such hostile conditions, these organisms flourish because of certain structural, physiological, and molecular variations that are associated with it. Adaptations related to the cell membrane, enzymes, transporters, chaperones, antifreeze proteins, osmolytes, and cold- and heat-shock proteins help the organisms in thriving under such situations. In the present chapter, we discussed various microbial adaptations in detail to throw light on the lifestyle microorganisms thriving under low temperature. Understanding such adaptations may assist us with investigating the prospects for advancement in various novel biotechnological applications.

Keywords

Psychrophiles · Chaperones · Cryoprotectants · Cold-shock proteins

D. Singh · S. Thapa · M. K. Das · H. Chakdar (✉)
ICAR-National Bureau of Agriculturally Important Microorganisms, Kushmaur, Uttar Pradesh, India

K. Pandiyan
ICAR-Central Institute for Research on Cotton Technology, Nagpur, Maharashtra, India

A. Pramanik
Central University of Haryana, Mahendragarh, Haryana, India

R. Goel et al. (eds.), *Survival Strategies in Cold-adapted Microorganisms*,
https://doi.org/10.1007/978-981-16-2625-8_9

9.1 Introduction

Much of the Earth's habitat, both marine and terrestrial, is either periodically or permanently cold. More than three quarters of the Earth's habitat is permanently frozen, having temperature below 5 °C. It comprises primarily the oceans, polar ice caps, mountains, and man-made habitats such as fridges and cooling towers. Microorganisms are one of the most fascinating forms of life which have an outstanding capability to adapt under such cold environments. Their adaptive capabilities have evolved in such a way that many of them can tolerate very low temperatures, while many like psychrophiles show their optimal growth under low temperatures only. Habitats that are most often frozen and generally considered to be inhospitable to life are home to psychrophilic microorganisms. Overall, low-temperature habitats are most prevalent where the psychrophiles inhabit (De Maayer et al. 2014). Majority of known psychrophilic microorganisms belong to varieties of archaea, bacteria, yeast, fungi, and algae. That is due to the capability of microorganisms to adapt for optimum functioning in their normal physiological environments. Any extreme change in environmental conditions (transient or permanent) from the optimum induces stress on the dwelling microorganism. Under such stress, the extent of stress determines the adaptability and growth, which can be ceased under extreme stress. Cold-adaptive microorganisms manage to survive such a low-temperature environment by adapting and combating the cold-induced stress in their own unique way. Various ingenious physiological and metabolic adaptations have empowered these tiny bugs to survive low to ultralow temperatures (Fig. 9.1). Such unique adaptations have several industrial implications as well, for example, cold-active lipases have multifarious applications in pharmaceutical industry, fine chemical synthesis, detergent industry, leather industry, environmental applications, etc. (Kavitha 2016). Apart from that various other cold-active biomolecules like proteases, amylases, and antifreeze proteins find similar industrial applications. Therefore, a detailed understanding of adaptive mechanisms has direct correlation with genetic or metabolic engineering of industrially important biomolecules. In this chapter, we have discussed major survival strategies employed by various cold-adaptive microorganisms.

9.2 Microbial Adaptations Under Low Temperature

9.2.1 Sensing the Temperature

The ability of microorganisms to adapt to environmental changes such as low temperature primarily depends on the ability to sense the changes in the environment. Biological membranes are present at the interface of the organism, and the environment could be the primary sensor to the changes in temperature. It now depends on the capability of the sensor to perceive the signal and transduce the signal to the genome for upregulation or downregulation of the adaptive response to cold adaptation. Model and biological membranes turn rigid, that is, reduction in

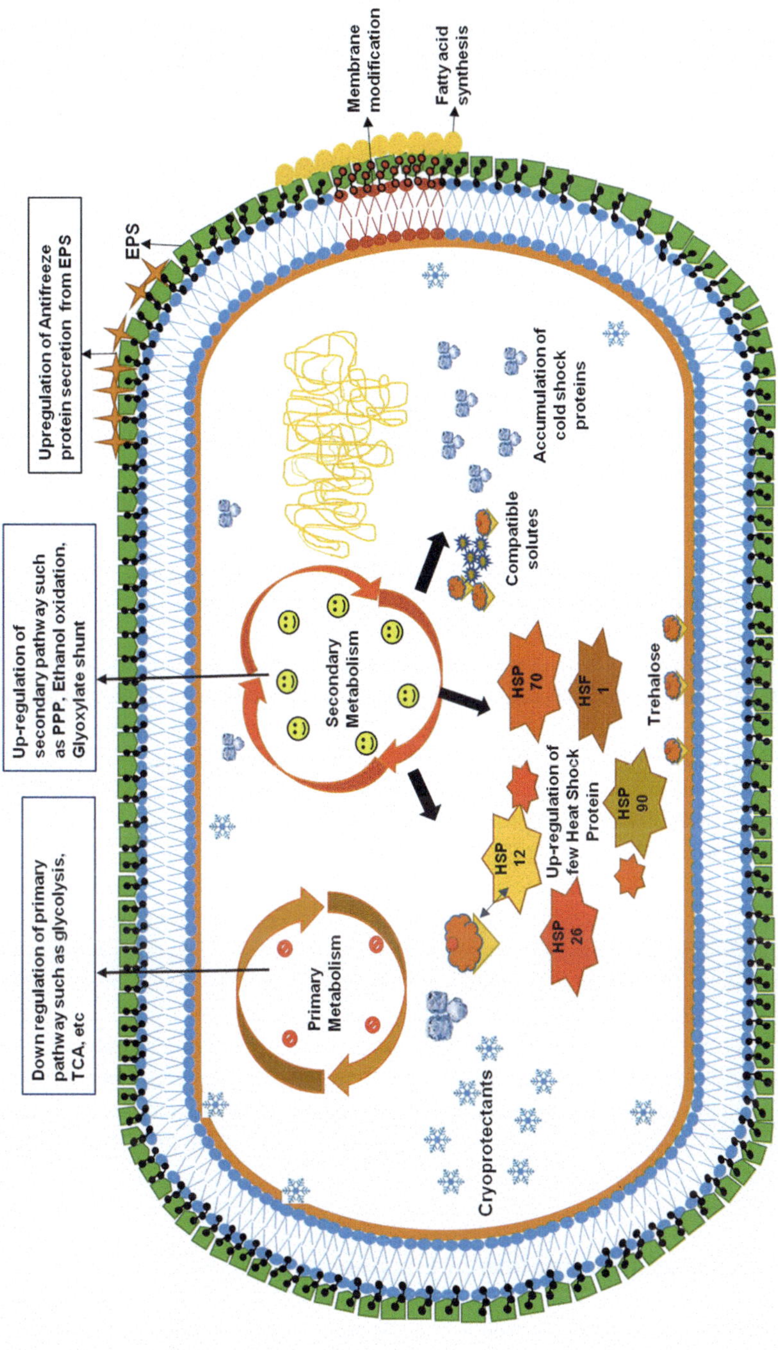

Fig. 9.1 Various adaptations deployed by microorganisms to survive under low temperature

membrane dynamics when exposed to low temperatures. This rigidification of biological membrane is poised to be a sensing signal of low-temperature environment. By deactivation of fatty acid desaturase, membrane rigidity was brought upon in an experiment by Inaba et al. Following microarray revealed a set of cold adaptation genes which were upregulated in response to membrane rigidification (Inaba et al. 2003). This asked for speculation that some sensor response regulator is yet to be identified that transduce the membrane rigidity signal to genome. Fatty acid desaturase activities are also upregulated in cold adaptation. The Pd-catalyzed hydrogenation of plasma membrane lipids induced the expression of the DesA fatty acid desaturase gene (Vigh et al. 1993). Such results suggest that increase in membrane rigidity serves as a primary signal for cold perception. To identify the response regulator, Suzuki et al. systematically disrupted all 43 putative genes for histidine kinases of *Synechocystis* sp. PCC 6803 in conjunction with screening by the transcriptional activity of the promoter of the desB gene for the omega-3 desaturase. This screen identified two histidine kinases and a response regulator as components of the perception and transduction of low-temperature signals for the expression of genes for fatty acid desaturases (Suzuki et al. 2000). The *Bacillus subtilis* DesK/DesR two-component system is believed to sense the decrease in membrane lipid fluidity (Albanesi et al. 2004).

Other than the membrane, nucleic acid present in the cell also responds to environmental temperature. In bacteria, the genome is present as a twisted superhelical state. The degree of this superhelicity varies with the environmental temperature (Dorman 1996). A number of bacterial proteins are involved in creating and maintaining superhelical states of the DNA. In *Shigella flexneri*, VirF controls VirB which in turn controls large 230 kb pathogenicity plasmid gene expression. The product of VirR, the H-NS, controls VirB transcription in a temperature-regulated manner. In in vivo studies, VirB transcription was decreased by inhibiting DNA gyrase, while the transcription was increased by mutating H-NS. Evidently the VirB transcription activation is DNA topology dependent (Tobe et al. 1993).

Theoretically RNA molecules can form excellent temperature sensors owing to their secondary and tertiary structure variability. Indeed, RNA thermometers have been reported in literature (Kortmann and Narberhaus 2012). LcrF is a transcriptional activator in *Yersinia pestis* and controls expression of several virulence genes. At lower temperature, LcrF production is low, while rate of transcription is similar to optimum temperature. LcrF mRNA secondary structure prediction revealed that the ribosomal binding Shine-Dalgarno sequence is sequestered in a stem loop structure at low temperature (Hoe and Goguen 1993). RNA thermometers are cis-acting riboregulators in bacteria. In *Shigella dysenteriae*, a RNA thermometer controls the translation of OmpA (Murphy et al. 2020).

9.2.2 Structural Adaptation of Enzymes

Enzymes run the metabolism inside a cell. Freezing temperature decreases the fluidity of cytoplasm which creates unfavorable conditions for enzymes to catalyze the metabolic functions. Also, low temperature alters enzyme structure; as a result, they could not attain their activation energy required to catalyze the reaction (Chandler 2018). Hence, a major adaptation to cold temperature is crucial for enzymes in their effective functioning inside a cell. Many structural adaptations are present in psychrophilic enzymes as compared to thermophilic or mesophilic ones, making it more flexible at low temperatures. The psychrophilic enzymes have increased specific activity and flexibility at low temperatures due to their following properties: decreased arginine/lysine ratio, low proline content in disulfide bridges and loops with more α-helices, more number of glycine residues, more nonpolar residues on protein surface resulting in higher-surface hydrophobicity, weaker hydrogen bonds, protein and other electrostatic interactions, and reduced secondary structures, with higher number and size of loops (Feller 2010; Cavicchioli et al. 2011; De Maayer et al. 2014).

Arrhenius best described the role of temperature in chemical reaction in his equation $k = Ae^{-Ea/RT}$, where A is the pre-exponential factor of reaction, k is the rate constant, R is the gas constant (8.31 kJ/mol), Ea is the activation energy, and T is the temperature in kelvins. It indicates that with the decrease in temperature, the decrease in the reaction rate will be exponential where the activation energy will provide the extent of the reaction. This could be the main drawback for lives at cold if not adapted. The basic difference between mesophilic and psychrophilic enzymes lies in the fact that the latter exhibit enzymes with higher specific activity at lower temperatures and the maximal activity for their enzymes are shifted toward low temperatures with lessened thermostability. Moreover, the displayed specific activity at lower temperatures though high is lower than their mesophilic counterpart, making it less efficient (D'Amico et al. 2002).

In a study of kinetics and structure of a cold-adapted hetero-octameric ATP phosphoribosyltransferase (ATPPRT), it was found that two forms of ATPPRT exist in nature depending on the species. The long form, $HisG_L$, has regulatory domain, while the short form, $HisG_S$, lacks regulatory domain which is supplemented by regulatory protein HisZ, constituting the STPPRT holoenzyme. The holoenzyme constituted with HisZ have twofold to threefold better K_{cat} than catalytic subunit $HisG_S$ (Stroek et al. 2017). In a study with cold-adapted Antarctic polyextremophilic *β*-galactosidase, it was found that six key amino acid replacements are responsible for higher K_{cat}/k_m of the enzyme at low temperature (Laye et al. 2017). Characterization of four novel cold-adapted *β*-galactosidases found in psychrophilic *Cryobacterium* sp. LW097 showed high level of enzyme activity at 5 °C and different optimal temperatures that ranged from 25 °C to 40 °C. Moreover, the enzyme kinetics of these novel enzymes having lower K_m to both ONPG and lactose at 5 °C provided evidence for their adaptation to low temperature (Wang et al. 2020). Psychrophilic enzymes are often thermo-sensitive. However, an interesting example of cold-active enzyme, superoxide dismutase isolated from a psychrophilic microbe, *Deschampsia antarctica*, had an unexpected high thermostability (Rojas-Contreras et al. 2015). Its optimal temperature was 20 °C, but its activity was not affected at 80 °C, and its half-life time was 35 min at 100 °C. Similarly, high activity of cold-active enzymes isolated from mesophilic organisms is unexpected. Interestingly, lipase enzyme from *Candida albicans* and *Staphylococcus epidermidis* had an optimal temperature of 15 °C and 25 °C, respectively (Lan et al. 2011; Kamarudin et al. 2014). More challenging task is to discover a thermophilic enzyme with high activity at low temperatures. But surprisingly, *β*-galactosidase isolated from *Pyrococcus furiosus* with optimal activity at 90 °C had retained 8% of its activity at 0 °C (Dong et al. 2014). Despite the decrease in activity than its optimal temperature, the lactase activity at 0 °C was 40% of its optimal activity. In an ingenious work on cold, adaptation of a mesophilic subtilisin-like protease by laboratory evolution k_{cat}/k_m increased 9.6 times with concomitant decrease of 3.3 times in half-life at 70 °C (Wintrode et al. 2000).

Enzymes run the metabolism inside a cell. Hence, a major adaptation to cold temperature working is by enzymes. Arrhenius best described the role of temperature in chemical reaction in his equation $k = Ae^{-Ea/RT}$, where A is the pre-exponential factor of reaction, k is the rate constant, R is the gas constant (8.31 kJ/mol), Ea is the activation energy, and T is the temperature in kelvins. It indicates that with the decrease in temperature, the decrease in the reaction rate will be exponential where the activation energy will provide the extent of the reaction. This could be the main drawback for lives at cold if not adapted. The basic difference between mesophilic and psychrophilic enzymes lies in the fact that the latter exhibit enzymes with higher specific activity at lower temperatures and the maximal activity for their enzymes are shifted toward low temperatures with lessened thermostability. Moreover, the displayed specific activity at lower temperatures though high is lower than their mesophilic counterpart, making it less efficient (D'Amico et al. 2002). In a study of structure and kinetics of a cold-adapted hetero-octameric ATP phosphoribosyltransferase, it was found that the holoenzyme complementation with HisZ provided only twofold to threefold better K_{cat} than catalytic subunit $HisG_S$ (Stroek et al. 2017). In a study with cold-adapted Antarctic polyextremophilic β-galactosidase, it was found that six key amino acid replacements are responsible for higher K_{cat}/k_m of the enzyme at low temperature (Laye et al. 2017). Characterization of four novel cold-adapted β-galactosidases found in the psychrophilic *Cryobacterium* sp. LW097 showed lower K_m to both ONPG and lactose at 5 °C, as their adaptation to low temperature (Wang et al. 2020). Psychrophilic enzymes are often thermo-sensitive. In an ingenious work, adaptation of a mesophilic subtilisin-like protease by laboratory evolution increased k_{cat}/k_m by 9.6 times with concomitant decrease of 3.3 times in half-life at 70 °C (Wintrode et al. 2000).

9.2.3 Membrane Fluidity

Cell membrane is the first line of defense against any changes coming from external environment. Extreme low temperature causes cell damage due to the transition of cell membrane lipids from liquefied crystalline to gel phase, resulting in disruption of membrane function (Crowe et al. 1987). Maintenance of membrane fluidity is a vital mechanism for any living beings to adapt to lower temperatures. With the drop in temperature, microbes undergo a number of changes in their fatty acid profile of lipid bilayer to maintain an optimum fluidity (Chattopadhyay 2006; Siddiqui et al. 2013). The cold-adapted microbes increase their membrane fluidity through synthesis of unsaturated fatty acids, methyl-branched fatty acids, anteiso-fatty acids to long-chain fatty acids, iso-fatty acids, cis-fatty acids, and shorter acyl-chain fatty acids at higher proportion (Chintalapati et al. 2004; Ji and Wei 2020). These fatty acids contribute in the disruption of packing order as well as reduction in packing density of phospholipid bilayer, leading to transition of liquid phase to gel phase to maintain the functional bilayer even at low temperatures (Collins and Margesin 2019). The long-chain polyunsaturated fatty acids (LC-PUFAs), mostly present in marine organisms, are frequently produced at lower temperatures at an increased concentration (Feng et al. 2014; Yoshida et al. 2016). They mainly help in quenching reactive oxygen species (ROS), acting as antioxidative agents, instead of agents in cold adaptation. LC-PUFAs prevent the membranes by forming more hydrophobic interfaces between lipid bilayers, thereby restricting the entry of ROS into the cells and acting as shields. Hence, its presence at low temperatures may not be a direct indication of cold response, per se, but a reaction to other stresses which are inherent at lower temperatures such as oxidative stress (Collins and Margesin 2019).

Besides, upregulation of various genes involved in the synthesis as well as desaturation of fatty acids (desaturases), production of branched fatty acids, and cis-isomerization (fatty acid cis-trans-isomerase) have been reported (De Maayer et al. 2014; Goordial et al. 2016). Cis-trans-isomerase, which converts cis-fatty acids into trans-fatty acids, is believed to facilitate the survival of bacteria at higher temperature, since the trans-fatty acids are well known to contribute for decrease in membrane fluidity (Chattopadhyay 2006). In addition, an increased number of genes involved in the degradation of membrane-rigidifying molecules have also been observed in the genome which may help in the reduction of membrane rigidity at low temperatures (Medigue et al. 2005; Collins and Margesin 2019). Few psychrophiles have reported an upregulation of membrane transport proteins which is responsible for counteracting the reduced diffusion rates and transport, inherently associated with low temperature (De Maayer et al. 2014). *Pseudomonas putida* reported an upregulation of genes associated with cold-shock proteins (CSPs) when temperature was decreased from 30 to 10 °C within a span of 2 h. CSPs are basically responsible for regulating various kinds of cellular processes like transcription, translation, protein-folding, and membrane fluidity (D'Amico et al. 2006; Casanueva et al. 2010).

Membrane fluidity is also modulated by pigments, particularly carotenoids (Collins and Margesin 2019). Pigmentation is commonly produced by psychrophiles, reported from glacier and ice cores to marine surface water (Shen et al. 2018; Dieser et al. 2010). A recent study recognized diminished pigmentation at low temperatures in Arctic bacteria (Singh et al. 2017). Interestingly, it is believed that polar carotenoids augment membrane rigidity which is beneficial to cold-adapted microorganisms at low temperatures (Collins and Margesin 2019). But it has been suggested that these pigments may offset the fluidity effects of unsaturated fatty acids and stabilizes the membrane (Jagannadham et al. 2000). Besides, these pigments also a play a number of other vital roles such as light harvesters, antioxidants, photoprotection, and antimicrobials (Pandey et al. 2018).

9.2.4 Metabolism at Low Temperatures

Psychrophiles growth and development have been better demonstrated through the metabolic activity taking place in microbes. Due to cold oxidative stress, ROS level increases such as, an upregulation of catalase gene depicting enhancement of ROS was observed in *Rhodotorula* sp. (USM-PSY62) (Chai et al. 2020). This leads to downregulation of primary metabolism such as glycolysis, TCA cycle, and electron transport chain (Huo et al. 2020). Alternative secondary metabolic pathway or metabolism of reserve compound was hence induced. Studying the major source of energy (glucose metabolism), 13C labeling showed bifurcation between glycolysis (catabolic, at -5 °C) and pentose phosphate (anabolic, at +5 °C and -20 °C) pathway (Bore et al. 2017). Hence, it was noted that antifreeze protein secretion from exopolysaccharide (EPS) increases via pentose phosphate pathway such as in bacterium *Colwellia psychrerythraea* 34H (Casillo et al. 2017). This helps the cells to become cryotolerant forming membranal plasticity. Low temperature results in lower production of energy than their utilization. Hence, energy was conserved by decreasing EPS synthesis and increasing antifreeze protein secretion. Also, *Psychrobacter arcticus* 273-4 grown at -10 °C shows lack in glycolysis genes but instead possesses alternative pathway of gluconeogenic genes resembling utilization of other carbon sources instead of sugar (Ayala-Del-Río et al. 2010). The lactose was broken down during cold adaptation in *Rahnella aquatilis* BS1 and *Buttiauxella* HS39 (Park et al. 2006). The transcriptomics and proteomics approaches in *Psychrobacter* sp. PAMC 21119 isolated from Antarctic soil recorded an upregulation in acetyl Co-A metabolism while downregulation in protein synthesis responsible for energy production was also seen (Koh et al. 2016). Similar transcriptomic finding was reported in *Pseudomonas extremaustralis* in which genes responsible for secondary metabolism such as ethanol oxidation (exaA, exaB, and exaC) were upregulated while primary metabolism genes were found to be depressed (Tribelli et al. 2015). *Flavobacterium* sp. (at 15 °C) and *Arthrobacter* sp. (at 5 °C and 10 °C) showed good oxidoreductase enzyme activity for ethanol oxidation (Araújo et al. 2011).

Polyhydroxyalkanoate (PHA) is a biodegradable bioplastic polymer produced in high content by many cold habitat microbes (such as bacteria, fungi, etc.) (Rogala et al. 2020; Mishra et al. 2020; Pärnänen et al. 2015). It functions in cryoprotection and cell motility and acts as dynamic reserves of nutrients, overcoming carbon and nitrogen uptake during low temperature (Tribelli and Lopez 2018). The proteomic analysis found increases in PHA depolymerase in -10 °C influencing polymer utilization under freezing and depressed metabolism condition (Nunn et al. 2015). A major metabolic modification can be studied during cold stress. As heme protein synthesis was downregulated, it was observed that putrescine synthesis was upregulated in *Psychrobacter* sp. PAMC 21119 (Zhu et al. 2015; Koh et al. 2016).

Besides, algZ for alginate production and spermidine synthesis was also upregulated during downregulation of iron synthesis (Tribelli and Lopez 2018; Kapse et al. 2017). Putrescine not only helps to overcome iron metabolism modification but also is responsible for polyamine synthesis (Mackelprang et al. 2011). The metagenomic analysis of microbial community reported that nitrogen fixation increases under permafrost condition despite available biological nitrogen while denitrification increases on short-term thawing (Mackelprang et al. 2011). Although different nitrogenases were responsible for nitrogen fixation, the genomic approach in a bacterium, *Rhodoferax antarcticus*, showed the presence of Mo-nitrogenase in abundance and no other nitrogenase (Baker et al. 2017). narGHJI, nirK, and qnor genes was responsible for denitrification, and sulfur oxidation (soxBXAZY) and sulfide dehydrogenases (FSD and soxF) genes were also detected in the same species. *R. fermentans* and *R. antarcticus* use acetate, lactate, succinate, and pyruvate as carbon sources (Jin et al. 2020). In anammox consortia, low temperature results in increased protein degradation and accumulation of amino acid (Huo et al. 2020; Lin et al. 2018). The increased degradation of misfolded protein is more energy-conserving than de novo synthesis of amino acid (Lawson et al. 2017). The metabolic activity of anammox consortia revealed the sequestration of glycine betaine, a cryoprotectant. The upregulation of glycine betaine by dehydrogenation of choline results in in vivo active protein-folding in chaperone-like manner during low temperature (Huo et al. 2020). The yeast *Candida antarctica* was found to produce Lipase B under cold adaptation which acts as biocatalyst. This also results in depolymerization of polyethylene terephthalate (PET) (Carniel et al. 2017). Also, enzyme such as protease was produced in high amount in yeast *Rhodotorula mucilaginosa* CBMAI 1528 under low and mild temperature (Lario et al. 2020). In view of these two literature, it was postulated that low temperature enhances the production of enzymes in certain microbes. In cold adaptation, trypsin was the major metabolic material, while xylanases (glycoside hydrolase) have catalytic property (Fornbacke and Clarsund 2013; Zheng et al. 2016). The psychrophilic microbes produce handful of enzymes, but due to slow rate of growth, there was a low expression of target protein.

To resist low temperature, psychrophilic microbes also produce nanostructure called outer membrane vesicles (OMV). OMVs formed by bacteria *Shewanella* sp. HM13 are reported for proteolytic activity which helps bacteria degrade proteins and survive in harsh condition (Casillo et al. 2019). In *Nesterenkonia* sp. AN1, enzymes responsible for glyoxylate cycle were induced at 5 °C to form intermediate carbon compound (Habibu et al. 2016). During hypoxia condition at low temperatures, *G. antarctica* PI12 shows upregulation of nitrogenase reductase gene generating more nitrite as a terminal acceptor instead of oxygen (Gupta and Igamberdiev 2016). Desaturase and glycerol 3-phosphate dehydrogenase were upregulated in psychrophilic fungus *Mrakia psychrophila* to produce unsaturated fatty acid for cell membrane and glycerol as a compatible solute (Su et al. 2016; Yao et al. 2016).

9.2.5 Heat-Shock Proteins

Heat-shock protein acting as a chaperone regulates the folding and unfolding of proteins. Heat-shock protein (Hsp) could degrade the aggregated stress-damaged proteins and can be a useful indicator of stress (Riback et al. 2017). Gene expression in chaperones like SGT1 was increased in psychrophilic yeast *Glaciozyma antarctica* providing them the properties of heat shock proteins to deteriorate aggregated proteins (Yusof et al. 2016). GaSGT1 was also upregulated at 0 °C and prevents protein aggregation. Cold stress and protein degradation results in the accumulation of protein. These proteins bind to heat-shock protein such as GroEL-depleting Hsp. Hence, during low level, Hsp was synthesized by modulating and dissociating σ 32 and htpR gene regulator of Hsp gene expression (Roncarati and Scarlato 2017).

Hsp70 is a 70 kDa, ubiquitously found conserved protein that protects cells machinery regulating protein-folding. The mutation in DnaK, DnaJ, and grpE gene of *E. coli* results in expression of Hsp at low temperature. There are eight Hsp70 (DnaK) genes found in psychrophilic alga *Chlamydomonas* sp. ICE-L homologous to prokaryotic DnaK. CidnaK was highly transduced than other Cihsp 70 in freezing condition and prevents unfolded protein denaturation through better chaperone activity (Liu et al. 2018). Although in normal condition DnaK acts in protein-folding, during stress condition, they bind with hydrophobic surface of unfolded protein stabilizing it. CidnaK gene from sea ice symbiotic bacterium proves to be a low temperature-tolerant gene for *Chlamydomonas* sp. (Liu et al. 2018). Heat-shock proteins (Hsp70 and Hsp90) were reported to be overexpressed in psychrophilic bacterium *G. antarctica* PI12 and help in proper folding of misfolded protein (Wong et al. 2019). Heat-shock protein (DNAJ) was also upregulated in *Mrakia psychrophila* at psychrophilic condition (Su et al. 2016; Yao et al. 2016). A study reported that Hsp70 (DnaK) and Hsp90 (mitochondrial chaperones) were upregulated in *Rhodotorula mucilaginosa* AN5 at 10 °C preventing ROS-mediated unfolding (Kan et al. 2018).

Small heat-shock protein Hsp26 is an oligomeric complex which forms primary defense during extreme stress and aging condition. The yeast Hsp26 is downregulated during optimal condition, while it is upregulated in temperature stress and called temperature-regulated molecular chaperone (Lytras et al. 2017). At low temperature, carbon starvation revealed the activation of Hsf1p which induces Hsp26 for utilizing protein aggregates for energy generation (Amoros and Estruch 2001). Shotgun metagenomic approach found that Hsp60, Hsp70, Hsp90, and Hsp100 were recorded in abundance among bacteria and archaea in low temperature (0–9 °C) (Centurion et al. 2020).

It was also found that heat-shock factor protein 1 (HSF1) was downregulated in *Rhodotorula* sp. (USM-PSY62) at 0 °C (Chai et al. 2020). Another small heat-shock protein Hsp12 is a lipid-binding plasma membrane protein. During cold stress, glucose was utilized as energy, and depletion in its level results in overexpression of Hsp12 gene. Hence, Hsp12 protein was upregulated in freezing stress in *S. cerevisiae* at 4 °C (Qiu et al. 2019). Hsp12 was overexpressed in yeast, and because it was interchangeable with trehalose (sugar molecules), it maintains cell architecture (Tiwari et al. 2015). Hsp12 binds with lipids in the membrane instead of trehalose during cold (−20 °C) and maintains cellular integrity against desiccation (Pachec et al. 2009). Besides, Hsp12 also plays a major role in proteostasis by modulating and stabilizing protein aggregation activity during cold adaptation (Kim et al. 2018). Like Hsp12, Zmo0094 from anaerobic bacteria *Zymomonas mobilis* was also hydrophilic class of protein which helps in cold stress tolerance when expressed in *E. coli* (Yang et al. 2020).

Many heat-shock proteins, under cold adaptation, play a vital role in cellular homeostasis. It either binds with the denatured protein in folded state or accumulates the protein in reversible state. The recombinant yeast with Hsp17.7 expression shows 95% survival at 2 °C. Being molecular chaperone, it functions in protein solubility, substrate binding, and formation of reversible protein aggregates under stress (Ko et al. 2017). Hsp1 acts as cryotolerance to the cell, controlling and maintaining cell membranal structure under cold stress (Arena et al. 2019). Microorganisms from New Zealand and South Africa were also found to induce Hsp synthesis at 8–12 °C.

Although being chaperones and binding to reversible unfolded proteins, some Hsp like Hsp60 (GroEL) and Hsp10 (GroES) are chaperonins too. Chaperonins are oligomeric ring with active sites at the center. The genomic sequencing analysis among ten strains of *Exiguobacterium* reported the presence of GroES and GroEL families' protein which provides its temperature stability at psychrophilic condition (Kasana and Pandey 2017). Many psychrophiles like *Pseudoalteromonas haloplanktis* and strains of *Psychrobacter* recorded to have chaperonins GroES and GroEL as cold-adapted core genes (Bakermans 2018).

9.2.6 Cold-Shock Proteins

The determination of microbial metabolic processes such as gene expression, energy production, and biochemical synthesis under low temperature is called cold shock. Cold-shock proteins are enzymes and proteins responsible for adaptation of microbes at low temperature. The expression of *Colwellia psychrerythraea* Csps increases at −1 °C to 10 °C (Lee et al. 2018). These are chaperones that destabilize unwanted protein secondary structure under cold stress. Psychrophilic Csps are more heat-labile as compared to other Csps due to lesser ionic interaction which decreases production of secondary structures and increases molecular adaptation. According to a study, the secondary structures of RNA were destabilized by *CspA*, and RNA degradation was prevented with the binding of *CspE* (Koh et al. 2016; Barria et al. 2013).

A psychrophilic yeast, *Glaciozyma antarctica*, living at temperatures ranging from −2.2 °C to 4 °C can tolerate temperature up to 20 °C (Firdaus-Raih et al. 2018). It possesses cold-shock domain containing *Ga16676* gene. The overexpression of *Ga16676* gene is responsible for *E. coli* multiplication at low temperature (Charles et al. 2020). Among various classes of Csps, major Csp (*CspA*) level was upregulated at 15 °C in cold-adapted psychrotrophic *Pseudomonas* sp. (Awasti et al. 2019). The firmicute *Planococcus halocryophilus* which grows at −15 °C also contains *CspA* gene (White et al. 2019). A study also revealed that to adapt low temperature (at 0 °C), four out of six Csp genes were overexpressed in *Exiguobacterium antarcticum* (Dall'Agnol et al. 2014). Some of the cold sensor gene also play a role in upregulation of metabolic pathway of cold-shock protein. The cold sensor response gene *hpk4* regulates Csp in *Leuconostoc gelidum*, a lactic acid bacterium (Duru et al. 2021).

Cold-shock proteins identified so far include RNA helicase, peptidylprolyl isomerase (PPIase), subunit of DNA gyrase, and different transcription factors (like NusA). Among these, RNA helicase and peptidylprolyl isomerase (PPIase) were constitutively expressed in *Glaciozyma antarctica* PI2 under cold stress (Wong et al. 2019). Likewise, *Shewanella* sp. SIB1, growing at the range of 0–20 °C, contain PPIase which is responsible to prevent insulin aggregation (Budiman et al. 2021). Also, *Acidithiobacillus ferrivorans* possesses five genes of PPIase, and *gyrB* genes were upregulated in *Pseudomonas helmanticensis* at low temperature (Ccorahua-Santo et al. 2017; Kumar et al. 2020). Trehalose operon was also regulated in *A. ferrivorans* for iron oxidation. Psychrotolerant yeast, *Naganishia albida*, shows an increase in DNA methylation during cold stress at 4 °C (Turchetti et al. 2020). Cold-inducible protein like proline symporter gene (Psyc_1415) was reported in *Pseudomonas articus* as well as Csp D, and Cold acclimation protein B (second class of Csps) was found to be common in *P. psychrophila* strains (Abraham et al. 2020).

9.2.7 Cryoprotectants

Cryoprotectants are exopolymeric substances which are mainly produced by cold-inhabiting poikilothermic organisms to prevent cellular freezing and maintain the membrane fluidity under very cold environmental temperature (Hamdan 2018). Cryoprotectants include amino acids (proline, alanine), sugars (fructose, glucose), and sugar alcohols (glycerol, mannitol) (Chattopadhyay 2006). These cryoprotective substances are accumulated inside the cell of the microorganisms, lowering their cytoplasmic freezing point, and providing protection against hyper-osmolality and desiccation (Collins and Margesin 2019). It is hypothesized that cold-resistant organisms have low thermal hysteresis values but have high inhibition activity of ice recrystallization (Lorv et al. 2014). Different psychrophilic microorganisms produce different cryoprotectants depending on their nature of habitat and their metabolic properties. *Listeria monocytogenes* was the first food-borne bacterial pathogen in which the role of glycine betaine was shown as a cryoprotectant as well as growth enhancer. In this experiment, the growth of *L. monocytogenes* was only observed in betaine added media when incubated at 7 °C for 32 days (Ko et al. 1994). The role of betaine as an excellent cold stress molecule was also supported by Bashir et al. (2014) when they performed their experiment on *Bacillus subtilis*, giving dimethylglycine (intermediate product of glycine betaine) as a exogenous source to cope up with stressful cold temperature. Trehalose, which is an important naturally occurring sugar, consisting of two molecules of glucose having alpha-configuration at the anomeric carbon, have been synthesized by many species of bacteria, fungi, plant, and invertebrate animal for diverse purposes such as source of energy, scavenger for free radicals, and prevention of denaturation as well as aggregation of proteins during cold stress (Elbein et al. 2003). Robinson (2001) reported that for surviving in stressful low temperature, fungi accumulate trehalose in their hyphae and reproductive bodies. High rate of trehalose production has been observed in psychrotrophic ectomycorrhizal fungi such as *Hebeloma* spp. of temperate and arctic region when they were subjected to cold environmental condition (Tibbett et al. 1998). As per the report of Weinstein et al. (2000), at low temperature of maritime Antarctica, in *Humicola marvinii*, trehalose concentration increased inside the cell, and glycerol concentration increased outside the cell, while in another fungi *Mortierella elongate*, enhanced trehalose concentration was observed only inside the cell. An experiment, conducted on thermophilic fungus *Myceliophthora thermophila*, showed an increasing content of mannitol and trehalose and decreasing content of inositol when the fungi exposed to below its optimal growth temperature which indicates the combating role of mannitol and trehalose at cold climatic condition in fungal strains (Feofilova et al. 1994). The upregulation of trehalose biosynthetic genes (such as otsA and otsB) under low temperature was revealed through transcriptome analysis of *Escherichia coli* (Phadtare and Inouye 2004). Exopolysaccharides also act as a cryo-protectant in psychrophilic microorganisms. Exopolysaccharides are secreted outside the bacterial cell surface which helps in adapting harsh cold climatic conditions by modifying physiochemical state of cell, nutrient and water sequestration, cell surface adhesion and protection of extracellular enzymes (D'Amico et al. 2006). Ali et al. (2019) reported that the psychrotrophic bacterium *Pseudomonas* sp. BGI-2 contains 11 genes responsible for exopolysaccharide production.

9.2.8 Antifreeze Proteins

Antifreeze proteins (AFPs) are proteins that inhibit ice growth and recrystallization by binding to ice crystals. This non-colligative property leads to structurally diverse polypeptides and is also known as ice-binding proteins (IBPs) (Bialkowska et al. 2020). AFPs are mainly synthesized by psychrophilic organisms for surviving their life at subzero temperature (De Maayer et al. 2014). AFPs are hypothesized to function by adsorption inhibition mechanism depending on two activities such as thermal hysteresis and ice recrystallization inhibition. In thermal hysteresis, AFPs bind to ice and lower the freezing temperature of extracellular fluids through its non-colligative property, whereas in inhibition of ice recrystallization APFs adsorb ice crystals and change the ice crystal pattern from hexagonal to pyramid, thereby impeding large recombination of ice and generating small ice crystals (Lorv et al. 2014; Hassan et al. 2016). A study reports that bacteria *Rhodococcus erythropolis* and *Micrococcus cryophilus* show the presence of thermal hysteresis proteins (Duman and Olsen 1993) and in Antarctic region AFP (identified as 52 kDa lipoprotein) had been first time found in the bacterium *Moraxella* sp. (Yamashita et al.

2002). Muryoi et al. (2004) were able to isolate AFP gene denoted as afpA from plant growth-promoting rhizobacterium *Pseudomonas putida* GR12–2 and reported the gene had both the activities like antifreeze and ice nucleation. Xu et al. (1998) also supported the presence of afpA gene in the genome of arctic *P. putida* GR12–2 and characterized afpA as 164 kDa in size with lipid and sugar moieties. Singh et al. (2014) reported that AFPs isolated from *P. putida* from the Arctic region belongs to the AFP family IBP-1 which have an important physiological role for surviving the harsh cold condition. Due to freeze tolerance nature of psychrophilic microorganisms especially in bacteria, AFPs have lower thermal hysteresis value as compared to freeze-avoiding organisms (Chattopadhyay 2006; D'Amico et al. 2006). But some bacteria like *Marinomonas primoryensis* possess freeze avoidance strategy due to the presence of powerful AFP (Gilbert et al. 2005). Here in this bacterium, AFP showed a high value of thermal hysteresis and Ca^{2+}-dependent activity which is unique among bacterial AFPs. In the case of bacteria, thermal hysteresis value ranges from 0.1 to 0.35 °C (Duman and Olsen 1993), and in fungi, it ranges from 0.3 to 0.35 °C (Snider et al. 2000). AFPs have been investigated in snow molds also. Hoshino et al. (2003) reported that the fungi *Coprinus psychromorbidus* belonging to class *Basidiomycetes* produced three kinds of thermal hysteresis proteins in the extracellular space, having comparatively different N-terminal amino acid sequences and approximately 23 kDa molecular mass. As per the report of Hoshino (2005) among the different fungal strain isolated from Antarctica, seven strains were found to produce such substance that can alter the ice crystal nature, but the substance were not identified as AFP. However, in Antarctic ascomycetes, novel and purified form of AFP was recognized in *Antarctomyces psychrotrophicus* (Xiao et al. 2010). The novelty relies on that the AFP from *A. psychrotrophicus* showed high thermal hysteresis activity (ranging from 0.42 to 0.48 °C) similar to fish AFP and created bipyramidal pattern of ice crystal with rough facet.

9.3 Future Prospects

This chapter deals with the microorganisms living at cold habitats and the adaptations they share to cope with the challenging conditions present. Genetic and "omic" approaches have contributed significantly in improving our understanding of the molecular mechanisms taking place at such low-temperature regimes. Transcriptomics and metatranscriptomics have developed a connection between microbial activity and its associated role in polar habitats with its impact on the nature and on larger global processes including biogeochemical cycles and global warming. With an advancement in the technologies, novel methods and metabolic strategies will be unraveled, leading to a more comprehensive view of the overall processes being undertaken on a global scale. Further novel metabolites can be isolated and identified having a greater potential in various pharmaceutical and biotechnological fields. With the growing anthropological interventions in the polar regions, bioremediation of pollutants comprises an arduous task for which bioprospection of effective organisms is a major challenge to the scientists all across the globe. In all, our understanding of such pristine habitats is just a tip of the iceberg, and a lot needs to be substantiated.

References

Abraham WP, Raghunandanan S, Gopinath V, Suryaletha K, Thomas S (2020) Deciphering the cold adaptive mechanisms in *Pseudomonas psychrophila* MTCC12324 isolated from the Arctic at 79° N. Curr Microbiol 77:2345–2355. https://doi.org/10.1007/s00284-020-02006-2

Albanesi D, Mansilla MC, de Mendoza D (2004) The membrane fluidity sensor DesK of *Bacillus subtilis* controls the signal decay of its cognate response regulator. J Bacteriol 186(9):2655–2663. https://doi.org/10.1128/JB.186.9.2655-2663.2004

Ali P, Shah AA, Hasan F, Cai H, Sosa A, Chen F (2019) Draft genome sequence of a cold-adapted *Pseudomonas* sp. strain, BGI-2, isolated from the ice of Batura Glacier, Pakistan. Microbiol Resour Announc 8:e00320–e00319. https://doi.org/10.1128/MRA.00320-19

Amoros M, Estruch F (2001) Hsf1p and Msn2/4p cooperate in the expression of *Saccharomyces cerevisiae* genes HSP26 and HSP104 in a gene- and stress type-dependent manner. Mol Microbiol 39(6):1523–1532. https://doi.org/10.1046/j.1365-2958.2001.02339.x

Araújo LS, Kagohara E, Garcia TP, Pellizari VH, Andrade LH (2011) Screening of microorganisms producing cold-active oxidoreductases to be applied in enantioselective alcohol oxidation. An Antarctic survey. Mar Drugs 9(5):889–905. https://doi.org/10.3390/md9050889

Arena MP, Capozzi V, Longo A, Russo P, Weidmann S, Rieu GJ, Spano G, Fiocco D (2019) The phenotypic analysis of *Lactobacillus plantarum* shsp mutants reveals a potential role for hsp1 in cryotolerance. Front Microbiol 10:5. https://doi.org/10.3389/fmicb.2019.00838

Awasti S, Anjney S, Saxena P, Yadav J, Pandiyan K, Kumar M, Singh A, Chakdar H, Bhowmik A, Kashyap PL, Srivastava A, Saxena AK (2019) Molecular detection and in silico characterization of cold shock protein coding gene (cspA) from cold adaptive *Pseudomonas koreensis*. J Plant Biochem Biotechnol 28(4):405–413. https://doi.org/10.1007/s13562-019-00500-8

Ayala-Del-Río H, Chain PS, Grzymski JJ, Ponder MA, Ivanova N, Bergholz PW, Di Bartolo G, Hauser L, Land M, Bakermans C, Rodrigues D, Klappenbach J, Zarka D, Larimer F, Richardson P, Murray A, Thomashow M, Tiedje JM (2010) The genome sequence of *Psychrobacter arcticus* 273-4, a psychroactive Siberian permafrost bacterium, reveals mechanisms for adaptation to low-temperature growth. Appl Environ Microbiol 76(7):2304–2312. https://doi.org/10.1128/aem.02101-09

Baker JM, Riester CJ, Skinner BM, Newell AW, Swingley WD, Madigan MT, Jung DO, Asao M, Chen M, Loughlin PC, Pan H, Lin Y, Li Y, Shaw J, Prado M, Sherman C, Tang JK, Blankenship RE, Zhao T, Touchman JW, Sattley WM (2017) Genome sequence of *Rhodoferax antarcticus* ANT.BRT; a psychrophilic purple nonsulfur bacterium from an Antarctic microbial mat microorganisms. Microorganisms 5(1):8. https://doi.org/10.3390/microorganisms5010008

Bakermans C (2018) Adaptations to marine versus terrestrial low temperature environments as revealed by comparative genomic analyses of the genus *Psychrobacter*. FEMS Microbiol Ecol 94:7. https://doi.org/10.1093/femsec/fiy102

Barria C, Malecki M, Arraiano C (2013) Bacterial adaptation to cold. Microbiologica 159:2437–2443. https://doi.org/10.1099/mic.0.052209-0

Bashir A, Hoffmann T, Kempf B, Xie X, Smits SH, Bremer E (2014) Plant-derived compatible solutes proline betaine and betonicine confer enhanced osmotic and temperature stress tolerance to *Bacillus subtilis*. Microbiologica 160:2283–2294. https://doi.org/10.1099/mic.0.079665-0

Bialkowska A, Majewska E, Olczak A, Twarda-Clapa A (2020) Ice binding proteins: diverse biological roles and applications in different types of industry. Biomolecules 10:1–26. https://doi.org/10.3390/biom10020274

Bore EK, Apostel C, Halicki S, Kuzyakov Y, Dippold MA (2017) Microbial metabolism in soil at subzero temperatures: adaptation mechanisms revealed by position-specific 13C labeling. Front Microbiol 8:946. https://doi.org/10.3389/fmicb.2017.00946

Budiman C, Goh CKW, Arief II, Yusuf M (2021) FKBP22 from the psychrophilic bacterium *Shewanella* sp. SIB1 selectively binds to the reduced state of insulin to prevent its aggregation. Cell Stress Chaperones 26:377–386. https://doi.org/10.1007/s12192-020-01183-0

Carniel A, Valoni É, Nicomedes J, Gomes AC, Castro AM (2017) Lipase from *Candida antarctica* (CALB) and cutinase from *Humicola insolens* act synergistically for PET hydrolysis to terephthalic acid. Process Biochem 59:84–90. https://doi.org/10.1016/j.procbio.2016.07.023

Casanueva A, Tuffin M, Cary C, Cowan DA (2010) Molecular adaptations to psychrophily: the impact of 'omic' technologies. Trends Microbiol 18:374–381

Casillo A, Di GR, Carillo S, Chen C, Kamasaka K, Kawamoto J, Corsaro MM (2019) Structural elucidation of a novel lipooligosaccharide from the Antarctic bacterium OMVs producer *Shewanella* sp. HM13. Mar Drugs 17(1):34. https://doi.org/10.3390/md17010034

Casillo A, Parrilli E, Sannino F, Mitchell DE, Gibson MI, Marino G, Lanzetta R, Parrilli M, Cosconati S, Novellino E, Randazzo A, Tutino ML, Corsaro MM (2017) Structure-activity relationship of the exopolysaccharide from a psychrophilic bacterium: a strategy for cryoprotection. Carbohydr Polym 156:364–371. https://doi.org/10.1016/j.carbpol.2016.09.037

Cavicchioli R, Charlton T, Ertan H, Omar SM, Siddiqui K, Williams T (2011) Biotechnological uses of enzymes from psychrophiles. J Microbial Biotechnol 4:449–460. https://doi.org/10.1111/j.1751-7915.2011.00258.x

Ccorahua-Santo R, Eca A, Abanto M, Guerra G, Ramírez P (2017) Physiological and comparative genomic analysis of *Acidithiobacillus ferrivorans* PQ33 provides psychrotolerant fitness evidence for oxidation at low temperature. Res Microbiol 168(5):482–492. https://doi.org/10.1016/j.resmic.2017.01.007

Centurion VB, Lacerda-Júnior GV, Duarte AWF, Silva TR, Silva LJ, Rosa LH, Oliveira VM (2020) Dynamics of microbial stress responses driven by abiotic changes along a temporal gradient in Deception Island, maritime Antarctica. Sci Total Environ 758:143671. https://doi.org/10.1016/j.scitotenv.2020.143671

Chai CN, Yam HC, Rosli N, Azlan A, Azzam G, Halim MA, Najimudin N (2020) Transcriptomic response of an Antarctic yeast *Rhodotorula* sp. USM-PSY62 to temperature changes. bioRxiv 080796. https://doi.org/10.1101/2020.05.07.080796

Chandler S (2018) The effect of temperature on cell membranes. https://sciencing.com/effect-temperature-cell-membranes-5516866.html

Charles J, Masnoddin M, Nazaie F, Yusof NA (2020) Structure and function of a novel cold regulated cold shock domain containing protein from an obligate psychrophilic yeast, *Glaciozyma antarctica*. Adv Polar Sci 31(2):137–145. https://doi.org/10.13679/j.advps.2019.0030

Chattopadhyay MK (2006) Mechanism of bacterial adaptation to low temperature. J Biosci 31:157–165. https://doi.org/10.1007/BF02705244

Chintalapati S, Kiran MD, Shivaji S (2004) Role of membrane lipid fatty acids in cold adaptation. Cell Mol Biol (Noisy-le-Grand) 50(5):631–642

Collins T, Margesin R (2019) Psychrophilic lifestyles: mechanisms of adaptation and biotechnological tools. Appl Microbiol Biotechnol 103:2857–2871. https://doi.org/10.1007/s00253-019-09659-5

Crowe JH, Carpenter JF, Wistrom CA (1987) Stabilization of dry phospholipid bilayers and proteins by sugars. Biochem J 242:1–10. https://doi.org/10.1042/bj2420001

D'Amico S, Claverie P, Collins T, Georlette D, Gratia E, Hoyoux A, Meuwis M-A, Feller G, Gerday C (2002) Molecular basis of cold adaptation; Philos. Trans R Soc Lond 357:917–925. https://doi.org/10.1098/rstb.2002.1105

D'Amico S, Collins T, Marx J-C, Feller G, Gerday C (2006) Psychrophilic microorganisms: challenges for life. EMBO Rep 7:385–389. https://doi.org/10.1038/sj.embor.7400662

Dall'Agnol HP, Baraúna RA, de Sá PH, Ramos RT, Nóbrega F, Nunes CI, Graças A, Carneiro AR, Santos DM, Pimenta AM, Carepo MS, Azevedo V, Pellizari VH, Schneider MP, Silva A (2014) Omics profiles used to evaluate the gene expression of *Exiguobacterium antarcticum* B7 during cold adaptation. BMC Genomics 15(1):986. https://doi.org/10.1186/1471-2164-15-986

De Maayer P, Anderson D, Cary C, Cowan DA (2014) Some like it cold: understanding the survival strategies of psychrophiles. EMBO Rep 15:508–517. https://doi.org/10.1002/embr.201338170

Dieser M, Greenwood M, Foreman CM (2010) Carotenoid pigmentation in Antarctic heterotrophic bacteria as a strategy to withstand environmental stresses. Arc Antarc Alp Res 42(4):396–405. https://doi.org/10.1657/1938-4246-42.4.396

Dong Q, Yan X, Zheng M, Yang Z (2014) Characterization of an extremely thermostable but cold-adaptive β-galactosidase from the hyperthermophilic archaeon Pyrococcus furiosus for use as a recombinant aggregation for batch lactose degradation at high temperature. J Biosci Bioeng 117(6):706–710. https://doi.org/10.1016/j.jbiosc.2013.12.002

Dorman CJ (1996) Flexible response: DNA supercoiling, transcription and bacterial adaptation to environmental stress. Trends Microbiol 4:214–216. https://doi.org/10.1016/0966-842X(96)30015-2

Duman JG, Olsen MT (1993) Thermal hysteresis protein activity in bacteria, fungi, and phylogenetically diverse plants. Cryobiology 30:322–328. https://doi.org/10.1006/cryo.1993.1031

Duru IC, Ylinen A, Belanov S, Pulido AA, Paulin F, Auvinen P (2021) Transcriptomic time-series analysis of cold- and heat-shock response in psychrotrophic lactic acid bacteria. BMC Genomics 22:28. https://doi.org/10.1186/s12864-020-07338-8

Elbein AD, Pan YT, Pastuszak I, Carroll D (2003) New insights on trehalose: a multifunctional molecule. Glycobiology 13:17–27. https://doi.org/10.1093/glycob/cwg047

Feller G (2010) Protein stability and enzyme activity at extreme biological temperatures. J Phys Condens Matter 22:323101. https://doi.org/10.1088/0953-8984/22/32/323101

Feng S, Powell SM, Wilson R, Bowman JP (2014) Extensive gene acquisition in the extremely psychrophilic bacterial species *Psychroflexus torquis* and the link to sea-ice ecosystem specialism. Genome Biol Evol 6(1):133–148. https://doi.org/10.1093/gbe/evt209

Feofilova EP, Tereshina VM, Gorova IB (1994) Changes in carbohydrate composition of fungi during adaptation to thermostress. Microbiologica 63:442–445

Firdaus-Raih M, Hashim NHF, Bharudin I et al (2018) The *Glaciozyma antarctica* genome reveals an array of systems that provide sustained responses towards temperature variations in a persistently cold habitat. PLoS One 13(1):e0189947. https://doi.org/10.1371/journal.pone.0189947

Fornbacke M, Clarsund M (2013) Cold-adapted proteases as an emerging class of therapeutics. Inf Dis Ther 2(1):15–26. https://doi.org/10.1007/s40121-013-0002-x

Gilbert JA, Davies PL, Laybourn-Parry J (2005) A hyperactive Ca2+ –dependent antifreeze protein in an Antarctic bacterium. FEMS Microbiol Lett 245:67–72. https://doi.org/10.1016/j.femsle.2005.02.022

Goordial J, Raymond-Bouchard I, Zolotarov Y, de Bethencourt L, Ronholm J, Shapiro N, Woyke T, Stromvik M, Greer CW, Bakermans C, Whyte L (2016) Cold adaptive traits revealed by comparative genomic analysis of the eurypsychrophile Rhodococcus sp. JG3 isolated from high elevation McMurdo Dry Valley permafrost, Antarctica. FEMS Microbiol Ecol 92(2): fiv154. https://doi.org/10.1093/femsec/fiv154

Gupta KJ, Igamberdiev AU (2016) Reactive nitrogen species in mitochondria and their implications in plant energy status and hypoxic stress tolerance. Front Plant Sci 7:369. https://doi.org/10.3389/fpls.2016.00369

Habibu A, Pieter DM, Don C (2016) The genome of the Antarctic polyextremophile *Nesterenkonia* sp. AN1 reveals adaptive strategies for survival under multiple stress conditions. FEMS Microbiol Ecol 92(4):fiw032. https://doi.org/10.1093/femsec/fiw032

Hamdan A (2018) Psychrophiles: ecological significance and potential industrial application. S Afr J Sci 114:1–6. https://doi.org/10.17159/sajs.2018/20170254

Hassan N, Rafiq M, Hayat M, Shah AA, Hasan F (2016) Psychrophilic and psychrotrophic fungi: a comprehensive review. Rev Environ Sci Biotechnol 15:147–172. https://doi.org/10.1007/s11157-016-9395-9

Hoe NP, Goguen JD (1993) Temperature sensing in *Yersinia pestis*: translation of the LcrF activator protein is thermally regulated. J Bacteriol 175(24):7901–7909. https://doi.org/10.1128/jb.175.24.7901-7909.1993

Hoshino T (2005) Ecophysiology of snow mold fungi. Plant Biol 6:27–35

Hoshino T, Kiriaki M, Nakajima T (2003) Novel thermal hysteresis proteins from low temperature basidiomycete, *Coprinus psychromorbidus*. CryoLetters 24:135–142

Huo T, Zhao Y, Tang X, Zhao H, Ni S, Gao Q, Liu S (2020) Metabolic acclimation of anammox consortia to decreased temperature. Environ Int 143:105915. https://doi.org/10.1016/j.envint.2020.105915

Inaba M, Suzuki I, Szalontai B, Kanesaki Y, Los DA, Hayashi H, Murata N (2003) Gene-engineered rigidification of membrane lipids enhances the cold inducibility of gene expression in *Synechocystis*. J Biol Chem 278(14):12191–12198. https://doi.org/10.1074/jbc.M212204200

Jagannadham MV, Chattopadhyay MK, Subbalakshmi C, Vairamani M, Narayanan K, Rao CM, Shivaji S (2000) Carotenoids of an Antarctic psychrotolerant bacterium, *Sphingobacterium antarcticus*, and a mesophilic bacterium, *Sphingobacterium multivorum*. Arch Microbiol 173(5–6):418–424. https://doi.org/10.1007/s002030000163

Ji X, Wei Y (2020) Mechanism of microbial adaptation and survival within psychrophilic habitat. In: Singh RP, Manchanda G, Maurya IK, Wei Y (eds) Microbial versatility in varied environments. Springer, Singapore, pp 67–84. https://doi.org/10.1007/978-981-15-3028-9_5

Jin CZ, Zhuo Y, Wo X, Ko SR, Li T, Jin FJ, Ahn CY, Oh HM, Lee HG, Jin L (2020) Genomic and metabolic insights into denitrification, sulfur oxidation, and multidrug efflux pump mechanisms in the bacterium *Rhodoferax sediminis* sp. nov. Microorganisms 8(2):262. https://doi.org/10.3390/microorganisms8020262

Kamarudin NHA, Rahman RNZRA, Ali MSM, Leow TC, Basri M, Salleh AB (2014) A new cold-adapted, organic solvent stable lipase from mesophilic *Staphylococcus epidermidis* AT2. Protein J 33(3):296–307. https://doi.org/10.1007/s10930-014-9560-3

Kan G, Wang X, Jiang J, Zhang C, Chi M, Ju Y, Shi C (2018) Copper stress response in yeast *Rhodotorula mucilaginosa* AN5 isolated from sea ice, Antarctic. Microbiol Open 3:e00657. https://doi.org/10.1002/mbo3.657

Kapse N, Singh P, Roy U, Singh SM, Dhakephalkar PK (2017) Insights into the psychrophilic and sea ice-specific lifestyle of *Marinobacter* sp. strain AC-23: a genomic approach. Genome Announc 5(15):e00134. https://doi.org/10.1128/genomea.00134-17

Kasana RC, Pandey CB (2017) *Exiguobacterium*: an overview of a versatile genus with potential in industry and agriculture. Crit Rev Biotechnol 38(1):141–156. https://doi.org/10.1080/07388551.2017.1312273

Kavitha M (2016) Cold active lipases – an update. Front Life Sci 9(3):226–238

Kim SX, Çamdere G, Hu X, Koshland D, Tapia H (2018) Synergy between the small intrinsically disordered protein Hsp12 and trehalose sustain viability after severe desiccation. Elife 7:e38337. https://doi.org/10.7554/eLife.38337

Ko E, Kim M, Park Y, Ahn YJ (2017) Heterologous expression of the carrot hsp17.7 gene increased growth, cell viability, and protein solubility in transformed yeast (*Saccharomyces cerevisiae*) under heat, cold, acid, and osmotic stress conditions. Curr Microbiol 74(8):952–960. https://doi.org/10.1007/s00284-017-1269-z

Ko R, Smith LT, Smith GM (1994) Glycine betaine confers enhanced osmotolerance and cryotolerance on *Listeria monocytogenes*. J Bacteriol 176:426–431. https://doi.org/10.1128/jb.176.2.426-431.1994

Koh HY, Park H, Lee JH, Han SJ, Sohn YC, Lee SG (2016) Proteomic and transcriptomic investigations on cold-responsive properties of the psychrophilic Antarctic bacterium *Psychrobacter* sp. PAMC 21119 at subzero temperatures. Environ Microbiol 19(2):628–644. https://doi.org/10.1111/1462-2920.13578

Kortmann J, Narberhaus F (2012) Bacterial RNA thermometers: molecular zippers and switches. Nat Rev Microbiol 10(4):255–265. https://doi.org/10.1038/nrmicro2730

Kumar S, Suyal DC, Yadav A, Shouche Y, Goel R (2020) Psychrophilic *Pseudomonas helmanticensis* proteome under simulated cold stress. Cell Stress Chaperones 25:1025–1032. https://doi.org/10.1007/s12192-020-01139-4

Lan DM, Yang N, Wang WK, Shen YF, Yang B, Wang YH (2011) A novel cold-active lipase from *Candida albicans*: cloning, expression and characterization of the recombinant enzyme. Int J Mol Sci 12:3950–3965. https://doi.org/10.3390/ijms12063950

Lario LD, Pillaca-Pullo OS, Sette LD, Converti A, Casati P, Spampinato C, Pessoa A (2020) Optimization of protease production and sequence analysis of the purified enzyme from the cold adapted yeast *Rhodotorula mucilaginosa* CBMAI 1528. Biotechnol Rep 28:e00546. https://doi.org/10.1016/j.btre.2020.e00546

Lawson CE, Wu S, Bhattacharjee AS, Hamilton JJ, McMahon KD, Goel R, Noguera DR (2017) Metabolic network analysis reveals microbial community interactions in anammox granules. Nat Commun 8:1–12. https://doi.org/10.1038/ncomms15416

Laye VJ, Karan R, Kim JM, Pecher WT, DasSarma P, DasSarma S (2017) Key amino acid residues conferring enhanced enzyme activity at cold temperatures in an Antarctic polyextremophilic β-galactosidase. Proc Natl Acad Sci 114(47):12530–12535. https://doi.org/10.1073/pnas.1711542114

Lee Y, Kwak C, Jeong K-W, Durai P, Ryu K-S, Kim E-H, Cheong C, Ahn H-C, Kim HJ, Kim Y (2018) Tyr51: key determinant of the low thermostability of the *Colwellia psychrerythraea* cold-shock protein. Biochem 57(26):3625–3640. https://doi.org/10.1021/acs.biochem.8b00144

Lin X, Wang Y, Ma X, Yan Y, Wu M, Bond PL, Guo J (2018) Evidence of differential adaptation to decreased temperature by anammox bacteria. Environ Microbiol 20(10):3514–3528. https://doi.org/10.1111/1462-2920.14306

Liu C, Zhao X, Wang X (2018) Identification and characterization of the psychrophilic bacterium CidnaK gene in the Antarctic *Chlamydomonas* sp. ICE-L under freezing conditions. J Appl Phycol 30:3519. https://doi.org/10.1007/s10811-018-1492-4

Lorv JS, Rose DR, Glick BR (2014) Bacterial ice crystal controlling proteins. Forensic Sci 2014:1–20. https://doi.org/10.1155/2014/976895

Lytras G, Zacharioudakis I, Tzamarias D (2017) Asymmetric inheritance of the yeast chaperone Hsp26p and its functional consequences. Biochem Bioph Res 491(4):1055–1061. https://doi.org/10.1016/j.bbrc.2017.08.009

Mackelprang R, Waldrop MP, DeAngelis KM, David MM, Chavarria KL, Blazewicz SJ, Rubin EL, Jansson JK (2011) Metagenomic analysis of a permafrost microbial community reveals a rapid response to thaw. Nature 480:368–371. https://doi.org/10.1038/nature10576

Medigue C, Krin E, Pascal G, Barbe V, Bernsel A, Bertin PN, Cheung F, Cruveiller S, D'Amico S, Duilio A, Fang G, Feller G, Ho C, Mangenot S, Marino G, Nilsson J, Parrilli E, Rocha EP, Rouy Z, Sekowska A, Tutino ML, Vallenet D, von Heijne G, Danchin A (2005) Coping with cold: the genome of the versatile marine Antarctica bacterium *Pseudoalteromonas haloplanktis* TAC125. Genome Res 15(10):1325–1335. https://doi.org/10.1101/gr.4126905

Mishra PK, Joshi S, Gangola S, Khati P, Bisht JK, Pattanayak A (2020) Psychrotolerant microbes: characterization, conservation, strain improvements, mass production, and commercialization. In: Goel R, Soni R, Suyal D (eds) Microbiological advancements for higher altitude agro-ecosystems & sustainability. Rhizosphere biology. Springer, Singapore, pp 227–246. https://doi.org/10.1007/978-981-15-1902-4_12

Murphy ER, Roßmanith J, Sieg J, Fris ME, Hussein H, Kouse AB, Gross K, Zeng C, Hines JV, Narberhaus F, Coschigano PW (2020) Regulation of OmpA translation and *Shigella dysenteriae* virulence by an RNA thermometer. Infect Immun 88(3):e00871–e00819. https://doi.org/10.1128/IAI.00871-19

Muryoi N, Sato M, Kaneko S, Kawahara H, Obata H, Yaish MW, Griffith M, Glick BR (2004) Cloning and expression of afpA, a gene encoding an antifreeze protein from the arctic plant growth-promoting rhizobacterium *Pseudomonas putida* GR12-2. J Bacteriol 186:5661–5671. https://doi.org/10.1128/JB.186.17.5661-5671.2004

Nunn BL, Slattery KV, Cameron KA, Timmins-Schiffman E, Junge K (2015) Proteomics of *Colwellia psychrerythraea* at subzero temperatures—a life with limited movement, flexible membranes and vital DNA repair. Environ Microbiol 17:2319–2335. https://doi.org/10.1111/1462-2920.12691

Pachec A, Pereira C, Almeida MJ, Sousa MJ (2009) Small heat-shock protein Hsp12 contributes to yeast tolerance to freezing stress. Microbiologica 155(6):2021–2028. https://doi.org/10.1099/mic.0.025981-0

Pandey N, Jain R, Pandey A, Tamta S (2018) Optimisation and characterisation of the orange pigment produced by a cold adapted strain of *Penicillium* sp. (GBPI_P155) isolated from mountain ecosystem. Mycologia 9(2):81–92. https://doi.org/10.1080/21501203.2017.1423127

Park JW, Oh YS, Lim JY, Roh DH (2006) Isolation and characterization of cold-adapted strains producing beta-galactosidase. J Microbiol 44(4):396–402

Pärnänen K, Karkman A, Virta M, Eronen-Rasimus E, Kaartokallio H (2015) Discovery of bacterial polyhydroxyalkanoate synthase (PhaC)-encoding genes from seasonal Baltic Sea ice and cold estuarine waters. Extremophiles 19(1):197–206. https://doi.org/10.1007/s00792-014-0699-9

Phadtare S, Inouye M (2004) Genome-wide transcriptional analysis of the cold shock response in wild-type and cold-sensitive, quadruple-csp-deletion strains of *Escherichia coli*. J Bacteriol 186: 7007–7014. https://doi.org/10.1128/JB.186.20.7007-7014.2004

Qiu X, Zhang J, Zhou J, Fang Z, Zhu Z, Li J, Du G (2019) Stress tolerance phenotype of industrial yeast: industrial cases, cellular changes, and improvement strategies. Appl Microbiol Biotechnol 103:6449–6462. https://doi.org/10.1007/s00253-019-09993-8

Riback JA, Katanski CD, Kear-Scott JL, Pilipenko EV, Rojek AE, Sosnick TR, Drummond DA (2017) Stress-triggered phase separation is an adaptive, evolutionarily tuned response. Cell 168: 1028–1040. https://doi.org/10.1016/j.cell.2017.02.027

Robinson CH (2001) Cold adaptation in Arctic and Antarctic fungi. New Phytol 151:341–353. https://doi.org/10.1046/j.1469-8137.2001.00177.x

Rogala MM, Gawor J, Gromadka R, Kowalczyk M, Grzesiak J (2020) Biodiversity and habitats of polar region polyhydroxyalkanoic acid-producing bacteria: bioprospection by popular screening methods. Gene 11(8):873. https://doi.org/10.3390/genes11080873

Rojas-Contreras JA, de la Rosa AP, De León-Rodríguez A (2015) Expression and characterization of a recombinant psychrophilic cu/Zn superoxide dismutase from *Deschampsia antarctica* E. Desv.[Poaceae]. Appl Biochem Biotechnol 175(7):3287–3296. https://doi.org/10.1007/s12010-015-1496-3

Roncarati D, Scarlato V (2017) Regulation of heat-shock genes in bacteria: from signal sensing to gene expression output. FEMS Microbiol Rev 41(4):549–574. https://doi.org/10.1093/femsre/fux015

Shen L, Liu Y, Wang N, Jiao N, Xu B, Liu X (2018) Variation with depth of the abundance, diversity and pigmentation of culturable bacteria in a deep ice core from the Yuzhufeng glacier, Tibetan plateau. Extremophiles 22(1):29–38. https://doi.org/10.1007/s00792-017-0973-8

Siddiqui KS, Williams TJ, Wilkins D, Yau S, Allen MA, Brown MV, Lauro FM, Cavicchioli R (2013) Psychrophiles. Annu Rev Earth Planet Sci 41(1):87–115. https://doi.org/10.1146/annurev-earth-040610-133514

Singh A, Krishnan KP, Prabaharan D, Sinha RK (2017) Lipid membrane modulation and pigmentation: a cryoprotection mechanism in Arctic pigmented bacteria. J Basic Microbiol 57(9):770–780. https://doi.org/10.1002/jobm.201700182

Singh P, Hanada Y, Singh SM, Tsuda S (2014) Antifreeze protein activity in Arctic cryoconite bacteria. FEMS Microbiol Lett 351:14–22. https://doi.org/10.1111/1574-6968.12345

Snider CS, Hsiang T, Zhao G, Griffith M (2000) Role of ice nucleation and antifreeze activities in pathogenesis and growth of snow molds. Phytopathology 90:354–361. https://doi.org/10.1094/PHYTO.2000.90.4.354

Stroek R, Ge Y, Talbot PD, Glok MK, Bernaś KE, Thomson CM, Gould ER, Alphey MS, Liu H, Florence GJ, Naismith JH (2017) Kinetics and structure of a cold-adapted hetero-octameric atp phosphoribosyltransferase. Biochem 56(5):793–803. https://doi.org/10.1021/acs.biochem.6b011

Su Y, Jiang X, Wu W, Wang M, Hamid MI, Xiang M, Liu X (2016) Genomic, transcriptomic and proteomic analysis provide insights into the cold adaptation mechanism of the obligate psychrophilic fungus *Mrakia psychrophila*. G3: genes, genomes. Genetics 6(11):3603–3613. https://doi.org/10.1534/g3.116.033308

Suzuki I, Los DA, Kanesaki Y, Mikami K, Murata N (2000) The pathway for perception and transduction of low-temperature signals in *Synechocystis*. EMBO J 19(6):1327–1334. https://doi.org/10.1093/emboj/19.6.1327

Tibbett M, Grantham K, Sanders FE, Cairney JWG (1998) Induction of cold active acid phosphomonoesterase activity at low temperature in psychrotrophic ectomycorrhizal *Hebeloma* spp. Mycol Res 102:1533–1539. https://doi.org/10.1017/S0953756298006753

Tiwari S, Thakur R, Shankar J (2015) Role of heat-shock proteins in cellular function and in the biology of fungi. Biotechnol Res Int 2015:1–11. https://doi.org/10.1155/2015/132635

Tribelli PM, Lopez NI (2018) Reporting key features in cold-adapted bacteria. Life (Basel) 8:1. https://doi.org/10.3390/life8010008

Tribelli PM, Venero EC, Ricardi MM, Gómez-Lozano M, Iustman LJR, Molin S, López NI (2015) Novel essential role of ethanol oxidation genes at low temperature revealed by transcriptome analysis in the Antarctic bacterium *Pseudomonas extremaustralis*. PLoS One 10:e0145353. https://doi.org/10.1371/journal.pone.0145353

Vigh L, Los DA, Horvath I, Murata N (1993) The primary signal in the biological perception of temperature: Pd-catalyzed hydrogenation of membrane lipids stimulated the expression of the desA gene in *Synechocystis* PCC6803. Proc Natl Acad Sci 90(19):9090–9094. https://doi.org/10.1073/pnas.90.19.9090

Tobe T, Yoshikawa M, Mizuno T, Sasakawa C (1993) Transcriptional control of the invasion regulatory gene virB of *Shigella flexneri*: activation by virF and repression by H-NS. J Bacteriol 175(19):6142–6149. https://doi.org/10.1128/jb.175.19.6142-6149.1993

Turchetti B, Marconi G, Sannino C, Buzzini P, Albertini E (2020) DNA methylation changes induced by cold in psychrophilic and Psychrotolerant *Naganishia* yeast species. Microorganisms 8(2):296. https://doi.org/10.3390/microorganisms8020296

Wang L, Mou Y, Guan B, Hu Y, Zhang Y, Zeng J, Ni Y (2020) Genome sequence of the psychrophilic *Cryobacterium* sp. LW097 and characterization of its four novel cold-adapted β-galactosidases. Int J Biol Macromol 163:2068–2083. https://doi.org/10.1016/j.ijbiomac.2020.09.100

Weinstein RN, Montiel PO, Johnstone K (2000) Influence of growth temperature on lipid and soluble carbohydrate synthesis by fungi isolated from fellfield soil in the maritime Antarctic. Mycologia 92(2):222–229. https://doi.org/10.1080/00275514.2000.12061148

White RA, Soles SA, Gavelis G, Gosselin E, Slater GF, Lim DSS, Leander B, Suttle CA (2019) The complete genome and physiological analysis of the eurythermal firmicute *Exiguobacterium chiriqhucha* strain rw2 isolated from a freshwater microbialite, widely adaptable to broad thermal, pH, and salinity ranges. Front Microbiol 9:3189. https://doi.org/10.3389/fmicb.2018.03189

Wintrode PL, Miyazaki K, Arnold FH (2000) Cold adaption of a mesophilic subtilisin-like protease by laboratory evolution. J Biol Chem 275:31635–31640. https://doi.org/10.1074/jbc.M004503200

Wong CMVL, Boo SY, Voo CLY, Zainuddin N, Najimudin N (2019) A comparative transcriptomic analysis provides insights into the cold-adaptation mechanisms of a psychrophilic yeast, *Glaciozyma antarctica* PI12. Polar Biol 42:541–553. https://doi.org/10.1007/s00300-018-02443-7

Xiao N, Suzuki K, Nishimiya Y, Kondo H, Miura A, Tsuda S, Hoshino T (2010) Comparison of functional properties of two fungal antifreeze proteins from *Antarctomyces psychrotrophicus* and *Typhula ishikariensis*. FEBS J 277:394–403. https://doi.org/10.1111/j.1742-4658.2009.07490.x

Xu H, Griffith M, Patten CL, Glick BR (1998) Isolation and characterization of an antifreeze protein with ice nucleation activity from the plant growth promoting rhizobacterium *Pseudomonas putida* GR12-2. Can J Microbiol 44:64–73. https://doi.org/10.1139/cjm-44-1-64

Yamashita Y, Nakamura N, Omiya K, Nishikawa J, Kawahara H, Obata H (2002) Identification of an antifreeze lipoprotein from *Moraxella* sp. of Antarctic origin. Biosci Biotechnol Biochem 66: 239–247. https://doi.org/10.1271/bbb.66.239

Yang J, Kim HE, Jung YH, Kim J, Kim DH, Walmsley AR, Kim KH (2020) Zmo0994, a novel LEA- like protein from *Zymomonas mobilis*, increases multi- abiotic stress tolerance in *Escherichia coli*. Biotechnol Biofuels 13:151. https://doi.org/10.1186/s13068-020-01790-0

Yao S, Xianzhi J, Wenping W, Manman W, Hamid MI, Meichun X, Xingzhong L (2016) Genomic, transcriptomic, and proteomic analysis provide insights into the cold adaptation mechanism of the obligate psychrophilic fungus *Mrakia psychrophila*. G3 Genes Genomes Genet 11(6):3603–3613. https://doi.org/10.1534/g3.116.033308

Yoshida K, Hashimoto M, Hori R, Adachi T, Okuyama H, Orikasa Y, Nagamine T, Shimizu S, Ueno A, Morita N (2016) Bacterial long-chain polyunsaturated fatty acids: their biosynthetic genes, functions, and practical use. Mar Drugs 14(5):94. https://doi.org/10.3390/md14050094

Yusof NA, Hashim NHF, Beddoe T, Mahadi NM, Illias RM, Bakar FDA, Murad AMA (2016) Thermotolerance and molecular chaperone function of an SGT1-like protein from the psychrophilic yeast, *Glaciozyma antarctica*. Cell Stress Chaperones 21(4):707–715. https://doi.org/10.1007/s12192-016-0696-2

Zheng Y, Li Y, Liu W, Chen CC, Ko TP, He M, Xu Z, Liu M, Luo H, Guo RT, Yao B (2016) Structural insight into potential cold adaptation mechanism through a psychrophilic glycoside hydrolase family 10 endo-β-1,4-xylanase. J Struct Biol 193(3):206–211. https://doi.org/10.1016/j.jsb.2015.12.010

Zhu X, Li Q, Yin C, Fang X, Xu X (2015) Role of spermidine in overwintering of cyanobacteria. J Bacteriol 197:2325–2334. https://doi.org/10.1128/JB.00153-15

Molecular Mechanisms of Cold-Adapted Microorganisms

10

Anupam Singh, Ashish Kumar Rai, Nupur Sinha, and Prachi Bhargava

Abstract

Most of the ecosystems have cold environment, be it high mountains or deep oceans. The microbes residing in these ecosystems have well-developed metabolism which helps them to thrive in extreme climatic conditions. This chapter focuses on the adaptations and variations at present at molecular level in cold-adapted microorganisms. The adaptations can be mimicked in industries like food, tannery, dairy, and healthcare for the betterment of mankind.

Keywords

Psychrophiles · Molecular adaptations · Cold-shock proteins · Cold-adapted enzymes

10.1 Introduction

The competence of low-temperature-loving psychrophilic microorganisms to live and replicate at cold temperatures indicates that they have been victorious in conquering the hurdles of extremely cold environments which include decreased enzyme activity; reduced membrane fluidity; abated rates of transcription,

A. Singh · A. K. Rai
Department of Biotechnology, Bansal Institute of Engineering and Technology, Lucknow, India

N. Sinha
Amity Institute of Biotechnology, AUUP, Noida, Uttar Pradesh, India

P. Bhargava (✉)
Institute of Agricultural Sciences and Technology, Sri Ramswaroop Memorial University, Barabanki, Uttar Pradesh, India
e-mail: prachi.bio@srmu.ac.in

R. Goel et al. (eds.), *Survival Strategies in Cold-adapted Microorganisms*,
https://doi.org/10.1007/978-981-16-2625-8_10

translation, and thereby cell division; inappropriate protein-folding; protein cold denaturation; and formation of intracellular ice flakes (D'Amico et al. 2006). Cold-adapted microorganisms have various features to sustain the negative effects of extreme low temperatures and help them to grow in the extreme environmental conditions. The various adaptations which have helped the cold-loving microorganisms to thrive in harsh conditions are summarized in this chapter.

10.2 Cold-Adapted Enzymes

In permanently cold conditions and at subzero levels in supercooled liquid water, psychrophiles flourish (Miteva 2008). To maintain the cell cycle, psychrophiles survive and thrive with the help of cold-active enzymes. In order to maintain adequate translation and proper protein-folding under cold conditions, genome sequences and proteomic and transcriptomic studies indicate various adaptive features. The work of enzymes in psychrophiles plays a central role in living organisms' cold adaptation. Enzyme function in psychrophiles occupies a central role in cold adaptation of living organisms. There has been comprehensive analysis of cold-active enzymes isolated through metagenomic approaches (Cavicchioli et al. 2011; Vester et al. 2015). Several strategies have been explored which are known to encourage proper expression and folding of cold-active enzymes expressed in heterologous host, improving their solubility, activity, and yield with the help of molecular chaperons and cold-active promotors.

Molecular chaperones: Proteins that enable newly manufactured polypeptides and denatured proteins to attain their native conformation are called molecular chaperones. They are ubiquitous in nature and are present in bacteria, yeast, plants, and animals (Evstigneeva et al. 2001). Chaperons like Cpn60 and Cpn10 have been isolated from psychrophilic bacterium *Oleispira antarctica*, and RB8 from *E. coli* has shown to decrease the optimum growth temperature of microbes below 15 °C. It was observed that the low temperature enhanced the enzyme's proper folding, increasing its activity 180-fold compared to the enzyme purified from the normal *E. coli* strain grown at 37 °C (Ferrer et al. 2003, 2004a, 2004b). Today, a professional *E. coli* strain that co-expresses Cpn60 and Cpn10 cold-active chaperones is marketed under the name Arctic Express by Agilent Technologies.

Cold active promoters: Scientists have developed vectors of cold-shock expression (pColdl-IV) that harbors the CspA promoter from CspA, which is *E. coli*'s main cold-shock protein, allowing for high expression by cold shock of certain genes upon induction (Qing et al. 2004). They stated that for the expression of 38 genes, pCold vectors are highly complementary to the commonly used pET vectors. For functionally expressing different proteins in *E. coli*, pCold vectors have been used which were hard to get at low temperatures, mainly from mesophilic species (Hayashi and Kojima 2008).

Shuo-Shuo et al. (2011) worked on a psychrotrophic bacterium, *Psychrobacter* species, and cloned the cold-active lipase gene Lip-948 to the pColdl plasmid and converted it into *E. coli* BL21, which obtained substantive lipase Lip-948 expression with a yield of 39% of the total protein, the majority of which was present as inclusion bodies.

The ability of unicellular organisms to survive in cold environments requires a wide variety of adaptations at all stages, enabling the disruptions stressed by these harsh environments to be compensated for (Russell and Fukunaga 1990). These adaptations range from changes in the lipid composition of the cell membrane to changes in the sequence and structure of enzymes that ensure that all biochemical reactions are successful (Gerday et al. 2000). In order to compensate for freezing effects of low temperatures, structurally, enzymes often involve modification of their three-dimensional structure while at the same time preventing disastrous cold-induced unfolding events that hinder proper operation (Ramírez-Sarmiento et al. 2013). Cold-adapted enzymes act as base of the strategy developed in microbes to respond well at low and freezing temperatures (Singh et al. 2009; Sundareswaran et al. 2010; Singh and Shivaji 2010) and help them to compensate for the drastic decrease in low-temperature-induced biochemical reaction rates. Psychrophilic species increase catalytic efficiency by increasing Kcat and decreasing Km catalytic efficiency to cope with this limitation or by modifying both parameters as a low-temperature adaptive response (Siddiqui and Cavicchioli 2006). The presence, at low temperatures, of high psychrophilic enzyme activity, is already recorded (Georlette et al. 2000; Russell 2000). Most of the cold-adapted enzymes are thermo-labile and show peak activity at 20 °C (Choo et al. 1998; Kulakova et al. 1999). Psychrophilic enzymes isolated from various sources often have dramatic variations in optimal temperature for their activity (Yasuda et al. 2013). Commercially important psychrophilic alkaline proteases have been classified and studied to explore the modifications at their active sites. A wide opening of the catalytic cleft is observed (Jung et al. 2008) in many proteases and has been seen apart from substitution of bulky side chains for smaller groups, separating conformation of the loops bordering the active site or minor deletions in these loops (Russell et al. 1998). Besides, there is a presence of an additional ion which pulls the backbone of the site entrance and increases the accessibility exponentially in psychrophilic counterparts (Aghajari et al. 2003). As a result of this increased accessibility, cold-active enzymes can accommodate substrates at lower-energy costs regarding conformation changes, thus reducing the activation energy needed for the formation of complex enzyme substrates (Bjelic et al. 2008). The larger active site may also promote the release and exit of products more easily and thus mitigate the impact on the reaction rate of a rate-limiting step (De Vos et al. 2006, 2007). In addition, variations in the electrostatic potential of psychrophilic enzymes in and around the active site tend to be a critical parameter for low-temperature activity. Interestingly in comparison to their mesophilic or thermophilic equivalents, cold-active citrate synthase (Russell et al. 1998), malate dehydrogenase (Kim et al. 2015), uracil-DNA glycosylase (Leiros et al. 2003; Olufsen et al. 2008; Raeder et al. 2010), elastase (Papaleo et al. 2007), and trypsin (Gorfe et al. 2000; Smalås et al. 2000; Brandsdal et al. 2001) are distinguished by marked differences in electrostatic potential near the active site area. The enhanced positive potential at and around the binding site for oxaloacetate and the slightly diminished negative surface potential at the binding area for NADH facilitate the interaction between the oppositely charged ligands and the enzyme surface in malate dehydrogenase (Kim et al. 2015). Hence, it can be interpreted that the variations in non-conserved charged residues were caused by distinct substitutions, resulting in local electrostatic potential varying in both sign and magnitude (Moe et al. 2004).

Mesophilic beta-galactosidases are homo-tetrameric enzymes bearing four active sites (Skalova et al. 2005). However, the cold-active beta-galactosidase crystal structure showed that it is a homohexamer, thus possessing six active sites. With two additional active sites, this unique quaternary structure contributes to enhancing activity at low temperatures. Similarly, psychrophilic cellulases have exceptionally long binders of more than 100 residues of amino acids, about five times longer than mesophilic cellulases to bind with substrates (Garsoux et al. 2004). The long linker adopts a large number of conformations between the fully extended and bent conformations, and when compared to a mesophilic cellulase with a much shorter linker, it was found that the catalytic domain has a 40-fold greater accessible substrate surface area which will increase the activity of this enzyme at low temperatures (Violot et al. 2005). Most psychrophilic enzymes maximize high activity at low temperatures, at the cost of substrate affinity, thereby reducing the free energy barrier of the transition state. The optimization of activity at low temperature in these enzymes is attained by destabilizing the structures bearing the active site or by destabilizing the entire molecule by reducing the number and strength of all forms of weak interactions or the loss of stability factors, leading to improved dynamics of cold residues in the active site.

10.3 Modifications in Transcription and Translation

Any decrease in temperature has two major effects on the cellular environment directly; firstly, it exponentially affects the rate of biochemical reactions, and secondly, it increases the viscosity of aqueous environments. Both these changes affect proteins which plays a major role in the structure and function of a cell as they regulate the harmony between substrates and products, nutrient influx and efflux, waste product outflow, formation of macromolecular, and metabolism as a whole. Key biological activities such as replication of DNA, transcription, and translation into protein are directly affected by cold temperatures. Both transcription and translation are tightly coupled in bacterial cells, and enzymes involved in these processes work beautifully in cold-adapted microorganisms at low temperatures. For example, in most of the psychrophilic microorganisms, a ribosomal extract, RNA polymerase, elongation factor, and peptidyl-prolyl cis-trans isomerase retain activity at a temperature near 0 °C. Indeed, this latter enzyme catalyzes cis-trans prolyl isomerization and is critical for maintaining protein-folding rates at low temperatures due to its high activity and overexpression at low temperatures. CspA-related proteins and RNA helicases which are nucleic acid-binding proteins play an important role in destabilization of secondary structures of DNA and RNA (Berger et al. 1996). The secondary structures in mRNA pose an even greater problem for the initiation process under cold shock as the transcript is too far away from the nearest ribosome to be kept linear (Lim et al. 2000). Therefore, cold-shock proteins (CSPs) and cold-induced RNA helicases (CSHs) come into picture to prevent the formation of secondary mRNA structures and enhance the transcription and translation coupling (Hunger et al. 2006; Walid and Graumann 2007). The proximity of CSPs and CSHs to their presumed substrate, mRNA, suggests such a cooperative relationship. The transient growth of bacteria as a result of blockage in translation due to the highly stable secondary structure of low-temperature mRNAs is prevented by cold shock (Broeze et al. 1978; Das and Goldstein 1968). The yfiA gene encodes protein Y (PY), which in *E coli* is a cold-shock protein, and mediates significant repression of protein synthesis during cold shock (Di Pietro et al. 2013), and such cold-inducible genes have more than 100 bp of untranslated region (5'-UTR) in various bacteria and cyanobacteria (Fang et al. 1998; Graumann and Marahiel 1998). The cold box containing genes are regulated during cold shock by effective transcriptional and translational machinery, and this cold box element works as an effective repressor binding site (Phadtare et al. 1999; Nogueira and Springer 2000). At low temperatures, the efficiency of transcription is retarded by improved DNA duplex (enhances negatively supercoiled) stability, low efficiency of RNA polymerase promoter melting, and slow diffusion of enzymes and substrates, which can be seen in psychrophilic *Pseudomonas syringae* (Lz4W) whose RNA polymerase is active at a low temperature (Uma et al. 1999). Psychrophilic bacteria's RNA polymerase also has a typical eubacterial composition consisting of subunits β, β, σ, and α and prefers supercoiled DNA at low temperatures for gene transcription. It is understood that gyrase A is associated with RNA polymerase and helps to unwind DNA at a low temperature during transcription.

10.4 Role of Polyhydroxyalkanoates (PHA)

Polyhydroxyalkanoates (PHAs) are highly reduced bacterial storage compounds that increase fitness in changing environments. Polyhydroxyalkanoate (PHA) is a family of polymers of esters produced by few prokaryotes as carbon source under abnormal growth conditions and occurs as water-insoluble aggregates in the cells (Laycock et al. 2014). Lemoigne in 1926 isolated them from *Bacillus megaterium*. These polyesters have gained interest globally because they are biodegradable (Ho et al. 2002; Lenz and Marchessault 2005; Lim et al. 2005) and biocompatible (Hazer and Steinbüchel 2007) and are produced by fermentation of bacteria. That is why for traditional petrochemical-based plastics, PHA has potential as an alternative material. They minimize pollution caused by the rising global demand for polymers because of their unique properties. PHAs are bacterial storage compounds that are highly reduced, increasing fitness in changing environments. For the initiation of the oxidative stress response induced by cold treatment, PHA was required. The metabolism of PHA modulates the availability of reducing equivalents, helping to reduce low-temperature oxidative stress. PHA accumulation and degradation endow bacteria with increased longevity, competitiveness, resistance to stress, and increasing fitness of bacteria in evolving environments (Pham et al. 2004; Kadouri et al. 2005).

It is possible to classify microorganisms capable of synthesizing PHAs into two classes. Under nutrient restrictions, such as oxygen, nitrogen, or phosphorus, bacteria belonging to the first category can synthesize PHAs and do not accumulate them in their growth process. Microorganisms that accumulate biopolymers during the growth process and do not need to reduce nutrients are submitted to the second category (Muhammadi et al. 2015). Only one species of *Pseudomonas*, i.e., *P. extremaustralis*, of Antarctic origin has been found to be capable of synthesizing poly(3-hydroxybutyrate) (PHB) from octanoate but not from glucose (Ayub et al. 2004; López et al. 2009; Diard et al. 2002). Some studies on non-Antarctic bacteria have shown that the accumulation of reserve polymers such as PHA could help bacteria survive hunger and harsh environmental conditions (Wang and Bakken 1998). Therefore, the accumulation of PHA in Antarctic bacteria in the extreme environments and low nutrient availability may also increase the survival potential of these bacteria by helping them to withstand oxidative stress and dramatic temperature shifts, which are common stress factors found in Antarctic environments (Ayub et al. 2009).

Six major proteins including two polymerases (PhacC1, PhaC2), one depolymerase (PhaZ), two phasins (PhaI, PhaF), and two regulatory proteins (PhaD, PhaG) have been found to play a major role in PHA synthesis and aggregation. The Pha cluster is well-preserved in *Pseudomonas* species and has two operons, PhaC1ZC2D and PhaF1. The genes transcribed in the same direction encode the proteins PhacC1, PhaC2, PhaZ, and PhaD, while the proteins PhaF and PhaI are transcribed in the opposite direction (Sandoval et al. 2007). PhaD is the only protein known that is not bound to the surface of PHA granules. It encodes the TetR family's putative transcription regulator, which plays an important role in the biosynthesis of PHA. The PhaD gene prevent expression or binding to major granule proteins that lead to accumulation of mcl-PHA (Klinke et al. 2000). PhaF and PhaI are associated with granule formation, and during cell division, they are assumed to promote the segregation of the granules that act as an interface between the granule and the cytoplasm (Galán et al. 2011). Besides, PhaF phasin helps in regulating the shape and length of PHA. Global transcriptional regulators such as the Crc, RpoS, PsrA, and GacS/GacA systems are in charge of the expression of the Pha genes. However, according to one of the findings, PhaG expression is correlated with oleic acid (a structurally related carbon source) synthesis of mcl-PHA, indicating that PhaG can also play a role in this process. Możejko-Ciesielska et al. (2018) stated that PhaG expression is associated with oleic acid (a structurally related carbon source) synthesis of mcl-PHA, implying that PhaG can also play a role in this process. Using the *Pseudomonas putida* KT2440 relA/spot mutant, an association between increased PhaG expression and mcl-PHA synthesis on oleic acid was also established. This complexity in PHA synthesis is likely to arise not only from the effects of PhaD, PhaF, or PhaI but also from other proteins that are still undiscovered that influence the function of the promoters of the Pha gene cluster. *Pseudomonas* genus seems to be a factory for production of biopolyesters and can become a platform bacterium capable of accumulating established structures of PHAs and constant biopolymer properties. In addition, a deeper understanding of the regulation of genes involved in the synthesis and aggregation of PHAs would help to enhance their bacterial cell content and help isolate new biopolymer films that are useful in many applications.

10.5 Cryoprotectant and Cold-Shock Proteins

Around 80% of the biosphere of our world is permanently cold, that is, at temperatures below 5 °C. This covers much of the seas of the planet, which occupy 70% of the surface of the Earth, the Polar Regions that form Antarctica, and portions of the Arctic Circle in North America and Europe. Microorganisms that live in Polar Regions such as Antarctica must be cold-tolerant as well as desiccation-tolerant and have unfavorable environmental stress factors for adaptation with the mechanisms involved in the process (Rohini 2010).

Freezing of cells induce the formation of ice flakes which creates a disbalance in osmotic pressure causing severe damage to the cell. Cryoprotectants like glycine, betaine, sucrose, and mannitol decrease the cytoplasmic freezing point and combat freezing, desiccation, and hyper-osmolality (Cowan 2009; Klähn and Hagemann 2011). Trehalose disaccharide prevents protein denaturation and aggregation, scavenges free radicals, and stabilizes cell membranes under cold conditions (Kandror et al. 2002). *E. coli* transcriptome analysis has shown that under cold conditions the trehalose biosynthetic genes *otsA* and *otsB* are induced (Phadtare and Inouye 2004). Some psychrophiles generate antifreeze or ice-binding proteins (IBP) that bind to regulate the growth and recrystallization of ice crystals by reducing the freezing point (thermal hysteresis) (Celik et al. 2013). Ice-nucleating (IN) proteins can prevent water from supercooling by promoting the formation of ice crystals at temperatures near the melting point (Kawahara 2002). Depending on the climate and microbial community structure, the cryoprotective mechanisms used may differ as demonstrated by a metagenomic analysis of temperate lakes that revealed a predominance of isolates with high cytoplasmic osmolyte material with negligible phenotypes of ice association (IN/IBP) while half of the epiphytic isolates from a frost-exposed chrysanthemum phyllosphere communication (Wilson et al. 2012; Wu et al. 2012). Another possible cryoprotection mechanism is the production of exopolysaccharides (EPS), and psychrophiles generate high amounts of EPS under cold conditions (Feng et al. 2014). The high polyhydroxyl content of EPS decreases the temperature of the freezing point and ice nucleation of water. In addition, EPS can trap and promote surface adhesion, cell aggregation, and biofilm formation by trapping water, nutrients, and metal ions and can also play a role in defending extracellular enzymes against cold denaturation and autolysis (De los Ríos et al. 2004). The psychrophilic diatom *Melosira arctica* and cold-tolerant bacterium *Colwellia psychrerythraea* contain exopolymeric substances which cause changes in desalination and microstructure of rising ice by increasing ice crystal disorder and pore density. This results in a decrease in ice permeability which ultimately contributes to salt retention. Therefore, by reducing ice growth due to increased salinity, biologically active EPS can affect the colonization and survival of organisms in the sea ice habitat (Ewert and Deming 2011; Krembs et al. 2011).

Cold-shock proteins (CSPs) play an important role among the numerous biomolecules responsible for developing cold tolerance for microorganisms. CSPs are protein domains that have been found in prokaryotic and eukaryotic DNA-binding proteins of around 70 amino acids. These so-called CSPs are thought to assist the cell, likely through condensation of the chromosome and prokaryotic nucleoid organization, to survive at temperatures below optimum growth temperature. In plants, animals, and bacteria, CSPs are nucleic acid-binding proteins and are well maintained. In order to counteract adverse cold-shock effects, prokaryotes and eukaryotes exhibit cold-shock response with the development of cold-shock proteins. In cell physiology, CSPs play a significant role, and studies have documented they bind mRNA and control ribosomal translation (Barria et al. 2013). Multifunctional RNA/DNA binding proteins distinguished by the inclusion of one or more cold-shock domains (CSDs) are CSPs. Among the most evolutionary preserved proteins are CSPs (Wolffe et al. 1992; Wolffe 1994). Their distinctive feature is the presence of one or more domains of cold shock that possess binding properties of nucleic acid. In bacteria, a sudden drop in temperature (from 37 °C to 10 °C) induced a 200-fold increase in the expression of cold-shock protein A (CspA) within minutes, which was independent of transcriptional activity (Jones and Inouye 1994; Gottesman 2018). Among species, this rapid inducibility is conserved. A recent research revisited the original observation to examine the global changes occurring in bacteria during the cold-shock response using genome-wide methods (Graumann and Marahiel 1998; Zhang et al. 2018). The authors listed the major players as RNase R and CspA. The degradation of misfolded RNAs appears to be the responsibility of RNase R, while CspA melts double-stranded RNAs to allow translation. Bacterial CSPs are primarily induced to control the adaptation to cold stress after a rapid temperature downshift but are also present to regulate other biological functions under normal conditions. A simple collection of gene products involved in the metabolic processes to respond to cold was defined by *B. subtilis* cells. A subject of systems biology that is not yet resolved is the complete explanation of cold-shock response and cold adaptation within the cell (Kaan et al. 2002; Beckering et al. 2002; Budde et al. 2006). Cold-adapted microorganism research not only seeks to understand the molecular mechanisms of cold adaptation but also projects the important and future use of organisms in various applications in the fields of biotechnology, manufacturing, and healthcare (Casanueva et al. 2010).

10.6 Role of RNA Degradosome

In most bacteria, the RNA degradosome is a multiprotein complex contributing in the processing of ribosomal RNA. It includes the RNA helicase B, RNase E, and polynucleotide phosphorylase proteins (Carpousis 2002). The structure of the RNA degradosome is like a molecular domain in which RNA interacts with each other of the components as a substrate, and when this occurs, it is very difficult for RNA to escape from the complex (Carpousis 2007). Degradosome is the major determinant of cellular RNA stability having many ribonucleases in it. Endoribonuclease RNase E, an RNA helicase, and exoribonuclease, called RNase R, are known to be associated with degradosomes. These enzymes are known to play an important role in maintaining the regulation of rRNA quality. Therefore, by removing the need for helicase-induced ATP, it can enable the cell to conserve energy at low temperatures (Purusharth et al. 2005). There are several alternative types of RNA degradosome that have been identified with different proteins. PcnB (poly A polymerase) and RNA helicases (RhlE and SrmB) are supplementary alternative degradosome components. RNA helicase, CsdA, contains other alternative components during cold shock. RNase R and the putative RNA helicase Hrp A are additional alternative degradosome components during the stationary process. Another constituent stated to be part of the complex is polyphosphate kinase (Ppk), like RNA chaperone Hfq, prostatic acid phosphatase (PAP), and other forms of chaperones and ribosomal proteins. These were discovered from *E. coli* in cell-extracted degradosome preparations. The method of RNA destruction is very complex. We use the mRNA degradation method in *E. coli* as an example to make it easier to understand, since it is the best-known process. It is primarily mediated by endonucleases and ribonucleases. In a 3′ by 5′ manner, the enzymes RNase II and polynicleotide phosphorylase degrade mRNA. There are four compartments in the degradosome, which have multiple ribonucleases. Initially, a polyphosphate framework is the synthesized RNA. Therefore, dephosphorylation is necessary by the action of RNA pyrophosphohydrolase (Pph) in order to obtain monophosphate terminal as an end and a stem-loop structure. RNase E cleavages the P-terminal endoribonucleolytically, while RNA helicases digest the stem loop. The efficiency of polymerase PAP is important to simplify the reduction of exoribonucleases such as PNPase, if there are any secondary structures. Finally, oligoribonucleases are processed to manage the scraps. In other organisms, the mechanism is similar and differs only in the enzymatic machinery. For example, *Bacillus subtilis* uses RNase Y or RNase J instead of using RNase E as endoribonuclease (Górna et al. 2012). *P. syringae*, a psychrophilic microorganism, is a part of the novel degrading RNase R, a multisubunit RNA-degrading complex that also includes endoribonuclease RNase E and RhlE helicase (Purusharth et al. 2005). Both RNase RPs and RNase Rec enzymes can degrade secondary organized RNAs without the aid of RNA helicases using Mg^{+2} and Mn^{+2} divalent cations. The enzymes, however, differed in their thermal stability and temperature-dependent operations from each other. Intriguingly, *M. genitalium* RNase R has been shown to contain mostly trinucleotides as end products (and only a small amount of dinucleotides) (Lalonde et al. 2007). Although the molecular basis for the development of different end-product lengths of the RNA chain by different RNase R homologues is not known, the proposed catalytic pocket residues in the enzyme's RNB domain could play an important role, as evidenced by RNase II and RNase R from *E. coli* (Barbas et al. 2008; Domingues et al. 2009). RNase REc structural models based on the crystal structures of RNase II (Frazao et al. 2006; Zuo et al. 2006) have positioned the catalytic site deep in a putative RNB domain funnel-like structure into which the 3′ ends of the RNA chains are threaded for cleavage (Barbas et al. 2008; Vincent and Deutscher 2006). It is therefore probable that the strength of RNA 3′ anchoring ends in the RNase R active site domain of various species may differ, leading to the release of various end products that do not remain bound to the enzyme. The RNase RP possibility of preserving certain characteristics of RNase II such as the generation of 4nts as a limit product, likely due to the sequence preservation at the RNB site, was not evident from the sequence alignment of RNase RPs, RNase Rec, and RNase II. *Pseudomonas* species contain only one member of the RNB/RNR exoribonuclease family that has greater identity to RNase R (Matos et al. 2011). Any consequence of the variation of the limit product size or its role in environmental adaptation of the RNB/RNR enzymes is currently an open issue. The essential difference appears to lie in the optimum temperature of action between the psychrophilic RNase RPs and the mesophilic RNase REc. At 22 °C, the *P. syringae* enzyme showed the highest activity, while in *E. coli* at 37 °C, the enzyme demonstrated optimum activity (Cheng and Deutscher 2002). The enzymes also varied in their thermal stability, with the heat-labile nature of the psychrophilic protein shown by the RNase RPs. Thus, during evolution, *P. syringae* RNase R appears to have been a cold-loving enzyme, presumably from a mesophilic ancestor, as previously indicated based on several other enzymes that have maximum activity at 37 °C (Ray et al. 1998). The maximum activity shown at 22 °C by RNase R is noteworthy as it points out a facet of the multicomponent protein complex evolution. While the entire RNA degradation complex consisting of RNase E, RNase R, and Rhl R, as well as the isolated *P. syringae* RNase E enzyme, exhibits optimum activity at 37 °C, RNase R has been modified to operate optimally at a lower temperature (Purusharth 2006, 2007). This indicates that in low-temperature physiology, when the enzyme operates alone outside the complex, the psychrophilic nature of RNase RPs is very significant, as demonstrated by the lack of cold sensitivity in the RNase E C-terminal deleted *P. syringae* strain in which degradosomal proteins are not associated. However, it may also reflect that it has evolved faster for cold adaptation than RNase E and/or that RNase R has been recruited to the degrading complex in Antarctic *P. syringae* at a later stage of evolution.

10.7 Changes in Membrane Fluidity

As temperature decreases, so does phospholipid bilayer fluidity, and an increase in the viscosity of membrane is seen. Dropping of mercury decreases the kinetic energy of the phospholipids in the bilayer asking them to cluster together more closely, thereby increasing intermolecular interactions and decreasing membrane fluidity. In order to preserve the membrane in a liquid crystalline phase that is necessary for its proper function, many bacteria modify the composition of their membrane's fatty acids. To assist this phenomenon, a set of cold-inducible proteins responsible for altering the mechanism of protein translation is upregulated. It has been observed in *Bacillus subtilis*, a model Gram-positive organism, that the fatty acid profile is characterized by the iso- and anteiso-branched fatty acids and branched amino acids are used as precursors for their synthesis: isoleucine for anteiso-branched fatty acids and valine and leucine for iso-branched fatty acids. It is noteworthy to know that the melting point of fatty acids that are anteiso-branched is considerably smaller than that of their iso-branched isomers (Jones and Inouye 1994). Mesophilic cells have more saturated fatty acid content (laurate) which decreases in psychrophiles and is replaced by unsaturated fatty acid (palmitoleate), which helps in enhancing membrane fluidity and decreases membrane phase separation. An enzyme called fatty acid desaturase is known to regulate unsaturation of fatty acids in phospholipids of cell membranes which is regulated by a thermosensor called DesK kinase. This sensor induces the transcriptional activators DesR at low temperature (Aguilar et al. 2001).

Membrane fluidity is very important so as to allow the cells to sustain its basic membrane functions such as bacterial proliferation, cell division, cell growth, signal transduction, energy generation, and transport. Another known mechanism which helps in sustaining the membrane fluidity is by changing the polar head groups of glycerophospholipids that modify the packing of the membrane in two. Similarly by changing the magnitude of fatty acid desaturation or fatty acid chain length and by changing the ratio of isocoagulant and anticoagulant fatty acids or changing the chain length, bacteria try to maintain the viscosity of their membrane. Some generations have also established a preferential mechanism to resolve membrane rigidity connected to changes at low temperatures. *Shewanella livingstonensis* Ac10 (Antarctic bacterium) is developed at 4 °C and produces eicosapentaenoic acid (EPA). EPA-less mutants, on the other side, are cold-sensitive and filamentous, display phenotypes of several nucleoids, and demonstrate defective cell division at low temperatures (Kawamoto et al. 2009).

10.8 Fatty Acid Desaturation

The polymers of polysaccharide and the polypeptide residues are carbohydrates and protein. In comparison, lipids are comprised of a large variety of substances with enormous structural variations and a scarcity of building blocks. Fatty acids are, thus, the primary constituents of different types of lipids, such as glycerides and sphingolipids. Fatty acids are organic compounds mostly at the end of an aliphatic chain that comprise a group of carboxylic acids and are classified into saturated and unsaturated. The majority of fatty acids, in particular, the so-called polyunsaturated fatty acids (PUFAs) and their derivatives, play a crucial biological role in inflammatory reaction, cell division, lipid metabolism regulation, signaling molecules, energy supply, and defense of the structure and function of biological membranes.

Since temperatures below the optimum specifications for organisms induce rigidity of membrane lipids and irregular cellular behaviors by remodeling the lipid content of their membrane, poikilothermic organisms such as bacteria, plants, and fish have to respond to such environmental changes in order to survive in abiotic stress (Mc Elhaney 1984). A membrane fluidity is known as the sum of molecular disturbance and molecular displacement in a fluid bilayer. Stress at temperature tends to influence an organism's membrane fluidity as cold temperature typically results in lower membrane flow, and the fatty acids that are already linked to fatty acid desaturations can be rescued by desaturation of the membrane lipids. Phospholipids rich in unsaturated fatty acids also have a slightly lower-transition temperature than phospholipids containing high concentrations of saturated fatty acids, which is partly attributed to the assumption that saturated acyl chains are closely assembled into the membrane lipids, while unsaturated fatty acids cause stearic impairment due to the rigid twisting of cis-double ties, allowing the unsaturated acids chain to be bundled far worse. Therefore, optimal membrane lipid fluidity is recovered, and normal cellular activities at lower temperatures are restored (Los et al. 2013). In several species such as bacteria, cyanobacteria, and fish, the correlation between membrane permeability and changes in temperature has been studied by a polarization of 1,6-diphenyl-1,3,5-hexatriene (DPH) by fluorescence wherein parallel to the membrane lipid acyl chains, and the DPH was added experimentally. As the DPH interacted stably with rigidized membranes upon fluorescence, weak depolarization was detected, but depolarization was found to increase with fluidized membranes. Heat shock enhances membrane permeability and is known to induce lipid bilayer as well as to disintegrate.

Biological membranes occur in the liquid crystalline form and transit to a gel phase after exposure to low temperatures. The modification of fatty acid amounts (saturated to unsaturated) could effectively change the fluidity of the membrane. By elongating palmitoleic acid at low temperatures, *Escherichia coli* controls its membrane fluidity by growing the quantity of cis-vaccine acid (Garwin and Cronan 1980). In contrast to mesophilic bacteria, a greater proportion of unsaturated fatty acids has been recorded in psychrophilic bacteria. Adaptation to low-temperature tension by certain bacteria takes place by growing the volume of unsaturated fatty acids integrated into the phospholipids of the membrane (Prabagaran et al. 2005). Saturated fatty acids reduce the fluidity of the membrane owing to the close packing of the saturated fatty acids' acyl chains. In comparison, unsaturated fatty acids improve the fluidity owing to inadequate packaging and kinks in unsaturated fatty acid induced by the involvement of cis-double bonds (Mansilla et al. 2004).

It has been shown that *Bacillus subtilis* and *B. megaterium* transform current saturated fatty acids into unsaturated fatty acids in reaction to low temperatures. The desaturase enzyme (encoded by the des gene), which is transiently induced by temperature downshift, catalyzes the transformations of certain fatty acids (Diaz et al. 2002). The cold induction of the gene regulates two component signal transductions, which consist of a cytoplasmic reaction regulator (DesR) and a membrane-associated kinase sensor (DesK) (Shivaji and Prakash 2010). The nonphotosynthetic bacterium has been identified with cold-inducible desaturase (*Bacillus* sp.). Cold-shock responses have been studied using DNA microarrays in *Bacillus subtilis*, and findings have shown that *des* is the strongest gene inducible by cold (Beckering et al. 2002). The absence of isoleucine and *des* gene deletion causes a clear cold-sensitive phenotype. There are studies that show that *des* expression can substitute the isoleucine-dependent fatty acid branching pathways during cold adaptation (Weber et al. 2001).

10.9 Branching of Fatty Acids

Antarctica is known to harbor microbes who thrive in several extremes, viz., nutrient shortage, high or lower pH, desiccation and osmotic stress, high UVB radiation levels, and a remarkably variable photoperiod, i.e., from no light to continuous light for 24 h as the cold, most inaccessible continent in the world besides its low temperatures (Feller 2017). This inevitably implies that they have a variety of adaptive strategies in order to retain the essential functions of the cellular cells even in certain prohibitive conditions to allow psychrophilic prokaryotes to live and proliferate in the Antarctica (Pearce 2012). *Proteobacteria*, a large phylum of Gram-negative bacteria, constitute the most abundant phylum along with *Actinobacteria* in the above said region (Nunez-Montero and Barrientos 2018). Lipopolysaccharides are amphiphile macromolecules necessary for viability and longevity, since they provide the structural stabilization and protection for the whole bacterial surface, in a complex interplay with the outside world (Aislabie et al. 2008). Bacteria fight hostile environments under unfavorable circumstances by altering their primary structure of the LPS to enhance the cell envelope in a way that provides more cushioning defense and adaptation.

A fluidizing influence on the membrane is exerted by the introduction of low-melting point fatty acids (unsaturated, short-chain, and branched chain fatty acids) into lipids. These adjustments are termed homeoviscous membrane fluidity modifications. The production of anteiso-fatty acids rises among the branched chain of fatty acids in preference to iso-fatty acid syntheses. This is a natural shift triggered by a temperature drop. Branching happens in an anteiso-fatty acid from the penultimate carbon atom farthest from the functional group; thus, branching takes place in an iso-fatty acid from the farthest carbon atom. There is evidence that anteiso-fatty acids play a significant role in bacteria's cold adaptation (Suutari and Laakso 1992).

The integration of branched chain fatty acids (iso- and anteiso-fatty acids) into the lipid bilayer has a fluidizing effect on the membrane. In *Brevibacterium fermentans* and *Bacillus subtilis*, an increase of anteiso-branched fatty acids (methyl branching from the third last ante-penultimate carbon in the chain) was also observed when the growth temperature decreased, and the concomitant decrease of iso-branched fatty acids (methyl branching from the second last carbon fatty acid) was observed. *Listeria monocytogenes* (food-borne disease pathogen) may spread at cooling temperature in response to cold adaptation and mostly synthesizes alpha-C15:0 in order to promote homeoviscous adaptation (Annous et al. 1997).

10.10 Cis- and Trans-Fatty Acids

Different techniques for adaptation to low temperatures have been established for psychrophilic and psychrotolerant bacteria. The capacity of the cell to modulate membrane fluidity is an effective technique that is necessary for the survival of the cell at low temperatures (Heipieper et al. 2003). Bacteria modulate membrane fluidity in general by modifying the structure of their fatty acids. However, through changing the lipid head group, the protein content of the membrane, the form of carotenoids synthesized, the length of the fatty acid chain, and the proportion of *cis*- to *trans*-fatty acids, bacteria could also accomplish the same by numerous other techniques (Diefenbach et al. 1992). In addition to bacteria, there is a two-component signal transduction mechanism consisting of a diaphragm-binding sensor and a soluble, low-temperature cytoplasmic reaction regulator.

The fatty acids regulate membrane fluidity in particular through isomers (*cis* and *trans*) (Loffeld and Keweloh 1996). With a rise in growth temperature, solvent, and salt shock exposure, an increase in the trans-fatty acid content is seen.

10.11 Amino Acid: Composition and Length Variation

A spectrum of adapted proteins, including less salt bridges and less proline residue, lower hydrogen bands, a decreased arg/(Arg + Lys) ratio, and an improved serine and glycine content, are observed in psychrophiles (Molinaro et al. 2015). As a result of less salt bridges, less closely packed hydrophobic cores, and a decrease in proline residues in surface loops, the elongation factor 2 protein of eurypsychrophilic archaea *Methanococcoides burtonii*, isolated from the chronically cold Ace Lake in Antarctica, exhibits greater structural versatility (Metpally and Reddy 2009). The same features of reduced arginine residues, the small hydrophobic base, and a very low number of aromatic-aromatic interactions, containing small amounts of salt and proline residues, were found in lipase isolated from *Psychrobacter immobilis* which is a psychrophile abundant in Antarctic region (Thomas and Cavicchioli 2000). Variations in the kcat and Km values, corresponding with increasing catalytic efficiency in cold temperatures, are a characteristic feature of psychrophilic enzymes (Metpally and Reddy 2009). Comparative genomic analysis between cold-loving microbes and mesophiles has helped to explore the amino acid structure and composition role in cold adaptations (Mykytczuk et al. 2013). It is also important to ascertain whether psychrophiles which are capable of growth under null temperatures (cryophiles) are associated with different adaptations compared to psychrophiles with narrower range of growth that are not adaptable to the challenges associated with under freezing temperature. As we expect these members to have near shared origin, comparisons using species of the same genus or adjacent genus offer a more comprehensive study and thus reported amino acid discrepancies between psychrophiles and their close mesophilic associates are more possible to be the consequence of cold adaptation. As amino acids are building blocks of proteins, this adaptation directly affects the working efficiency of all the biocatalysts.

It gives psychrophiles an exceptional ability to transcribe and read the templates at cold temperatures very efficiently. It improves the stability of DNA and also affects the production of RNA polymerase which works efficiently at low temperatures (Uma et al. 1999).

There is little awareness about how RNA polymerase overcomes these psychrophile barriers. However, according to some studies, RNA polymerase is involved at low temperatures of the psychrophilic *Pseudomonas syringae* (Lz4W). In psychrophilic bacteria, a standard eubacterial composition of RNA polymerase consists of many subunits like alpha, beta, etc. which favors supercoiled DNA at low temperatures for gene transcription (Broeze et al. 1978). The cold shock stops temporary developments from being triggered by the blockage during the translation owing to the extremely stable secondary mRNA structure at low temperatures (Di Pietro et al. 2013). Cold-inducible genes in some bacteria and cyanobacteria have been reported to be greater than 100 bp untranslated region (5'-UTR) (Singh and Shivaji 2010). In *Escherichia coli*, cold shock and 5'-UTR (cspA, cspB, and csdA) overexpression results in the corresponding genes being de-repressed (prolonged synthesis) (Fang et al. 1998). It is claimed that the cold box containing genes are regulated during cold shock by successful transcriptional and translational machinery. As a putative repressor binding site, the cold box portion works. In order to allow productive low-temperature translation, the posttranscriptional regulator, such as the CspA cold-shock protein family, has the capacity to destabilize the secondary mRNA structures (Nogueira and Springer 2000).

10.12 Modifications at Protein-Folding Stage

Before being fully functional, the freshly synthesized embryonic polypeptide must be specifically folded (at tertiary and quaternary structure level). Protein synthesis and protein-folding have been considered over many years as temperature-sensitive cellular processes which, with no unique adaptations, seriously limit microbial growth in low temperatures (Yang et al. 2015). Cold adaptation is proposed to overlap preexisting cellular structure, and thus, adaptive strategies may be different among different psychrophilic species (Piette et al. 2011; Shin et al. 2012). However, some general patterns have been observed as regards protein synthesis and folding.

One phenomenon commonly observed is the overexpression of cold-shock proteins (CSP) which work as chaperons and aid in proper folding of proteins at low temperature (Rabus et al. 2004). In addition, some Csps are known to control unsaturated fatty acid synthesis and have already been documented in the genomes of *Desulfotalea psychrophila*, *Colwellia psychrerythraea* 34H, and *Pseudoalteromonas haloplanktis* (Medigue et al. 2005). In *E. coli* during cold shock, molecular chaperones such as caseinolytic proteases (Clps), trigger factor (TF), GroEL, DnaK, and GroES are upregulated. It has already been identified that during cold adaptation, folding and refolding of cold-damaged proteins are important. In order to monitor co-translational protein-folding, several experiments have often indicated that chaperones and embryonic polypeptides interact together. One of the studies confirmed that when expressed in trans, cpn60 (encoding GroEL) and cpn10 (encoding GroES) of the Antarctic bacterium *Oleispira antarctica* could promote the growth of *E. coli* at 4 °C, thus demonstrating the significance of folding the chaperone-mediated protein during low-temperature development (Yoshimune et al. 2005).

10.12.1 Translation

Proteomic analysis of cold temperature microbes have interpreted that more 30% of the upregulated proteins were specifically associated with protein synthesis at 4 °C. Protein synthesis becomes a rate-limiting stage for almost all cellular reactions which is normalized by enhancement of synthesis of ribosomal proteins and RNA chaperones and the genes used in preserving RNA at low temperatures (Bowman 2008). The comparatively large number (up to 106 genes, often arranged in long sequences with repetitive sequences) of rRNA genes and of tRNA genes at least in *P. haloplanktis*, *C. psychrerythraea*, and *P. ingrahamii* has proven their importance in cold adaptations (Uh et al. 2010). In many psychrophiles like *P. haloplanktis*, *S. denitrificans*, *P. arcticus*, *S. alaskensis*, *E. sibiricum*, and *M. burtonii*, overexpression of RNA helicases at low temperatures that allow the secondary RNA structures to be disassembled for effective cold translation has proved their mettle (Riley et al. 2008). These examples show that psychrophiles have also evolved adaptive pathways for the optimization of low-temperature protein synthesis.

10.12.2 Folding Assistance

Several interesting studies have shown that cold-adjusting chaperones, including DnaK and GroEL expressed in *E. coli*, give cold tolerance and increase the rate of growth, thus illustrating the critical role played by chaperones in microorganism temperature adaptation (Cartier et al. 2010). The ribosome-bound trigger factor (TF) involved in protein-folding is the first chaperone to interact with nearly all ribosome-synthesized nascent polypeptides, and most small proteins (~70% of the total) can easily be synthesized without further support. Interestingly enough, TF is a cold-shock protein of *E coli*, while others are well-known proteins of the heat-shock protein family (HSPs) (Hartl and Hayer-Hartl 2009). Overexpression of TF and controlled synthesis of HSP in several cold-adapted microorganisms suggested that in view of these imbalances in chaperone machinery, TF tends to rescue the chaperone function (Kandror and Goldberg 1997). The decreasing influence of HSP chaperones in cold shows that these folding aids are synthesized at elevated temperatures of the environment (Piette et al. 2010). In comparison, TF synthesis repression or HSP-chaperonin upregulation has been documented as well. *S. alaskensis* (thermophile) also has two sets of genetic clusters dnaK-dnaJ-grpE (DnaK and its cochaperones). Quantitative proteomics indicated that one set acts as a chaperone system of low temperatures, while the other set operates at higher temperatures of growth. Psychrophiles have clearly implemented separate techniques to help better fold protein.

10.13 Conclusion

All the adaptations can be used as methods to induce the ability to thrive in cold temperatures which find its importance in many industries like milk industry, detergent industry, and leather industry. Many cold-loving enzymes which have been isolated from psychrophiles are being used for commercial purpose. The recent science of metagenomics is also trying to explore and identify many psychrophiles which may boost up the rate of industrial growth.

References

Aghajari N, Van Petegem F, Villeret V (2003) Crystal structures of a psychrophilic metalloprotease reveal new insights into catalysis by cold-adapted proteases. Proteins 50(4):636–647

Aguilar PS, Hernandez-Arriaga AM, Cybulski LE, Erazo AC, de Mendoza D (2001) Molecular basis of thermosensing: a two-component signal transduction thermometer in Bacillus subtilis. EMBO J 20(7):1681–1691

Aislabie JM, Jordan S, Barker GM (2008) Relation between soil classification and bacterial diversity in soils of the Ross Sea region, Antarctica. Geoderma 144(1-2):9–20

Annous BA, Becker LA, Bayles DO, Labeda DP, Wilkinson BJ (1997) Critical role of anteiso-C15: 0 fatty acid in the growth of Listeria monocytogenes at low temperatures. Appl Environ Microbiol 63(10):3887–3894

Ayub ND, Pettinari MJ, Ruiz JA, López NI (2004) A polyhydroxybutyrate-producing *Pseudomonas* sp. isolated from Antarctic environments with high stress resistance. Curr Microbiol 49:170–174

Ayub ND, Tribelli PM, López NI (2009) Polyhydroxyalkanoates are essential for maintenance of redox state in the Antarctic bacterium *Pseudomonas* sp. 14-3 during low temperature adaptation. Extremophiles 13:59–66

Barbas A, Matos RG, Amblar M, Lopez-Vinas E (2008) Gomez-Puertas, Arraiano CM. New insights into the mechanism of RNA degradation by ribonuclease II: identification of the residue responsible for setting the RNase II end product. J Biol Chem 283:13070–13076

Barria C, Malecki M, Arraiano CM (2013) Bacterial adaptation to cold. Microbiol 159:2437–2443

Beckering CL, Steil L, Weber MH, Volker U, Marahiel MA (2002) Genomewide transcriptional analysis of the cold shock response in Bacillus subtilis. J Bacteriol 184:6395–6402

Berger F, Morellet N, Menu F, Potier P (1996) Cold shock and cold acclimation proteins in the psychrotrophic bacterium *Arthrobacter globiformis* SI55. J Bacteriol 178:2999–3007

Bjelic S, Brandsdal BO, Åqvist J (2008) Cold adaptation of enzyme reaction rates. Biochemistry 47 (38):10049–10057

Bowman JP (2008) Genomic analysis of psychrophilic prokaryotes. In: Psychrophiles: from biodiversity to biotechnology. Springer, Cham, pp 265–284

Brandsdal BO, Smalås AO, Åqvist J (2001) Electrostatic effects play a central role in cold adaptation of trypsin. FEBS Lett 499(1-2):171–175

Broeze RJ, Solomon CJ, Pope DH (1978) Effects of low temperature on in vivo and in vitro protein synthesis in Escherichia coli and Pseudomonas fluorescens. J Bact 134:861–874

Budde I, Steil L, Scharf C, Volker U, Bremer E (2006) Adaptation of Bacillus subtilis to growth at low temperature: a combined transcriptomic and proteomic appraisal. Microbiology 152:831–853

Carpousis AJ (2002) The *Escherichia coli* RNA degradosome: structure, function and relationship in other ribonucleolytic multienzyme complexes. Biochem Soc Trans 30(2):150–155

Carpousis AJ (2007) The RNA degradosome of *Escherichia coli*: an mRNA-degrading machine assembled on RNase E. Annu Rev Microbiol 61(1):71–87

Cartier G, Lorieux F, Allemand F, Dreyfus M, Bizebard T (2010) Cold adaptation in DEAD-box proteins. Biochemistry 49(12):2636–2646

Casanueva A, Tuffin M, Cary C, Cowan DA (2010) Molecular adaptations to psychrophile: the impact of 'omic' technologies. Trends Microbiol 18:374–381

Cavicchioli R, Charlton T, Ertan H, Omar SM, Siddiqui K, Williams T (2011) Biotechnological uses of enzymes from psychrophiles. Microb Biotechnol 4:449–460

Celik Y, Drori R, Petraya-Braun N, Altan A, Barton T, Bar-Dolev M, Groisman A, Davies PL, Braslavsky I (2013) Microfluidic experiments reveal that antifreeze proteins bound to ice crystals suffice to prevent their growth. Proc Natl Acad Sci U S A 110:1309–1314

Cheng ZF, Deutscher MP (2002) Purification and characterization of the Escherichia coli exoribonuclease RNase R: comparison with RNase II. J Biol Chem 277:21624–21629

Choo DW, Kurihara T, Suzuki T, Soda K, Esaki N (1998) A cold-adapted lipase of an Alaskan psychrotroph, Pseudomonas sp. strain B11-1: gene cloning and enzyme purification and characterization. Appl Environ Microbiol 64:486–491

Cowan DA (2009) Cryptic microbial communities in Antarctic deserts. Proc Natl Acad Sci U S A 106:19749–19750

D'Amico S, Collins T, Marx JC, Feller G, Gerday C (2006) Psychrophilic microorganisms: challenges for life. EMBO Rep 7(4):385–389

Das HK, Goldstein A (1968) Limited capacity for protein synthesis at zero degrees centigrade in Escherichia coli. J Mol Biol 31:209–226

De los Ríos A, Wiezchos J, Sancho LG, Ascaso C (2004) Exploring the physiological state of continental Antarctic endolithic microorganisms by microscopy. FEMS Microbiol Ecol 50:143–152

De Vos D, Collins T, Nerinckx W (2006) Oligosaccharide binding in family 8 glycosidases: crystal structures of active-site mutants of the β-1,4-xylanase pXyl from *Pseudoalteromonas haloplanktis* TAH3a in complex with substrate and product. Biochemistry 45(15):4797–4807

De Vos D, Xu Y, Hulpiau P, Vergauwen B, Van Beeumen JJ (2007) Structural investigation of cold activity and regulation of aspartate carbamoyltransferase from the extreme psychrophilic bacterium *Moritella profunda*. J Mol Biol 365(2):379–395

Di Pietro F, Brandi A, Dzeladini N, Fabbretti A, Carzaniga T, Piersimoni L, Pon CL, Giuliodori AM (2013) Role of the ribosome-associated protein PY in the cold-shock response of Escherichia coli. Microbiology 2:293–307

Diard S, Carlier JP, Ageron E, Grimont PAD, Langlois V, Guérin P (2002) Accumulation of poly (3-hydroxybutyrate) from octanoate in different *Pseudomonas* belonging to the rRNA homology group. J Syst Appl Microbiol 25:183–188

Diaz AR, Mansilla MC, Vila AJ, de Mendoza D (2002) Membrane topology of the acyl-lipid desaturase from Bacillus subtilis. J Biol Chem 277(50):48099–48106

Diefenbach R, Keweloh H, Rehm HJ (1992) Fatty acid impurities in alginate influence the phenol tolerance of immobilized Escherichia coli. Appl Microbiol Biotechnol 36(4):530–534

Domingues S, Matos RG, Reis FP, Fialho AM, Barbas A, Arraiano CM (2009) Biochemical characterization of the RNase II family of exoribonucleases from the human pathogens Salmonella typhimurium and Streptococcus pneumoniae. Biochemistry 48:11848–11857

Evstigneeva Z, Solov'eva N, Sidel'nikova L (2001) Structures and functions of chaperones and chaperonins (review). Appl Biochem Microbiol 37:1–13

Ewert M, Deming JW (2011) Selective retention in saline ice of extracellular polysaccharides produced by the cold-adapted marine bacterium *Colwellia psychrerythraea* strain 34H. Ann Glaciol 52:111–117

Fang L, Hou Y, Inouye M (1998) Role of the cold-box region in the 5′ untranslated region of the cspA mRNA in its transient expression at low temperature in Escherichia coli. J Bacteriol 180:90–95

Feller G (2017) Cryosphere and psychrophiles: insights into a cold origin of life? Life 7(2):25

Feng S, Powell SM, Wilson R, Bowman JP (2014) Extensive gene acquisition in the extremely psychrophilic bacteria species *Psychroflexus torques* and the link to sea-ice ecosystem specialism. Genome Biol Evol 6:133–148

Ferrer M, Chernikova TN, Timmis KN, Golyshin PN (2004b) Expression of a temperature-sensitive esterase in a novel chaperone-based Escherichia coli strain. Appl Environ Microbiol 70:4499–4504

Ferrer M, Chernikova TN, Yakimov MM, Golyshin PN, Timmis KN (2003) Chaperonins govern growth of Escherichia coli at low temperatures. Nat Biotechnol 21:1266–1267

Ferrer M, Lünsdorf H, Chernikova TN, Yakimov M, Timmis KN, Golyshin PN (2004a) Functional consequences of single: double ring transitions in chaperonins: life in the cold. Mol Microbiol 53:167–182

Frazao C, McVey CE, Amblar M, Barbas A, Vonrhein C, Arraiano CM, Carrondo MA (2006) Unravelling the dynamics of RNA degradation by ribonuclease II and its RNA-bound complex. Nature 443:110–114

Galán B, Dinjaski N, Maestro B, de Eugenio LI, Escapa IF, Sanz JM, García JL, Prieto MA (2011) Nucleoid-associated PhaFphasin drives intracellular location and segregation of polyhydroxyalkanoate granules in Pseudomonas putida KT2442. Mol Microbiol 79:402–418

Garsoux G, Lamotte J, Gerday C, Feller G (2004) Kinetic and structural optimization to catalysis at low temperatures in a psychrophilic cellulase from the Antarctic bacterium *Pseudoalteromonas haloplanktis*. Biochem J 384(2):247–253

Garwin JL, Cronan JE (1980) Thermal modulation of fatty acid synthesis in Escherichia coli does not involve de novo enzyme synthesis. J Bacteriol 141(3):1457–1459

Georlette D, Jonsson ZO, Petegem FV, Chessa J, Beeumen JV, Hubscher U, Gerday C (2000) A DNA ligase from the psychrophile Pseudoalteromonas haloplanktis gives insights into the adaptation of proteins to low temperatures. Eur J Biochem 267:3502–3512

Gerday C, Aittaleb M, Bentahir M, Chessa JP, Claverie P, Collins T (2000) Cold-adapted enzymes: from fundamentals to biotechnology. Trends Biotechnol 18:103–107

Gorfe AA, Brandsdal BO, Leiros HKS, Helland R, Smalås AO (2000) Electrostatics of mesophilic and psychrophilic trypsin isoenzymes: qualitative evaluation of electrostatic differences at the substrate binding site. Proteins 40(2):207–217

Górna MW, Carpousis AJ, Luisi BF (2012) From conformational chaos to robust regulation: the structure and function of the multi-enzyme RNA degradosome. Q Rev Biophys 45(2):105–145

Gottesman S (2018) Chilled in translation: adapting to bacterial climate change. Mol Cell 70 (2):193–194

Graumann PL, Marahiel MA (1998) A superfamily of proteins that contain the cold-shock domain. Trends Biochem Sci 23:286–290

Hartl FU, Hayer-Hartl M (2009) Converging concepts of protein folding in vitro and in vivo. Nat Struct Mol Biol 16(6):574

Hayashi K, Kojima C (2008) pCold-GST vector: a novel cold-shock vector containing GST tag for soluble protein production. Protein Expr Purif 62:120–127

Hazer B, Steinbüchel A (2007) Increased diversification of polyhydroxyalkanoates by modification reactions for industrial and medical applications. Appl Microbiol Biotechnol 74:1–1

Heipieper HJ, Meinhardt F, Segura A (2003) The cis–trans isomerase of unsaturated fatty acids in Pseudomonas and Vibrio: biochemistry, molecular biology and physiological function of a unique stress adaptive mechanism. FEMS Microbiol Lett 229(1):1–7

Ho YH, Gan SN, Tan IKP (2002) Biodegradation of a medium-chain-length polyhydroxyalkanoate in tropical river water. Appl Biochem Biotechnol 10:337–347
Hunger K, Beckering CL, Wiegeshoff F, Graumann PL, Marahiel MA (2006) Cold-induced putative DEAD box RNA helicases CshA and CshB are essential for cold adaptation and interact with cold shock protein B in Bacillus subtilis. J Bacteriol 188:240–248
Jones PG, Inouye M (1994) The cold-shock response--a hot topic. Mol Microbiol 11(5):811–818
Jung SK, Jeong DG, Lee MS (2008) Structural basis for the cold adaptation of psychrophilic M37 lipase from *Photobacterium lipolyticum*. Proteins 71(1):476–484
Kaan T, Homuth G, Mader U, Bandow J, Schweder T (2002) Genome-wide transcriptional profiling of the Bacillus subtilis cold-shock response. Microbiology 148:3441–3455
Kadouri D, Jurkevitch E, Okon Y, Castro-Sowinski S (2005) Ecological and agricultural significance of bacterial polyhydroxyalkanoates. Crit Rev Microbiol 31:55–67
Kandror O, Goldberg AL (1997) Trigger factor is induced upon cold shock and enhances viability of Escherichia coli at low temperatures. Proc Natl Acad Sci 94(10):4978–4981
Kandror OA, DeLeon A, Goldberg AL (2002) Trehalose synthesis in induced upon exposure of *Escherichia coli* to cold and is essential for viability at low temperatures. Proc Natl Acad Sci U S A 99:9727–9732
Kawahara H (2002) The structures and functions of ice crystal controlling proteins from bacteria. J Biosci Bioeng 94:492–496
Kawamoto J, Kurihara T, Yamamoto K, Nagayasu M, Tani Y, Mihara H et al (2009) Eicosapentaenoic acid plays a beneficial role in membrane organization and cell division of a cold-adapted bacterium, Shewanella livingstonensis Ac10. J Bacteriol 191(2):632–640
Kim H, Wi AR, Jeon BW, Lee JH, Shin SC, Park H (2015) Cold adaptation of a psychrophilic chaperonin from Psychrobacter sp. and its application for heterologous protein expression. Biotechnol Lett 37:1887–1893
Klähn S, Hagemann M (2011) Compatible solute biosynthesis in cyanobacteria. Environ Microbiol 13:551–562
Klinke S, de Roo G, Witholt B, Kessler B (2000) Role of phaD in accumulation of medium-chain-length poly (3-hydroxyalkanoates) in Pseudomonas oleovorans. Appl Environ Microbiol 66 (9):3705–3710
Krembs C, Eicken H, Deming JW (2011) Exopolymer alteration of physical properties of sea ice and implications for ice habitability and biogeochemistry in a warmer Arctic. Proc Natl Acad Sci U S A 108:3653–3658
Kulakova L, Galkin A, Kurihara T, Yoshimura T, Esaki N (1999) Cold-active serine alkaline protease from the psychrotrophic bacterium Shewanella strain ac10: gene cloning and enzyme purification and characterization. Appl Environ Microbiol 65:611–617
Lalonde MS, Zuo Y, Zhang J, Gong X, Wu S, Malhotra A, Li Z (2007) Exoribonuclease R in mycoplasma genitalium can carry out both RNA processing and degradative functions and is sensitive to RNA ribose methylation. RNA 13:1957–1968
Laycock B, Arcos-Hernandez M, Langford A, Pratt S, Werker A, Halley P, Lant P (2014) Crystallisation and fractionation of selected polyhydroxyalkanoates produced from mixed cultures. New Biotechnol 31(4):345
Leiros I, Moe E, Lanes O, Smalås AO, Willassen NP (2003) The structure of uracil-DNA glycosylase from Atlantic cod (*Gadus morhua*) reveals cold-adaptation features. Acta Crystallogr D 59(8):1357–1365
Lenz RW, Marchessault RH (2005) Bacterial polyesters: biosynthesis, biodegradable plastics and biotechnology. Biomacromolecules 6:1–8
Lim J, Thomas T, Cavicchioli R (2000) Low temperature regulated DEAD-box RNA helicase from the Antarctic archaeon, *Methanococcoides burtonii*. J Mol Biol 297:553–567
Lim SP, Gan SN, Tan IKP (2005) Degradation of medium-chain-length polyhydroxyalkanoates in tropical forest and mangrove soils. Appl Biochem Biotechnol 126:23–33
Loffeld B, Keweloh H (1996) Cis/trans isomerization of unsaturated fatty acids as possible control mechanism of membrane fluidity in Pseudomonas putida P8. Lipids 31(8):811–815

López NI, Pettinari MJ, Stackerbrandt E, Tribelli PM, Pōtter M, Steinbüchel A (2009) *Pseudomonas extremaustralis* sp. nov., a poly(3-hydroxybutyrate) producer isolated from an Antarctic environment. Curr Microbiol 59:514–519

Los DA, Mironov KS, Allakhverdiev SI (2013) Regulatory role of membrane fluidity in gene expression and physiological functions. Photosynth Res 116(2):489–509

Mansilla MC, Cybulski LE, Albanesi D, de Mendoza D (2004) Control of membrane lipid fluidity by molecular thermosensors. J Bacteriol 186(20):6681–6688

Matos RG, Barbas A, Gómez-Puertas P, Arraiano CM (2011) Swapping the domains of exoribonucleases RNase II and RNase R: conferring upon RNase II the ability to degrade dsRNA. Proteins 79:1853–1867

Mc Elhaney RN (1984) The relationship between membrane lipid fluidity and phase state and the ability of bacteria and mycoplasmas to grow and survive at various temperatures. In: Membrane fluidity. Springer, Boston, MA, pp 249–278

Medigue C, Krin E, Pascal G, Barbe V, Bernsel A, Bertin PN, Danchin A (2005) Coping with cold: the genome of the versatile marine Antarctica bacterium Pseudoalteromonas haloplanktis TAC125. Genome Res 15(10):1325–1335

Metpally RPR, Reddy BVB (2009) Comparative proteome analysis of psychrophilic versus mesophilic bacterial species: insights into the molecular basis of cold adaptation of proteins. BMC Genomics 10(1):1–10

Miteva V (2008) Bacteria in snow and glacier ice. In: Margesin R, Schinner F, Marx J-C, Gerday C (eds) Psychrophiles: from biodiversity to biotechnology. Springer, Berlin, pp 31–50

Moe E, Leiros I, Riise EK (2004) Optimisation of the surface electrostatics as a strategy for cold adaptation of uracil-DNA N-glycosylase (UNG) from Atlantic cod (*Gadus morhua*). J Mol Biol 343(5):1221–1230

Molinaro A, Holst O, Di Lorenzo F, Callaghan M, Nurisso A, D'Errico G et al (2015) Chemistry of lipid a: at the heart of innate immunity. Chem Eur J 21(2):500–519

Możejko-Ciesielska J, Pokój T, Ciesielski S (2018) Transcriptome remodeling of Pseudomonas putida KT2440 during mclPHAs synthesis: effect of different carbon sources and response to nitrogen stress. J Ind Microbiol Biotechnol 45:433–446

Muhammadi S, Afzal M, Hameed S (2015) Bacterial polyhydroxyalkanoates-eco-friendly next generation plastic: production, biocompatibility, biodegradation, physical properties and applications. Green Chem Lett Rev 8:56–77

Mykytczuk NC, Foote SJ, Omelon CR, Southam G, Greer CW, Whyte LG (2013) Bacterial growth at– 15 C; molecular insights from the permafrost bacterium Planococcus halocryophilus Or1. ISME J 7(6):1211–1226

Nogueira T, Springer M (2000) Post-transcriptional control by global regulators of gene expression in bacteria. Curr Opin Microbiol 3:154–158

Nunez-Montero K, Barrientos L (2018) Advances in Antarctic research for antimicrobial discovery: a comprehensive narrative review of bacteria from Antarctic environments as potential sources of novel antibiotic compounds against human pathogens and microorganisms of industrial importance. Antibiotics 7(4):90

Olufsen M, Smalås AO, Brandsdal BO (2008) Electrostatic interactions play an essential role in DNA repair and cold adaptation of uracil DNA glycosylase. J Mol Model 14(3):201–213

Papaleo E, Olufsen M, De Gioia L, Brandsdal BO (2007) Optimization of electrostatics as a strategy for cold-adaptation: a case study of cold- and warm-active elastases. J Mol Graph Model 26 (1):93–103

Pearce DA (2012) Extremophiles in antarctica: life at low temperatures. In: Adaption of microbial life to environmental extremes. Springer, Vienna, pp 87–118

Phadtare S, Alsina J, Inouye M (1999) Cold-shock response and cold-shock proteins. Curr Opin Microbiol 2:175–180

Phadtare S, Inouye M (2004) Genome-wide transcriptional analysis of the cold shock response in wild-type and cold-sensitive, quadruple-*csp*-deletion strains of *Escherichia coli*. J Bacteriol 186:7007–7014

Pham TH, Webb JS, Rehm BHA (2004) The role of polyhydroxyalkanoate biosynthesis by Pseudomonas aeruginosa in rhamnolipid and alginate production as well as stress tolerance and biofilm formation. Microbiology 150:3405–3413

Piette F, D'Amico S, Struvay C, Mazzucchelli G, Renaut J, Tutino ML, Feller G (2010) Proteomics of life at low temperatures: trigger factor is the primary chaperone in the Antarctic bacterium Pseudoalteromonas haloplanktis TAC125. Mol Microbiol 76(1):120–132

Piette F, Struvay C, Feller G (2011) The protein folding challenge in psychrophiles: facts and current issues. Environ Microbiol 13(8):1924–1933

Prabagaran SR, Suresh K, Manorama R, Delille D, Shivaji S (2005) Marinomonas ushuaiensis sp. nov., isolated from coastal sea water in Ushuaia, Argentina, sub-Antarctica. Int J Syst Evol Microbiol 55(1):309–313

Purusharth RI (2006) Characterization of the RNA degrading machinery (degradosome) from the Antarctic psychrotrophic bacterium Pseudomonas syringae. Ph D thesis, Jawaharlal Nehru University, New Delhi, India

Purusharth RI, Klein F, Sulthana S, Jager S, Jagannadham MV, Evguenieva-Hackenberg E, Ray MK, Klug G (2005) Exoribonuclease R interacts with endoribonuclease E and an RNA-helicase in the psychrotrophic bacterium Pseudomonas syringae LZ 4W. J Biol Chem 280:14572–14578

Purusharth RI, Madhuri B, Ray MK (2007) Exoribonuclease R in *Pseudomonas syringae* is essential for growth at low temperature and plays a novel role in the 3′ end processing of 16 and 5 S rRNA. J Biol Chem 282:16267–16277

Qing G, Ma L, Khorchid A, Swapna G, Mal TK, Takayama MM (2004) Cold-shock induced high-yield protein production in Escherichia coli. Nat Biotechnol 22:877–882

Rabus R, Ruepp A, Frickey T, Rattei T, Fartmann B, Stark M, Klenk HP (2004) The genome of Desulfotalea psychrophila, a sulfate-reducing bacterium from permanently cold Arctic sediments. Environ Microbiol 6(9):887–902

Raeder ILU, Moe E, Willassen NP, Smalås AO, Leiros I (2010) Structure of uracil-DNA N-glycosylase (UNG) from *Vibrio cholerae*: mapping temperature adaptation through structural and mutational analysis. Acta Crystallograph F 66(2):130–136

Ramírez-Sarmiento CA, Baez M, Wilson CA, Babul J, Komives EA, Guixé V (2013) Observation of solvent penetration during cold denaturation of E. coli phosphofructokinase-2. Biophys J 104:2254–2263

Ray MK, Seshu Kumar G, Janiyani K, Kannan K, Jagtap P, Basu MK, Shivaji S (1998) Adaptation to low temperature and regulation of gene expression in Antarctic psychrotrophic bacteria. J Biosci 23:423–435

Riley M, Staley JT, Danchin A, Wang TZ, Brettin TS, Hauser LJ, Thompson LS (2008) Genomics of an extreme psychrophile, Psychromonas ingrahamii. BMC Genomics 9(1):1–19

Rohini K (2010) Molecular evolution of cell division proteins FtsA, FtsL, and FtsZ in bacteria: a phylogenetic analysis. Malays J Microbiol 6:94–98

Russell N, Fukunaga N (1990) A comparison of thermal adaptation of membrane lipids in psychrophilic and thermophilic bacteria. FEMS Microbiol Lett 75:171–182

Russell NJ (2000) Towards a molecular understanding of cold activity of enzymes from psychrophiles. Extremophiles 4:83–90

Russell RJM, Gerike U, Danson MJ, Hough DW, Taylor GL (1998) Structural adaptations of the cold-active citrate synthase from an Antarctic bacterium. Structure 6(3):351–362

Sandoval A, Arias-Barrau E, Arcos M, Naharro G, Olivera ER, Luengo JM (2007) Genetic and ultrastructural analysis if different mutants of Pseudomonas putida affected in the poly-3-hydroxy-n-alkanoate gene cluster. Environ Microbiol 9(3):737–751

Shin SC, Kim SJ, Lee JK, Ahn DH, Kim MG, Lee H, Park H (2012) Transcriptomics and comparative analysis of three Antarctic notothenioid fishes. PLoS One 7(8):e43762

Shivaji S, Prakash JS (2010) How do bacteria sense and respond to low temperature? Arch Microbiol 192(2):85–95

Shuo-Shuo C, Xue-Zheng L, Ji-Hong S (2011) Effects of co-expression of molecular chaperones on heterologous soluble expression of the cold-active lipase Lip-948. Protein Expr Purif 77:166–172

Siddiqui KS, Cavicchioli R (2006) Cold-adapted enzymes. Annu Rev Biochem 75:403–433

Singh AK, Pindi PK, Dube S, Sundareswaran VR, Shivaji S (2009) Importance of trmE for growth of the psychrophile *Pseudomonas syringae* at low temperatures. Appl Environ Microbiol 75:4419–4426

Singh AK, Shivaji S (2010) A cold-active heat-labile t-RNA modification GTPase from a psychrophilic bacterium *Pseudomonas syringae* (Lz4W). Res Microbiol 161:46–50

Skalova T, Dohnalek J, Spiwok V (2005) Cold-active beta-galactosidase from *Arthrobacter* sp. C2-2 forms compact 660 kDa hexamers: crystal structure at 1. 9 Å resolution. J Mol Biol 353:282–294

Smalås AO, Leiros HK, Os V, Willassen NP (2000) Cold adapted enzymes. Biotechnol Annu Rev 6:1–57

Sundareswaran VR, Singh AK, Dube S, Shivaji S (2010) Aspartate aminotransferase is involved in cold adaptation in psychrophilic *Pseudomonas syringae*. Arch Microbiol 192:663–672

Suutari M, Laakso S (1992) Changes in fatty acid branching and unsaturation of Streptomyces griseus and Brevibacterium fermentans as a response to growth temperature. Appl Environ Microbiol 58(7):2338–2340

Thomas T, Cavicchioli R (2000) Effect of temperature on stability and activity of elongation factor 2 proteins from Antarctic and thermophilic methanogens. J Bacteriol 182(5):1328–1332

Uh JH, Jung YH, Lee YK, Lee HK, Im H (2010) Rescue of a cold-sensitive mutant at low temperatures by cold shock proteins from Polaribacter irgensii KOPRI 22228. J Microbiol 48 (6):798–802

Uma S, Jadhav RS, Kumar GS, Shivaji S, Ray MK (1999) A RNA polymerase with transcriptional activity at 0 degrees C from the Antarctic bacterium *Pseudomonas syringae*. FEBS Lett 453:313–317

Vester JK, Glaring MA, Stougaard P (2015) Improved cultivation and metagenomics as new tools for bioprospecting in cold environments. Extremophiles 19:17–29

Vincent HA, Deutscher MP (2006) Substrate recognition and catalysis by the exoribonuclease RNase R. J Biol Chem 281:29769–29775

Violot S, Aghajari N, Czjzek M (2005) Structure of a full length psychrophilic cellulase from *Pseudoalteromonas haloplanktis* revealed by X-ray diffraction and small angle X-ray scattering. J Mol Biol 348(5):211–1224

Walid ME-S, Graumann PL (2007) Cold shock proteins aid coupling of transcription and translation in Bacteria. Sci Prog 90(1):15–27

Wang JG, Bakken LR (1998) Screening of soil bacteria for poly-β-hydroxybutyric acid production and its role in the survival of starvation. Microb Ecol 35:94–101

Weber MH, Klein W, Müller L, Niess UM, Marahiel MA (2001) Role of the Bacillus subtilis fatty acid desaturase in membrane adaptation during cold shock. Mol Microbiol 39(5):1321–1329

Wilson SL, Frazer C, Cumming BF, Nuin PA, Walker VK (2012) Cross-tolerance between osmotic and freeze-thaw stress in microbial assemblages from temperate lakes. FEMS Microbiol Ecol 82:405–415

Wolffe AP (1994) Structural and functional properties of the evolutionarily ancient Y-box family of nucleic acid binding proteins. BioEssays 16(4):245–251

Wolffe AP, Tafuri S, Ranjan M, Familari M (1992) The Y-box factors: a family of nucleic acid binding proteins conserved from Escherichia coli to man. New Biol 4(4):290–298

Wu Z, Kan FW, She YM, Walker VK (2012) Biofilm, ice recrystallization inhibition and freeze-thaw protection in an epiphyte community. Appl Biochem Microbiol 48:363–370

Yang LL, Tang SK, Huang Y, Zhi XY (2015) Low temperature adaptation is not the opposite process of high temperature adaptation in terms of changes in amino acid composition. Genome Biol Evol 7(12):3426–3433

Yasuda W, Kobayashi M, Takada Y (2013) Analysis of amino acid residues involved in cold activity of monomeric isocitrate dehydrogenase from psychrophilic bacteria, *Colwellia maris* and *Colwellia psychrerythraea*. J Biosci Bioeng 116:567–572

Yoshimune K, Galkin A, Kulakova L, Yoshimura T, Esaki N (2005) Cold-active DnaK of an Antarctic psychrotroph Shewanella sp. Ac10 supporting the growth of dnaK-null mutant of Escherichia coli at cold temperatures. Extremophiles 9(2):145–150

Zhang Y, Burkhardt DH, Rouskin S, Li GW, Weissman JS, Gross CA (2018) A stress response that monitors and regulates mRNA structure is central to cold shock adaptation. Mol Cell 70 (2):274–286

Zuo Y, Vincent HA, Zhang J, Wang Y, Deutscher MP, Malhotra A (2006) Structural basis for processivity and single-strand specificity of RNase II. Mol Cell 24:149–156

Microbe-Mediated Plant Functional Traits and Stress Tolerance: The Multi-Omics Approaches

11

Vinay Kumar, Lata Jain, Swati Jain, and Ravindra Soni

Abstract

Plants are sessile in nature and have to cope up with adverse effect of environmental conditions and biotic stresses using their intrinsic biological mechanisms. Microorganisms are considered as one of the important natural inhabitants of diverse environments and have tremendous metabolic capabilities to alleviate biotic and abiotic stresses. These plant-associated microorganisms assist in the plant growth and development and in providing immunity to the host by producing hormones and enhancing the uptake of nutrients and synthesis of secondary metabolites, small peptides, and molecules for defense response. The interaction between plant and associated microbes is key component of the living ecosystem, considered as natural partner which modulates several functional traits including systemic defense response and providing tolerance to host several abiotic stresses in plants. This chapter will comprehend the various types of microorganisms associated with plants (phyllosphere and rhizosphere), role of plant-microbe interaction in governing the plant traits, and stress (biotic and abiotic) tolerance mechanism. Recent concept that proved the essential role of microbes in governing plant traits includes holobionts and DefenseBiome. Applications of recent biotechnological and multi-omics approaches for deciphering the role of microbes governing plant traits and mitigating stresses.

V. Kumar (✉) · L. Jain
ICAR-National Institute of Biotic Stress Management, Raipur, Chhattisgarh, India

S. Jain
Department of Computer Science, Government J. Yoganandam Chhattisgarh College, Raipur, India

R. Soni
Department of Agricultural Microbiology, Indira Gandhi Krishi Vishwavidyalaya, Raipur, India

R. Goel et al. (eds.), *Survival Strategies in Cold-adapted Microorganisms*,
https://doi.org/10.1007/978-981-16-2625-8_11

Keywords

Microbiome · Plant stress · Endophytes · Omics approaches · Plant-microbe interaction

11.1 Introduction

Plants host diverse microbial communities called plant microbiota, which colonize a variety of plant tissues. Microbiota are comprised of all living organisms including archaea, algae, bacteria, fungi, and small protists forming the microbiome (Berg et al. 2020). Plant microbiomes are generally shaped by both factors, namely, plant genotype, organ, tissues, species, and health status and plant's environment, such as soil, soil condition, and climatic factors. Plant microbiome consists of rhizosphere and phyllosphere, which are involved in the interaction with the host plants. Soil and plant microbiomes have been considered as crucial determinants of plant fitness and productivity and microbial diversity, and balance is essential for maintaining plant health (Yan et al. 2008; Singh et al. 2018).

11.1.1 Rhizosphere Microbiome

Rhizosphere is a complex and dynamic environment where a massive number of interactions between roots, microorganisms, organic compounds, gases, and minerals occurred to drive the biogeochemical cycling of elements (Oburger and Schmidt 2016). The rhizosphere refers to the portion of soil where microorganism-mediated processes are under the influence of the root system. Rhizosphere basically considered as a biologically active zone of the soil around plant roots which harbor soil-borne microorganisms. These microbes are known to play a crucial role in various vital processes of the plants including plant health, nutrition, carbon sequestration, and nutrient cycling in terrestrial ecosystems (Berg and Smalla 2008). The microbial communities present in the rhizosphere and their structural and functional diversity are highly influenced by plant genotypes, species, soil types, edaphic factors, and an array of biotic and abiotic stresses imposed on the plants (Mendes et al. 2011; Mendes et al. 2013; Goel et al. 2017). Several studies have been conducted to explore the hidden bacterial world and diversity of rhizospheric microbiome including rhizoplane and endosphere using culture-based and culture-independent approach and metagenomics-based approach (Soni et al. 2017; Armanhi et al. 2018; Goel et al. 2018b; Alawiye and Babalola 2019). The rhizospheric microbes have strong impact in plant health and development as they facilitate nutrient solubilization and mobilization, helping plant to tolerate abiotic stresses (Pérez-Jaramillo et al. 2015; Goel et al. 2018a).

11.1.2 Phyllosphere Microbiome

The phyllosphere refers to aboveground plant tissues which provide habitats for diverse microbial communities (Lei 2020). Microbes (bacteria and fungi) live in the symbiotic relationship within the aerial plant tissues including stems, leaves, buds, flowers, and seeds of the host plants (Carvalho and Castillo 2018). This is an open system and greatly influenced by the variable environmental factors, availability of nutrients, and microbial species have to manage to colonize under the prevalent environmental conditions. Thus, phyllosphere microbiota is an important component to study the diversity, interrelationship, and flow of resources and energy within and between microbial communities and the host plants (Lindow and Brandl 2003). Successful interactions play a vital role in maintaining the homeostasis of plants and confers several benefits to the plant host including growth promotion, nutrient uptake defense against pathogens, and tolerance to different environmental stresses including cold stresses (Venkatachalam et al. 2016; Saleem et al. 2017; Dash et al. 2019; Dubey et al. 2020; Kumar et al. 2020c). Among the phyllosphere, the microbes can be categorized as epiphytes and endophytes. Microorganisms that live on the surface of plants and organs such as leaves, flowers, and seeds are called epiphytes, while endophytes are the microbes that have potential to live inside the plant and their tissues and play a significant role in the growth and health of the host plant.

11.1.2.1 Endophytic Microbes

Endophytes are microbes (bacterial and fungi) which live within the plant tissues, internally causing any disease symptoms, and are known to possess beneficial effect on host plants. It was reported that the structure and composition of microbial communities varied among plant tissues, genotypes, developmental stages, soil conditions, altitude level, and other edaphic factors (Hallmann et al. 1997; Hardoim et al. 2015; Kumar et al. 2019). Studies have identified the bacterial and fungal endophytes from different crop plants, and their diversities have been explored. Bacterial endophytes have been isolated from different plant tissues, namely, root, leaf, stem, and seed tissues of legume crops *Lathyrus* and rice (Kumar et al. 2018; Kumar et al. 2020a). These microbes have different roles in the promotion of plant growth and protection of plant from various biotic and abiotic stresses, production of phytoharmones, and bioactive compounds (Santoyo et al. 2016; Kumar et al. 2020b). Seed microbiome is considered as very crucial for the host, as it is the first stage of plant's life cycle (Truyens et al. 2015). These microbes are different from the epiphytes as they have the ability to transfer vertically from seed to the seedling and to the next generation. Endophytes have additional benefits over other phyllosphere microbes as they live inside and grow with the plant; hence, their growth is not influenced by the external environmental factors, and they produce various bioactive compounds for survival and adaptation of host in the extreme environmental conditions. These microbes actively respond to various biotic and environmental stresses which hamper the overall agricultural growth and productivity.

The plant-microbe interaction can be positive, negative, and neutral interactions. The positive interaction refers to symbiotic/mutual and negative interactions to productive (either one or both the participants harmed by each other's), while neutral interactions are the irrelevant interactions (Carvalho and Castillo 2018). The proper balance of microbial diversity is essential to maintain the health of host plants; microbial species have the potential to contribute to the host in various ways including (1) germination of seeds and growth promotion, nutrient supply by nitrogen fixation, and solubilization and mobilization of phosphorus and minerals, (2) resistance/tolerance to various biotic and abiotic stresses, and (3) influencing physiology and production of bioactive metabolites (Truyens et al. 2015; Berg et al. 2017). The contribution of plant-associated microbes in such a way that any disturbance in the microbial composition/microbial dysbiosis/imbalance led to the appearance of disease.

11.2 Plant-Associated Microbial Communities and their Functional Roles

Plants select a specific group of microbes to live outside and inside their roots which provide an important functions related to the growth and health of the plant host (Song et al. 2020). These microbes include fungi, bacteria, archaea, viruses, protists, mycorrhizal candidates, and communities. The soil and plant microbiomes have been considered as a crucial determinant of plant fitness and productivity (Singh et al. 2018). The fitness of plants is the result of physical and physiological functions governed by the plant genome as well as the associated microbiome, which collectively called as a "plant holobiont" (Vandenkoornhuyse et al. 2015). Plant as a holobiont governs its associated microbiota/microbiome both in aboveground (phyllosphere) and belowground (rhizosphere) of the plant (Kaiser et al. 2015; Hacquard 2016; Berg et al. 2017). It has been reported that the development and appearance of phenotypes of plant and association of microbiota are determined by both the host plant and environmental factors (Wagner et al. 2016). The plant microbiome jointly acts as a collection for genes, which may improve the total genetic potential of the plant by enhancing plant growth and survival under stress conditions (Wang and Haney 2020). In addition, plant-associated microbial communities have the potential to contribute in numerous aspects which comprise beneficial effects on the health, fitness, and evolution of the host including vital functions (Bulgarelli et al. 2013). Studies have reported the plant microbiome has an important role in the development and establishment of plant diseases (Trivedi et al. 2012; Erlacher et al. 2014). The good balance of microbes decides the impact of microbes in plants such as disbalance of microbes, leading to disease development and pathogenesis. Several reports showed the role of phyllosphere and rhizosphere microbes to produce growth regulators, namely, auxins and cytokinins, which enhance plant growth and increase nutrient uptake via root elongation and enhancing photosynthesis (Mwajita et al. 2013).

11.2.1 Seed Germination and Nutrient Acquisition

Microbes associated with seeds are very crucial as they help in the processes of seed germination and promotion of seedling growth through the production of growth hormones and may enable seedlings to survive and thrive under adverse environmental conditions as well (Kandel et al. 2017; Verma and White 2019). The seed microbiome has special role as it confirms the dissemination and conservation of coevolved microbial communities to the next generation which may be necessary for seed germination in different plant phyla (Jacquemyn et al. 2015; van der Heijden et al. 2016). These microbes assist the plant in the acquisition and availability of nutrients from the soil by processes of nitrogen fixation and solubilization and mobilization of phosphate and other minerals. They have been found to promote growth and decrease stress damage in different plant species (Compant et al. 2005; Mastretta et al. 2009).

11.2.2 Phytoharmones/Bioactive Compound Production

Plant hormones/phytoharmones, namely, auxins, cytokinins, gibberellins, ethylene, and abscisic acid, are known to play a crucial role in plant growth and development, lateral root formation, and flower morphogenesis and assist plants to survive in adverse environmental conditions (Aloni et al. 2006). Microbes have the ability to produce or modulate production of phytohormone level impacting plant hormone homeostasis. Rhizosphere microbes have been reported to modulate phytohormones and thereby influence the hormonal balance of plants and their responses to stressful environment (Glick et al. 2007). In addition, microbes have the ability to produce various types of bioactive compounds/metabolites like alkaloids, aromatic compounds, terpenoids, polypeptides, enzymes, lipo-peptides, and other enzymes with the activity to hydrolyze cellulose, hemicellulose, chitin, proteins, etc. These compounds provide protection to the plants from various pathogens and abiotic stresses. It has been reported that the *Fusarium* spp. was found to promote growth of *Euphorbia pekinensis* and enhance the level of terpenoids (Yang et al. 2009). Recent studies have further proved the role of endophytic microbe in the synthesis of phytoharmones (Turbat et al. 2020).

11.2.3 Tolerance to Biotic and Abiotic Stresses

The plant-associated microbiome plays a significant role in maintaining the health of both plant and ecosystem. The beneficial interaction of plant with the associated microbiota is solely responsible for maintaining the health of the holobiont. Any disbalance in the same or microbial symbiosis often correlated with the appearance and development of disease (Berg et al. 2017). Microbial communities have been reported to play a central role in the functioning of the plants through modulating the physiological and developmental processes (Mendes et al. 2013). Furthermore, studies have reported that plant microbiome plays a key role in epigenetic and phenotypic plasticity and evolution of plants (Partida-Martınez and Heil 2011). Plants generally overcome the adverse effect of stresses by adjusting or modulating their metabolism.

11.2.3.1 Resistance Against Biotic Stress

Plant-associated microbial communities develop a network which influences the microbial colonization and invasion. These networks of microbes provide an opportunity for enhancing disease resistance/tolerance and biocontrol agent. Plant-associated microbes assist the plants to suppress disease development by inhibiting the pathogen growth and stimulation of growth to occupy space that would otherwise be available to pathogenic microbes. Recent studies reported that the microbes associated with plants, and whose abundance is increased during the plant stress conditions called as 'DefenseBiome' which could benefits the plant health through various ways (Liu et al. 2020b). In addition, they also promote stress tolerance and influence crop yield by nutrient acquisition and solubilization (Lugtenberg and Kamilova 2009). Microbes assist the host plants to cope up with the adverse effects of pathogens and provide tolerance mechanisms through biocontrol of plant pathogens in the rhizospheric region (root zone) through antagonism, synthesis of antimicrobial compounds, siderophore production, and induction of systemic acquired resistance (SAR) in the host. Biocontrol of pathogens may be controlled through competition for nutrients, production of antimicrobial metabolites, and activation of immune responses in plant (Islam et al. 2015; Desgarennes et al. 2020). Endosphere and rhizosphere microbes were found to suppress plant diseases caused by *Gaeumannomyces graminis* and damping off caused by *Rhizoctonia solani* fungal pathogens (Duran et al. 2018; Carrión et al. 2018). Similarly, endophytes (bacteria and fungi) have been identified from different plants and used as plant growth-promoting and biocontrol agent for controlling plant pathogens (Berg et al. 2017; Kumar et al. 2020a, 2020b). *Trichoderma* spp. is a beneficial fungus that essentially lives in the soil and/or soil ecosystems and is widely used for plant growth promotion and biopesticides to control the pathogenic microbes, more specifically soil-borne fungal pathogen including *Rhizoctonia* (Chaudhary et al. 2020).

11.2.3.2 Resistance Against Abiotic Stresses

An abiotic stress includes extreme conditions such as drought, high and low temperatures, salinity, and high rainfall which affect the productivity of agricultural crops. Plant adopts diverse machinery to mitigate the adverse effects of extreme climatic conditions and abiotic stresses (Pareek et al. 2020). Among them, plant-associated microbial communities have been reported to mitigate the impact of abiotic stresses in most effective manner (Meena et al. 2017). Plant growth-promoting (PGP) beneficial bacteria found to assist plant to cope up with abiotic stresses and promote their normal physiological functions (Ojuederie et al. 2019).

11.2.3.3 Cold Stress Tolerance

Mountain or hill ecosystems are known for the uniqueness of the agro-forestry and agricultural practices. Plants grown in these conditions are reservoir of microbial communities which have the potential to grow and colonize at low temperature or cold environments (Pandey et al. 2018). Cold-adapted microbes which have been isolated from high altitudes in the Indian Himalayan region (IHR) were found to possess different functional attributes like nutrient mobilization or mineralization at low temperatures (Kumar et al. 2019). Plants have complex regulatory mechanism to direct crop tolerance under cold and other environmental stresses and induce several biochemical and physiological changes in plants which enable them to acclimatize and survive in changed temperature condition. Natural mechanism of plants is to cope up with the environmental stresses; on the basis of time tree, it was believed that flowering plants made to fight the cold. Another mechanism involved by plants includes dropping their leaves seasonally for shutting down the pathways that may normally transport water between roots and leaves. Cold tolerance is a very multifarious trait which comprises the ability plant to tolerate the adverse effect of low temperature and formation of ice within and surrounding of the plant tissue to maintain the proper functioning of plants (Preston and Sandve et al. 2013). In addition, cold-adapted microorganisms are gaining importance due to their ability to colonize and inhabit in extreme low-temperature environments. As these microbes have the ability to produce cold-active enzymes with higher activity at low temperatures, they produce various compounds to protect themselves against intracellular freezing.

Microbe-mediated plant cold tolerance involves various physiological and molecular mechanisms underlying plant-microbe interaction (Kumar and Verma 2018). These microorganisms are emerging as promising source of bioactive metabolites and cold-active enzymes, secondary metabolites, natural pigments, recombinant proteins, and biofuels (Pandey et al. 2018). Antifreeze proteins have the ability to modify the ice crystal structure and inhibit the growth of ice (Margesin and Miteva 2011). Acuña-Rodríguez et al. (2020) showed the functional roles of microbial symbionts in plant tolerance to cold and freezing stresses modulating the hormonal signaling, antioxidant activity, and osmotic balance in host. It was found that bacterial endosymbionts *Azospirillum*, *Burkholderia*, and *Pseudomonas* affected the photosynthesis leading to the accumulation of trehalose and raffinose which helped in maintaining cell osmotic pressure and integrity of plasma membrane. Several studies have confirmed the involvement of bacterial and fungal endophytes in mitigating environmental stresses by the production of certain antioxidants, metabolites, enzymes, synthesis of stress hormones like ethylene, ABA, and compatible solutes in the host plants. They also found to induce or enhance the expression of genes related to stress-mitigating pathways (Pandey et al. 2018; Kumar et al. 2020b; Dubey et al. 2020).

11.3 Multi-Omics Approaches for Characterization of Plant Microbiome

Recent developments in the field of molecular biology and genomics led to the development of multiple "-omics" technologies which enabled us to gain insights into the genomics characteristics and structural and functional attributes of plant-associated microbes (Qin et al. 2016). These approaches hold a huge potential for unlocking fundamental and applied insights in microbial plant disease ecology and deciphering the genetic basis for plant pathogen resistance and susceptibility and role of microbial community in influencing the gene expression and cascade of metabolic pathway operated under stress conditions/infections (Crandall et al. 2020). With the availability of genome sequence data, it became possible to characterize the plant microbiome and identify the links connecting to the development and management of diseases (Berg et al. 2017). Comparative genomics aims to identify structural and functional genomics elements conserved within the species or across different species. Multi-omics approaches used for the identification of stress-induced elements are depicted in Fig. 11.1. The different omics approaches are the following:

- Genomics: It enables the structural and functional analysis of genes, genome analysis, and identification of genetic variants and genetic compositions.
- Transcriptomics: It provides information about the differentially expressed gene under stress conditions and global gene expression profiling different developmental stages and identification of functional regulators/factors.
- Proteomics: It involves the qualitative and quantitative analysis of proteins, identification of proteins and stress-induced proteins/enzymes, and posttranslational modifications.
- Metabolomics: This allows the profiling of metabolites, hormones, and signaling molecule produced by the plants and during the plant-microbe interactions.
- Ionomics: It enables identification, characterization, and distribution of elements in the plant and associated microbiota, signaling molecule.
- Interactomics: It offers information related to protein-protein interaction and molecule interaction to identify the compatible and noncompatible interactions.
- Phenomics: Morphological, molecular, physiological, and biochemical characterization.

Due to large-scale/huge amount of genomics data generated by these techniques, bioinformatics and computational approaches are essentially required to derive the meaningful results and get the useful information. The multi-omics resources need to have integrated approaches for efficient utilization of resources and in-depth understanding of the molecular processes.

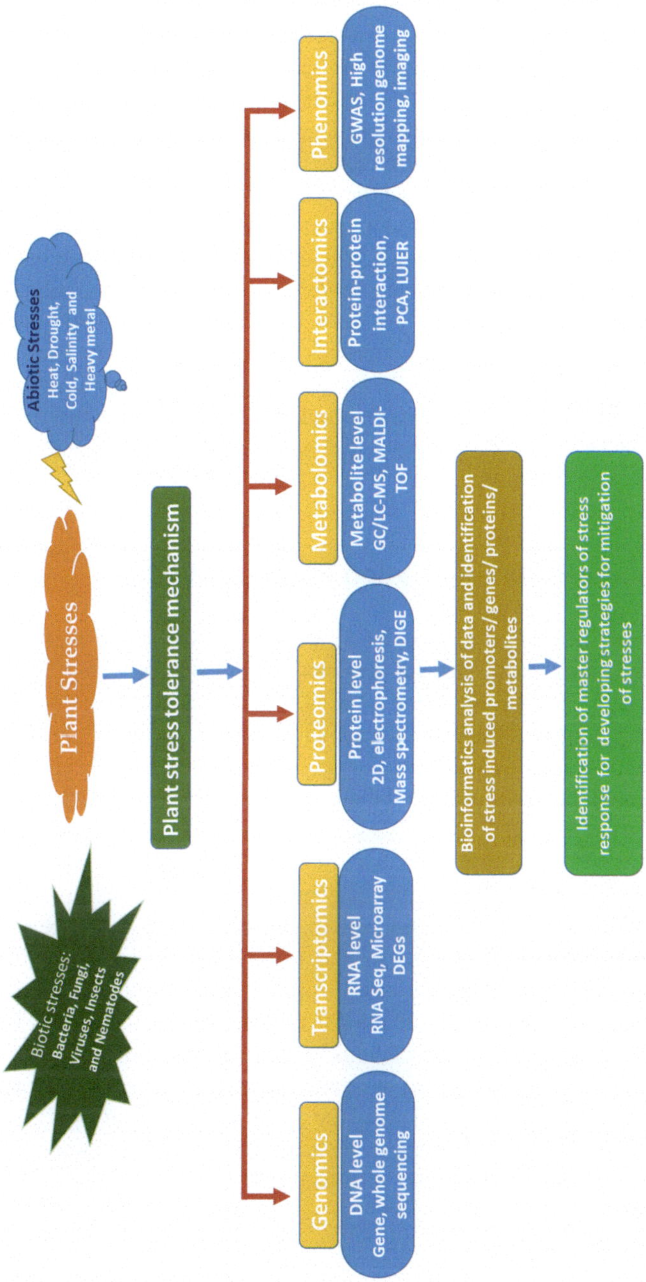

Fig. 11.1 A schematic representation of *-omics* approach for identification of master regulator of stress response in plants

11.4 Bacterial Genome Sequencing

The first whole genome of bacteria, *Haemophilus influenzae*, was sequenced in the year 1995 (Fleischmann et al. 1995). Thereafter, with the development of modern genomics technologies like next-generation sequencing (NGS), it became possible to sequence the whole genome of any organism within a short time (Mehla et al. 2011). Nowadays, sequencing of bacterial genome sequences appears to have a standard procedure to get the information related to the genetic composition of microbes and their functional role, and the availability of massive bacterial genome sequences in public domain had a major impact on bacterial world (Land et al. 2015). Studies have reported the sequencing of rhizospheric and endophytic bacteria and characterized the genes involved in the various biochemical and metabolic pathways including the genes and enzymes involved in mitigating the biotic and abiotic stresses. A list of major bacterial endophytes with plant growth-promoting (PGP) attributes and antimicrobial activities against pathogen is presented in Table 11.1. Abundance of genomics data helps in understanding the process and factors involved in mutualism and parasitism and evolution of traits (Kahlke et al. 2012).

11.5 Genome-Wide Association Studies (GWAS)

Genome-wide association studies (GWAS) are very powerful and useful approach that can be used to identify genetic variants responsible for naturally occurring variation in phenotypic and quantitative traits. GWAS approach is mainly used to identify gene function as it involves screening of large numbers of genetic variants for deletions and insertions (indels) and/or single-nucleotide polymorphisms (SNPs) within a population of an organism and their significant associations with the specific phenotype (Corvin et al. 2010). GWAS in bacteria offers a new opportunity due to the advances in whole-genome sequencing (WGS) approach and provides a complement to the approaches for analyzing gene function using de novo mutations (Read and Massey 2014; Epstein et al. 2018). This approach is also being used in analyzing the impact of host genetics on the associated microbiomes in organisms (Beilsmith et al. 2019). GWAS approach is highly useful for identification of plant genetic variation associated with microbes in a community perspective. This technique has enabled the testing of associations between genetic variation at a particular locus and organismal phenotypes to several loci across the genome (Brachi et al. 2011). Epstein et al. (2018) conducted the association analyses to 20 traits in agricultural and ecological important bacteria, *Ensifer meliloti*, as it involves in the fixation of nitrogen during symbiosis with legume crops.

Table 11.1 List of endophytic bacteria with their functional characteristics whose whole genome has been sequenced

Sr. no.	Name of host plant and tissues/ samples	Endophyte species sequenced	Genome size (Mb)	Plant growth-promoting traits/ activities	References
1.	Rice	*Azoarcus* sp. BH72	4.37	Nitrogen fixation	Krause et al. (2006)
2.	Rice	*Pseudomonas stutzeri* A1501	4.50	Nitrogen fixation	Yan et al. (2008)
3.	Maize, wheat	*Klebsiella pneumoniae* 342	5.90	Nitrogen fixation	Fouts et al. (2008)
4.	Poplar	*Pseudomonas putida* W619	5.77	IAA synthesis, ACC deaminase	Taghavi et al. (2009)
5.	Sugarcane, rice, coffee, tea	*Gluconacetobacter diazotrophicus* Pal5	3.90	Nitrogen fixation, auxin synthesis	Bertalan et al. (2009)
6.	Poplar	*Enterobacter* sp. 638	4.67	Siderophore, IAA, acetoin and 2,3-butanediol synthesis, antifungal action (indirect PGP)	Taghavi et al. (2009)
7.	Soybean	*Serratia proteamaculans* 568	5.50	IAA synthesis, ACC deaminase, acetoin and 2,3-butanediol synthesis	Taghavi et al. (2009)
8.	Poplar	*Stenotrophomonas maltophilia* R551–3	4.57	IAA synthesis, ACC deaminase	Taghavi et al. (2009)
9.	Rice	*Azospirillum* sp. B510	7.60	Nitrogen fixation, phytohormone secretion	Kaneko et al. (2010)
10.	Poplar (*Populus trichocarpa*, stem	*Enterobacter* sp. 638	4.51	PGP activities	Taghavi et al. (2010)
11.	Rice, maize, wheat	*Azospirillum lipoferum* 4B	6.85	Nitrogen fixation, phytohormone secretion	Wisniewski-Dye et al. (2011)
12.	Potato, tomato, maize, barley, onion, canola, grapevine	*Burkholderia phytofirmans* PsJN	8.2	IAA synthesis, ACC deaminase	Weilharter et al. (2011)
13.	Rice root	*Stenotrophomonas maltophilia*	4.66	Volatile organic compounds with antifungal activity	Zhu et al. (2012)

(continued)

Table 11.1 (continued)

Sr. no.	Name of host plant and tissues/ samples	Endophyte species sequenced	Genome size (Mb)	Plant growth-promoting traits/ activities	References
14.	Rice	*Burkholderia* spp. KJ006	6.60	ACC deaminase, nif gene cluster, antifungal action (indirect PGP)	Kwak et al. (2012)
15.	Pepper	*Enterobacter cloacae* ENHKU01	4.70	Role in PGP activities	Liu et al. (2020a)
16.	Rapeseed root	*Serratia plymuthica* strain AS13	5.44	Plant growth-promoting bacteria with antagonistic effect on plant pathogens	Neupane et al. (2012)
17.	Rice	*Pantoea ananatis* strain AMG521	4.89	Plant growth-promoting bacterial endophyte	Megías et al. (2016)
18.	*Butea monosperma*, root tissues	*Enterobacter* sp. MR1	4.58	Plant growth-promoting (PGP) activity	Parakhia et al. (2016)
19.	Potato, endosphere	*Bacillus mycoides* M2E15	6.08	Genes for proteins involved in phosphate utilization, iron acquisition	Yi et al. (2016)
20.	*Miscanthus giganteus*	*Pseudomonas fluorescens*	6.22	Potential biofertilizer strains	Moreira et al. (2016)
21.	*Santiria apiculata*, stem tissues	*Paenibacillus tyrfis* strain SUK123	8.04	Plant growth-promoting hormones and antimicrobial peptide production	Haruna et al. (2017)
22.	Cover, *Trifolium repens*, root tissues	*Paraburkholderia* sp. strain A27	7.39	Contains genes for host colonization	Laugraud et al. (2017)
23.	*Festuca rubra*, aerial tissues	*Paenibacillus amylolyticus* strain GM1FR	7.30	Plant growth promotion and biocontrol	Poehlein et al. (2018)
24.	*Panax quinquefolius*, Root tissues	*Chryseobacterium indologenes* PgBE177	5.00	Potential to stimulate plant growth	Hong et al. (2018)
25.	Plant tissues of *Lolium perenne*	*Bacillus mycoides* strain GM6LP	6.20	Plant growth-promoting bacterium	Wemheuer et al. (2018a)

(continued)

Table 11.1 (continued)

Sr. no.	Name of host plant and tissues/ samples	Endophyte species sequenced	Genome size (Mb)	Plant growth-promoting traits/ activities	References
26.	*Festuca rubra*, plant tissues	*Paenibacillus* sp. strain GM2FR	7.40	Contains genes for PGP and siderophore and bacillibactin	Wemheuer et al. (2018b)
27.	*Solanum lycopersicum*, leaf tissues	*Bacillus pumilus* SCAL1	3.75	Heat-tolerant plant growth-promoting bacterium	Mukhtar et al. (2018)
28.	*Pellaea calomelanos*, leaf tissues	*Enterobacter hormaechei* strain MHSD6	4.81	Defense of plant and secondary metabolite production	Tshishonga and Serepa-Dlamini (2019)
29.	Sugarcane, *Saccharum officinarum*, stem	*Kosakonia radicincitans* UYSO10	6.58	Plant growth-promoting bacterium of	Beracochea et al. (2019)
30.	*Dicoma anomala*, leaf tissues	*Bacillus* sp. strain MHSD28	5.57	PGP activities	Makuwa and Serepa-Dlamini (2019)
31.	Switchgrass plant, Leaf tissues	*Microbacterium* sp. strain LKL04	2.90	Isolated from coal mining site and PGP activities	Sahib et al. (2019)
32.	Apple fruit	*Bacillus velezensis* PG12	3.99	Inhibition of growth of broad spectrum fungal pathogens	Zeng et al. (2019)
33.	*Citrus sinensis*, branches	*Streptomyces* sp. strain CBMAI 2042	8.21	Inhibits the growth of *Xylella fastidiosa*	Gonzaga de Oliveira et al. (2019)
34.	Wheat, root tissues	*Bacillus* sp. strain WR11	4 0.32	Abiotic stress-alleviating properties and PGP activities	Chen et al. (2020)
35.	*Echinocystis lobata*, seed	*Acinetobacter* sp. strain *Enterobacter hormaechei*	4.760 and 4.75	Antagonism against soil-borne fungal pathogen	Khalaf and Raizada (2020a)
36.	Sugarcane, stem	*Pantoea ananatis*	5.17	Growth-promoting effects and IAA production abilities	Zeng et al. (2020)
37.	*Dicoma anomala*, leaf tissues	*Stenotrophomonas pavanii* strain MHSD12		Important genes for an endophytic lifestyle	Maela and Serepa-Dlamini (2020)

(continued)

Table 11.1 (continued)

Sr. no.	Name of host plant and tissues/ samples	Endophyte species sequenced	Genome size (Mb)	Plant growth-promoting traits/ activities	References
38.	*Luffa acutangula*, seeds	*Bacillus* sp. strain EKM601B	4.19	Antagonism against soil-borne pathogens in vitro	Khalaf and Raizada (2020b)
39.	*Zea mays*, roots	*Endobacterium cerealis*	6.11	Genes for colonization and hydrolytic enzymes	Menendez et al. (2020)
40.	Potato, healthy tubers	*Leifsonia* sp. strain *PS1209*	4.09	Potato endophyte with PGP activities	Liu et al. (2020b)

11.6 Conclusion

Plant-associated microbial communities, "plant microbiome," play a crucial role in plant health and growth and development. These microbes seem to be very essential for the vital functioning of plants ranging from seed germinations, seedling growth, and protecting plants from adverse effect of environmental stresses, pathogens, and insect herbivores. Microbial diversity is a crucial factor for maintaining plant health as it enables the recruitment of beneficial microbes. Plant-associated microbes called hologenome are involved in the appearance of plant traits and mitigation of stress conditions. Recent studies have proved the microbial dysbiosis could lead to disease development which offers the possibilities to engineer the microbial composition for improving plant traits. The microbiome could be engineered to make a combination of potential beneficial microbes of rhizospheric and phyllospheric microbes governing the growth and development and providing protection to the biotic and abiotic stresses alone or in combination. Recent advances in microbial biotechnology have been identified as microbes are the integral part of the plant genome and can be used for microbial-assisted breeding.

References

Acuña-Rodríguez IS, Newsham KK, Gundel PE, Torres-Díaz C, Molina-Montenegro MA (2020) Functional roles of microbial symbionts in plant cold tolerance. Ecol Lett 23(6):1034–1048

Alawiye TT, Babalola OO (2019) Bacterial diversity and community structure in typical plant rhizosphere. Diversity 11:179

Aloni R, Aloni E, Langhans M, Ullrich CI (2006) Role of auxin in regulating Arabidopsis flower development. Planta 223(2):315–328

Armanhi JSL, de Souza RSC, Damasceno, N. de B., de Araújo, L. M., Imperial, J., and Arruda, P. (2018) A community-based culture collection for targeting novel plant growth-promoting

Bacteria from the sugarcane microbiome. Front Plant Sci 8:1–17. https://doi.org/10.3389/fpls.2017.02191

Beracochea M, Taulé C, Battistoni F (2019) Draft genome sequence of *Kosakonia radicincitans* UYSO10, an endophytic plant growth-promoting bacterium of sugarcane (*Saccharum officinarum*). Microbiol Resour Announc 8:e01000

Berg G, Köberl M, Rybakova D, Müller H, Grosch R, Smalla K (2017) Plant microbial diversity is suggested as the key to future biocontrol and health trends. FEMS Microbiol Ecol 93:1–9

Berg G, Rybakova D, Fischer D et al (2020) Microbiome definition re-visited: old concepts and new challenges. Microbiome 8:103

Bertalan M, Albano R, de Pádua V, Rouws L, Rojas C, Hemerly A, Teixeira K, Schwab S, Araujo J, Oliveira A et al (2009) Complete genome sequence of the sugarcane nitrogen-fixing endophyte *Gluconacetobacter diazotrophicus* Pal5. BMC Genom 23(10):450

Bulgarelli D, Schlaeppi K, Spaepen S, van Themaat EVL, Schulze-Lefert P (2013) Structure and functions of the bacterial microbiota of plants. Annu Rev Plant Biol 64:807–838

Carvalho SD, Castillo JA (2018) Influence of light on plant–Phyllosphere interaction. Front Plant Sci 9:1–16

Chaudhary S, Sagar S, Lal M, Tomar A, Kumar V, Kumar M (2020) Biocontrol and growth enhancement potential of *Trichoderma spp.* against *Rhizoctonia solani* causing sheath blight disease in Rice. J Environ Biol 41:1034–1045

Chen C, Yue Z, Chu C, Ma K, Li L, Sun Z (2020) Complete genome sequence of *Bacillus sp.* WR11, an endophyte isolated from wheat root providing genomic insights into its plant growth-promoting effects. Mol Plant-Microbe Interact 33:876–879

Compant S, Reiter B, Sessitsch A, Nowak J, Clément C, Ait Barka E (2005) Endophytic colonization of *Vitis vinifera* L. by plant growth-promoting bacterium *Burkholderia sp.* strain PsJN. Appl Environ Microbiol 71(4):1685–1693

Corvin A, Craddock N, Sullivan PF (2010) Genome-wide association studies: a primer. Psychol Med 40:1063–1077

Crandall SG, Gold KM, Jiménez-Gasco MM, Filgueiras CC, Willett DS (2020) A multi-omics approach to solving problems in plant disease ecology. PLoS One 15(9):e0237975

Dash B, Soni R, Kumar V, Suyal DC, Dash D, Goel R (2019) Mycorrhizosphere: microbial interactions for sustainable agricultural production. In: Varma A, Choudhary D (eds) Mycorrhizosphere and pedogenesis. Springer, Singapore. https://doi.org/10.1007/978-981-13-6480-8_18

Dubey P, Kumar V, Ponnusamy K, Sonwani R, Singh AK, Suyal DC, Soni R (2020) Microbe assisted plant stress management. In: De Mandal S, Bhatt P (eds) The recent advancements in microbial diversity, 1st edn. Academic Press, Cambridge, MA, pp 351–378. https://doi.org/10.1016/B978-0-12-821265-3.00015-3

Epstein B, Abou-Shanab RAI, Shameldsin A, Taylor MR, Guhlin J, Burghardt LT, Nelson M, Sadowsky MJ, Tiffin P (2018) Genome-wide association analyses in the model rhizobium *Ensifer meliloti*. mSphere 3:e00386. https://doi.org/10.1128/mSphere.00386-18

Fleischmann RD, Adams MD, White O, Clayton RA, Kirkness EF, Kerlavage AR, Bult CJ, Tomb J-F, Dougherty BA, Merrick JM (1995) Whole-genome random sequencing and assembly of *Haemophilus influenzae* Rd. Science 269:496–512

Fouts DE, Tyler HL, DeBoy RT, Daugherty S, Ren Q, Badger JH, Durkin AS, Huot H, Shrivastava S, Kothari S, Dodson RJ, Mohamoud Y, Khouri H, Roesch LF, Krogfelt KA, Struve C, Triplett EW, Methé BA (2008) Complete genome sequence of the N2-fixing broad host range endophyte Klebsiella pneumoniae 342 and virulence predictions verified in mice. PLoS Genet 4(7):e1000141

Glick BR, Cheng Z, Czarny J, Duan J (2007) Promotion of plant growth by ACC deaminase-producing soil bacteria. Eur J Plant Pathol 119:329–339

Goel R, Kumar V, Suyal DC, Narayan SR (2018b) Toward the un-culturable microbes for sustainable agricultural production. In: Meena V (ed) Role of rhizospheric microbes in soil. Springer, Singapore. https://doi.org/10.1007/978-981-10-8402-7_4

Goel R, Kumar V, Suyal DK, Dash B, Kumar P, Soni R (2017) Root-associated bacteria: rhizoplane and endosphere. In: Singh D, Singh H, Prabha R (eds) Plant-microbe interactions in agro-ecological perspectives. Springer, Singapore. https://doi.org/10.1007/978-981-10-5813-4_9

Goel R, Suyal DC, Kumar V, Jain L, Soni R (2018a) Stress-tolerant beneficial microbes for sustainable agricultural production. In: Panpatte D, Jhala Y, Shelat H, Vyas R (eds) Microorganisms for green revolution. Microorganisms for sustainability, vol 7. Springer, Singapore. https://doi.org/10.1007/978-981-10-7146-1_8

Gonzaga de Oliveira L, Sigrist R, Sachetto Paulo B, Samborskyy M (2019) Whole- genome sequence of the endophytic *Streptomyces sp.* strain CBMAI 2042, isolated from *Citrus sinensis*. Microbiol Resour Announc 8:e01426

Hacquard S (2016) Disentangling the factors shaping microbiota composition across the plant holobiont. New Phytol 209:454–457

Hallmann J, Quadt-Hallmann A, Mahaffee WF, Kloepper JW (1997) Bacterial endophytes in agricultural crops. Can J Microbiol 43:895–914

Hardoim PR, van Overbeek LS, Berg G, Pirttilä AM, Compant S, Campisano A et al (2015) The hidden world within plants: ecological and evolutionary considerations for defining functioning of microbial endophytes. Microbiol Mol Biol Rev 79(3):293–320. https://doi.org/10.1128/MMBR.00050-14

Haruna E, Zin NM, Adams JM (2017) Draft genome sequence of an endophytic bacterium, *Paenibacillus tyrfis* strain SUK123, isolated from *Santiria apiculata* stem. Genomics Data 14:44–46

Hong CE, Kim JU, Lee JW, Bang KH, Jo I-H (2018) Complete genome sequence of the endophytic bacterium *Chryseobacterium indologenes* PgBE177, isolated from *Panax quinquefolius*. Microbiol Resour Announc 7:e01234–e01218

Jacquemyn H, Waud M, Merckx VS et al (2015) Mycorrhizal diversity, seed germination and long-term changes in population size across nine populations of the terrestrial orchid Neottia ovata. Mol Ecol 24:3269–3280

Kahlke T, Goesmann A, Hjerde E, Willassen NP, Haugen P (2012) Unique core genomes of the bacterial family Vibrionaceae: insights into niche adaptation and speciation. BMC Genomics 13:179

Kaiser C, Kilburn MR, Clode PL, Fuchslueger L, Koranda M, Cliff JB et al (2015) Exploring the transfer of recent plant photosynthates to soil microbes: mycorrhizal pathway vs direct root exudation. New Phytol 205:1537–1551

Kandel SL, Joubert PM, Doty LS (2017) Bacterial endophyte colonization and distribution within plants. Microorganisms 5:77

Kaneko T, Minamisawa K, Isawa T, Nakatsukasa H, Mitsui H, Kawaharada Y, Nakamura Y, Watanabe A (2010) Complete genomic structure of the cultivated rice endophyte Azospirillum sp. B510. DNA Res 17(1):37–50

Khalaf EM, Raizada MN (2020a) Draft genome sequences of *Acinetobacter sp.* strain EKM10A, Enterobacter hormaechei EKM10E, and *Enterobacter hormaechei* EKM11E (phylum Proteobacteria) colonizing the seed surface biogel of *Echinocystis lobata* (wild cucumber). Microbiol Resour Announc 9:e00184–e00120

Khalaf EM, Raizada MN (2020b) Draft genome sequence of *Bacillus sp.* strain EKM601B (phylum Firmicutes), living inside the seeds of *Luffa acutangula* (Chinese okra). Microbiol Resour Announc 9:e00180–e00120

Krause A, Ramakumar A, Bartels D, Battistoni F, Bekel T, Boch J, Böhm M, Friedrich F, Hurek T, Krause L, Linke B (2006) Complete genome of the mutualistic, N2-fixing grass endophyte *Azoarcus* sp. strain BH72. Nat Biotechnol 24(11):1385–1391

Kumar A, Verma JP (2018) Does plant—microbe interaction confer stress tolerance in plants: a review? Microbiol Res 207:41–52

Kumar V, Jain L, Jain SK, Chaturvedi S, Kaushal P (2020a) Bacterial endophytes of rice (*Oryza sativa* L.) and their potential for plant growth promotion and antagonistic activities. S Afr J Bot 34:50–63

Kumar V, Jain L, Jain SK, Kaushal P (2018) Metagenomic analysis of endophytic and Rhizospheric bacterial diversity of Lathyrus sativus using Illumina based sequencing. In: proceeding of international conference on microbiome research 2018, Pune with abstract ID 2018EVM0152

Kumar V, Jain L, Kaushal P, Soni R (2020c) Fungal endophytes and their applications as growth promoters and biological control agents. In Fungi bio-prospects in sustainable agriculture, environment and nano-technology Sharma VK, Shah M. P, Parmar S, and Kumar A by Academic Press pp 315–337. DOI: doi:https://doi.org/10.1016/B978-0-12-821394-0.00012-3

Kumar V, Jain L, Soni R, Kaushal P, Goel R (2020b) Bio-prospecting of endophytic microbes from higher altitude plants: recent advances and their biotechnological applications. In: Goel R, Soni R, Suyal D (eds) Microbiological advancements for higher Altitude Agro-Ecosystems & Sustainability. Springer, Singapore. https://doi.org/10.1007/978-981-15-1902-4_18

Kumar V, Soni R, Jain L, Dash B, Goel R (2019) Endophytic fungi: recent advances in identification and explorations. In: Singh B (ed) Advances in endophytic fungal research. Springer, Cham. https://doi.org/10.1007/978-3-030-03589-1_13

Kwak MJ, Song JY, Kim SY, Jeong H, Kang SG, Kim BK, Kwon SK, Lee CH, Yu DS, Park SH, Kim JF (2012) Complete genome sequence of the endophytic bacterium *Burkholderia* sp. strain KJ006. J Bacteriol 194(16):4432–4433

Land M, Hauser L, Jun SR, Nookaew I, Leuze MR, Ahn TH et al (2015) Insights from 20 years of bacterial genome sequencing. Funct Integr Genomics 15:141–161

Laugraud A, Young S, Gerard E, O'Callaghan M, Wakelin S (2017) Draft genome sequence of the clover (*Trifolium repens* L.) root endophyte *Paraburkholderia sp.* strain A27. Genome Announc 5:e00466–e00417

Lei L (2020) Phyllosphere dysbiosis. Nat Plants 6:434. https://doi.org/10.1038/s41477-020-0674-7

Liu H, Brettell LE, Qiu Z, Singh BK (2020a) Microbiome-mediated stress resistance in plants. Trends Plant Sci 25(8):733–743

Liu Y, Ma X, Helmann TC, McLane H, Stodghill P, Swingle B, Filiatrault MJ, Perry KL (2020b) Complete genome sequence of a gram- positive bacterium, *Leifsonia sp.* strain PS1209, a potato endophyte. Microbiol Resour Announc 9:e00447

Lugtenberg B, Kamilova F (2009) Plant-growth promoting rhizobacteria. Annu Rev Microbiol 63:541–556

Maela MP, Serepa-Dlamini MH (2020) Draft genome sequence of *Stenotrophomonas pavanii* strain MHSD12, a bacterial endophyte associated with *Dicoma anomala*. Microbiol Resour Announc 9:e00550

Makuwa SC, Serepa-Dlamini MH (2019) Data on draft genome sequence of *Bacillus* sp. strain MHSD28, a bacterial endophyte isolated from *Dicoma anomala*. Data Brief 26:104524

Margesin R, Miteva V (2011) Diversity and ecology of psychrophilic microorganisms. Res Microbiol 162:346–361

Mastretta C, Taghavi S, van der Lelie D, Mengoni A, Galardi F, Gonnelli C, Barac T, Boulet J, Weyens N, Vangronsveld J (2009) Endophytic bacteria from seeds of Nicotiana tabacum can reduce cadmium phytotoxicity. Int J Phytoremediation 11:251–267

Meena KK, Sorty AM, Bitla UM, Choudhary K, Gupta P, Pareek A, Singh DP, Prabha R, Sahu PK, Gupta VK, Singh HB, Krishanani KK, Minhas PS (2017) Abiotic stress responses and microbe-mediated mitigation in plants: the omics strategies. Front Plant Sci 8:172. https://doi.org/10.3389/fpls.2017.00172

Megías E, Megías M, Ollero FJ, Hungria M (2016) Draft genome sequence of *Pantoea ananatis* strain AMG521, a rice plant growth-promoting bacterial endophyte isolated from the *Guadalquivir marshes* in southern Spain. Genome Announc 4(1):e01681–e01615

Mehla K, Chaudhary S, Kumar A, Kumar V, Chauhan P, Gupta S et al (2011) Advances in DNA sequencing: challenges and limitations of personal sequencing. Afr J Agric Res 6(6):1277–1280

Mendes R, Garbeva P, Raaijmakers JM (2013) The rhizosphere microbiome: significance of plant beneficial, plant pathogenic, and human pathogenic microorganisms. FEMS Microbiol Rev 37 (5):634–663

Mendes R et al (2011) Deciphering the rhizosphere microbiome for disease-suppressive bacteria. Science 332:1097–1100

Menéndez E, Flores-Félix JD, Ramírez-Bahena MH, Igual JM, García-Fraile P, Peix A, Velázquez E (2020) Genome analysis of *Endobacterium cerealis*, a novel genus and species isolated from *Zea mays* roots in North Spain. Microorganisms 8(6):939

Moreira AS, Germaine KJ, Lloyd A, Lally RD, Galbally PT, Ryan D, Dowling DN (2016) Draft genome sequence of three endophyte strains of *Pseudomonas fluorescens* isolated from *Miscanthus giganteus*. Genome Announc 4(5):e00965–e00916

Mukhtar T, Afridi MS, McArthur R, Van Hamme JD, Rineau F, Mahmood T, Amna S, Zahid M, Salam A, Khan MN, Ali F, Mehmood S, Bangash N, Chaudhary HJ (2018) Draft genome sequence of *Bacillus pumilus* SCAL1, an endophytic heat-tolerant plant growth-promoting bacterium. Genome Announc 6:e00306–e00318

Neupane S, Finlay RD, Kyrpides NC, Goodwin L, Alström S, Lucas S, Land M, Han J et al (2012) Complete genome sequence of the plant-associated *Serratia plymuthica* strain AS13. Stand Genomic Sci 7(1):22–30

Ojuederie OB, Olanrewaju OS, Babalola OO (2019) Plant growth promoting rhizobacterial mitigation of drought stress in crop plants: implications for sustainable agriculture. Agronomy 9 (11):712. https://doi.org/10.3390/agronomy9110712

Pandey A, Dhakar K, Jain R, Pandey N, Gupta VK, Kooliyottil R et al (2018) Cold adapted fungi from Indian Himalaya: untapped source for bioprospecting. Proc Natl Acad Sci India Sect B Biol Sci 89:1125–1132. https://doi.org/10.1007/s40011-018-1002-0

Parakhia MV, Tomar RS, Vala AG, Rathod VM, Kheni JV, Kothari VV, Golakiya BA (2016) Whole genome sequencing of drought stress tolerance endophytic bacterium *Enterobacter sp.* Mr1. J Bacteriol Mycol 3:207–209

Pareek A, Dhankher OP, Foyer CH (2020) Mitigating the impact of climate change on plant productivity and ecosystem sustainability. J Exp Bot 71:451–456. https://doi.org/10.1093/jxb/erz518

Pérez-Jaramillo JE et al (2015) Impact of plant domestication on rhizosphere microbiome assembly and functions. Plant Mol Biology 90:635–644. https://doi.org/10.1007/s11103-015-0337-7

Poehlein A, Hollensteiner J, Granzow S, Wemheuer B, Vidal S, Wemheuer F (2018) First insights into the draft genome sequence of the endophyte *Paenibacillus amylolyticus* strain GM1FR, isolated from *Festuca rubra* L. Genome Announc 6:e01516

Qin Y, Druzhinina IS, Pan X, Yuan Z (2016) Microbially mediated plant salt tolerance and microbiome-based solutions for saline agriculture. Biotechnol Adv 34(7):1245–1259

Read TD, Massey RC (2014) Characterizing the genetic basis of bacterial phenotypes using genome-wide association studies: a new direction for bacteriology. Genome Med 6:1–11

Sahib MR, Yang P, Bokros N, Shapiro N, Woyke T, Kyrpides NC, Xia Y, DeBolt S (2019) Improved draft genome sequence of *Microbacterium sp.* strain LKL04, a bacterial endophyte associated with switchgrass plants. Microbiol Resour Announc 8:e00927

Santoyo G, Loeza PD, Villegas J, Farias R, Contreras M (2016) A glimpse of the endophytic bacterial diversity in roots of blackberry plants (*Rubus fruticosus*). Genet Mol Res 15 (3):15038542

Singh BK, Trivedi P, Singh S, Macdonald CA, Verma JP (2018) Emerging microbiome technologies for sustainable increase in farm productivity and environmental security. Microbiology 39:17–23

Soni R, Kumar V, Suyal DC, Jain L, Goel R (2017) Metagenomics of plant rhizosphere microbiome. In: Singh R, Kothari R, Koringa P, Singh S (eds) Understanding host-microbiome interactions - an omics approach. Springer, Singapore. https://doi.org/10.1007/978-981-10-5050-3_12

Taghavi S, Garafola C, Monchy S, Newman L, Hoffman A, Weyens N, Barac T, Vangronsveld J, van der Lelie D (2009) Genome survey and characterization of endophytic bacteria exhibiting a beneficial effect on growth and development of poplar trees. Appl Environ Microbiol 75 (3):748–757

Taghavi S, van der Lelie D, Hoffman A, Zhang YB, Walla MD, Vangronsveld J, Newman L, Monchy S (2010) Genome sequence of the plant growth promoting endophytic bacterium Enterobacter sp. 638. PLoS Genet 6(5):e1000943

Truyens S, Weyens N, Cuypers A et al (2015) Bacterial seed endophytes: genera, vertical transmission and interaction with plants. Environ Microbiol Rep 7:40–50

Tshishonga K, Serepa-Dlamini MH (2019) Draft genome sequence of *Enterobacter hormaechei* strain MHSD6, a plant endophyte isolated from medicinal plant *Pellaea calomelanos*. Microbiol Resour Announc 8:e01251

Turbat A, Rakk D, Vigneshwari A, Kocsubé S, Thu H, Szepesi Á, Bakacsy LD, Škrbić B, Jigjiddorj EA, Vágvölgyi C, Szekeres A (2020) Characterization of the plant growth-promoting activities of endophytic fungi isolated from *Sophora flavescens*. Microorganisms 8(5):683

Van der Heijden MG, de Bruin S, Luckerhoff L et al (2016) A widespread plant-fungal-bacterial symbiosis promotes plant biodiversity, plant nutrition and seedling recruitment. ISME J 10:389–399

Vandenkoornhuyse P, Quaiser A, Duhamel M, Le Van A, Dufresne A (2015) The importance of the microbiome of the plant holobiont. New Phytol 4:1196–1206

Verma SK, White JF (2019) Seed endophytes: biology and biotechnology. Springer, Cham. https://doi.org/10.1007/978-3-030-10504-4

Wagner MR, Lundberg DS, Tijana G, Tringe SG, Dangl JL, Mitchell-Olds T (2016) Host genotype and age shape the leaf and root microbiomes of a wild perennial plant. Nat Commun 7:1–15

Wang NR, Haney CH (2020) Harnessing the genetic potential of the plant microbiome. Biochem 42 (4):20–25

Weilharter A, Mitter B, Shin MV, Chain PS, Nowak J, Sessitsch A (2011) Complete genome sequence of the plant growth-promoting endophyte *Burkholderia phytofirmans* strain PsJN. J Bacteriol 193(13):3383–3384

Wemheuer F, Hollensteiner J, Poehlein A, Liesegang H, Daniel R, Wemheuer B (2018a) Draft genome sequence of the endophyte *Bacillus mycoides* strain GM6LP isolated from *Lolium perenne*. Genome Announc 6:e00011

Wemheuer F, Wemheuer B, Hollensteiner J, Daniel R, Poehlein A (2018b) Draft genome sequence of the endophyte *Paenibacillus sp.* strain GM2FR isolated from *Festuca rubra*. Genome Announc 6:e00017

Wisniewski-Dyé F, Borziak K, Khalsa-Moyers G, Alexandre G, Sukharnikov LO, Wuichet K et al (2011) *Azospirillum* genomes reveal transition of bacteria from aquatic to terrestrial environments. PLoS Genet 7(12):e1002430

Yan Y, Yang J, Dou Y, Chen M, Ping S, Peng J, Lu W, Zhang W, Yao Z, Li H, Liu W, He S, Geng L, Zhang X, Yang F, Yu H, Zhan Y, Li D, Lin Z, Wang Y, Elmerich C, Lin M, Jin Q (2008) Nitrogen fixation island and rhizosphere competence traits in the genome of root-associated *Pseudomonas stutzeri* A1501. Proc Natl Acad Sci U S A 105(21):7564–7569

Yang J, Kloepper JW, Ryu CM (2009) Rhizosphere bacteria help plants tolerate abiotic stress. Trends Plant Sci 14:1–4

Yi Y, de Jong A, Spoelder J, Elzenga JTM, van Elsas JD, Kuipers OP (2016) Draft genome sequence of *Bacillus mycoides* M2E15, a strain isolated from the endosphere of potato. Genome Announc 4(1):e00031–e00016

Zeng Q, Shi GY, Nong ZM, Ye XL, Hu CJ (2020) Complete genome sequence of *Pantoea ananatis* strain NN08200, an endophytic bacterium isolated from sugarcane. Curr Microbiol 77:1864–1870

Zeng Q, Xie J, Li Y, Chen X, Wang Q (2019) Draft genome sequence of an endophytic biocontrol bacterium, *Bacillus velezensis* PG12, isolated from apple fruit. Microbiol Resourc Announc 8: e00468

Zhu B, Liu H, Tian WX, Fan XY, Li B, Zhou XP, Jin GL, Xie GL (2012) Genome sequence of *Stenotrophomonas maltophilia* RR-10, isolated as an endophyte from rice root. J Bacteriol 194 (5):1280–1281

Omic Technologies and Cold Adaptations 12

Poonam Sehgal and Pankaj Chaturvedi

Abstract

Psychrophiles or cold-adapted microorganisms are one of the most ecologically diverse groups of life-forms, touted to hold the key for the sustainable future of mankind. Even after centuries of microbial research, significant gaps about the microbial diversity and their intricate interactions with the surroundings remain. Such studies of psychrophiles from extreme environments are plagued by our inability to apply conventional culture methods to elucidate their physiological and metabolic properties. This inability stems from our limited understanding of in vitro growth requirements for these microbes. The advent of several high-throughput and sensitive "-omic" technologies have ushered a new era for microbial ecology. As these methods rely on the detection of macromolecules isolated directly from environmental samples, the focus is primarily on community composition rather than an isolated organism. They reveal extensive knowledge of dynamic interactions of the individual microbial cell with their surroundings. Reliance on these "-omic" studies has provided a framework of testable hypothesis to understand metabolic requirements of axenic microbial cultures. Such concerted efforts are expected to unlock novel microbial properties that were previously unavailable by standalone approaches. The current chapter aims to elucidate the different available "-omic" technologies and their generalized workflow for analyzing the diversity of environmental samples with particular emphasis on cold environments. Few recent studies are also presented in relevant sections to guide the reader to more specialized applications of these "-omic" methods.

P. Sehgal · P. Chaturvedi (✉)
Department of Cell and Developmental Biology, University of Illinois at Urbana-Champaign, Urbana, IL, USA

R. Goel et al. (eds.), *Survival Strategies in Cold-adapted Microorganisms*,
https://doi.org/10.1007/978-981-16-2625-8_12

Keywords

Psychrophiles · High throughput · Culture independent · Metagenomic · Metatranscriptomic · Global warming

12.1 Introduction

Psychrophiles (cryophiles or cold-adapted microbes) are one of the most diverse and abundant groups of organisms representing all three domains of life (*Bacteria*, *Archaea*, and *Eukarya*). In general, the psychrophiles thrive at ambient temperatures of −20 °C to +4 °C in deep oceans, natural caves, aquifers, high altitudes, cold deserts, and permafrost (Mackelprang et al. 2017; Dhakar and Pandey 2020) (Fig. 12.1). The largest diversity of psychrophiles is found in oceans at depths greater than 1000 m. Although per unit microbial biomass of psychrophiles (23–47 million microbial cells per g dry soil weight) from the Arctic and sub-Arctic permafrost is about 10–100 times lower than the corresponding values from temperate soils, they are critical for sequestering carbon and maintenance of geological balance (Mackelprang et al. 2016; Burkert et al. 2019). The exploration of cold-adapted microorganisms comes with their own unique set of challenges such as inability to grow under standard laboratory conditions, difficulties in sample collection from remote sites, and reproducibility of such studies. Some of these issues and how the advent of "-omic" approaches have helped us answer these challenges have been discussed in this chapter.

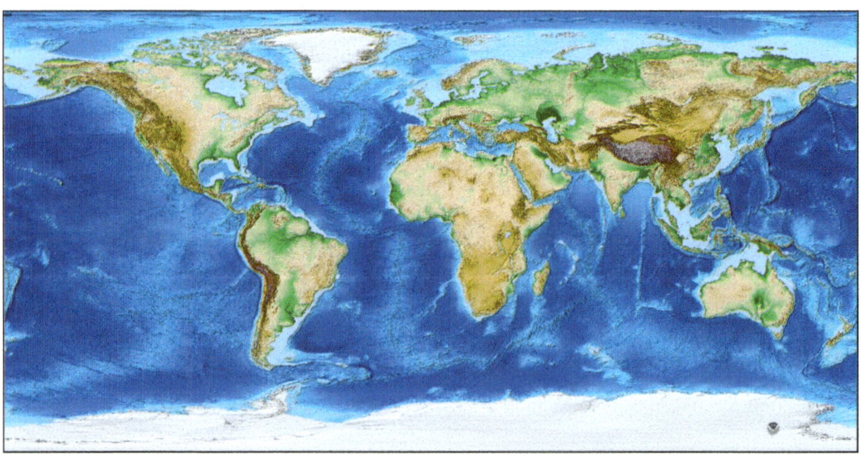

Fig. 12.1 Ubiquitous distribution of cold-adapted microbes and their habitats on the planet. A global relief model of Earth's surface indicating the distribution of glacial ice (white), mountainous (dark brown and purple), and deep marine (dark blue) habitats of cold-adapted microbes. Source map (ETOPO1) courtesy of National Oceanic and Atmospheric Administration (NOAA), USA

12.2 Need for Identifying Psychrophiles and their Biosynthetic Potential

Microorganisms have been drivers of geological transformation for billions of years. In modern times, microorganisms, in particular, the ones from extreme environments, are being explored as they hold high inherent value both for sustainable ecology and microbial biotechnology. These microbial applications are manifold that include but are not limited to diagnostics, pharmaceuticals, nutritional supplements, biocatalysis, carbon capture and cycling, biofertilization, and bioremediation (Dhakar and Pandey 2020; Edwards et al. 2020).

12.2.1 Ecological Impacts

Initial psychrophilic microbial ecology studies were mainly academic, aiming to know more about their physiology and survival strategies. However, there is renewed interest in psychrophilic diversity as they are important for carbon sequestration and cycling in polar environments. Rising temperatures in the polar regions due to global warming have led to a shift in microbial ecology. The shifted microbial balance can impact the fate of carbon reservoirs leading to the runoff effect of greenhouse gases such as CH_4 and CO_2 (Mackelprang et al. 2016; Edwards et al. 2020). The abundance of nitrogen-fixing microbes (*nif*H gene) in permafrost and the adverse impact of thaw have been demonstrated by metagenomic methods underscoring the importance of cold environments in nitrogen assimilation and cycling (Hultman et al. 2015; Mackelprang et al. 2016). Multiple lines of evidence have also revealed iron reduction and uptake potential in permafrost microbes (Hultman et al. 2015). Cumulatively, the cold-adapted microbes are vital for life-sustaining biogeochemical transformations such as carbon and nitrogen cycling, iron and sulfur assimilation, methane generation and oxidation, and organic matter decomposition.

The past decade has witnessed unprecedented warming of glaciers and permafrost regions that wasn't even predicted to happen till the 2050s, and the models suggest at least four times faster warming in polar regions compared to the rest of the planet (Legendre et al. 2014). The climate change models further imply that at the current rate the Arctic regions will warm by around 10 °C by the end of this century (Edwards et al. 2020). These Arctic and other cold environments (e.g., permafrost, brine channels in sea ice, cold deserts, glacial crusts, englacial conduits, nival levels of mountains, etc.) are a niche for a diverse array of psychrophilic microbes critical for biogeochemical cycles on a global scale (Fig. 12.1). Apart from creating an imbalance in greenhouse gas levels, the gradual increase in temperature is also expected to induce a change in microbial diversity where the true psychrophiles are replaced by psychrotolerant/psychrotrophic organisms. The appearance of darkened or pigmented patches due to such abnormal microbial growth on the glacial ice surface can reduce its solar reflectivity and can enhance the melting of ice (Edwards et al. 2020). Few studies have even predicted that thaw in polar regions can "reawaken" dormant pathogens and may pose threat to human and animal health. One of many such examples is the discovery of a giant *Pandoravirus* named *Pithovirus sibericum* from 30,000-year-old Siberian ice. The virus has many unusual traits compared to other known DNA viruses. This virus was found to retain infectivity to its host amoeba (*Acanthamoeba castellanii*) after the thaw and interestingly contains the least compacted (610 kb genome in ~1.5 μm in length, 500 nm diameter capsid) and unusually AT-rich (64%) genomic DNA compared to other viruses (Legendre et al. 2014).

Literature in the field is replete with examples of psychrophilic adaptation for surviving oligotrophic conditions, freezing temperatures, low water availability, high salinity, and background radiations. Only recently their metabolic properties and adaptive strategies for subzero survival are being unraveled. Expression of cold-shock proteins, chaperones, stress-related proteins, or strategies promoting membrane fluidity, or increasing protein flexibility by low incorporation of proline and arginine amino acids, and pigmentation seem to be common survival methods in psychrophiles (Mackelprang et al. 2016). Lateral gene transfer to acquire newer traits also appears to be a prominent adaptive strategy to improve genetic diversity. A recent example of such gene transfer has been documented in a metagenome-assembled sequence of Arctic sea planktonic *Chloroflexi* bacteria. The marine *Chloroflexi* seems to have recently acquired aromatic compound degradation genes from terrestrial bacteria and can now utilize terrestrial humic-rich organic material as a source of carbon (Colatriano et al. 2018). Detailed molecular, genetic, and physiological strategies for survival by these psychrophiles are beyond the scope of this chapter.

12.2.2 Economic Importance

Apart from the ecological contributions, the cold-adapted microbes are also keenly studied as a sustainable bioresource for biotechnological, industrial, and biomedical applications. Just the immediate share of novel microbial products of industrial importance such as enzymes, antibiotics, and pigments has been estimated to be worth 6.20 billion dollars (Bernard et al. 2018).

Pigments: The quest for biodegradable and environmentally sustainable pigments for the pharmaceutical, textile, cosmetic, and food industries has led researchers to the pigments produced by psychrophiles. The cold-adapted bacteria and fungi produce these pigments as an adaptive strategy to cope with temperature, oxidative, and photo-damage by ultraviolet radiations (Sajjad et al. 2020). Few such identified psychrophiles and pigments produced by them are Antarctic fungal species *Friedmanniomyces endolithicus* (melanin), Antarctic bacterium *Sphingobacterium antarcticus* (carotenoids), and Alaskan soil isolate *Janthinobacterium lividum* (prodigiosin). Further, several psychrophilic pigments also have antimicrobial potential and are valuable candidates for pharmaceutical industries (Sajjad et al. 2020). Violacein, an indole derivative produced by the bacterium of *Janthinobacterium* genus, has also been reported for antimicrobial, antitumor, and antiparasitic properties. Similarly, the pigments derived from *Micrococcus luteus* can inhibit the growth of pathogens such as *Staphylococcus* sp., *Klebsiella* sp., and *Pseudomonas* sp. Similarly, carotenoids sourced from *Halomonas* sp. are effective against antibiotic-resistant *S. aureus*, *Klebsiella* sp., *Pseudomonas aeruginosa*, *Streptococcus pyogenes*, and *E. coli* (Ravikumar et al. 2016; Sajjad et al. 2020).

Bioactive molecules: The uncultured microbes represent a largely untapped resource for bioactive molecules, secondary metabolites of profound relevance to human health and well-being. Direct secretion of bioactive molecules such as essential vitamins, short-chain fatty acids, metabolized complex sugars, or antimicrobial substances by gut commensal microbiota plays a pivotal role in host health and metabolism. Additionally, environmental microbes have been found as one of the vital sources of other bioactive molecules possessing antioxidant, anti-inflammatory, or antimicrobial properties. Lantibiotics, a class of antimicrobial peptides, are attractive options against multidrug-resistant bacteria (also called superbugs) that have evolved as a major threat to public health. Novel antimicrobial compounds have also been sourced from marine strains as *Verrucosispora* sp. (abyssomicins), *Salinispora tropica* (Salinosporamide A), and *Salinispora pacifica* (Cyanosporaside) (Monciardini et al. 2014). A potential biotechnological application of psychrophiles is to express heat-labile compounds or proteins that are susceptible to aggregation during normal culture conditions used in the large-scale fermenters.

Bioremediation: A considerable fraction of environmental pollutants are petroleum hydrocarbons. Being hydrophobic, these molecules need to be emulsified and mobilized into sediments for microbial degradation. Production of biosurfactants and degradation of such hydrocarbons by cold-adapted microbes are an emergent strategy as a large fraction of oil reservoirs, drilling, and transport are performed in cold environments. The activity of enzymes derived from temperate counterparts in such conditions is limiting or inadequate. For example, comparable kerosene-emulsifying activity at 4 °C was observed by Antarctic bacterial isolates (*Psychrobacter* sp. PL19 and *Janthinobacterium* sp. CG23.3) to that observed at 37 °C by mesophilic bacterial strains (*Pseudomonas aeruginosa* or *Bacillus subtilis*) (Trudgeon et al. 2020). In addition to the activity at a lower temperature, these isolates were also reported to utilize diesel, motor oil, and crude oil as carbon sources and present a promising aspect of cold-adapted microbes for petroleum hydrocarbon degradation.

12.3 Challenges in Studying Psychrophiles by Conventional Approaches

A vast majority of representative strains of estimated ~1300 prokaryotic phyla have not been cultured in laboratory (Yarza et al. 2014; Steen et al. 2019), thereby constraining our ability to understand the ecological impact of microorganisms. The unculturability of bacterial cells was first observed in the nineteenth century by Robert Koch during the early ages of microbiology. However, the first well-documented study about "viable but nonculturable (VBNC)" microbial population on routinely used culture media was made only a few decades earlier (Xu et al. 1982).

12.3.1 Viable but Nonculturable (VBNC) State

The VBNC state represents microbial cells that are viable yet are not able to multiply to form colonies on the applied culture conditions. This state may be either due to suboptimal culture conditions or due to an induced state of dormancy caused by a change in salinity, pH, osmotic stress, reactive oxygen species, nutrient starvation, heavy metals, or other abiotic stress (Oliver 2010). In contrast to the dead cells, the cells or spores in the VBNC state have intact cell membrane and genetic material and reduced but active metabolic processes. Frequently, in response to limiting metabolic activity, many bacterial species adopt sporulation, coccoid, or smaller cell shapes to achieve a lower surface area to volume ratio (Zhao et al. 2017). Survival of protozoa in soils for >30,000 years as dormant cysts has also been observed (Legendre et al. 2014). The VBNC state has been predicted to be a long-term natural survival strategy to survive oligotrophy rather than an artificial state induced by in vitro culture conditions. However, recent permafrost studies have suggested that different strategies are adopted by psychrophiles: some endospore-forming classes as *Bacilli* undergo sporulation, while *Clostridia* choose to stay in a vegetative state and avoid an accumulation of DNA damage in endospores (Burkert et al. 2019). Favorable growth conditions stimulate these VBNC cells from the dormant state to resume regular growth activities. An interesting hypothesis of "scout cells" that relates to reactivation or awakening of dormant microbes has been proposed (Buerger et al. 2012) that predicts periodic sporadic activation of spores/cells from a dormant state. If the conditions are favorable, then the scout cells secrete inducers to promote activation of the entire population; otherwise, the scout cell dies, and another scout randomly gets activated at a different time point. This cycle of sensing favorable conditions by scout cells continues till the optimal growth conditions are found (Buerger et al. 2012).

12.3.2 Missing in Vivo Signaling Factor or Component in Media

Apart from the VBNC state, the limited growth in vitro can also be due to reasons such as nonavailability of specific nutrients required, presence of toxic or inhibitory substances in media, and dependency of target microbe on synergistic association in the community. Loss of viability of some pathogenic bacteria like *Salmonella enteritidis*, *Vibrio cholerae*, and *V. vulnificus* upon exposure to salt water, fresh water, and low temperatures, respectively, has been demonstrated (Amann et al. 1995). Bacterial signaling through secreted cytokines is also an essential factor determining the growth of other microbes in a complex population, and their absence on solid media can be another factor causing nonculturability. A recent study used such an approach of coculture and demonstrated a symbiotic relationship between two hypersaline Antarctic lake isolates *Candidatus Nanohaloarchaeum antarcticus* and its host haloarchaeon *Halorubrum lacusprofundi* (Hamm et al. 2019). The cell-to-cell contact is required for in vitro growth of *C. Nanohaloarchaeum antarcticus*. Assembled metagenomes from several nanohaloarchaea revealed that these archaea are deficient in phospholipid biosynthesis (mevalonate pathway), amino acids, nucleotides, and certain cofactors and have evolved for symbiotic association with host cells. The host *Hrr. lacusprofundi* in turn derives by-products like ammonium or acetate molecules from its symbiont partner (Hamm et al. 2019). Multiple growth factors or cytokines to promote culturability of novel isolates can also be provided by the transgenic approaches where the host symbiont is engineered to heterologously express one or more such factors. Simutaneous, stable expression of multiple transgenes is a promising method routinely employed in several model organisms such as bacteria, yeast and mammalian cells (Chaturvedi et al. 2018; Zhao et al. 2020).

12.3.3 Use of Growth Inducers

In the natural environment, bacterial cells secrete a myriad of different factors promoting differentiation or virulence, or quorum-sensing heat-stable factors for cell-to-cell signaling, or resuscitation-promoting factors (Rpf) (Piel 2011). Supplementation of Rpf, a small cytokine-like molecule from *Micrococcus luteus* in picomolar concentrations, has been shown to resuscitate dormant cells (Vartoukian et al. 2010). Siderophores are high-affinity iron scavengers that form soluble complexes with ferric iron (Fe^{3+}) and are then internalized by cells. Limiting the availability of iron has adverse effects on bacterial growth even in nutrient-rich media. Provision of siderophores from *E. coli* and *Micrococcus luteus* KLE1011 in coculture has been shown to induce growth of *Maribacter polysiphoniae* KLE1104, a previously unculturable bacterium from tidal ocean beach biofilm (Lewis et al. 2010).

12.4 Culture-Independent Microbial Identification: Molecular Analytical Methods

As discussed in Sect. 12.3, simulating natural habitat by coculture, growth inducers, or diffusion chambers has led to in vitro growth of many uncultivated strains. Hence, it should be noted that a vast majority of them are "uncultured" because of elusive media requirements and conditions, and should not be labeled as unculturable per se. A paradigm shift was achieved when microbes were directly studied from native habitats by the application of culture-independent methods. However, the initial approaches needed prior knowledge of molecular markers and remained low throughput.

12.4.1 Phylogenetic Marker Gene Sequencing

The most direct approach of identifying microbial diversity is through sequencing of a segment of the gene of interest and building a phylogenic relationship between the representative microbes in the sample. This marker gene-sequencing (16S rRNA) technology was pioneered by Carl Woese to delineate a new phylogenetic domain archaea that is different from bacteria (Woese and Fox 1977). Theoretically, any prokaryotic gene can be used as a genetic marker to identify microbes or related members of a population. However, for a universal or broad detection capability, the marker gene sequence should be ubiquitous in the population and serve as a molecular clock, i.e., the divergence of the sequence should be proportional to the evolutionary distance between the members in the population. Few such genetic markers are multicopy marker (16S rRNA gene) or single-copy markers (e.g., *rpo*B, *gyr*B, *rec*A) and functional genes such as *nif*H (nitrogenase reductase) gene for nitrogen-fixing bacteria or *amo*A (ammonia monooxygenase) gene for nitrifying bacteria, or *gac*A (response regulator) for *Pseudomonads* (van Elsas and Boersma 2011). Internal transcribed spacers (ITS) within ribosomal transcripts have been used to elucidate relationships among congeneric species and closely related genera for fungi (Schoch et al. 2012). The 1.5-kb-sized 16S rRNA gene consists of a region conserved throughout the bacterial domain and also has a genus-specific hypervariable region. These hypervariable regions are PCR-amplified and sequenced and are then assigned different operational taxonomic units (OTUs) based on conservation. The OTU concept is similar to species used in classical taxonomy as many of the genomes represented by OTUs do not have a representative culture strain (Yarza et al. 2014). Assignment of sequences to OTUs is frequently accomplished by clustering the sequences based on similarity and modeling the phylogenetic distances based on mutation rates to the evolutionary relationship for the marker. NGS-based approaches have proved to be a more robust and sensitive tool for de novo construction of OTUs in complex microbial consortia (Niu et al. 2017).

The advantage of phylogenetic marker sequencing is that it is inexpensive, simple in design, and fast and works well in the presence of contaminating DNA from other sources. However, this approach has low resolution and is prone to PCR biases. As the method relies on sequencing hypervariable segments, differential amplification of target sequences in a complex population can occur and lead to underrepresentation of the microbial community having very divergent sequences. Horizontal gene transfer, multiple copies of rRNA gene, chimeric PCR amplicons, type of variable region selected, and amplification protocols are a few other sources of inconsistency in results in this method. Furthermore, it should be noted that the marker gene analysis focuses on a specific gene(s) rather than the entire genome of microbes presenting a unidimensional view of the sample, and is not a metagenomic method per se.

12.4.2 Molecular Fingerprinting Techniques

Before the development of NGS approaches, several molecular methods to glean information were extensively used to probe the microbial community. The popular

PCR-based methods are denaturing and temperature gradient gel electrophoresis (DGGE/TGGE), PCR followed by terminal restriction fragment length polymorphism (PCR-T-RFLP), amplified ribosomal DNA restriction analysis (ARDRA), 16S rRNA gene clone library analysis, serial analysis of ribosomal sequence tags (SARST), and single-strand conformation polymorphism (SSCP) (van Elsas and Boersma 2011). The dependency of these methods on PCR amplification of target regions makes them vulnerable to the PCR biases, limits their applicability to the most abundant microbes, and cannot distinguish between viable or nonviable or dead microbes. PCR-independent method (such as fatty acid methyl esters (FAME)) analysis relies on the occurrence of signature fatty acids in microbial communities. Here, the phospholipids that constitute cellular membrane are extracted, converted to fatty acid methyl esters, and analyzed by gas chromatography. Despite the simple experimental requirement and setup, the method can lead to underestimation of diversity if the uncultured microbe has a similar profile to another known microbe. However, the method is very useful if the temporal or external factor-induced shift in community composition is to be analyzed.

Detailed information about molecular fingerprinting methods is not intended to be presented here and can be found in this review article (van Elsas and Boersma 2011) and the references cited in the article.

12.4.3 DNA Hybridization-Based Methods

Hybridization/Microarray – Nucleic acid hybridization has been a reliable approach to detect and identify noncultivable or rare microorganisms which are not easily achieved by traditional culture-based methods. The complementary sequences (probes) are designed based on a priori knowledge or estimation of genomic complexity of the sample being analyzed, and the probes are labeled with radionucleotide, fluorescent, or magnetic tags for detection. The approach is based on the premise that the rate of reannealing of DNA is indicative of its complexity (diversity). Faster reannealing is expected for a more diverse microbial community. However, the method has low throughput and sensitivity and is more suited for the detection of abundant cells or genes.

Microarray on the other hand is based on prefabricated chips spotted with thousands of oligonucleotide probes either for 16S rRNA (PhyloChip) or functional genes (GeoChip). The GeoChips represent microbial functional genes for biogeochemical, metabolic, stress response, metal homeostasis/resistance, and virulence pathways. Based on homology, the DNA extracted from the environmental sample is labeled with fluorophores, hybridized to the probes on a microarray chip, and the fluorescent signals are detected. The method provides high sensitivity and throughput but is prone to cross-hybridization with closely related sequences. Additionally, both of these hybridization-based approaches need prior knowledge of target sequences and are thus biased in detecting microbial cells or genes. The microarray-based direct analysis of samples is being widely used in clinical and bioremediation studies (Jo et al. 2020).

FISH – Fluorescence in situ hybridization (FISH) using labeled oligonucleotide probes against rRNA or other functional genes has been an invaluable molecular tool for the identification of rare or uncultured microorganisms. However, unique challenges are encountered when uncultured microbes from cold or oligotrophic habitats are being visualized. First, optimal probe stringency for validation and hybridization to uncultured strains has to be determined empirically. Also, under oligotrophic or unfavorable conditions, the cell size and number of ribosomal copies get reduced, restricting the detection sensitivity. Amplification of FISH signal (e.g., tyramide amplification, rolling circle amplification, hybridization chain reaction (HCR)) has been achieved for better detection sensitivity (Takahashi et al. 2020). Improvements in DNA synthetic technology have fostered high-throughput and cost-effective detection of microbes through thousands of custom-made unique barcode FISH probe libraries.

12.5 "-Omics"-Based Platforms for Microbial Identification

The contribution of tools and approaches used to characterize elusive microorganisms from the environmental sample during early metagenomics or "proto-metagenomics" era (van Elsas and Boersma 2011) has been a cornerstone of microbiology. However, the improved sensitivity and accuracy of next-generation sequencing (NGS) techniques have paved the way for ultrahigh-throughput mapping of the ecological diversity of thousands of environmental samples and unlocking novel microbial metabolites that were earlier not possible.

The biological complexity of "uncultured" microbes has been evaluated predominantly by one or more of these "-omics": metagenomics (DNA), metatranscriptomics (RNA), metaproteomics (proteins), metabolomics (metabolites), lipidomics (lipids), and glycomics (glycans) (Fig. 12.2). Interestingly, the informational diversity of these macromolecules keeps increasing in the same order, making their study progressively complex. The culture-independent nature of these "-omic" methods avoids potential unnatural phenotypes/genotypes that arise when the microbes are grown in axenic cultures in vitro (Morales and Holben 2011). Beyond ecological surveys, the NGS methods have also garnered interest in bioprospecting newer microbial products for industrial or commercial potentials. One such discovery made from then uncultured bacteria was cytotoxic compound pederin (Morales and Holben 2011). It is a polyketide produced by *Pseudomonas* symbiont of female rove beetles *Paederus sabaeus*. Similar compounds have later been identified from free-living marine *Alphaproteobacteria* or other symbiotic bacteria found in insects, lichen, and marine sponges (Piel 2011). Some of these NGS methods have been grouped based on the biomolecule of interest and discussed here. It should be noted that earlier these NGS approaches were developed as a standalone technique, but they are increasingly being used in a combinatorial manner to provide a holistic view of the microbial potential.

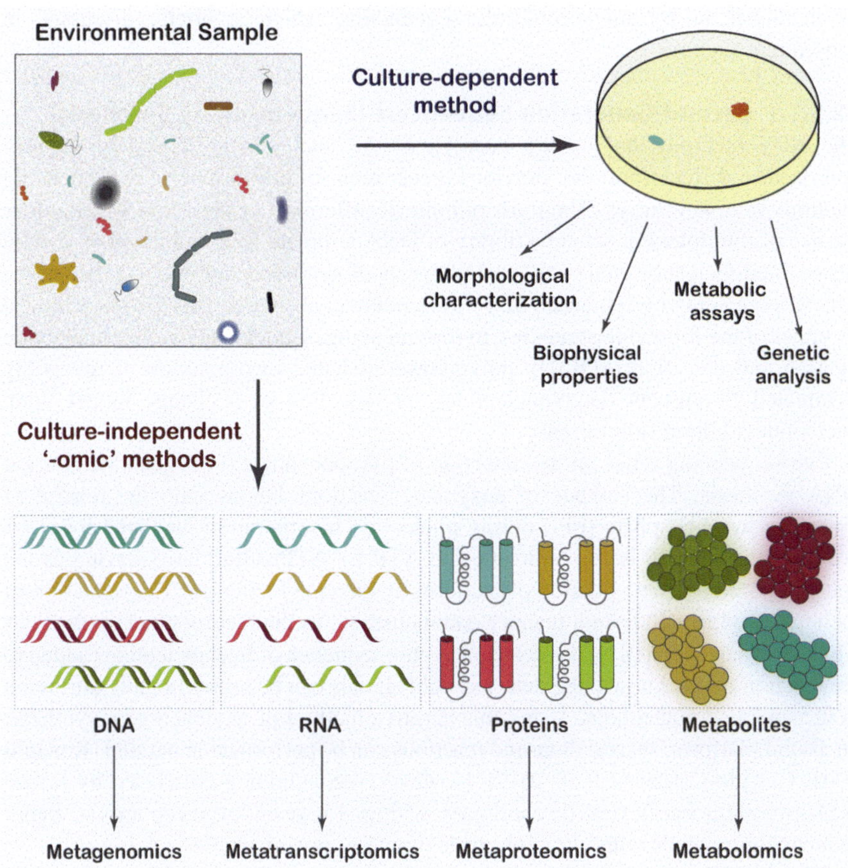

Fig. 12.2 Culture-dependent and culture-independent ("-omic") approaches to study microbial cells. A schematic representation of the commonly employed approaches to study microbes from the environmental samples. The culture-dependent methods (top right) rely on axenic culture for characterizing morphological, metabolic, biochemical, and genetic potential of microbes in vitro. The culture-independent methods (bottom mid) depend on extracted biomolecules (DNA, RNA, protein, and metabolites shown) to detect microbial composition of the environmental sample

12.5.1 Metagenomics: Evaluating DNA Sequences

An ensemble of microbial genes representing diverse microbial community and their response to perturbations in their surroundings is generally referred to as metagenome. The method captures all the DNA present in the sample, including viral and eukaryotic DNA (Fig. 12.2). This untargeted approach provides a wider genomic and taxonomic information that is not achieved by phylogenetic marker analysis alone. The metagenomic sequencing has allowed for unprecedented analysis of complex microbiomes, enabling refined predictions about the biosphere composition and function. The metagenomic approach relies on either sequencing by synthesis or by single-molecule sequencing, which is briefly discussed in subsequent sections.

12.5.1.1 Second-Generation Sequencers (Sequencing by Synthesis)

The NGS methods that form a bedrock of the nucleic acid-dependent "-omic" approaches differ from the previous generation of electrophoretic sequencing methods in several ways. The most prominent difference of these NGS methods is the use of multiplexing, where a library of DNA template is immobilized on a solid surface and then subjected to sequential cycles of synthesis and imaging, hence the name "sequencing by synthesis (SBS)" (Shendure et al. 2017). This SBS is achieved by either of the following strategies: pyrosequencing, sequencing by oligonucleotide ligation and detection (SOLiD), polymerase-mediated incorporation of reversibly terminated fluorescent deoxynucleotides, or detection of a change in pH upon nucleotide addition (ion torrent).

Pyrosequencing relies on the detection of pyrophosphate (PPi) generated during DNA synthesis. The cascade of enzymatic reactions begins with the release of inorganic pyrophosphate (PPi) during nucleotide incorporation mediated by DNA polymerase. The PPi is then converted to ATP by ATP sulfurylase, serving as the energy source for the oxidation of luciferin by enzyme luciferase. The oxidation of luciferin produces light in a manner commensurate to the incorporated nucleotides which are then captured by the detector. As the sequence of free nucleotides added to the reaction mix is known, the detected light signals can be proportionally attributed to incorporated nucleotides. Using this iterative method of sequencing by synthesis on a solid platform, several thousand reactions can be performed in parallel (Ronaghi 2001; Garrido-Cardenas et al. 2017). However, the technology marketed by Roche (454 pyrosequencer) was discontinued within 15 years of inception as better sequencing methods with ultrahigh-throughput became available.

The SOLiD sequencing platform made commercially available by Applied Biosystems (now Thermo Fisher Scientific) is a DNA ligase-based sequence detection method (Shendure et al. 2017). The single fragments or mate-paired fragments are ligated to adapters at either end, clonally amplified onto a bead in an emulsion PCR. Millions of such beads representing individual DNA from the library are then covalently bound to a solid glass surface. The DNA fragments on beads are then hybridized to a universal sequencing primer of length n and exposed to a library of 8-mer probes. The probes have two bases of known sequence at the first two positions followed by degenerate bases thereafter and either of the four different fluorescent dyes conjugated at the 5′ end. This ensures that only complementarity by the dinucleotide determines binding to the target sequence. The 8-mer probe gets ligated to the adjacent primer by DNA ligase, and the fluorescence is detected. In the next cycle, this fluorophore and the degenerate bases are cleaved-off, exposing the 5'hydroxyl group for the next cycle of ligation/detection. After one round of sequencing, the DNA strands are melted and then hybridized again to a different universal primer. Successive rounds of sequencing using incremental lengths ($n + 1$, $n + 2$, $n + 3$, and so on) ensure that the entire sequence is mapped. The choice of fluorophores is not random, and each fluorophore is shared by four dinucleotide pairs. So at least two rounds of sequencing are needed to correctly assign the bases in the target DNA. Further, as each base of target DNA is sequenced twice, the error rates are lower ($< 0.1\%$) than other contemporary methods (Garrido-Cardenas et al. 2017). Despite being low cost and high precision, the method has major drawbacks of longer run time and smaller read length.

The reversible terminator sequencing (marketed by Illumina/Solexa) method relies on the controlled incorporation of fluorescently labeled nucleotide analogs by DNA polymerase. The method differs from conventional Sanger's sequencing by substituting the 3'-OH group with a 3′-o-azidomethyl group which prevents incorporating more than one nucleotide per amplification cycle (Garrido-Cardenas et al. 2017). In each amplification cycle of immobilized DNA on a solid support, the fluorescent reversibly terminated nucleotides are added, and signals from added nucleotides are detected. The 3'-OH group is regenerated by removing the 3′-o-azidomethyl group and fluorescent moiety using TCEP (tris(2-carboxyethyl) phosphine). Similar cycles of nucleotide addition, imaging, and 3'-OH retrieval are iteratively continued till the entire DNA fragment (typically 100–150 bp) is sequenced. These platforms can generate over a billion reads, with very high accuracy (99.9%) in less than 3-day run time (Shendure et al. 2017).

The ion torrent sequencer (Thermo Fisher Scientific) is essentially a miniaturized solid-state pH meter that can detect the release of a proton (H^+ ion) when a nucleotide is incorporated into the growing strand of template DNA. Hence, the method directly transforms chemical information to binary digital data that can be converted to sequence information without the use of any specific modification of nucleotides. By utilizing a layer of ion-sensitive field-effect transistor (ISFET) coated under microwells, thousands of reactions in parallel can be measured simultaneously (Garrido-Cardenas et al. 2017). The method can provide reads of average length ~ 150 bp at a lower cost per sample, although like other single-nucleotide incorporation-based methods (e.g., pyrosequencing) detection of homopolymeric DNA stretches is error-prone.

12.5.1.2 Third-Generation Sequencers (Real-Time Single-Molecule Sequencing)

Notwithstanding the prowess of these NGS approaches, the dependence on template amplification makes them vulnerable to sequence-dependent biases and occasionally higher error rates. On the contrary, single-molecule real-time (SMRT) sequencing methods such as polymerase-based synthesis in real time (PacBio) and nanopore sequencing (Oxford Nanopore Technologies, ONT) have gained widespread attention (Shendure et al. 2017). The single-molecule methods have the potential to extend beyond DNA to RNA and even protein sequencing in real time. The biggest advantage of these real-time sequencing methods is that several kilobase long reads can be produced without the use of complicated library preparation steps or modified sequencing reagents.

The PacBio platform was the first system of its kind to use the SMRT sequencing approach. The DNA polymerase is immobilized on a solid surface and template DNA to be sequenced slides along with the polymerase. The PacBio sequencers rely on the use of nanophotonic zero-mode waveguides (ZMWs) for signal detection. The ZMWs constrain signal detection in a zeptoliter (10^{-21} liter) volume ensuring that only the signal from the incorporation of a cognate fluorescent nucleotide at the active site of DNA polymerase is detected. The bulk fluorescent nucleotides remain out of the detection range, manifesting as a constant low background distinct from fluorescence pulse generated by incorporated nucleotide. The sequence of these fluorescence pulses over time is used to derive the reads by instrument (Eid et al. 2009). The prime selling points of this platform are long reads (average length of 10 kb, occasionally up to 100 kb), higher fidelity, and independence of GC bias or repetitive sequences for accurate and contiguous assemblies (Shendure et al. 2017).

The Oxford Nanopore Technologies (ONT) is a sequencing approach of monitoring changes in electric potential when a single DNA or RNA polymer is tunneled through a proteinaceous nanopore rachet. The major benefit of the ONT sequencer is its extreme portability (Shendure et al. 2017). As the instrument depends on direct measurement of the composition of target DNA through the altered flow of ions, it can be miniaturized to the size of a USB stick and used in remote ecological sampling locations. The processivity rates of the instrument can be up to 400 bases per second, with an average read length greater than 100 kb and the longest read of ~900 kb in a single run (Shendure et al. 2017). These ultra-long reads have enabled de novo assembly of repeat-rich regions containing satellite or tandem repeats.

12.5.1.3 Historical Perspective of Metagenomic Studies

Early attempts at identification of soil metagenome were labor-intensive and low-throughput, wherein cloned libraries were prepared from DNA isolated from available soil samples and then screened for genes of interest (Myrold et al. 2014). The approaches were succeeded by Sanger's sequencing-based method in which random DNA fragments from soil samples were cloned into phage library (Scholz et al. 2016). Later, approaches relied on direct shotgun sequencing of isolated DNA, improving the throughput and efficacy of metagenome analysis. The shotgun sequencing provides reads of 400–500 bp average length with a typical metagenome size of up to 0.5 Gbp. However, only a fraction of these reads could be assembled into a draft sequence due to the small read length (Myrold et al. 2014). Soil metagenomes derived through the Illumina platforms have succeeded in producing larger assemblies of about 4.0 Gbp with reads ranging from 0.4 to 29 million. The success of this approach was demonstrated by the assembly of a draft genome of the dominant methanogens such as *Methanobacterium* (Mackelprang et al. 2017). In the metagenome assembly approach, the sensitivity of the method is directly dependent on the number of reads (sequencing depth). With the high enough number of reads (deep sequencing), taxonomic resolution to species or strain level is also possible (Scholz et al. 2016).

A practical application of metagenomic analysis is to identify patterns of nutrient cycling in the microbial community. A metagenomic study on the impact of thawing on Alaskan cryosols (soil containing permafrost in cold regions) identified shifts in C and N cycling. The study found that after the thaw, anaerobic oxidation of methane through methane monooxygenases (*pmoA* and *mmoX*) and ammonification and denitrification through nitrate reductase I (*narG*) increased dramatically (Hultman et al. 2015; Mackelprang et al. 2016).

12.5.1.4 Shotgun Metagenomic Workflow

A comprehensive reading about the shotgun metagenomic approach and useful computational tools can be found in these references (Knight et al. 2012; Niu et al. 2017; Quince et al. 2017). As several steps of sequence identification are common between metagenomic and metatranscriptomic methods, a detailed description is provided here, while only salient features are described for the metatranscriptomic workflow section:

1. *Experimental design, sampling, and metadata.* Although the downstream sample processing steps of the metagenomic study are streamlined, the experimental design requires very careful consideration. Primary factors for investigation of largely unexplored microbial life in a sample hugely depend on the variables such as type of sample collected, abiotic composition of the sample, time/season of sampling, etc. (Knight et al. 2018). Along with these descriptors, the experimental metadata such as sample handling, nucleic acid extraction/detection protocols, and statistical and bioinformatic analysis tools should also be recorded. This can have a huge impact on results as some environments such as soil are very heterogeneous in composition, while marine or frozen isolates are low in microbial density. Therefore, a careful forethought about the effect of intrinsic and extrinsic factors on the dynamic association of microbial community should be made. These parameters gain more significance for a hypothesis-based comparison and less so if a discovery-oriented study of novel microorganisms is envisioned. However, rigorous standards for the acquisition of experimental samples and careful documentation of metadata describing variables that can influence results should be made to ensure the reproducibility of the study. Inclusion of decontamination protocols and use of tracers during sampling and handling of ancient DNA studies have been suggested as a way to avoid contamination of permafrost samples (Saidi-Mehrabad et al. 2020).
2. *DNA extraction from samples.* DNA extraction is the quintessential part of metagenomic methods. Several factors such as DNA yield, shear of DNA fragments, extraction reproducibility, and proportional representation of rare microorganisms can introduce bias in the outcome of the study. Studies have shown that the observed-to-expected ratio of representative microbes in the population differ with the DNA extraction method, PCR amplification, and data analysis protocol used (Yuan et al. 2012). Although as low as 1 ng DNA can be used to prepare libraries using commercial kits such as Nextera XT (Illumina), low yield and purity of DNA from permafrost samples are still tricky for metagenomic evaluation. It has been observed that the most effective DNA recovery from diverse microbial taxa can be achieved by mechanical lysis than chemical DNA extraction methods (Yuan et al. 2012). Additionally, a unique set of challenges for nucleic acid-based microbial studies are encountered while using different soil or permafrost samples (Saidi-Mehrabad et al. 2020).

Extracellular DNA from dead microbes can get adsorbed to soil colloids and get coextracted with lysed DNA, leading to overestimation of ecological diversity. The presence of humic acid and other inhibitory contaminants in extracted DNA (Mackelprang et al. 2016) can also adversely affect the enzymatic library preparation steps.

3. *Library preparation and sequencing.* The method and depth of sequencing depend on the question to be addressed. Deep sequencing is advantageous when rare microbes in the sample are the intended targets of the study. Such an approach will yield detailed genomic data of rare microbes that can be assembled and interpreted for their functional roles in the complex environment. However, rare microorganisms in one environmental isolate may be abundant in another environment. Shallow sequencing of multiple samples, on the other hand, can provide a broad picture of abundant microbial community and dynamics in one go. Occasionally, shallow sequencing is preferred in conjunction with metatranscriptomic or metaproteomic methods. However, even with shallow sequencing depth, the detailed profiling of microbial dynamics in the environmental sample can be achieved by iterative sampling at regular intervals (Gilbert et al. 2012).
4. *Data processing*: Initial step of all sequencing methods is the removal of short- or low-quality reads and correcting for errors. Frequently, the sequences are trimmed after a predetermined length to get rid of adapter sequences or low-quality reads. The error rates can vary from 0.5% to 2.0% with more frequent errors at the 3′ ends of reads. Few preferred algorithms for these purposes are Trimmomatic, PrinSeq, Cutadapt, and Skewer (Lott et al. 2017). Some of the analytical platforms developed for metagenomic data analysis are QIIME, UPARSE, mothur, dada2, and minimum entropy decomposition (MED) (Niu et al. 2017; New and Brito 2020). The metagenomic data can even be mined to derive strain-specific information using tools as PanPhlAn, Kraken, CLARK, MG-RAST, and ConStrains (Scholz et al. 2016; Niu et al. 2017).

The sequencing reads from shotgun sequencing technology are typically less than 400 bp. This provides a huge volume of sequence information in a cost-efficient way. But the shorter reads contain lesser information and need to have higher coverage to derive meaningful overlaps. The assembly of these reads therefore can be computationally more challenging than SMRT sequencing. Initial shotgun sequence assembly tools were based on a greedy algorithm of pairwise comparison of every sequenced read in the dataset. This approach originally developed for assembling Sanger sequencing data was computationally taxing and less efficient. In the improved algorithm, the sequencing reads are decomposed into smaller subsequence (*k*-mers) of prespecified length *k*. These *k*-mers represent all possible combinations of nucleotides A, T, G, and C, and each *k*-mer serves as a node for the graph. Two nodes are considered connected by an edge if the *k*-mers have an overlap at the *k*-1 position. The algorithm then finds a Eulerian path derived from a De Bruijn graph of connected *k*-mer edges, which is then converted to contigs with minimal computational memory requirements (New and Brito 2020). Popular examples of such algorithms are SOAPdenovo, SPAdes, EULER, and Velvet (Shendure et al. 2017). Nevertheless, this approach of *k*-mer-based fast assembly can lead to different assemblies based on the choice of *k*-mer length and may fail for long repetitive sequences losing the genomic context. The method also falls short in the assembly of mobile genetic elements and diverse SNPs in the mapped strains. Other tools such as Genovo use Bayesian probabilistic models for assembly of reads (Carvalhais et al. 2012). A qualitative evaluation of different genome assemblies made by the algorithms mentioned above can be made through software tools as QUAST, MetaQUAST, Ikarus, etc.

Iterative clustering of thousands of such contigs (a process called binning) then leads to larger assemblies termed scaffolds. In the majority of instances, these contigs remain fragmented and incomplete or are explored for the presence of certain taxonomic markers only. Further binning of contigs into larger refined scaffolds is made by considering several factors as depth of sequencing, DNA composition, GC content, etc. These binned assemblies may be either a draft genome or complete MAG (metagenome-assembled genome). Application of single long-read methods (SMRT sequencing) can bypass many of these computational assembly steps, and further improvements are expected to lower the sequencing cost in the near future. The portability offered by ONT has been heralded as a breakthrough for an on-site metagenomic sampling of remote locations when sample quality can be adversely affected during transport to storage facilities (Edwards et al. 2020).

Non-assembly-Based Analysis (*Resequencing*).

Cataloging genetic variants by aligning metagenomic reads to the reference genome has also been referred to as resequencing. As the end of resequencing is to discover common variants, the tools employed for this goal differ from those for genome assembly. An important requirement for calling bona fide variants from sequencing errors is high redundant coverage (~30-folds or more). Data compression techniques as Bowtie or Burrows-Wheeler Aligner (BWA) align these short reads to larger genomes, allowing mismatches which are then utilized by analytical tools as SAMtools, ATLAS, AntCaller, and GATK to identify variants (Shendure et al. 2017). Furthermore, newer RNA-seq analysis packages (e.g., Cufflinks, TopHat, etc.) can directly generate statistics for transcript abundance, differential expression, number of splice junctions, and de novo transcript assemblies. Additionally, these non-assembly-based analysis is useful in discovery-based metagenomic or metatranscriptomic studies to predict and annotate multigene biosynthetic pathways or secondary metabolite gene clusters in the sequenced library. Some of the specialized non-assembly-based pipelines for such analysis are antiSMASH (antibiotics and secondary metabolite analysis shell), CLUSEAN (cluster sequence analyzer), and SMURF (secondary metabolite unknown region finder) (Medema et al. 2011).

5. *Post-processing analysis*: The quality of metagenomic assemblies is adjudged by factors as maximum contig length, N50 value (shortest contig length in assembly needed to cover 50% of the representative genome) and presence of chimeric contigs. To ensure a uniform application of standards for quality, completeness,

and contamination of assembled metagenomes, the genomic standard consortium (GSC) has proposed guidelines termed minimum information about any (x) sequence (MIxS). The MIxS descriptors contain information about the geolocation of sampling site, type of habitat (e.g., soil, water, permafrost, human gut, etc.), and sequencing method. These parameters for genomes (MIGS), metagenomes (MIMS), and environmental marker sequence (MIMARKS) are required during submission for publications or public databases. Public nucleotide accession databases (DDBJ/EMBL/GenBank) have also added these MIxS on their submission pages (Field et al. 2011).

12.5.2 Metatranscriptomics

The first several years of microbial ecology studies were focused on metagenomic and 16S rRNA-based sequencing approaches. These approaches revealed a rich trove of information regarding the composition and functional potential of microbial niches, but the temporal aspect of this potential in the active community was missing. Even selective depletion of DNA from nonviable sources by reagents such as DNA intercalating agent propidium monoazide (PMA) has limited success in the identification of viable microbial population in sample (New and Brito 2020). Owing to inherent instability compared to DNA or proteins, RNA-based analysis is more suited to study immediate cellular responses to environmental stimuli. Therefore, in recent years, metatranscriptomics, the NGS-based profiling of expressed genes comparing total cellular RNA to the total rRNA genes in the sample, has gained prominence (Fig. 12.2).

Before the advent of RNA-seq in 2008, the earlier environmental transcriptomic studies relied on microarrays or sequencing of expressed sequence tag (EST) clone libraries. But recent improvements in sequencing technologies have enabled faster and cheaper transcriptomic identification of active microbial diversity in environmental samples (Jo et al. 2020; Kaster and Sobol 2020). Several versions of RNA-seq are in use depending on the outcome anticipated, e.g., differential RNA-seq (dRNA-seq) for global mapping of transcriptional start sites, dual RNA-seq to evaluate roles of small RNA (sRNA) in host-pathogen interaction, and metatranscriptomic RNA-Seq (metaRNA-Seq) are to analyze highly complex microbiomes (Lott et al. 2017).

The average half-life of prokaryotic mRNA is typically in the range of few seconds to minutes, although the ones with housekeeping function tend to be more stable. The stability also varies with microbial species and nutritional status of individual cells (Edwards et al. 2020). Therefore, any sample intended for metatranscriptomic analysis must be snap-frozen in liquid nitrogen or supplemented with RNase inhibitors as soon as possible. The frozen or aqueous samples from cold regions need to be thawed or filtered over extended periods for preprocessing and hence present a bigger challenge to the chemical preservation of transcripts. Multiple sampling has been proposed as an alternate to average out the loss of transcripts due to natural decay. A generalized workflow for the metatranscriptomic experiment involves these four steps: RNA extraction, mRNA enrichment, sequencing library construction, and annotation of sequenced libraries:

1. *Total RNA extraction*: Metatranscriptomic approach warrants careful preparation of RNA, avoiding inhibitory organic molecules such as humic substances (e.g., humic acid, fulvic acid) and tannic acids (Wang et al. 2012; Uchii et al. 2019). Compared to genomic studies, these contaminants are more problematic as higher amounts of RNA are required for cDNA and library preparation. Specialized methods such as Sephadex gel column or polyethylene glycol (PEG)-based purification or adsorption to activated charcoal have been used to get rid of such inhibitory chemicals (Carvalhais et al. 2012). Further, adsorption of RNA to soil particles and the presence of RNases in the sample present a bigger challenge for transcriptomic studies (Tveit et al. 2014).
2. *mRNA enrichment*: As prokaryotic mRNA is devoid of 3′-end poly(A) tail and more than 95% of cellular RNA is composed of rRNA and tRNA (Sorek and Cossart 2010), an essential step before transcriptome sequencing is depletion of rRNAs through subtractive hybridization or selective enrichment of mRNA. Although mRNA enrichment can introduce biases in the sequenced library, the increased depth of sequencing by enriched mRNA compensates for the biases and improves the overall resolution of the transcriptome. Enrichment of mRNAs can be achieved by subtractive hybridization of cellular 16S and 23S bacterial rRNAs to complementary probes derived from constant regions of rRNAs. This approach is used in commercial kits such as Ribo-Zero Plus rRNA Depletion Kit (Illumina) and MICROBExpress kit (Thermo Fisher Scientific) and has been extensively used in different studies (Gilbert et al. 2008; Petrova et al. 2017). Selective hydrolysis of prokaryotic rRNAs and tRNAs, but not mRNA, has also been achieved using 5′-monophosphate-dependent ribonuclease (RNase E). The RNase E preferentially targets prokaryotic RNA molecules that have 5-′-monophosphates (5'-P), while mRNA molecules containing 5′-triphosphates (5'-PPP) are spared (Mackie 1998). Another approach employed for mRNA enrichment is through polyadenylation of only prokaryotic mRNAs, followed by oligo(dT) primer capture. A commercially available version of this enzyme is Terminator™ 5′-phosphate-dependent exonuclease kit (Lucigen). Selective enrichment of rare transcripts has also been achieved through the targeted degradation of dsDNA through a duplex-specific nuclease (DSN)-based kit (Trimmer-2 cDNA normalization kit, Evrogen). However, all the methods used above can act only on full-length RNA molecules rather than sheared RNA in samples. The rRNA depletion approaches also require higher input amounts of total RNA which can be difficult for environmental isolates. Studies have even used co-immunoprecipitation (Co-IP) using antibodies of interest to purify the subset of mRNAs.
3. *Sequencing library construction and data preprocessing*: The purified or enriched mRNA is then fragmented to a certain size range (~500 bp) either by alkaline solutions or enzymatically. Alternatively, the RNA is reverse transcribed to cDNA and then fragmented by sonication, enzymatically, or through

transposition (tagmentation). Because of the size limitations of NGS methods, this fragmentation is performed for both metagenomic and metatranscriptomic samples. Regardless of the fragmentation approach, the next step requires ligation of sequencing platform-specific adapters to generate a sequencing library. The amplified cDNA library is sequenced, and the quality of reads based on Phred score is visualized through FastQC or NGS QC toolkits (Lott et al. 2017). A typical RNA-seq library generates millions of raw reads corresponding to the RNA present in the metatranscriptome. The raw reads are first trimmed (adapter or low-quality reads), demultiplexed, and aligned to the reference genome to get normalized reads. Irrespective of the source of RNA, few initial steps of data filtering (e.g., removal of duplicates, filtering out low-quality and non-mRNA reads) are common to all analysis protocols. Analysis suites as Bowtie2, BWA, or TopHat2 for mapping, Trimmomatic for quality filtering, and DESeq2 or CuffDuff56 for differential gene expression are popular bioinformatics tools used to improve dataset quality (Lott et al. 2017). To eliminate sequencing biases arising from the length of transcript or sequencing depth, the raw data is usually normalized by converting to reads per kilobase of exon per million mapped sequence reads (RPKM) or transcripts per million sequenced reads (TPM). The RPKM value is represented as follows:

$$RPKM = \frac{(\text{number of mapped mRNA reads})\ \text{X}\ 10^9}{(\text{Total reads in the sample})\ \text{X}\ (\text{sum of exons in base pairs})}$$

No true consensus pipeline for RNA-seq data processing exists. However, biologically relevant information from these reads can be inferred through several comprehensive analysis tools which have been developed such as SAMSA, MetaTrans, and HUMAnN2 (Niu et al. 2017). Occasionally, high-quality and deep-sequenced datasets are used to assemble reads to contigs. Such an approach is useful when reference genome or gene annotation is unavailable. Here, the assembled contigs are then compared to large annotated transcript or protein databases, and the biological functionality is inferred based on homology with similar hits in the database. Further, an integrated analysis of metagenome to metatranscriptome from the same sample can be a more useful approach to identify expressed vs. non-expressed genes.

4. *Annotation of sequenced reads*: For metatranscriptomics, an additional step after quality filtering is the identification and removal of reads corresponding to rRNA by comparing the reads to the rRNA gene database. Additionally, if a polyadenylation approach was used to enrich mRNA, then the artificial polyA sequence also needs to be eliminated from the final reads. The filtered reads are aligned to public databases such as the National Center for Biotechnology Information (NCBI) nonredundant (nr) database (http://www.ncbi.nlm.nih.gov), the Integrated Microbial Genomes database (IMG/M; http://img.jgi.doe.gov), or the Metagenomics Analysis Server (MG-RAST, http://metagenomics.anl.gov). The alignment to public access databases let the researcher know the identity and

frequency of reads mapping to reference genomes, which in turn are normalized against gene abundance. This normalization provides the metatranscriptomic reads that are upregulated or downregulated in sample (Carvalhais et al. 2012). Sequence homology-dependent functional annotation of differentially expressed transcripts is made through databases that contain genes segregated on basis of functional pathways (e.g., methanogenesis, nitrification, etc.). Some of these databases are the Kyoto Encyclopedia of Genes and Genomes (KEGG), the Clusters of Orthologous Groups (COGs), MetaCyc, TIGRFAM, and evolutionary genealogy of genes: nonsupervised orthologous group (eggNOG) databases (Carvalhais et al. 2012; New and Brito 2020) (Table 12.1). Different statistical models (redundancy analysis (RDA), canonical correspondence analysis (CCA), permutational multivariate analysis of variance (PERMANOVA), or analysis of similarities (ANOSIM)) are applied to reveal the relationship between microbial transcripts in the environmental sample. The statistical interpretations also help identify the phylogenetic relatedness and diversity in the sampled library.

Detailed reviews on the application of microbial metatranscriptomics can be found in these references (Sorek and Cossart 2010; Bashiardes et al. 2016; Lott et al. 2017; Niu et al. 2017). Owing to space constraints, only two examples of ecological investigations relying on metatranscriptomics are listed here. The first example emphasizes the importance of understanding active microbial fraction in a complex environment, while the second example shows how metatranscriptomics can be leveraged to identify functional changes in microbial composition. *Verrucomicrobia* is an abundant bacterial phylum in soil that has been considered as an indicator of soil health. However, a recent transcriptomic study demonstrated that *Verrucomicrobia* is transcriptionally inactive and slow-growing (New and Brito 2020), raising questions about utility of these bacteria for tracking soil fertility. Another environmental metatranscriptomic study on peat soil samples from the Arctic region demonstrated a gradual shift of functional microbial activity from aerobic to anaerobic metabolism with increasing depth (Tveit et al. 2014).

12.5.3 Metaproteomics

Proteome, an ensemble of proteins present in the sample, represents active biological processes at a given time and provides a snapshot of the ecological diversity of microorganisms. The term metaproteomics was coined in 2004 by Wilmes and Bond as "the large-scale characterization of the entire protein complement of environmental microbiota, at a given point in time" (Kunath et al. 2019). Although metatranscriptomics has established its utility, RNA is not always the best predictor to illustrate potential biochemical functions in the microbial community. Cold or oligotrophic environments may have basal levels of gene expression that are not efficiently extracted from limiting samples. Also, the stability and turnover of RNA are influenced by ambient conditions.

Table 12.1 Resources for microbial data annotation and analysis

Name	Webpage	Purpose
BioCyc	https://biocyc.org/	A subscription-based curated database for omic data analysis and metabolic path searching
EMP (earth microbiome project)	https://www.earthmicrobiome.org/	Establish a global catalogue of the microbial distribution across the planet
Ensembl Bacteria	https://bacteria.ensembl.org/index.html	A specific web portal for bacterial and archaeal genomes only
GOLD (genomes OnLine database by joint genome institute, DOE-USA)	https://gold.jgi.doe.gov/index	Comprehensive resource for genome and metagenome sequencing projects
Integrated microbial genomes and microbiomes database	https://img.jgi.doe.gov/	For annotation, analysis, and distribution of microbial genome and microbiome datasets sequenced at DOE's joint genome institute (JGI)
Kbase (the Department of Energy Systems Biology Knowledgebase)	https://www.kbase.us/	An open-source software and data platform for data sharing, integration, and analysis
KEGG (Kyoto encyclopedia of genes and genomes)	https://www.kegg.jp/	Integrated functional annotation database for genomic, systems, chemical, and health information
List of prokaryotic names with standing in nomenclature	https://lpsn.dsmz.de/	A resource for bacterial and archaeal nomenclature and classification
MBGD (microbial genome database for comparative analysis)	http://mbgd.genome.ad.jp/	For ortholog identification, paralog clustering, motif analysis, and gene order comparison
MGnify	https://www.ebi.ac.uk/metagenomics/	Resource for analysis and archiving of microbiome data
NIH human microbiome project	Http://nihroadmap.nih.gov/hmp	Project themed at study of the microbial communities in human bodies
Patric (the Pathosystems resource integration center)	https://www.patricbrc.org/	Data and analysis tools for multiple omic data types
SILVA	https://www.arb-silva.de/	A comprehensive ribosomal RNA database

Since then, improved detection methods and analytical tools have presented metaproteomics as a complementary method to metagenomic and metatranscriptomic approaches for probing the dynamic interaction of microbial communities (Fig. 12.2). Recently ecological studies have embraced the concept of reporting the composition of microbial samples in terms of biomass (total protein per biomass protein), instead of counts per gene or genome, reflecting a changing perspective about metaproteomics (Kleiner et al. 2017). UniProtKB database contains ~195 million sequence entries (October 2020), representing 1,228,222 species, and more than 70% of the sequences are of bacterial origin (https://www.ebi.ac.uk/uniprot/TrEMBLstats).

In the earlier days of proteomics, 2D gel electrophoresis followed by mass spectrometry (MS) detection of proteins was the method of choice. Due to the limited information and resolution, these approaches have been replaced with one-dimensional gel-based or gel-free approaches for mass spectrometry. In-gel digestion followed by MS analysis of methanotrophs in cold seep sediments in one such study demonstrated expression of cold and oxidative stress response proteins in anaerobic methanotrophic archaea, suggesting adaptations to cold anaerobic marine environment (Stokke et al. 2012). Currently, rapid advances in the field have led to more efficient methods of systematic identification of proteins in samples through liquid chromatography coupled to tandem mass spectrometry. As a major goal of proteomic studies on environmental samples is the detection and identification of novel proteins in microbial consortia (discovery proteomics), the workflow relies heavily on mass spectrometry of enzymatically digested peptides. Broadly, these proteomic approaches can be segregated into two groups, untargeted shotgun (bottom-up) proteomics and targeted proteomics, which can further be accomplished by label-based or label-free methods. Although the label-based approaches such as SILAC (stable isotope labeling by amino acids in cell culture) and iTRAQ (isobaric tags for relative and absolute quantitation) are accurate and can be multiplexed (Ong et al. 2002), they are limited by expensive and tedious protocols and high sample requirement. On the contrary, the label-free methods are popular for their simplistic design even when metabolic labeling is not possible (Cox et al. 2014; Bostanci et al. 2020). The label-free proteomic quantitation is performed by spectral counting or by measuring peak intensities.

Here, the broad outlines of sample preparation, data acquisition, and interpretations for the metaproteomic approach are discussed. Detailed methods and required computational tools vary with the type of information sought from the samples and can be found in articles cited in these publications (Kunath et al. 2019; Sajulga et al. 2020; Wang et al. 2020):

1. *Protein extraction and preparation of peptide pool*: The profound challenge faced by environmental metaproteomics is the heterogeneity of sample types and their compositions. Oligotrophic sites such as permafrost, marine, or glacial sites have a lower abundance of microorganisms, and the samples need to be concentrated over several orders of magnitude to yield sufficient material. On the other hand, a smaller volume of samples from environments (e.g., soil, sewage) may be required for proteomic studies. Regardless, the initial step for the proteomic study is lysis using moderate concentrations of detergents (up to 5% SDS) and/or mechanical shearing. For shotgun proteomics, efficient digestion of proteins into peptides for mass spectrometry is necessary. This digestion is performed either on a gel, in solution, on filter (FASP), or through suspension trap (S-Trap). Subsequently, the peptides are desalted and cleaned up for fractionation by high-performance liquid chromatography (HPLC) or are further concentrated by precipitation with chemicals as trichloroacetic acid (TCA),

acetone, or methanol/ammonium acetate. As in the case of DNA/RNA extracted from soil samples, the peptides may also contain humic acids as contaminants. The common extraction protocols for removing humic acids from peptides exploit one or both of these properties: humic acid is insoluble at low pH (2–3) of peptide solvents, and humic acid molecules are larger than peptides. Therefore, a simple 10 kDa filtration of low pH peptide solution is preferred before HPLC separation.

2. *Detection of representative proteome*: The separated peptides are then identified through liquid chromatography coupled to high-performance mass spectrometry (LC-MS). The identification and quantification are done by data-dependent acquisition (DDA) or data-independent acquisition (DIA) methods (Kunath et al. 2019). In the DDA experiment (e.g., TopN), the top N number of abundant precursor ions is selected for fragmentation, which is not an ideal scenario for metaproteomics as rare or low-abundance peptides will be missed. The DIA method (e.g., SWATH) on the contrary provides better coverage of peptides in the sample by multiplexing of MS/MS spectra. Here, the instrument focuses on narrow mass windows per cycle and systematically scans through the entire range of peptide masses leading to high-accuracy spectra.
3. *Annotation of peptide spectra*: A single run for 24-h LC-MS/MS can yield several million MS/MS peptide fingerprints. Protein identification from the MS/MS spectra can be achieved either by comparison to reference databases or by deriving peptide sequence from the spectra accompanied by peptide sequence-based similarity search or finally through direct comparison to reference spectral libraries (Kunath et al. 2019). Assigning peptides to corresponding proteins of heterogeneous organisms need high-quality reference databases, preferably representing a similar microbial community or environment for high chances of discovery. However, only a small fraction of spectral fingerprints can be assigned to peptides due to a lack of comprehensive public databases, undermining the prowess of shotgun metaproteomics. Several proteomic data analysis platforms and tools are in use. Unfortunately, very few "meta-' oriented proteomic tools are available. A few of the available analysis tools are Unipept, Prophane, Pipasic, and MetaProteomeAnalyzer (MPA). The general purpose proteomic analytical search engines as Mascot, Andromeda, and X!Tandem produce an exhaustive list of peptides based on spectral matches. De novo sequencing software as PepNovo + and PEAKS directly infer an amino acid sequence from MS/MS spectra. This approach allows protein identification via error-tolerant similarity search on platforms as BLAST (Basic Local Alignment Search Tool) but is lower in throughput. An additional benefit of such an approach is the ability to identify unexpected posttranslational modifications (PTMs) that are missed by other methods. Spectral library search engines such as X!Hunter, SpectraST, and BiblioSpec rely on the fact that the peptide from the sampled population has been identified by MS before and is therefore restricted in applicability to metaproteomic studies.
4. *Functional annotation of metaproteome*: The functional annotation of identified proteins is more streamlined where standard homology-based methods as BLAST

and hidden Markov model-based HMMER are available. Publicly available tools as UniProt, COG (Clusters of Orthologous Groups), Kyoto Encyclopedia of Genes and Genomes (KEGG), MetaCyc (metabolic pathway database), Pfam, etc. are used in functional annotation or gene ontology (GO)-dependent subgrouping of annotated proteins in the metaproteome (Sajulga et al. 2020). Apart from functional curation, the single amino acid polymorphisms in the identified homologous proteins can also be used to detect different strains in a community.

In addition to label-free shotgun metaproteomics, recently active microbial communities in soil or marine environment have been probed through bioorthogonal noncanonical amino acid tagging (BONCAT) method (Hatzenpichler et al. 2016; Couradeau et al. 2019). The method relies on metabolic labeling of nascent proteins within the environmental sample with a methionine analog homopropargylglycine (HPG). The labeled proteins are then conjugated to fluorescent tags by "click chemistry" reaction. The fluorescent cells representing translationally active cells in the sample are then flow-sorted. As the method is dependent on the labeling of newly synthesized proteins, it has the sensitivity to identify nondividing or slow-growing cells in the sample. Furthermore, the metabolic labeling approach used in this method is rapid (requiring incubations as low as a few minutes), is cost-effective, and works even with low amounts of sample. A successful demonstration of this approach was made on two soil samples from Oak Ridge (Tennessee, USA), where up to 70% of cells enriched by flow cytometry were found to be metabolically active (Couradeau et al. 2019).

12.5.4 Metabolomics

Metabolites are the central players in all the cellular regulatory pathways integrating the cellular response to environmental perturbations. As regulatory metabolites induce rapid response toward external cues, they provide a direct link between genotype and phenotype. Therefore, a thorough evaluation of metabolites is important to understand active biosynthetic pathways and microbial function. Metabolomics is the measure of all such low-molecular-weight biosynthetic compounds at a given time. Although the concept of metabolic profiling and metabolomics has been proposed more than 20 years ago (Bostanci et al. 2020), the non-genomic approach of investigating secreted metabolites in the environment has recently gained momentum (Fig. 12.2).

The diversity of metabolites is a major challenge for both measurement and characterization methods. Unlike DNA, RNA, or proteins that are polymers made of specific nucleotides or amino acids, the metabolites are chemically distinct from each other. The challenge is further compounded by rapid fluctuations in the cell due to the ongoing biochemical activities. Metabolites involved in signal transduction (e.g., cyclic adenosine monophosphate (cAMP)) are produced in very low amounts compared to the ones required for metabolic activities (e.g., adenosine triphosphate

(ATP)). Hence, for reproducible metabolic profiling, it is imperative to quench the enzymatic activities instantaneously so that the composition of samples is maintained. Given the range of metabolites and complex cellular components, it has been estimated that even the best extraction methods fall short in the complete recovery of metabolites from samples (Baidoo 2019).

Metabolic profiling can be either global or targeted. In global profiling, a more diverse array of metabolite signatures (typically ~1500) are detected but not quantified, while in the targeted metabolomic studies a smaller subset of metabolites 200–500 across different samples are measured and quantified (Bostanci et al. 2020). These analytical metabolite signatures are detected either through NMR (nuclear magnetic resonance) or conventional mass spectrometry in conjunction with liquid chromatography (LC-MS) or gas chromatography (GC-MS). NMR-based metabolic profiling has the advantages of being quantitative, reproducible, and have simpler sample preparation steps. The method can generate high-throughput structural fingerprints in a nondestructive manner. However, NMR-based detection is better suited for abundant metabolites (μM concentrations) and is more expensive than MS-based approaches. Mass spectrometric methods have better sensitivity and can readily detect femtomolar concentrations. But MS-based measurements rely on ionization of sample and are, therefore, a destructive approach of profiling. It has been observed that only ~40–50% of metabolites can ionize in a chemical library. The nonionized components not only impact the quantitative estimation but also introduce artifacts called ion suppression. Ion suppression by nonvolatile compounds (e.g., salts, ion-pairing agents, coeluting compounds, plasticizers from tubes) impacts the efficiency of droplet formation, affecting the number of charged ions detected by the instrument. These ionization related problems make MS-measurements non-quantitative and can also limit the resolution of the method.

Additionally, both NMR and MS methods need complex analysis of chemical shifts in signatures or comparison to profiles of the library of compounds (Powers and Riekeberg 2017). Currently, the metabolic databases have limited NMR or MS metabolic spectra (a few thousand compared to millions for peptide databases) and allow simplistic search algorithms. The redundant and standalone nature of these databases further makes the analysis cumbersome, allowing one spectral search per query (Marshall and Powers 2017). Distinct NMR and MS spectral patterns also prohibit cross-comparisons of spectra and limit any statistical correlation between the two. These obstacles have led to a growing consensus that a combination of these two complementary analytical methods should be used to enhance the accuracy of metabolomic studies.

12.5.5 Other Developing "-Omic" Methods

Glycomics refers to a collective study of glycans and glycoconjugates present in a given organism. Both proteins and lipids can undergo covalent linkage of the diverse array of glycans at different positions, thereby providing innumerable combinations of such modifications. To put the diversity in numerical perspective, theoretical combinations of hexanucleotides are 4096, hexapeptides are 6.4×10^7, and hexasaccharides are 1.9×10^{11} (Kveton et al. 2020). Given the complexity and dynamic nature of glycoconjugates in a cell or an organism, the technology for the detailed appreciation of glycome is still in its infancy. The hurdles faced in the study of glycome are exemplified by limited entries in the available databases such as Carbohydrate Structure Database (CSDB), Glyco 3D, Glycan Reader, Unilectin3D, and GlycosciencesDB. The primary focus of microbial glycomics so far has been toward glycoimmunology and the glycobiology of infectious diseases. The host-pathogen recognition in such studies is elucidated through glycan microarrays. However, the diversity of these glycan arrays constructed by shotgun or synthetic methods is still very low (~300 microbial glycan antigens) compared to the potential diversity of natural glycoproteins (Geissner et al. 2019).

Lipidomics, an offshoot of metabolomics, is the study of structural and functional aspects of all the lipids found in cells, organisms, or the environment. Although the prokaryotic lipids are significantly distinct from their eukaryotic counterparts, they perform similar roles in energy storage, signal transduction, and membrane fluidity. In a manner analogous to other metabolomic studies, lipidomics also relies heavily on MS-based analyses. Both targeted and untargeted analytical MS methods have been used to obtain systemic insights. Strain-level bacterial fatty acid profiling for clinically relevant bacterial strains has been performed for "ESKAPE" pathogens (*Enterococcus faecium*, *Staphylococcus aureus*, *Klebsiella pneumoniae*, *Acinetobacter baumannii*, *Pseudomonas aeruginosa*, and *Enterobacter* species), establishing the potential of the approach for environmental lipidomics (Appala et al. 2020). However, the paucity of available databases for organism-specific lipids has hindered the wider application of lipidomics for environmental probing.

12.6 The Holistic Picture of "-Omic" Methods

The environmental microbiome is complex and diverse and is the second most abundant life-form on Earth in terms of global carbon content. The unique niches occupied by microorganisms have been successfully probed by several "meta-omic" approaches, revealing a hitherto unknown biochemical potential. However, it is now being widely acknowledged that an integrative mapping approach is required for better understanding of microbial diversity. Several recent studies have focused on this holistic microbial profiling approach; only a few of them are cited here.

The detailed study of permafrost soil samples through combinatorial "-omics," encompassing 16S rRNA gene sequencing, metagenomics, metatranscriptomics, and metaproteomics, revealed that >99.5% microbial population was made up of bacteria and archaea, and less than 0.5% corresponded to fungi (Hultman et al. 2015). The study sampled microbial diversity in three different environments, an intact permafrost, a seasonally thawed active layer, and a thermokarst bog, and found a high fraction of expressed cold-shock proteins in the former two environments of microbes compared to the bog. Additionally in permafrost soil, the ratio of transcripts to genes measured by the "-omics" approach was greater than 1.5 for phyla such as *Proteobacteria*, *Firmicutes*, and *Acidobacteria*, suggesting that these phyla are more acclaimed to survive in subzero temperatures (Hultman et al. 2015). Apart from bacterial sequences, the metagenomic datasets also include sequence data of fungi, viruses, and host DNA. The viral metagenome often gets ignored as representative genomes are limiting or co-identified with the host genome. These caveats are echoed in marine viral contigs assembled from several marine metagenomic studies which revealed that the lysogenic viruses are ubiquitous and abundant across oceans (New and Brito 2020). The analysis also indicated that in cold, aphotic marine water, the phages use their auxiliary metabolic genes to enhance nutrient uptake and nucleotide biosynthesis favoring synthesis of new viral particles in nutrient-depleted conditions (Coutinho et al. 2017).

12.7 Conclusion

With the ever-increasing affordability, optimized data collection, and analytical methods, the microbial universe has expanded at a previously unimaginable resolution. The culture-independent investigations have demonstrated the existence of giant viruses (>1.5 um) and symbiotic nanoarchaea or nanobacteria (<0.2) that lack key biosynthetic pathways (Legendre et al. 2014; Hamm et al. 2019). Such unusual biological features would never have been expected barely a decade ago. However, as the overarching goal of both the traditional and NGS methods is to characterize the elusive microbial "dark matter" from different environments, a polyphasic approach to analyze diversity through NGS methods and then culturing these underrepresented microbes by knowledge-driven approaches (e.g., cocultures, specific nutritional supplements) will be fruitful. As of December 2020, in less than 1.5 decades of the advent of nucleic acid-based "-omic" methods, ~38,000 metagenomic and ~ 12,000 metatranscriptomic environmental projects have been accomplished (refer to the Genomes Online Database at http://genomesonline.org/index.html). This sheer volume of data itself unquestionably has a huge impact on the way ecological diversity is interpreted. However, to develop deeper understandings of these novel organisms/assemblies, functional interpretations must be made in a holistic manner where the community composition, available metabolites, and inhibitors and changes over time are factored in. Instead of exploratory "-omics" from ecological samples to collect billions of nucleotide sequences, somewhat akin to stamp-collecting, a hypothesis-driven approach of microbial discovery can herald a new era of sustainable microbial ecology.

References

Amann RI, Ludwig W et al (1995) Phylogenetic identification and in situ detection of individual microbial cells without cultivation. Microbiol Rev 59:143–169

Appala K, Bimpeh K, Freeman C, Hines KM (2020) Recent applications of mass spectrometry in bacterial lipidomics. Anal Bioanal Chem 412:5935–5943. https://doi.org/10.1007/s00216-020-02541-8

Baidoo EEK (2019) Microbial metabolomics: a general overview. Methods Mol Biol 1859:1–8. https://doi.org/10.1007/978-1-4939-8757-3_1

Bashiardes S, Zilberman-Schapira G, Elinav E (2016) Use of metatranscriptomics in microbiome research. Bioinform Biol Insights 10:19–25. https://doi.org/10.4137/BBI.S34610

Bernard G, Pathmanathan JS, Lannes R et al (2018) Microbial darkmatter investigations: Howmicrobial studies transform biological knowledge and empirically sketch a logic of scientific discovery. Genome Biol Evol 10(3):707–715. https://doi.org/10.1093/gbe/evy031

Bostanci N, Grant M, Bao K et al (2020) Metaproteome and metabolome of oral microbial communities. Periodontol 2000:1–36. https://doi.org/10.1111/prd.12351

Buerger S, Spoering A, Gavrish E et al (2012) Microbial scout hypothesis, stochastic exit from dormancy, and the nature of slow growers. Appl Environ Microbiol 78:3221–3228. https://doi.org/10.1128/AEM.07307-11

Burkert A, Douglas TA, Waldrop MP, Mackelprang R (2019) Changes in the active, dead, and dormant microbial community structure across a pleistocene permafrost chronosequence. Appl Environ Microbiol 85:e02646. https://doi.org/10.1128/AEM.02646-18

Carvalhais LC, Dennis PG, Tyson GW, Schenk PM (2012) Application of metatranscriptomics to soil environments. J Microbiol Methods 91:246–251. https://doi.org/10.1016/j.mimet.2012.08.011

Chaturvedi P, Zhao B, Zimmerman D, Belmont AS (2018) Stable and reproducible transgene expression independent of proliferative or differentiated state using BAC TG-EMBED. Gene Ther 25(5):376–391. https://doi.org/10.1038/s41434-018-0021-z

Colatriano D, Tran PQ, Guéguen C et al (2018) Genomic evidence for the degradation of terrestrial organic matter by pelagic Arctic Ocean Chloroflexi bacteria. Commun Biol 1:90. https://doi.org/10.1038/s42003-018-0086-7

Couradeau E, Sasse J, Goudeau D et al (2019) Probing the active fraction of soil microbiomes using BONCAT-FACS. Nat Commun 10:2770. https://doi.org/10.1038/s41467-019-10542-0

Coutinho FH, Silveira CB, Gregoracci GB et al (2017) Marine viruses discovered via metagenomics shed light on viral strategies throughout the oceans. Nat Commun 8:15955. https://doi.org/10.1038/ncomms15955

Cox J, Hein MY, Luber CA et al (2014) Accurate proteome-wide label-free quantification by delayed normalization and maximal peptide ratio extraction, termed MaxLFQ. Mol Cell Proteomics 13(9):2513–2526. https://doi.org/10.1074/mcp.M113.031591

Dhakar K, Pandey A (2020) Microbial ecology from the himalayan cryosphere perspective. Microorganisms 8:1–17. https://doi.org/10.3390/microorganisms8020257

Edwards A, Cameron KA, Cook JM et al (2020) Microbial genomics amidst the arctic crisis. Microb Genomics 6:1–20. https://doi.org/10.1099/mgen.0.000375

Eid J, Fehr A, Gray J et al (2009) Real-time DNA sequencing from single polymerase molecules. Science 323:133–138. https://doi.org/10.1126/science.1162986

Field D, Amaral-Zettler L, Cochrane G, Cole JR, Dawyndt P, Garrity GM et al (2011) The genomic standards consortium. PLoS Biol 9(6):e1001088. https://doi.org/10.1371/journal.pbio.1001088

Garrido-Cardenas JA, Garcia-Maroto F, Alvarez-Bermejo JA, Manzano-Agugliaro F (2017) DNA sequencing sensors: an overview. Sensors 17:1–15. https://doi.org/10.3390/s17030588

Geissner A, Reinhardt A, Rademacher C et al (2019) Microbe-focused glycan array screening platform. Proc Natl Acad Sci U S A 116(6):1958–1967. https://doi.org/10.1073/pnas.1800853116

Gilbert JA, Field D, Huang Y et al (2008) Detection of large numbers of novel sequences in the metatranscriptomes of complex marine microbial communities. PLoS One 3(8):e3042. https://doi.org/10.1371/journal.pone.0003042

Gilbert JA, Steele JA, Caporaso JG et al (2012) Defining seasonal marine microbial community dynamics. ISME J 6(2):298–308. https://doi.org/10.1038/ismej.2011.107

Hamm JN, Erdmann S, Eloe-Fadrosh EA et al (2019) Unexpected host dependency of Antarctic Nanohaloarchaeota. Proc Natl Acad Sci U S A 116:14661–14670. https://doi.org/10.1073/pnas.1905179116

Hatzenpichler R, Connon SA, Goudeau D et al (2016) Visualizing in situ translational activity for identifying and sorting slow-growing archaeal - bacterial consortia. Proc Natl Acad Sci U S A 113:E4069–E4078. https://doi.org/10.1073/pnas.1603757113

Hultman J, Waldrop MP, Mackelprang R et al (2015) Multi-omics of permafrost, active layer and thermokarst bog soil microbiomes. Nature 521:208–212. https://doi.org/10.1038/nature14238

Jo J, Oh J, Park C (2020) Microbial community analysis using high-throughput sequencing technology: a beginner's guide for microbiologists. J Microbiol 58:176–192. https://doi.org/10.1007/s12275-020-9525-5

Kaster AK, Sobol MS (2020) Microbial single-cell omics: the crux of the matter. Appl Microbiol Biotechnol 104:8209–8220. https://doi.org/10.1007/s00253-020-10844-0

Kleiner M, Thorson E, Sharp CE et al (2017) Assessing species biomass contributions in microbial communities via metaproteomics. Nat Commun 8:1558. https://doi.org/10.1038/s41467-017-01544-x

Knight R, Jansson J, Field D et al (2012) Unlocking the potential of metagenomics through replicated experimental design. Nat Biotechnol 30:513–520

Knight R, Vrbanac A, Taylor BC et al (2018) Best practices for analysing microbiomes. Nat Rev Microbiol 16:410–422. https://doi.org/10.1038/s41579-018-0029-9

Kunath BJ, Minniti G, Skaugen M et al (2019) Metaproteomics: sample preparation and methodological considerations. Adv Exp Med Biol 1073:187–215. https://doi.org/10.1007/978-3-030-12298-0_8

Kveton F, Blsakova A, Kasak P, Tkac J (2020) Glycan nanobiosensors. Nano 10:1–30. https://doi.org/10.3390/nano10071406

Legendre M, Bartoli J, Shmakova L et al (2014) Thirty-thousand-year-old distant relative of giant icosahedral DNA viruses with a Pandoravirus morphology. Proc Natl Acad Sci U S A 111:4274. https://doi.org/10.1073/pnas.1320670111

Lewis K, Epstein S, D'Onofrio A, Ling LL (2010) Uncultured microorganisms as a source of secondary metabolites. J Antibiot (Tokyo) 63:468–476

Lott SC, Wolfien M, Riege K et al (2017) Customized workflow development and data modularization concepts for RNA-sequencing and metatranscriptome experiments. J Biotechnol 261:85–96. https://doi.org/10.1016/j.jbiotec.2017.06.1203

Mackelprang R, Burkert A, Haw M et al (2017) Microbial survival strategies in ancient permafrost: insights from metagenomics. ISME J 11:2305–2318. https://doi.org/10.1038/ismej.2017.93

Mackelprang R, Saleska SR, Jacobsen CS et al (2016) Permafrost Meta-omics and climate change. Annu Rev Earth Planet Sci 44:439–462. https://doi.org/10.1146/annurev-earth-060614-105126

Mackie GA (1998) Ribonuclease E is a 5′-end-dependent endonuclease. Nature 395 (6703):720–723. https://doi.org/10.1038/27246

Marshall DD, Powers R (2017) Beyond the paradigm: combining mass spectrometry and nuclear. Physiol Behav 176:139–148. https://doi.org/10.1016/j.pnmrs.2017.01.001.Beyond

Medema MH, Blin K, Cimermancic P et al (2011) AntiSMASH: rapid identification, annotation and analysis of secondary metabolite biosynthesis gene clusters in bacterial and fungal genome sequences. Nucleic Acids Res 39:W339–W346. https://doi.org/10.1093/nar/gkr466

Monciardini P, Iorio M, Maffioli S et al (2014) Discovering new bioactive molecules from microbial sources. Microb Biotechnol 7(3):209–220. https://doi.org/10.1111/1751-7915.12123

Morales SE, Holben WE (2011) Linking bacterial identities and ecosystem processes: can "omic" analyses be more than the sum of their parts? FEMS Microbiol Ecol 75:2–16. https://doi.org/10.1111/j.1574-6941.2010.00938.x

Myrold DD, Zeglin LH, Jansson JK (2014) The potential of metagenomic approaches for understanding soil microbial processes. Soil Sci Soc Am J 78:3–10. https://doi.org/10.2136/sssaj2013.07.0287dgs

New FN, Brito IL (2020) What is metagenomics teaching us, and what is missed? Annu Rev Microbiol 74:117–135. https://doi.org/10.1146/annurev-micro-012520-072314

Niu SY, Yang J, McDermaid A et al (2017) Bioinformatics tools for quantitative and functional metagenome and metatranscriptome data analysis in microbes. Brief Bioinform 19:1415–1429. https://doi.org/10.1093/bib/bbx051

Oliver JD (2010) Recent findings on the viable but nonculturable state in pathogenic bacteria. FEMS Microbiol Rev 34:415–425. https://doi.org/10.1111/j.1574-6976.2009.00200.x

Ong SE, Blagoev B, Kratchmarova I et al (2002) Stable isotope labeling by amino acids in cell culture, SILAC, as a simple and accurate approach to expression proteomics. Mol Cell Proteomics 1(5):376–386. https://doi.org/10.1074/mcp.M200025-MCP200

Petrova OE, Garcia-Alcalde F, Zampaloni C, Sauer K (2017) Comparative evaluation of rRNA depletion procedures for the improved analysis of bacterial biofilm and mixed pathogen culture transcriptomes. Sci Rep 7:1–15. https://doi.org/10.1038/srep41114

Piel J (2011) Approaches to capturing and designing biologically active small molecules produced by uncultured microbes. Annu Rev Microbiol 65:431–453. https://doi.org/10.1146/annurev-micro-090110-102805

Powers R, Riekeberg E (2017) New frontiers in metabolomics: from measurement to insight. F1000Res 6:1148

Quince C, Walker AW, Simpson JT et al (2017) Shotgun metagenomics, from sampling to analysis. Nat Biotechnol 35:833–844. https://doi.org/10.1038/nbt.3935

Ravikumar S, Uma G, Gokulakrishnan R (2016) Antibacterial property of halobacterial carotenoids against human bacterial pathogens. J Sci Ind Res 75(4):253–257

Ronaghi M (2001) Pyrosequencing sheds light on DNA sequencing. Genome Res 11:3–11. https://doi.org/10.1101/gr.11.1.3

Saidi-Mehrabad A, Neuberger P, Cavaco M et al (2020) Optimization of subsampling, decontamination, and DNA extraction of difficult peat and silt permafrost samples. Sci Rep 10:14295. https://doi.org/10.1038/s41598-020-71234-0

Sajjad W, Din G, Rafiq M et al (2020) Pigment production by cold-adapted bacteria and fungi: colorful tale of cryosphere with wide range applications. Extremophiles 24:447–473. https://doi.org/10.1007/s00792-020-01180-2

Sajulga R, Easterly C, Riffle M et al (2020) Survey of metaproteomics software tools for functional microbiome analysis. PLoS One 15:1–20. https://doi.org/10.1371/journal.pone.0241503

Schoch CL, Seifert KA, Huhndorf S et al (2012) Nuclear ribosomal internal transcribed spacer (ITS) region as a universal DNA barcode marker for Fungi. Proc Natl Acad Sci U S A 109 (16):6241–6246. https://doi.org/10.1073/pnas.1117018109

Scholz M, Ward DV, Pasolli E et al (2016) Strain-level microbial epidemiology and population genomics from shotgun metagenomics. Nat Methods 13(5):435–438. https://doi.org/10.1038/nmeth.3802

Shendure J, Balasubramanian S, Church GM et al (2017) DNA sequencing at 40: past, present and future. Nature 550:345–353. https://doi.org/10.1038/nature24286

Sorek R, Cossart P (2010) Prokaryotic transcriptomics: a new view on regulation, physiology and pathogenicity. Nat Rev Genet 11:9–16. https://doi.org/10.1038/nrg2695

Steen AD, Crits-Christoph A, Carini P et al (2019) High proportions of bacteria and archaea across most biomes remain uncultured. ISME J 13:3126–3130. https://doi.org/10.1038/s41396-019-0484-y

Stokke R, Roalkvam I, Lanzen A et al (2012) Integrated metagenomic and metaproteomic analyses of an ANME-1-dominated community in marine cold seep sediments. Environ Microbiol 14:1333–1346. https://doi.org/10.1111/j.1462-2920.2012.02716.x

Takahashi H, Horio K, Kato S et al (2020) Direct detection of mRNA expression in microbial cells by fluorescence in situ hybridization using RNase H-assisted rolling circle amplification. Sci Rep 10:9588. https://doi.org/10.1038/s41598-020-65864-7

Trudgeon B, Dieser M, Balasubramanian N et al (2020) Low-temperature biosurfactants from polar microbes. Microorganisms 8(8):1183. https://doi.org/10.3390/microorganisms8081183

Tveit AT, Urich T, Svenning MM (2014) Metatranscriptomic analysis of arctic peat soil microbiota. Appl Environ Microbiol 80:5761–5772. https://doi.org/10.1128/AEM.01030-14

Uchii K, Doi H, Okahashi T et al (2019) Comparison of inhibition resistance among PCR reagents for detection and quantification of environmental DNA. Environ DNA 1:359–367. https://doi.org/10.1002/edn3.37

van Elsas JD, Boersma FGH (2011) A review of molecular methods to study the microbiota of soil and the mycosphere. Eur J Soil Biol 47:77–87. https://doi.org/10.1016/j.ejsobi.2010.11.010

Vartoukian SR, Palmer RM, Wade WG (2010) Strategies for culture of "unculturable" bacteria. FEMS Microbiol Lett 309:1–7. https://doi.org/10.1111/j.1574-6968.2010.02000.x

Wang Y, Hayatsu M, Fujii T (2012) Extraction of bacterial RNA from soil: challenges and solutions. Microbes Environ 27:111–121. https://doi.org/10.1264/jsme2.ME11304

Wang Y, Zhou Y, Xiao X et al (2020) Metaproteomics: a strategy to study the taxonomy and functionality of the gut microbiota. J Proteome 219:103737. https://doi.org/10.1016/j.jprot.2020.103737

Woese CR, Fox GE (1977) Phylogenetic structure of the prokaryotic domain: the primary kingdoms. Proc Natl Acad Sci U S A 74(11):5088–5090. https://doi.org/10.1073/pnas.74.11.5088

Xu HS, Roberts N, Singleton FL et al (1982) Survival and viability of nonculturable Escherichia coli and Vibrio cholerae in the estuarine and marine environment. Microb Ecol 8:313–323. https://doi.org/10.1007/BF02010671

Yarza P, Yilmaz P, Pruesse E et al (2014) Uniting the classification of cultured and uncultured bacteria and archaea using 16S rRNA gene sequences. Nat Rev Microbiol 12:635–645. https://doi.org/10.1038/nrmicro3330

Yuan S, Cohen DB, Ravel J et al (2012) Evaluation of methods for the extraction and purification of DNA from the human microbiome. PLoS One 7(3):e33865. https://doi.org/10.1371/journal.pone.0033865

Zhao B, Chaturvedi P, Zimmerman D, Belmont AS (2020) Efficient and reproducible multigene expression after single-step transfection using improved BAC transgenesis and engineering toolkit. ACS Syn Biol 9(5):1100–1116. https://doi.org/10.1021/acssynbio.9b00457

Zhao X, Zhong J, Wei C et al (2017) Current perspectives on viable but non-culturable state in foodborne pathogens. Front Microbiol 8:580

Use of Proteomics and Transcriptomics to Identify Proteins for Cold Adaptation in Microbes

13

Aparna Anand and Abhinay Sharma

Abstract

Limitations with the conventional methods to study the behavior of microorganisms at molecular level forced the scientific world to develop advanced tools and techniques. After unfolding of genome sequences, array of "omics" tools (like genomics, bioinformatics, and proteomics) have been invented. These tools are being applied to study the structural as well as physiological response of microorganisms at molecular level and provide the snapshot of global response toward the environment they are inhabiting. It is always interesting to study the structural and physiological processes occurring in extremophiles to understand the evolution of life and new species. In this chapter, we discuss about different transcriptomics and proteomic techniques and how these have been employed in case of psychrophiles to understand the molecular basis of the stability and adaptability to low temperature. Furthermore, we discuss about their applications to identify genes and proteins which might play important roles in cold adaptation in psychrophiles.

Keywords

Proteomics · Transcriptomics · Metagenomics · Cold adaptation

A. Anand · A. Sharma (✉)
Department of Microbiology and Molecular Genetics, The Institute for Medical Research, Israel-Canada (IMRIC), The Hebrew University of Jerusalem, Faculty of Medicine, Jerusalem, Israel

R. Goel et al. (eds.), *Survival Strategies in Cold-adapted Microorganisms*, https://doi.org/10.1007/978-981-16-2625-8_13

13.1 Introduction

Life originated in hot oceanic soup in the form of primitive acellular and unicellular microorganisms with the metabolic machinery to withstand high temperature. As the time passed, our planet cooled down and organisms also evolved to live at different temperature conditions. Theoretically, the lowest and highest temperatures for the biological processes have been predicted to be +140 °C (nucleic acids spontaneously hydrolyze) and − 50 °C (cells vitrify), respectively (Clarke et al. 2013; Cowen 2004; Nguyen et al. 2017). Life exists in all possible temperature gradient along this scale, including extreme hyperthermophile (*Pyrolobus fumarii* grows at 121 °C) and psychrophiles (*Psychrobacter arcticus* living at −20 °C) (Deming 2002; Blöchl et al. 1997; Nguyen et al. 2017). Microbial life is limited by liquid water associated with thin films or saline solutions and solubility of lipids in water, protein stability at the lower and higher temperatures, respectively (McKay 2014).

Habitats for psychrophiles are either natural or man-made which range below +5 °C temperature and spread over a large proportion of the earth's area. Majority of ocean covers approximately 70% of our planet (averaging −1 to +5 °C) (Cowan et al. 2007). Polar regions (Antarctica and parts of North America and Europe within the Arctic Circle) constitute almost 20%, and Alpine regions including the Alps, Himalayas, and Rocky Mountains constitute a further 5% of the world's land surface area (Cowan et al. 2007). Man-made habitats (fridges and freezer) constitute only a small proportion of habitats. Cold-adapted organisms are classified into two groups depending on the cardinal growth temperatures. Psychrophiles show growth at minimum to maximum in the range of <0 to <20 °C while psychrotrophs (or psychrotolerants) between >0 and > 30 °C (Casanueva et al. 2010).

13.2 Physiological Adaptations of Psychrophiles

Psychrophiles remain at complete equilibrium with their cold habitat with the help of their characteristic structural and functional components. In case of terrestrial psychrophiles, they need to withstand desiccation, nutrient scarcity, and high radiation impacts with the fluctuating temp (-20 °C to +25 °C), whereas deep marine psychrophiles are adapted for constant low temperatures. The particular set of mechanistic features of adaptation has evolved to suit the particular needs of different ecological groups of psychrophiles. Some of the evolved features have been described here in brief. Cell membranes protect the cellular integrity and regulate transport processes in cell. The main constituents of membranes are phospholipids and proteins. As the growth temperature goes down, unsaturation of fatty acid gets increased by reduction in average chain length and increase in methyl branching. These changes give rise to reduced temperature of the transition from the liquid crystalline to the gel phase, also defined as membrane fluidity. The number of specialized proteins plays important roles to neutralize the damaging effects of cytoplasmic freezing. Antifreeze proteins and ice-binding proteins (IBPs) bind to ice, which inhibits ice-crystal growth and recrystallization. Ice-nucleating proteins induce the crystallization of ice at temperatures close to the melting point so as to prevent supercooling of water. These proteins can bring the freezing point down by over 2 °C and help in cellular cryopreservation (Perutz and Raidt 1975; Weaver et al. 1976; Kumar et al. 2000; Jaenicke 2000; Goldstein 2007). It is a well fact that some salts and solutes help in depression of freezing points. Psychrophiles smartly adopted this strategy to protect themselves from freezing at low temperature condition. The cells of these microorganisms accumulate compatible solutes such as glycine, glycerol, betaine, trehalose, mannitol, and sorbitol which results in a reduction of the freezing point of the cytoplasmic aqueous phase. Furthermore, they might help in stabilization of cytoplasmic macromolecules.

A sudden lowering of environmental temperature triggers a "cold-shock response" in many microorganisms resulting in higher expression of cold-shock proteins (Csps). Csps are highly conserved small proteins that bind to cold-shock domain (CSD) of single-stranded nucleic acids (Moon et al. 2019). The number of cold-induced RNA helicases and Csps works in tandem to stabilize DNA replication and protein synthesis (Hawwa et al. 2009). Catalysis at low temperature is a challenging situation for microorganisms because it has been suggested by Arrhenius that reaction rates go down by 10- to 60-fold when the temperature drops from 30 °C to 0 °C. But the microbial physiology evolved to develop psychrophilic enzymes which are adapted to operate efficiently at low temperatures. The backbone polypeptides of the active site in psychrophilic enzymes are more flexible than other portions of the protein (Friedrich and Antranikian 1996). This high flexibility enables increased complementarity between active site and substrate, at a low energy cost, resulting in high specific activities at low temperatures. The adaptation strategies and features have been detailed out in a recently published review (Kohli et al. 2020).

13.3 Omics to Explore Structural and Physiological Features in Psychrophiles

It is always interesting to study the structural and physiological processes occurring in extremophiles to understand how the life evolved. Some of the factors like the isolation, maintenance, and preservation of these microorganisms at their optimum extrinsic temperatures pose difficulty to study them. Furthermore, the stability and functionality of all the macromolecules of the cell such as DNA, RNA, proteins, etc. should be ensured (Orellana et al. 2018; Kohli et al. 2020). Lot of research work by using conventional as well as advanced methods has been published on different aspects of survival and adaptation of extremophiles. A combination of advanced tools like bioinformatics, genomics, transcriptomics, and proteomics, also called as "omics," has been employed to understand the molecular basis of their stability and mechanistic insights into fundamental biological processes important for adaptability to temperature variations. In this chapter, we discuss about diverse proteomic and transcriptomic technologies and their applications to identify proteins which might play important roles in cold adaptation in psychrophiles.

13.3.1 Brief about Genomics and Metagenomics

The genome sequences are complete array of data belonging to every possible gene. By examining the genetic content as well as transcription and translation profiles of any extremophile, a more detailed understanding of its unique structural and physiological features may be achieved at molecular level. Lot of work has been done in the last three decades to finish genome sequencing of thousands of prokaryotes after first bacterial (*Haemophilus influenzae*) genome sequence was published (http://www.ebi.ac.uk/genomes/bacteria.html, https://gold.jgi.doe.gov/cgi-bin/GOLD/bin/sequencing_status_distribution.cgi). More than one genome sequences are now available for each major extremophile, and currently, hundreds of genome sequences have been completed with many more ongoing (https://gold.jgi.doe.gov/cgi-bin/GOLD/bin/organisms?Organism.Domain=ARCHAEAL&Organism.Is%20Public=Yes). Being advantageous over conventional approaches, genomics and its modifications are highly efficient and have been adopted in recent past to study the unique sets of differentially expressed genes across whole genome of extremophiles. Furthermore, available multi-genome sequences of same species can be used to identify core set of genes as well as set of genes differentially expressed in extremophiles in comparison to mesophiles. Genome-based approaches can be applied to cultivable as well as non-cultivable microorganisms. The last decade has witnessed several technological advances that have the potential to identify every possible unique gene which are important for cold adaptation in psychrophiles.

13.3.1.1 Comparative Genomics

The sudden wealth of sequence data has permitted whole genome alignments to compare the evolution of extremophiles. Comparative genomics involves the alignment of sequences of available genomic data using computer-based tools. Such comparative strategies have been linked to conservation of DNA sequences (encoding the proteins and RNAs responsible for functions) over evolutionary time. Therefore, comparative genomic approaches begin with alignment of genome sequences and searching for orthologous sequences (sequences that share a common ancestry) in the aligned genomes and examining the conserved sequences. Genome comparison allows to analyze the phylogenetically relatedness of different species and facilitates the identification of the essential genes necessary for adaptation to the extreme environments. BLASTN (www.ncbi.nlm.nih.gov/BLAST/), GWFASTA (http://www.imtech.res.in/raghava/gwfasta), Microbial Genome Database (http://mbgd.genome.ad.jp/), MicrobesOnline ((http://www.microbesonline.org), Integrated Microbial Genomes with Microbiome Samples (IMG/M) (https://img.jgi.doe.gov), EzBioCloud (https://www.ezbiocloud.net/), etc. are popular databases used for comparative analyses of microbes on gene level.

The procedure for performing comparative genomic studies of psychrophilic bacteria involves collection of samples (such as water, soil sediment samples, permafrost, and ground ice samples) and storing them at 4 °C (water and soil samples) or − 20 °C (permafrost and ground ice samples) until sampling and isolation of different strains, followed by serial dilution and plating. Once the isolates are obtained, the pure cultures are preserved using 20% glycerol at −80 °C for further studies. The genomic DNA is isolated, and the 16S rRNA gene is amplified by PCR using universal primers. Purified PCR products are subjected to sequencing using the forward, reverse, and internal primers with suitable kits and a sequence analyzer. The generated sequences are utilized to perform Basic Local Alignment Search Tool (BLAST); this helps to identify the close relatives of queried 16S rRNA gene sequences (Mukhia et al. 2021, Raymond-Bouchard et al. 2018).

Comparative genomic analyses of fully sequenced genomes play a vital role in describing the entire core- and pan-genomes of different isolates from the same species. The core-genome includes the entire repertoire of translated genes conserved among all isolates; they are essential for its major phenotypic traits and existence. The pan-genome is the sum of the core genes (genes common to all strains of a species) and the accessory genes (set of unique genes shared by some but not all strains of the species as well as strain-specific genes). The accessory genome contributes to the diversity of species; they confer selective advantages such as antibiotic resistance, colonization of new hosts, and niche adaptation (Medini et al. 2005; Tettelin et al. 2008). To define the entire core- and pan-genomes of an organism, a large quantity of fully sequenced genomes is essential. The presence of a pan-genome is linked to environment adaptations, intra-species diversity, inter-species diversity, etc. (Tettelin et al. 2008).

13.3.1.2 Functional Genomics

Functional genomics refers to the systemic study of analysis of the functions and interactions of the genes and gene products encoded by the genome. The function of genes can be determined using sequence database that is annotated, and the functions of specific genes and their role in cellular processes can be studied. Functional genomics utilizes several approaches, like microarrays, total gene expression analysis, differential display polymerase chain reaction (PCR), etc. These techniques have been used for analyzing the expression of thousands of genes in parallel.

Over the last three decades, methodological breakthroughs have repeatedly revolutionized transcriptome profiling, and integration of transcriptomic data with other omics is giving an increasingly integrated view of cellular complexities. Transcriptome includes all transcripts present in a cell including mRNA, miRNA, small RNAs, and noncoding RNAs. Transcriptomics is a global approach that identifies RNA quantity and measures the differential transcript expression levels, spatially and temporally under varying physiological conditions. It gives information on diversity, noncoding RNAs, and the arrangement of transcriptional units in coding regions. However, one of the main drawbacks of transcriptomics is that information generated is highly variable, as transcript levels are potentially affected by a wide range of factors. To eliminate the trivial causes of variation, transcriptomic data must be generated under a wide variety of conditions. In brief, cDNA is the target of the sequencing technology which is obtained by the reverse transcription of the microbial RNA; RNA extraction is achieved by using lab-assembled protocols or commercial kits. Extracted RNA is then transcribed into cDNA prior to next-generation sequencing protocols, and data analysis is performed by bioinformatic tools (Wang et al. 2009). Modern transcriptomics involves the usage of high-throughput methods to analyze multiple transcript expression levels in different physiological conditions, mostly hybridization-based microarray, or sequence-based RNA-sequencing approaches are widely used. To generate data for functional annotation for psychrophiles, the following techniques are used:

Microarrays

Microarrays involve the determination of transcript levels by hybridization of fluorescently labelled transcripts to probes that arrayed on a solid substrate. On the array, the fluorescence intensity at each probe location suggests the transcript level of gene or RNA. The advantage of microarrays is that they have high throughput and are relatively inexpensive. However, the limitations of microarray include high background levels owing to cross-hybridization, dependence upon existing knowledge about genome sequence, and a limited detection range due to background and saturation of signals.

Quantitative PCR (qPCR)

Quantitative PCR (qPCR), also called real-time PCR, is a PCR-based technique that combines amplification of a target DNA sequence with measurement of the concentration of that DNA type in the reaction (Dymond 2013). qPCR uses either intercalating fluorescent dyes such as SYBR Green or fluorescent probes (TaqMan) and specific primers to measure the accumulation of DNA amplicons in real time. The increase of amplicons is proportional to an increase in fluorescence intensity, which is measured at each PCR cycle. When preceded by reverse-transcription PCR, qPCR is a powerful tool to measure mRNA expression; it is the standard method for confirmation of microarray gene expression data.

RNA-Seq

RNA-seq employs deep sequencing approach to investigate the transcriptome. The protocol for RNA-seq involves conversion of extensive RNA into cDNA after DNase treatment. Next, reverse transcription with random hexamers is performed, which results in cDNA libraries. Sequencing the cDNA library produces a standard RNA-seq dataset. The resulting reads are either aligned to a reference genome or reference transcripts to generate a genome-scale transcription map that consists of both the transcriptional structure and gene expression level. Some of the commonly used RNA-Seq platforms include ABI SOLiD, Illumina Solexa, Roche, 454 PacificBio, or combination of them. Although they are centered on the same principle, i.e., to generate RNA sequencing data, they are different based on read length, throughput, price, and generation and error rate (Loman et al. 2012).

RNA-seq is used to mainly determine RNA expression levels in different conditions at different time points. RNA-seq has several advantages which include large dynamic range of transcript expression detection, cost-effective high-throughput analysis of transcriptome, higher sensitivity, etc. The main challenges of RNA-seq are as follows: the manipulations done during cDNA library preparation can restrict its usage in whole transcript profiling, the construction of strand-specific library construction is laborious, and due to incompletely processed RNAs, background noise can be present.

13.3.1.3 Metagenomics

Metagenomic analysis involves the use of techniques to investigate the collective genomic content of microorganisms in a particular environment. It is commonly used to study phylogenetic relationship and metabolic potential of mixed populations, without the need to isolate and cultivate each member of the community (Aliyu et al. 2017; Sharon and Banfield 2013). The sample collection (water and sediment samples) for the metagenomic analysis from psychrophiles involves preservation of samples at −20 °C in 70% ethanol, first at −20 °C and then at −80 °C (Koo et al. 2018). The methodology of metagenomic analysis starts with isolation of genomic DNA; construction of a DNA library from the isolated DNA by fragmentation with appropriate sizes followed by cloning of DNA fragments into the cloning vector such as plasmid, cosmid, phage, etc.; and screening the available library for a target gene of a particular function. There are two common metagenomic strategies used. The first one includes the usage of marker genes such as the ribosomal genes 16S rRNA and 18S rRNA that are commonly used to study the composition of the microbial community in a certain environment; this is referred to as targeted metagenomics. The second approach deals with analysis of genomic DNA sequences using high-throughput next-generation sequencing to determine the entire taxonomic structure or functional capacity of microbial communities; this is called the shotgun metagenomics.

The advantage of metagenomics is that it is less biased than PCR and it gives information about relative abundance of different organisms and the community composition. It also helps in detecting different variants present in various environmental conditions. The disadvantage of metagenomics is that the microorganisms that are less abundant are difficult to detect. In addition, the sequencing projects are expensive, and analysis of metagenomic data constitutes evident bottlenecks.

13.3.2 Application of Genomic Tools on Psychrophiles

The advent of genome sequencing and latest techniques has greatly boosted our understanding of psychrophile biology. In this section, we discuss and highlight recent works that will enable to understand the molecular strategies underlying cold adaptation. Various genomic tools and high-throughput methods such as comparative genomics, microarrays, RNA-seq, quantitative PCR, and metagenomics have been used by researchers across the world for genomic studies and analysis of differential gene expression levels of psychrophiles.

13.3.2.1 Comparative Genomics of Psychrophiles

Comparative genomics provides comprehensive view of evolutionary and taxonomic relationships of psychrophiles. In addition, it plays an important role in determining the cold adaptation-related traits present in psychrophiles and relationship with the ecosystem due to these traits. Comparative and functional genomic studies have been completed using complete and draft genome sequences of single psychrophilic strain and multiple genomes from closely related species. These types of studies have been carried out on different groups of species of archaea growing in the cold-adapted regions that include *Methanococcoides burtonii*, *Methanogenium frigidum*, and *Methanolobus psychrophilus* R15 (Allen et al. 2009; Saunders et al. 2003; Chen et al. 2012). The presence of psychrophilic algae in such as *Chlorella vulgaris* NJ-7, *Fragilariopsis cylindrus*, *Tetrabaena socialis*, and *Chlamydomonas* sp. ICE-l has been reported (Li et al. 2009; Mock et al. 2017; Zhang et al. 2019; Zhang et al. 2020). *Colwellia psychrerythraea* strain 34H was the first psychrophilic bacterial genome to be sequenced (Methé et al. 2005; Nunn et al. 2015). Genomic studies of many bacterial species belonging to genera *Psychroflexus*, *Psychrobacter*, *Glaciimonas*, *Rhodoferax*, *Colwellia*, *Synechococcus*, *Hymenobacter*, and *Arthrobacter* have been reported (Feng et al. 2013; Moghadam et al. 2016; Margesin et al. 2016; Baker et al. 2017;Zhang et al. 2018; Tang et al. 2019; Roldán et al. 2020; Mukhia et al. 2021). Intensive genome studies on specific psychrophilic and psychrotolerant strains such as *Luteimonas abyssi*, *Nesterenkonia* sp. AN1, *Psychrobacter* sp. PAMC 21119, *Exiguobacterium antarcticum* B7, *Marinobacter gelidimuriae* sp., *Hymenobacter nivis* $P3^T$, *Labilibaculum antarcticum* sp., and *Pseudomonas psychrophila* MTCC12324 have been studied by various researchers (Fan et al. 2014; Aliyu et al. 2016; Koh et al. 2017; Dias et al. 2018; Chua et al. 2018; Terashima et al. 2019; Watanabe et al. 2020; Abraham et al. 2020). Psychrophilic and psychrotolerant fungi belonging to the genera *Alternaria*, *Aspergillus*, *Mucor*, *Cladosporium*, *Penicillium*, *Geomyces*, and *Rhizopus* have been reported (Rafiq et al. 2020; Schipper 1967; Ma et al. 2018a; Porto et al. 2020; Durán et al. 2019). *Mucor strictus*, *Cladosporium ossifragi*, *Mrakia psychrophile*, *Alternaria consortialis*, and *Penicillium griseofulvum*, etc. are fungal species reported from various cold regions (Martorell et al. 2021; Su et al. 2016; Schipper 1967; Rafiq et al. 2020).

The genomes of various species have been investigated in terms of cold adaptation strategies, lifestyle, ecology, and metabolic pathways. Taking hints from insights of the genomic data, several studies based on comparative molecular adaptation analysis have been reported considering psychrophiles as a query against mesophilic counterparts. In a study by Mukhia et al. (2021), genome-wide comparative analyses were performed to compare the psychrotrophic bacteria *Arthrobacter* sp. ERGS1:01 and ERGS4:06 genomes with the genomes of 13 mesophilic *Arthrobacter* strains. Genome analyses results revealed that both the *Arthrobacter* strains possessed vital genes for cold adaptation such as endonuclease multi-enzyme complex involved in DNA repair, cold shock proteins, chaperone proteins and stress response proteins, etc. Zhang et al. (2019) studied the genomic features that may aid the psychrophilic lifestyle of the sea ice alga *Chlamydomonas* sp. ICE-l. Genome analysis revealed that ecological success of this alga was enabled by gene families associated with DNA repair, photoprotection, ionic homeostasis, osmotic homeostasis, unsaturated fatty acid biosynthesis, and reactive oxygen species detoxification.

13.3.2.2 Functional Genomics of Psychrophiles

The vast sequence database that is now available can enable high-throughput exploration aiming at the determination of the function of the genes. This huge sequence information and the development of new technologies for analyzing gene expression have now made it possible to analyze gene expression on a genomic scale. To this end, microarrays have been widely used. Ghobakhlou et al. (2015) performed microarray printing of 5760 Arctic *Mesorhizobium* strain N33 genomic clones for partial analysis of transcripts of the strain grown under different temperature conditions. At low temperatures (4 and 10 °C), the genes that were significantly upregulated encode genes involved in various functions such as transcription regulation, protein turnover, metabolite transport, cryoprotection, fatty acid metabolism, oxidoreductase activity, and membrane fluidity. The genes that were downregulated included those which were involved in functions such as energy production and conversion, secretion, amino acid transport, cell envelope, and outer membrane synthesis and cell motility. To validate the microarray results, qRT-PCR on the genes identified within 54 selected clones was performed; the difference in gene expression profiles among the 54 selected transcripts was statistically significant.

For gene discovery, global gene expression analyses are used to understand cellular and molecular mechanisms with respect to the culture environment and stresses. At colder temperatures, the transcriptional profiles of psychrophiles undergo several changes and change the gene regulation in most cellular processes, including transcription and translation, functions of chaperones, metabolic and biosynthetic pathways, cell wall and peptidoglycan biosynthesis, cell membrane composition, lipid biosynthesis, and stress responses (Raymond-Bouchard et al. 2018). In a study published by Peng et al. (2017), RNA-seq-based analysis was used to uncover the fact that strain *Acidithiobacillus ferrivorans* YL15 uses a set of strategies to acclimatize to low temperature. RNA-seq analysis for cells grown at 28 °C and at 6 °C revealed differences in RNA transcripts between cells grown at these temperatures; a total of 372 genes with significantly different RNA transcript counts were recognized, of which 199 and 173 had higher and lower RNA transcript amounts at 6 °C, respectively. Several genes involved in transcription showed increased number of RNA transcript counts in cold condition; most of the genes for RNA polymerase complex core enzyme subunits (alpha, beta, and beta') had higher RNA transcripts at 6 °C, specifically the beta subunit coding gene which had 5.44-fold more RNA transcript counts. The gene coding for RpoD (RNA polymerase sigma factor) was stimulated by cold. Several transcription factor genes including those coding for transcriptional initiation protein Tat, transcription elongation factor GreA, transcription termination factor Rho, and transcription-repair coupling factor showed cold-enhanced RNA transcript levels. The increased counts of RNA transcripts of cellular components of the transcriptional machinery together with the genes involved in the transcriptional processes in strain YL15 showed that transcriptional regulation is crucial to cold adaptation. Raymond-Bouchard et al. (2018) used RNA-seq to perform transcriptomics on *Rhodococcus* sp. JG3 culture grown at −5 and 25 °C and *Polaromonas* sp. Eur3 1.2.1 grown at 0 and 20 °C, respectively. Overall, 515 and 359 transcripts were expressed differentially for

Rhodococcus sp. JG3 and *Polaromonas* sp. Eur3 1.2.1, respectively, between their respective higher and lower temperatures. In *Rhodococcus* sp. JG3, there was enhanced transcript expression at cold temperature that involved genes in amino acid transport and metabolism, iron transport, fatty acid synthesis modulation, and catabolism of alcohols/ethanolamine and sustaining redox potential. Conversely, *Polaromonas* sp. Eur3 1.2.1 induced transport and metabolism of carboxylates, energy metabolism relating to the ETC, oxidative phosphorylation, glycolysis, and global signal transduction mechanisms (Table 13.1).

13.3.2.3 Metagenomics of Psychrophiles

Metagenomics has offered unprecedented access to the structure and prospective function of various microbial communities found in cold regions. For example, evaluation of metagenomic sequencing data to study the taxonomic diversity and functional profile of psychrophilic and psychrotolerant microbial communities of Pangong Lake (a temperature of ± 10 °C) revealed that the *Proteobacteria* was the most prominent phylum, followed by *Firmicutes*, *Actinobacteria*, and *Ascomycota*, whereas the most abundant genera were *Marinobacter*, *Methylophaga*, and *Halomonas*, respectively (Rathour et al. 2020). The metagenomic dataset also included enzyme pathways responsible for nitrogen metabolism, sulfur reduction, methane metabolism, benzoate, and xylene degradation. The advantage of metagenomics is that it aids in the unearthing of metabolic pathways of unculturable microorganisms directly from the environment. The metagenomic data revealed the presence of stress response genes accountable for adaption to cold (*DesK*, *DesR* and *Des*), pH (*pstS*, *kdpA*, *dnaK*, *nhaR*, *atpF*), osmotic stress (*Ydp1*, *Chk1*, *Ssk1*), oxidative stress (*katG*, *SOD2*, *yncG*, *HspQ*, *cspA*, *dnaK*), as well as salt tolerance (*proP*, *nhaA*, *kdpA*, *trkA*, *otsB*, *putP*). In another study by Sherpa et al. (2020), 16S rRNA functional metagenomics of two glacier samples of Sikkim, India, disclosed presence of antibiotic-resistant genes [belonging to aminoglycoside (*aa6ic*, *aph3ia*, *ksgA*), bacitracin (*bacA*), beta-lactam (*bl1_sm*), quinolone (*qnrB*), and tetracycline (*tetC*, *tet41*)] that displayed maximum similarity with psychrotolerant Gram-negative bacteria. For example, the gene related to tetracycline resistance exhibited similarity with *Serratia marcescens*, whereas aminoglycoside showed similarity with *Escherichia coli*. Heavy metal-resistant genes [such as arsenic (*arsH*), cobalt (*mgtA*, *dmef*, *corD*, *corC*, *corB*, and *cnrA*), chromium (*yelf*, *ruv*B, *nfs*A, *chr*R, and *chr*A), copper (*cut*A, *cut*E, *cut*C, *cut*F, *cue*R, *cop*C, and *cop*B), iron (*yef*D, *yef*C, *yef*B, and *yef*A), and mercury (*merA*)] also were detected. This metagenomic study indicated the biological transmittance of resistant gene from microorganisms through lateral or horizontal gene transfer. Function-based metagenomic screening approaches have proven effective for the identification and subsequent characterization of a wide array of cold-active enzymes from cold environments. For instance, functional screening of Arctic metagenomes resulted in the characterization of novel cold-active esterases, e.g., Est-97 (active between 0.5 and 55 °C) obtained from an Arctic intertidal metagenomic library and EstM-N1 and EstM-N2 (active between 0 and 35 °C) acquired from Arctic soils (Fu et al. 2013; Yu et al. 2011).

Table 13.1 Summary of transcriptomic applications on psychrophiles

Organism	Type of technique used	Key findings		References
		Upregulated gene	Downregulated gene	
Methanolobus psychrophilus R15, archaea isolated from the Zoige wetland of the Tibetan plateau	RNA-seq and qRT-PCR performed at 4 °C and 18 °C	Genes for RNA polymerase complex, gene cluster for a putative exosome complex, super oxide dismutase, bacterioferritin, TRAM-domain proteins, chaperone genes including GroES/EL complex and thermosomes were upregulated at 4 °C	Genes for methanogenesis, biosynthesis, and protein synthesis were downregulated in R15 at 4 °C	Chen et al. (2012)
Methanococcoides burtonii, archaea isolated from ace Lake, Antarctica	Microarray platform following the growth of cultures at 4 °C and 23 °C, respectively	Increase of secretory and cell surface proteins, protein turnover, and maintenance of translation and initiation was observed at 4 °C. transcriptional regulation largely responsible for controlling gene expression was reported	Data not available	Campanaro et al. (2011)
Chlamydomonas sp. ICE-L, alga isolated from Antarctic Sea ICE	Differentially expressed genes (DEG) analysis and qRT-PCR performed at 7 °C and −20 °C, respectively	Genes encoding carbohydrate metabolism, lipid metabolism, proteins involved in protein phosphorylation, nitrogen transportation and metabolism, transporter proteins, redox equilibrium, heat shock protein 70 were upregulated at −20 °C	Ice-binding protein (IBP) was downregulated following long exposure at −20 °C	Liu et al. (2016)

(continued)

Table 13.1 (continued)

Organism	Type of technique used	Key findings		References
		Upregulated gene	Downregulated gene	
Pseudomonas extremaustralis, bacterium isolated from a temporary pond in Antarctica	RNA-seq and qRT-PCR performed at 8 °C and 30 °C, respectively	Regulatory genes (*rsmE*, *slyA*, *cpxR*, *algZ*, *rnk*, *cheY*, *cheC*), gene coding for major cold shock protein (*cspA*), transcription of genes involved in ethanol oxidation pathway (*exaA*, *exaB*, *exaC* and *erbR*), were upregulated in cold condition. Gene coding for agmatinase, a key enzyme for putrescine biosynthesis from arginine, was upregulated at 4 °C	Cytochrome coding genes (*azu*, *cyoA*, *cyoB*, *cyoC*, cytochrome c4 and B561), TCA-associated genes, genes coding for pyoverdine biosynthesis, iron uptake, molybdopterin biosynthesis (*mobA* and *mobB*), Pel exopolysaccharide biosynthesis, motility genes (*flG*, *flgJH*, *flgK*,*fleQ*, *fliM*, *flhA*), genes involved in nitrogen metabolism, genes encoding polyamine (putrescine and spermidine) transport and catabolism (*potABCD*, *potFGHI*, *gabT* and *gabD*), genes encoding chaperone function and heat shock proteins (*groEL*, *groES*, *ibpA*, *dnaJ* and *dnaK*) were downregulated at 4 °C	Tribelli et al. (2015)
Mesorhizobium strain N33, bacterium isolated from nodules of the legume *Oxytropis arctobia* in Canada's eastern Arctic region	Microarray and qRT-PCR were performed after growing N33 cells in YMB medium to mid-log phase under three steady-state growing temperature conditions (21 °C = control, GT21; GT4 = 4 °C, and	Cells treated under constant (4 °C and 10 °C) low temperatures expressed a prominent number of induced genes distinct from cells treated with short-term cold exposure (<60 min) but exhibited an intermediate expression profile	The genes that were significantly downregulated in cold conditions were those involved in energy production and conversion, amino acid transport, cell motility, secretion, cell envelope, and	Ghobakhlou et al. (2015)

	GT10 = 10 °C). For the cold stress treatments, mid-log phase cells grown at 21 °C (T0) were exposed to 4 °C for different time points (T1 = 2 min, T2 = 4 min, T3 = 8 min, T4 = 60 min, T5 = 240 min in a rotary shaker water bath (150 rpm)	when exposed to a prolonged cold exposure (240 min). The genes that were upregulated at cold temperatures were those involved in transcription regulation, metabolite transport, cryoprotection (mannitol, polyamines), oxidoreductase activity, fatty acid metabolism, and membrane fluidity	outer membrane biogenesis functions	
Nesterenkonia sp. AN1, bacterium isolated from Antarctic desert soil	RNA-seq was performed after growing the bacterium at 21 °C and the test temperature of 5 °C	Upregulation of genes involved in chaperone functions, cell membrane biogenesis, oxidative response, universal stress proteins, protein turnover, and induction of glyoxylate cycle at 5 °C	Genes involved in amino acid, lipid and inorganic ion transport and metabolism, cold shock protein gene (*cspA*), antioxidant thioredoxin gene (*trxA*) were downregulated at 5 °C	Aliyu et al. (2016)
Acidithiobacillus ferrivorans SS3 (DSM 17398), psychrotolerant bacterium was isolated from sediment sample of copper-nickel mining area in Norilsk Industrial Area, Russia	Samples were grown in continuous cultures at 8 °C and 20 °C, respectively. RNA transcript sequencing was performed	Transcripts coding for translation initiation factor IF3 (Acife_2538) and IF1 (Acife_2688) were 7.2- and 22.5-fold higher at 8 °C, respectively. Transcript counts for elongation factor EF-Tu (Acife_2712) and EF-G (Acife_2713) were increased 7.5- and 6.0-fold, respectively	A gene cluster responsible for sulfate assimilation, transcript coding for bacterial RNase P (Acife_R0052), gene cluster FtsYEX (Acife_0556–Acife_0558; FPKM 246–41) related to cell division, and septum formation inhibitor Maf (Acife_0089; FPKM 111–13) showed lower RNA transcript counts. The large subunit RuBisCo (Acife_2232) had significantly lower transcript counts at (FPKM 502–33) at 8 °C	Christel et al. (2016)

(continued)

Table 13.1 (continued)

Organism	Type of technique used	Key findings: Upregulated gene	Key findings: Downregulated gene	References
Polaromonas sp. Eur3 1.2.1 and *Rhodococcus* sp. JG3, bacteria isolated from permafrost and Antarctic dry valley permafrost, respectively	RNA-seq was carried following the growth of cultures of *Rhodococcus* sp. JG3 at −5 °C and 25 °C, respectively and *Polaromonas* sp. Eur3 1.2.1 at 0 °C and 20 °C, respectively	In *Polaromonas* sp. Eur3 1.2.1, abundant transcripts for energy production and translation and ribosomal proteins, osmotically inducible protein (*osmB*), and transporter transcript (*ompA*-*ompF* type porin) at 0 °C were noted In *Rhodococcus* sp. JG3, upshifts of genes involved in translation, amino acid and ion transport and metabolism, superoxide dismutase, siderophore biosynthesis, and transport at −5 °C were reported	In *Polaromonas* sp. Eur3 1.2.1, there was downregulation of genes encoding amino acid transport and metabolism at 0 °C In *Rhodococcus* sp. JG3, downregulation of transcripts involved in lipid transport, energy production and conversion, and metabolism at −5 °C was reported	Raymond-Bouchard et al. (2018)
Pseudogymnoascus destructans, psychrophilic fungus isolated from substrates in glebe mine, New Brunswick, Canada	DEG analysis and RNA-seq was performed by growing the culture on PDA versus SDA medium at 4 °C and 15 °C, respectively	Differentially expressed genes encoding putative secreted proteins were significantly enriched on SDA compared with PDA, and exudate production was higher on SDA	Downregulation of the RNA-interference pathway was identified	Donaldson et al. (2018)

Mrakia psychrophila NN053900, fungi isolated from permafrost on the Qinghai-Tibet plateau	DEG analysis and RNA-seq, *M. psychrophila* was cultured at 4 °C, 12 °C, and 20 °C, respectively. qPCR was performed to assess gene expression levels at 0, 10, and 30 min and 1, 2, 4, 6, and 8 hr. after cold shock (from 12 °C to 4 °C)	Genes involved in ribosome and energy metabolism and biosynthesis of unsaturated fatty acid and glycerol (desaturase and glycerol 3-phosphate dehydrogenase) were upregulated at 4 °C. genes involved in protein processing in the endoplasmic reticulum, proteasome, spliceosome, unfolded protein binding, and mRNA surveillance were upregulated at 20 °C	The desaturase gene DesC (MPSY2164) was downregulated at 10 min after which it was expressed at a low level from 1 to 8 hr. at 4 °C. A core gene encoding glycerol biosynthesis (Gpd MPSY651) and *Sod* (MPSY659) were downregulated immediately after cold shock. Catalase (*Cat*, MPSY4125), which degrades hydrogen peroxide, was expressed at low level after cold shock	Su et al. (2016)

13.3.3 Brief About Proteomic Techniques

The proteome (complete set of proteins) of every cell includes thousands of proteins with different and distinct locations, interactions, abundances, and biochemical activities. As the cell passes through its cycle, differentiates, ages, or responds to fluctuating internal or external environments, the proteome changes with it. Study of complete set of translated proteins from a genome at a particular growth or environmental condition is called proteomics. Genomics and associated techniques provide a lot of information related to gene content or transcribed part of whole genome of any organism. These techniques do not describe dynamic cellular processes (Humphery-Smith et al. 1997) and are unable to predict the translated set of genes because the level of mRNA expression does not represent the amount of active protein in a cell (Anderson and Seilhamer 1997). Furthermore, inability of these approaches to identify posttranslational modifications in any protein makes them compromised which could be resolved by using proteomics and associated techniques.

13.3.3.1 Protein Sample Preparation of Extremophiles

The preparation of the protein samples for proteomics analysis in case of extremophiles is little different and tricky in comparison to other cell types. Different kinds of extremophiles respond differently to the methods and reagent/chemicals used for protein preparation. Because of the extreme habitats, the proteins of extremophiles have evolved a range of chemical and biological properties. Therefore, protein preparation methodologies need to be evaluated on a case-by-case basis. For example, hyperthermophiles' proteins are extremely thermostable and are highly resistant to organic solvents, chemical denaturants, and proteolytic digestion.

13.3.3.2 Qualitative or Conventional Proteomics

When our sole purpose is to detect and identify any protein in a given sample, qualitative or conventional proteomic methods are enough to achieve our goal. By using the combination of 1D (SDS or native SDS-PAGE) or 2D (two-dimensional-PAGE) gel electrophoresis and Western blotting/liquid chromatography-mass spectrometry (LC-MS)/matrix-assisted laser desorption ionization-time of flight (MALDI-TOF), one can qualitatively characterize the proteins in a sample. Gel electrophoresis is used to resolve different proteins in a sample which are further identified by Western blotting, liquid chromatography, and MALDI-TOF.

Electrophoresis for Resolving Proteins in a Sample

The basic principle of electrophoresis is "the charged protein molecules travel through a solvent under electric field." It is a simple, rapid, and sensitive analytical tool. The mobility of a molecule depends on field strength, net charge on the molecule, size and shape of the molecule, and ionic strength. Furthermore, it also depends on properties of the matrix through which the molecule migrates (e.g., viscosity, pore size). Due to small pores size, polyacrylamide is considered ideal for separating majority of proteins and smaller nucleic acids, while agarose (with large pore size) is suitable for nucleic acids and large protein complexes. Several forms of polyacrylamide gel electrophoresis (PAGE) are applied to extract different types of information about proteins of interest. Denaturing and reducing sodium dodecyl sulfate PAGE (SDS-PAGE) is the most widely used technique that separates proteins primarily based on mass. Non-denaturing PAGE (or native PAGE) separates proteins according to their mass/charge ratio. Two-dimensional (2-D) PAGE separates proteins by native isoelectric point in the first dimension and by mass in the second dimension.

Identification of Proteins

Following the electrophoresis, resolved proteins are identified by Western blotting. The resolved proteins in the gel are transferred to an appropriate membrane which is then incubated with labelled antibodies specific to the protein of interest. The unbound antibody is washed off leaving only the bound antibody to the protein of interest. The bound antibodies are then detected by different methods. As the antibodies only bind to the protein of interest, only one band should be visible. The thickness of the band corresponds to the amount of protein present; thus, doing a standard can indicate the amount of protein present. Moreover, the amount of protein of interest in different samples may be compared. Although this approach is widely used to identify or quantitative comparison of a particular protein, but because of few limitations like it is time-consuming, demanding more calibrations, skilful handling and inaccuracy, it is imperative to develop more advanced mass spectrometry-based techniques.

MALDI-TOF-MS and LC-MS are also used to identify proteins in a particular band of interest visualized after electrophoresis. The band of interest is cut, gel is digested, and proteins are trypsinized to obtain peptide mixture. This mixture is dried and dissolved in an appropriate solvent. Now this peptide solution is mixed with a matrix in MALDI-TOF-MS to identify all possible peptides by mass spectrometry. The identified peptide sequences are analyzed by using BLAST against already available genome sequences to finally predict the proteins present in a sample. However, it is difficult to efficiently identify the proteins of hydrophobic nature by this method. Subsequently, 2D-LC-MS/MS was introduced, and that is helpful to identify proteins with high hydrophobicity or basicity, high molecular weight, and inadequately expressed and extreme isoelectric points (Bagnoli et al. 2011; Gygi et al. 1999; Chen et al. 2006). Thus, LC-MS based methods are helpful in qualitative analysis of any given sample for its identification.

13.3.3.3 Quantitative Proteomics

Previously, quantitative proteome analysis had been performed using two-dimensional (2D) gel separation of radioactive and stable isotope (^{15}N)-labelled metabolic proteins followed by scintillation counting or mass spectrometry (MS) (Gygi et al. 1999; Oda et al. 1999). But the problem with 2D gel separation from total cell lysates is that we can measure only highly abundant proteins. New and advanced methods like "shotgun proteomics" or "multidimensional protein identification technology" (MudPIT) have been developed. This is suitable for mapping those proteins which are expressed in lower abundance (Wu and Yates 2003). Further, it helps in the release of hydrophilic peptides from hydrophobic proteins; hence it has no biasness against proteins on basis of pI or mass. Although this method is good for absolute quantification of protein in a given sample, it is not perfect for comparative analysis unless label is added and identification error rates (false positives) were higher because of the high-throughput nature of the method.

The alternative strategy might be adopted to compare the proteome of different biological samples. Trypsinized peptides that have already been tagged or labelled with amino acid "tags" can be analyzed using 2-DLC-MS/MS. This methodology enables real-time peptide-peak comparisons resulting in efficient and accurate quantitative proteome analysis. Isotope-coded affinity tag (ICAT) method is essentially a tandem mass spectrometry that involves the use of a new class of chemical reagents termed as isotope-coded affinity tags (ICATs). It not only allows the accurate detection of almost every possible individual protein within complex mixtures but also is suitable for comparative quantitative proteomic analysis (Gygi et al. 1999). Furthermore, it does not require metabolic labelling because it is based on post-isolation stable isotope labelling of proteins and is therefore not limited to incompatibility of cells and tissues. The limitation of this method is binding of ICATs exclusively to cysteine. This amino acid is a relatively rare amino acid and is either poorly present or almost absent (10–20% of bacterial proteins have no cysteine). Therefore, ICAT experiments need to be replicated to provide greater statistical confidence in the generated results.

Another series of labels, iTRAQ reagents (Applied Biosystems), have been described to address these limitations. These are a set of four isobaric tags having three components: (1) an amine-specific reactive group that binds with peptide N-terminus and lysine residues, (2) a neutral linker group of mass ranging between 28–31 Da, and (3) a reporter region of mass ranging between 114 and 117 Da. Each label may be used for one sample; therefore, four samples can be tracked simultaneously in a single experiment (Choe et al. 2005). The tags have been designed in such a way that all tags show the same complete mass. Therefore, each peak detected in MS represents a single peptide from the combined four samples. MS/MS of each peptide releases the reporter allowing simultaneous identification and quantitation of the peptide. The iTRAQ neutralizes the limitation of exclusive binding with cysteine; therefore, it is capable to detect more number of proteins with greater statistical confidence in less number of replicates.

Quantitative proteomics analysis may be performed using stable isotope labelling with amino acids in cell culture (SILAC). In this method, control sample is cultured with an unlabelled amino acid, and cells belonging to test samples are grown in the presence of an essential amino acid with a stable isotopic nucleus (e.g., deuterium). The labelled or unlabelled essential amino acids are taken up by the cells for its growth and used up in protein synthesis allowing true quantitation (Ong et al. 2002, 2003; Kani 2017). The rest method remains the same as described in case of ICAT or iTRAQ. In the post-genomics era, there is a need of new tools for global, quantitative, and automatic measurement of gene translation in cells and tissues under different conditions. The ICAT and iTRAQ approaches have been proven to be

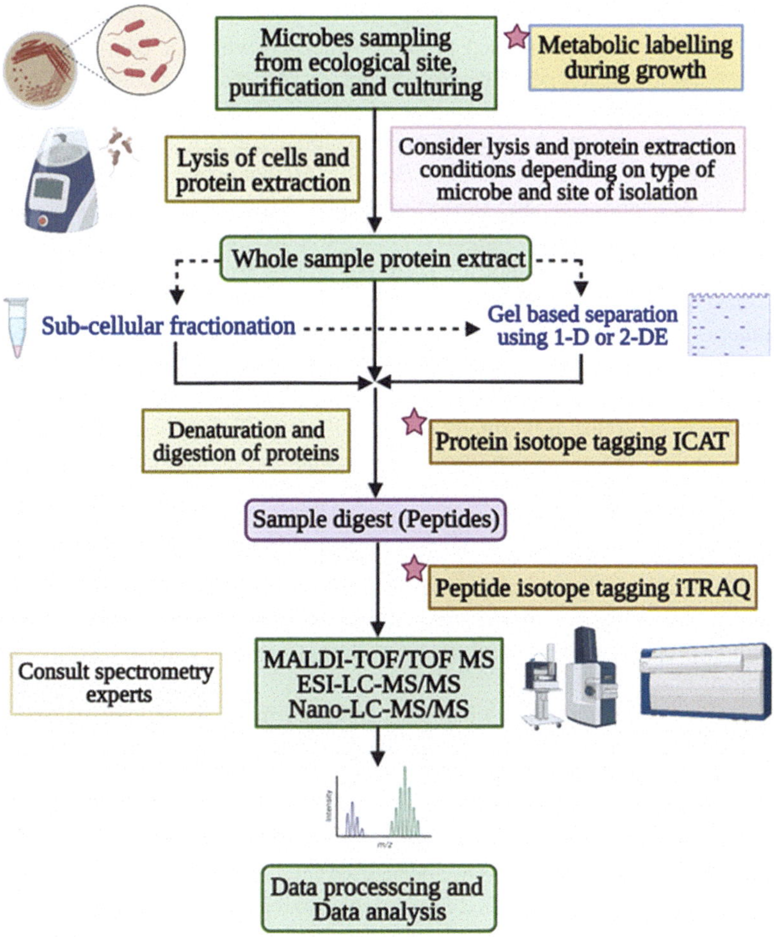

Fig. 13.1 Graphical representation of workflow for common proteomic studies

broadly applicable techniques for the quantitative cataloguing and comparison of protein expression in a different variety of biological samples.

In brief, after sampling, bacteria are purified and cultured under desired conditions, and then proteins are extracted. The proteins are denatured and enzymatically digested to peptides. Digested peptides are analyzed by LC-ESI-MS/MS or by MALDI-MS/MS. Starred text indicate points where labels can be introduced (Fig. 13.1).

13.3.3.4 Application of Proteomic Tools on Psychrophiles

Proteomic tools have been employed by several research groups on many different psychrophilic microorganisms to identify differentially expressed proteins at lower temperatures. It has been discussed with important examples in this section how these proteomics methodologies were applied on these microorganisms to explore the set of differentially expressed proteins which are vital for their survival and sustenance at lower temperature. According to the recently published reports, proteomes of psychrophilic microorganism species belonging to *Psychrobacter* and *Pseudomonas* genera have been most explored and discussed in comparison to other genera including *Shewanella*, *Bacillus*, *Methanococcoides*, *Rhodococcus*, *Pedobacter*, *Marinobacter*, and *Colwellia* (Table 13.2).

Psychrobacter are Gram-negative, heterotrophic, and rod-shaped bacteria. Although the temperature range (−10 to 42 °C) in which most species of this genus survive is very wide, they have mostly been isolated from cold environments. A number of species of this group have adopted several strategies at the molecular level to effectively adapt to cold habitats. *Psychrobacter cryohalolentis K5* is considered to survive in a 43,000-year-old burial of Siberian permafrost at −10 °C (Bakermans et al. 2006). Bakermans et al. (2007) used 2DE-LC-MS/MS to explore its proteome at different temperatures of 16, 4, and –4 °C (with constant salinity at 5%) and detected cold-induced proteins (CIPs) for the first time during growth at subzero temperatures. Out of 618, a total of 303 protein spots detected varied significantly with temperature. Five CIPs (AtpF, EF-Ts, TolC, Pcryo_1988, and FecA) were exclusively and additional 22 CIPs were abundantly detected during growth at –4 °C (relative to higher temperatures) suggesting specific stress on energy production, protein synthesis, and transport during growth at subzero temperatures (Bakermans et al. 2007).

Psychrobacter arcticus 273–4 had been isolated from Kolyma lowland region of Siberia having a 20,000- to 30,000-year-old continuously frozen permafrost soil. It is considered a model among microorganisms for life in freezing environments (Bakermans et al. 2006, 2009). It grows in range of – 10 °C to 28 °C temperature, with an optimum at 17 °C. It is the first cold-adapted bacterium from a terrestrial environment whose genome was sequenced (Ayala-del-Río et al. 2010). Genome analysis indicated that *P. arcticus* 273–4 adopts strategies for its survival at subzero temperature like changes in membrane composition, synthesis of cold shock proteins, and the use of acetate as an energy source. Comparative genomics showed reduced use of the acidic amino acids and proline and arginine in its proteome to increase protein flexibility at low temperatures. Interestingly, this feature was more common in proteins which are essential for cell growth and reproduction, indicating that *P. arcticus* evolved to grow at low temperatures (Ayala-del-Río et al. 2010). Omics analyses revealed that it follows the strategy of adopting a slow metabolic strategy rather than a cellular dormancy state which was evident by the fact that genes related to energy metabolism and carbon substrate incorporation were found to be downregulated and genes for maintenance of membranes, cell walls, and nucleic acid motion were upregulated. Therefore, it cannot only grow in temperatures as cold as - 10 °C, but also they can survive under limited-nutrient-availability conditions and low water activity (Bergholz et al. 2009; Ayala-del-Río et al. 2010; Ponder et al. 2005, 2008). Interestingly, *P. arcticus 273–4* does not overexpress either RNA or protein chaperones (helpful for folding of other proteins and RNA molecules), but it shows the overexpression of psychrophilic RNA helicase protein

Table 13.2 Summary of proteomic applications on psychrophiles

Microorganism (site of isolation and temperature conditions)	Type of technique used	Key findings		References
		Upregulation	Downregulation	
Psychrobacter sp. (a) P. cryohalolentis K5 (Siberian permafrost at -10 °C) (b) P. sp. PAMC 21119 (Antarctic permafrost soil at subzero temperature)	(a) 2D-MS/MS at 16, 4, and –4 °C (b) 2D-MALDI-TOF/TOF at -5 °C and 20 °C	(a) Exclusive detection of 5 CIPs (AtpF, EF-Ts, TolC, Pcryo_1988, and FecA) and other 22 CIPs. Proteins of gene expression, transport, energy production, amino acid metabolism (b) Proteins of metabolite transport, protein folding, membrane fluidity, acetyl-CoA metabolism, putrescine synthesis, and amino acid metabolism pathways	(a) Proteins of energy production/conversion and heme protein synthesis	Bakermans et al. (2007), Kohli et al. (2020)
Pseudomonas sp. (a) *P. migulae* S10724 (high alpine meadows in Chhiplakot, WIH) (b) *P. palleroniana* N26 (high alpine meadows in WIH) (c) *P. helmanticensis* (high altitude of Gangotri soil) (d) *P. psychrophila* MTCC12324 (Svalbard archipelago in the Arctic)	(a) 2D-MALDI-TOF/TOF at 5 and 10 °C (b) 2D-MS/MS (c) LC-MS/MS at 2 °C and 20 °C (d) LC-MS/MS at 4 °C and 25 °C	(a) Proteins of stress response, electron transport, ATP breakdown, nucleic acid metabolism, amino acid synthesis, and cofactor synthesis (b) Low molecular weight acidic proteins, general stress adaptation, protein synthesis and modifications, and energy metabolism (c) Molecular chaperons and cold shock proteins (Tif, Tig, DnaK, and Adk). Enzymes involved in proline,	(a) Proteins of energy-consuming processes including cell division, riboflavin biosynthesis (b) Proteins of biosynthetic processes (c) Proteins of energy generation and biosynthetic pathways (d) Basic survival pathways and factors involved in energy metabolism were found to be unaltered	Suyal et al. (2014), Soni et al. (2015), Kumar et al. (2020), Abraham et al. (2020)

(continued)

Table 13.2 (continued)

Microorganism (site of isolation and temperature conditions)	Type of technique used	Key findings		References
		Upregulation	Downregulation	
		polyamines, unsaturated fatty acid biosynthesis, ROS-neutralizing pathways, and arginine degradation (d) Stress response factors, enzymes involved in fatty acid elongation, CIPs, cold acclimation proteins (caps) including mRNA chaperones, cold shock proteins, proteins of cell division, DNA replication, transcription, and translation		
Shewanella livingstonensis Ac10 (Antarctic seawater at 0 °C)	2D-MALDI-TOF/TOF at 4 °C and 18 °C	Ratio of proteins to phospholipids in the inner membrane. 14 CIPs including AtoS, PspA, MreB, FtsY, Ald, SdhA, DegP, SurA, Dpp4, PykF, LeuB, SucB, RibP_PPkin, and Sliv_c417088 involved in protein folding, morphogenesis, sensing the environment, and other cellular processes	No significant data	Park et al. (2012)
Shewanella frigidimarina (Antarctic zone at 0 °C)	2D-MALDI-TOF/TOF at 4, 20, 28 °C	Several heat shock proteins as well as proteins of stress, redox homeostasis, or protein synthesis and degradation.	Trigger factor (CAP) and proteins of cell envelope like ton B-dependent siderophore receptors and porins	García-Descalzo et al. (2014)

		DnaK, GrpE, GroEL, GroES, ef-Tu, ef-G, FlgL		
Bacillus psychosaccharolyticus (soil and lowland marshes at 0–30 °C)	2D-ESI-LC-MS/MS at 0, 15, and 30 °C	Proteins related to glycolysis, translation, stress proteins, detoxication, energy production and conversion, nucleosides, and nucleotide metabolism and transport system component	No significant data	Seo et al. (2004)
Colwellia psychrerythraea (Arctic marine sediments at around 0 °C)	LC-MS/MS at -1^{0} and − 10 °C	Translation efTu, ribosomal protein S1, ribosomal protein L7/L12, DnaK, cold-shock DNA-binding domain protein, alanine dehydrogenase, DNA-directed RNA polymerase (beta subunit), translation elongation factor G, cold-shock DNA-binding domain protein, TonB-dependent receptor, ATP synthase F1 (alpha and beta subunit), ribosomal protein L9, DNA helicase II, SNF2 family protein, RNA helicase DeaD and ATP-dependent helicase DEAD-box family	At -100C, RNA polymerase sigma factor RpoS, D-alanine-D-alanine ligase, TonB-dependent receptor, flagellar proteins	Nunn et al. (2015)
Methanococcoides burtonii (Tibetan plateau wetland at 0 to 25 °C)	2D-MALDI-TOF/TOF at 30 °C and 4 °C	Methanol corrinoid methyltransferase (Mpsy_0909), methanol, corrinoid protein (Mpsy_0908, 3032), phosphoglycerate	TMA-derived methanogenesis, proteins of acetyl-CoA, gluconeogenesis, incomplete TCA cycle, energy, and biosynthetic precursors	Chen et al. (2012)

(continued)

Table 13.2 (continued)

Microorganism (site of isolation and temperature conditions)	Type of technique used	Key findings		References
		Upregulation	Downregulation	
		kinase, Rrp4, thermosome, group II chaperonin (Mpsy_1969, 2247, 3167), 3-Cys thioredoxin peroxidase (Mpsy_0762), and a superoxide reductase (Mpsy_2711)		
Oleispira Antarctica RB-8 (oil-enriched seawater from Rod Bay in southern Antarctica at 0–15 °C)	LC-MS/MS at 4 °C and 16 °C	CIPs, proteins for flagella structure/output, flagella rotation and proline metabolism to counteract oxidative stress, RNA metabolism, ribosome maturation and cold tolerance, oxidative stress response	No significant data	Gregson et al. (2020)
Pedobacter cryoconitis (alpine glacier cryoconite at 1 to 25 °C)	^{15}N metabolic labelling, 2D-LC-MS/MS at 1 and 20 °C (20 °C/1 °C)	Hsp70, 10-kDa chaperonin, chaperonin GroEL, and chaperone Hsp90, carbohydrate metabolism	Antioxidants, superoxide dismutase, aspartate-semialdehyde dehydrogenase, EF-Tu	Pereira-Medrano et al. (2012)
Rhodococcus qingshengii S10107 (Chhiplakot region of Kumaun Himalayas at 10 °C)	ESI- LC-MS/MS	nifH, nifD, and nifK, nifA, nifL, and nifB, biosynthesis of certain amino acids, viz., leucine, lysine, and alanine, peptidoglycan synthesis proteins, transketolase and transaldolase, CowN	Enzymes involved in lipid biosynthesis, biosynthetic processes and energy-consuming processes	Suyal et al. (2019)

Mrakia psychrophile (permafrost on the Qinghai-Tibet plateau)	iTRAQ labelling and NanoLC-MS/MS at 4, 12, and 20 °C	Transporter activity, nucleotide binding, nucleic acid biosynthetic processes, nucleobase transport, metallopeptidase, metal ion binding proteins, metabolic process, hydrolase	Unfolded protein binding, protein folding, protein binding, nucleic acid binding, helicase activity, cation binding	Su et al. (2016)

CsdA (Psyc_1082), which is considered as a key protein for life under freezing temperatures (Bergholz et al. 2009; Kuhn 2012).

In the most recent transcriptomics and proteomics study (Koh et al. 2017), the authors successfully identified and catalogued many transcribed genes and translated proteins which were upregulated or downregulated when *Psychrobacter sp. PAMC 21119* was grown at −5 °C in comparison to that of 20 °C. Out of the total 2906 transcripts identified by RNA-seq, 584 differentially expressed genes (belonging to different gene families like translation, ribosomal structure, and biogenesis) were upregulated. The genes related to lipid transport and metabolism showed downregulation at lower temperature. The combination of 2-DE-MALDI-TOF/TOF was employed to identify the proteins expressed differentially at lower temperature. A total of 60 spots on 2-DE were processed based on the differential expression, and the proteins were identified by mass spectrometry. Among the upregulated proteins in response to cold; many of them belonged to metabolite transport, protein folding and membrane fluidity. Furthermore, acetyl-CoA metabolism, putrescine synthesis, and amino acid metabolism pathways were also reported to be upregulated, while downregulation was reported in case of proteins involved in energy production/conversion and heme protein synthesis. Moreover, isoform exchange of cold-shock proteins (Csp) was detected at both temperatures. These results were similar as were detected in other species of *Psychrobacter* as described above.

Pseudomonas are Gram-negative, aerobic, non-spore-forming, and rod-shaped bacteria inhabiting a wide range of environment and temperatures. Many species have been reported to grow in psychrophilic temperature range (below 20 °C) like *Pseudomonas psychrophila MTCC12324*, *P. migulae S10724*, *P. helmanticensis*, and *P. palleroniana N26*. Several reports on proteomics studies on different psychrophilic pseudomonas species have been published which reveal the set of differentially expressed proteins important for survival at lower temperature conditions.

Pseudomonas migulae S10724, a Himalayan psychrophilic diazotroph to low temperature, is isolated from the rhizosphere of red kidney bean (*Phaseolus vulgaris L.*) from high alpine meadows in Chhiplakot (3290 m, 30.06_N, 79.01_E), Western Indian Himalaya (WIH). It showed the ability to fix nitrogen at 5–10 °C (Suyal et al. 2014). Global proteomic analysis of *Pseudomonas migulae S10724* was performed using 2-DE coupled with MALDI–TOF–MS to check the physiological response to low temperature diazotrophy. A total of 66 differentially expressed spots were identified, out of which 48 were found to be upregulated including 6 newly expressed proteins, while 18 were downregulated under nitrogen fixation conditions (NFC). Proteomic data suggested the fitness of bacteria for low temperature adaptation and nitrogen fixation, including general stress adaptation, protein and nucleic acid synthesis, energy metabolism, cell growth/maintenance, etc. Majority of upregulated proteins were stress proteins, while cell division proteins were downregulated (Suyal et al. 2014). One more 2-DE-LC-MS/MS-based proteomic study on Himalayan psychrophilic diazotroph, *Pseudomonas palleroniana N26*, was reported (Soni et al. 2015). A total of 53 differentially expressed protein spots were analyzed revealing many low temperature adaptation and nitrogen fixation mechanisms. Upregulation of low molecular weight acidic proteins and downregulation of proteins related to biosynthetic processes were reported. Another comparative proteomic study on Himalayan psychrophilic diazotroph, *Pseudomonas helmanticensis*, to explore differentially expressed proteome at low temperature growth conditions, was recently published (Kumar et al. 2020). It was previously isolated from high-altitude cold climatic Gangotri soil (altitude 3415 m, 30.98° N, 78.93° E). Its optimum growth temperature was reported 10 °C (Kumar et al. 2018, 2019). The proteomes were compared when it was grown at 2 °C and 20 °C. The research findings revealed that molecular chaperons and cold shock proteins were upregulated which provide cold stress resistance in bacteria. Upregulated enzymes were related to proline, polyamines, unsaturated fatty acid biosynthesis, ROS-neutralizing pathways, and arginine degradation. However, downregulated proteins were related to oxidative pathways of energy generation like most of the other studies.

Another proteomic study to explore the cold adaptive mechanisms in *Pseudomonas psychrophila MTCC12324* was recently published (Abraham et al. 2020). It is a facultative psychrophilic bacterium isolated from the Svalbard archipelago (79°55′N, 11°56′E) in the Arctic. LC-MS/MS-based comparative proteome analysis of *P. psychrophila* MTCC12324 was performed when grown at 4 °C and 25 °C and revealed unaltered expression of basic survival pathways and proteins related to energy metabolism. In contrast, stress response factors, enzymes involved in fatty acid elongation, and cold-inducible proteins (CIPs) and cold acclimation proteins (Caps) including nucleic acid chaperones and cold shock proteins were found to be upregulated under cold stress. It was evident that metabolic pathways and cellular processes were adjusted to withstand the cold environment (Abraham et al. 2020).

Shewanella livingstonensis Ac10 is a cold-adapted Gram-negative bacterium, isolated from Antarctic seawater, that grows well at temperatures close to 0 °C but does not grow at temperatures above 30 °C (Kulakova et al. 1999). Park et al. (2012) successfully established a method for separation of the inner and outer membranes and explored its inner membrane specific proteome by using 2-DE-MALDI-TOF/TOF and growing at 4 and 18 °C. They identified 14 CIPs (upregulated >two-fold and at 4 °C). These CIPs included proteins thought to be involved in membrane protein biogenesis (DegP, SurA, and FtsY), chemotaxis (AtoS and PspA), and morphogenesis (MreB) (Park et al. 2012). Another interesting proteomics study was performed on a typical psychrophilic bacterium, *Shewanella frigidimarina*, by growing it at 4, 20, and 28 °C temperatures to check the effects of climate change from cold to hot in Antarctic zone (García-Discalzo et al. 2014). They found overexpression of several heat shock proteins as well as other proteins related to stress, redox homeostasis, or protein synthesis and degradation. Downregulation of enzymes and components of the cell envelope was reported.

A classic example of SILAC has been described, where they applied this proteomic technique on *Edwardsiella tarda ATCC 15947*. It is a Gram-negative facultative anaerobic pathogen with a wide host range including humans. Low temperatures are conventionally employed to contain the growth of *E. tarda*. However, a prior study proving its retention of virulence at low temperature opened the scope of further exploration of proteome which might be playing significant roles in its virulence. Ma et al. (2018b) used SILAC technique to perform comparative proteomics of *E. tarda* ATCC 15947 under cold stress for 2 weeks and identified upregulated 72 proteins and 164 downregulated proteins in response to cold stress (Ma et al. 2018b).

An interesting study was published on exploration of proteome of *Psychroflexus torquis* under different salinity and light conditions (Feng et al. 2015). *P. torquis* is an extremely psychrophilic proteorhodopsin-containing bacterial species which is a model sea-ice microorganism with an epiphytic lifestyle. The comprehensive quantitative proteomic study was performed using nano LC-orbitrap tandem mass spectrometry (a gel-free label-free-based approach). The data suggested that *P. torquis* differentially responds to variation of salinity and illumination conditions. Electron-transfer chain proteins and TonB-dependent transporters were downregulated at a suboptimal salinity level, while Ton-B transporters upregulated under supra-optimal salinity. Furthermore, central metabolic pathways (TCA and glycolysis) were also induced by both salinity stress and illumination. The data suggested that response of *P. torquis* changed under both illumination and salinity by modulating membrane and central metabolic proteins that are involved in energy production as well as nutrient uptake processes (Feng et al. 2015).

Several other proteomic-based studies on different psychrophilic microorganisms like *Bacillus psychosaccharolyticus*, *Colwellia psychrerythraea*, *Halohasta litchfieldiae*, *Halorubrum lacusprofundi*, *Marinobacter gelidimuriae* sp., *Methanococcoides burtonii*, *Oleispira antarctica* RB-8, *Pedobacter cryoconitis*, and *Rhodococcus qingshengii* S10107 have been published (Seo et al. 2004; Nunn et al. 2015; Williams et al. 2017; Chua et al. 2018; Chen et al. 2012; Gregson et al. 2020; Pereira-Medrano et al. 2012; Suyal et al. 2019). The applications of proteomic tools have been summarized in Table 13.2.

13.4 Conclusion

Applications of transcriptomic and proteomic tools to study the cold adaptations in psychrophiles at genome level have revolutionized the ways of understanding the structural and physiological processes at molecular level.

References

Abraham WP, Raghunandanan S, Gopinath V, Suryaletha K, Thomas S (2020) Deciphering the cold adaptive mechanisms in *Pseudomonas psychrophila* MTCC12324 isolated from the Arctic at 79°N. Curr Microbiol 77(9):2345–2355. https://doi.org/10.1007/s00284-020-02006-2

Aliyu H, De Maayer P, Cowan D (2016) The genome of the Antarctic polyextremophile *Nesterenkonia sp.* AN1 reveals adaptive strategies for survival under multiple stress conditions. FEMS Microbiol Ecol 92(4):fiw032. https://doi.org/10.1093/femsec/fiw032

Aliyu H, De Maayer P, Sjöling S, Cowan DA (2017) Metagenomic analysis of low-temperature environments. In: Margesin R (ed) Psychrophiles: from biodiversity to biotechnology. Springer, Cham, pp 389–421

Allen MA, Lauro FM, Williams TJ, Burg D, Siddiqui KS, De Francisci D, Chong KW, Pilak O, Chew HH, De Maere MZ, Ting L, Katrib M, Ng C, Sowers KR, Galperin MY, Anderson IJ, Ivanova N, Dalin E, Martinez M, Lapidus A, Hauser L, Land M, Thomas T, Cavicchioli R (2009) The genome sequence of the psychrophilic archaeon, *Methanococcoides burtonii*: the role of genome evolution in cold adaptation. ISME J 3(9):1012–1035. https://doi.org/10.1038/ismej.2009.45

Anderson L, Seilhamer J (1997) A comparison of selected mRNA and protein abundances in human liver. Electrophoresis 18(3–4):533–537. https://doi.org/10.1002/elps.1150180333

Ayala-del-Río HL, Chain PS, Grzymski JJ, Ponder MA, Ivanova N, Bergholz PW, Di Bartolo G, Hauser L, Land M, Bakermans C, Rodrigues D, Klappenbach J, Zarka D, Larimer F, Richardson P, Murray A, Thomashow M, Tiedje JM (2010) The genome sequence of *Psychrobacter arcticus* 273-4, a psychroactive Siberian permafrost bacterium, reveals mechanisms for adaptation to low-temperature growth. Appl Environ Microbiol 76 (7):2304–2312. https://doi.org/10.1128/AEM.02101-09

Bagnoli F, Baudner B, Mishra RP, Bartolini E, Fiaschi L, Mariotti P, Nardi-Dei V, Boucher P, Rappuoli R (2011) Designing the next generation of vaccines for global public health. OMICS 15(9):545–566. https://doi.org/10.1089/omi.2010.0127

Baker JM, Riester CJ, Skinner BM, Newell AW, Swingley WD, Madigan MT, Jung DO, Asao M, Chen M, Loughlin PC, Pan H, Lin Y, Li Y, Shaw J, Prado M, Sherman C, Tang JK, Blankenship RE, Zhao T, Touchman JW, Sattley WM (2017) Genome sequence of *Rhodoferax antarcticus* ANT.BR^T; a psychrophilic purple nonsulfur bacterium from an Antarctic microbial mat. Microorganisms 5(1):8. https://doi.org/10.3390/microorganisms5010008

Bakermans C, Ayala-Del-Río HL, Ponder MA, Vishnivetskaya T, Gilichinsky D, Thomashow MF, Tiedje JM (2006) *Psychrobacter cryohalolentis* sp. nov. and *Psychrobacter arcticus* sp. nov., isolated from Siberian permafrost. Int J Syst Evol Microbiol 56:1285–1291. https://doi.org/10.1099/ijs.0.64043-0

Bakermans C, Bergholz PW, Ayala-del-Rıo H, Tiedje J (2009) Genomic insights into cold adaptation of permafrost bacteria. In: Margesin R (ed) Permafrost soils. Soil biology. Springer, Berlin, pp 159–168

Bakermans C, Tollaksen SL, Giometti CS, Wilkerson C, Tiedje JM, Thomashow MF (2007) Proteomic analysis of *Psychrobacter cryohalolentis* K5 during growth at subzero temperatures. Extremophiles 11(2):343–354. https://doi.org/10.1007/s00792-006-0042-1

Bergholz PW, Bakermans C, Tiedje JM (2009) *Psychrobacter arcticus* 273-4 uses resource efficiency and molecular motion adaptations for subzero temperature growth. J Bacteriol 191 (7):2340–2352. https://doi.org/10.1128/JB.01377-08

Blöchl E, Rachel R, Burggraf S, Hafenbradl D, Jannasch HW, Stetter KO (1997) *Pyrolobus fumarii*, gen. And sp. nov., represents a novel group of archaea, extending the upper temperature limit for life to 113 degrees C. Extremophiles 1(1):14–21. https://doi.org/10.1007/s007920050010

Campanaro S, Williams TJ, Burg DW, De Francisci D, Treu L, Lauro FM, Cavicchioli R (2011) Temperature-dependent global gene expression in the Antarctic archaeon *Methanococcoides burtonii*. Environ Microbiol 13(8):2018–2038. https://doi.org/10.1111/j.1462-2920.2010.02367.x

Casanueva A, Tuffin M, Cary C, Cowan DA (2010) Molecular adaptations to psychrophily: the impact of 'omic' technologies. Trends Microbiol 18(8):374–381. https://doi.org/10.1016/j.tim.2010.05.002

Chen EI, Hewel J, Felding-Habermann B, Yates JR 3rd (2006) Large scale protein profiling by combination of protein fractionation and multidimensional protein identification technology (MudPIT). Mol Cell Proteomics 5(1):53–56. https://doi.org/10.1074/mcp.T500013-MCP200

Chen Z, Yu H, Li L, Hu S, Dong X (2012) The genome and transcriptome of a newly described psychrophilic archaeon, *Methanolobus psychrophilus* R15, reveal its cold adaptive

characteristics. Environ Microbiol Rep 4(6):633–641. https://doi.org/10.1111/j.1758-2229.2012.00389.x
Choe LH, Aggarwal K, Franck Z, Lee KH (2005) A comparison of the consistency of proteome quantitation using two-dimensional electrophoresis and shotgun isobaric tagging in *Escherichia coli* cells. Electrophoresis 26(12):2437–2449. https://doi.org/10.1002/elps.200410336
Christel S, Fridlund J, Watkin EL, Dopson M (2016) *Acidithiobacillus ferrivorans* SS3 presents little RNA transcript response related to cold stress during growth at 8°C suggesting it is a eurypsychrophile. Extremophiles 20(6):903–913. https://doi.org/10.1007/s00792-016-0882-2
Chua MJ, Campen RL, Wahl L, Grzymski JJ, Mikucki JA (2018) Genomic and physiological characterization and description of *Marinobacter gelidimuriae* sp. nov., a psychrophilic, moderate halophile from blood falls, an antarctic subglacial brine. FEMS Microbiol Ecol 94(3): fiy021. https://doi.org/10.1093/femsec/fiy021
Clarke A, Morris GJ, Fonseca F, Murray BJ, Acton E, Price HC (2013) A low temperature limit for life on earth. PLoS One 8(6):e66207. https://doi.org/10.1371/journal.pone.0066207
Cowan D, Casanueva A, Stafford W (2007) Ecology and biodiversity of cold-adapted microorganisms. In: Gerday C, Glansdorff N (eds) Physiology and biochemistry of extremophiles. ASM Press, Washington, DC, pp 119–132
Cowen DA (2004) The upper temperature of life--where do we draw the line? Trends Microbiol 12 (2):58–60. https://doi.org/10.1016/j.tim.2003.12.002
Deming JW (2002) Psychrophiles and polar regions. Curr Opin Microbiol 5(3):301–309. https://doi.org/10.1016/s1369-5274(02)00329-6
Dias LM, Folador ARC, Oliveira AM, Ramos RTJ, Silva A, Baraúna RA (2018) Genomic architecture of the two cold-adapted genera Exiguobacterium and Psychrobacter: evidence of functional reduction in the *Exiguobacterium antarcticum* B7 genome. Genome Biol Evol 10 (3):731–741. https://doi.org/10.1093/gbe/evy029
Donaldson ME, Davy CM, Vanderwolf KJ, Willis CKR, Saville BJ, Kyle CJ (2018) Growth medium and incubation temperature alter the *Pseudogymnoascus destructans* transcriptome: implications in identifying virulence factors. Mycologia 110(2):300–315. https://doi.org/10.1080/00275514.2018.1438223
Durán P, Barra PJ, Jorquera MA, Viscardi S, Fernandez C, Paz C, Mora ML, Bol R (2019) Occurrence of soil fungi in Antarctic pristine environments. Front Bioeng Biotechnol 7:28. https://doi.org/10.3389/fbioe.2019.00028
Dymond JS (2013) Explanatory chapter: quantitative PCR. Methods Enzymol 529:279–289. https://doi.org/10.1016/B978-0-12-418687-3.00023-9
Fan X, Yu T, Li Z, Zhang XH (2014) *Luteimonas abyssi* sp. nov., isolated from deep-sea sediment. Int J Syst Evol Microbiol 64:668–674. https://doi.org/10.1099/ijs.0.056010-0
Feng S, Powell SM, Wilson R, Bowman JP (2013) Light-stimulated growth of proteorhodopsin-bearing sea-ice psychrophile *Psychroflexus torquis* is salinity dependent. ISME J 7 (11):2206–2213. https://doi.org/10.1038/ismej.2013.97
Feng S, Powell SM, Wilson R, Bowman JP (2015) Proteomic insight into functional changes of proteorhodopsin-containing bacterial species *Psychroflexus torquis* under different illumination and salinity levels. J Proteome Res 14(9):3848–3858. https://doi.org/10.1021/acs.jproteome.5b00241
Friedrich AB, Antranikian G (1996) Keratin degradation by *Fervidobacterium pennavorans*, a novel thermophilic anaerobic species of the order Thermotogales. Appl Environ Microbiol 62 (8):2875–2882. https://doi.org/10.1128/AEM.62.8.2875-2882.1996
Fu J, Leiros HK, de Pascale D, Johnson KA, Blencke HM, Landfald B (2013) Functional and structural studies of a novel cold-adapted esterase from an Arctic intertidal metagenomic library. Appl Microbiol Biotechnol 97(9):3965–3978. https://doi.org/10.1007/s00253-012-4276-9
García-Descalzo L, García-López E, Alcázar A, Baquero F, Cid C (2014) Proteomic analysis of the adaptation to warming in the Antarctic bacteria *Shewanella frigidimarina*. Biochim Biophys Acta 1844(12):2229–2240. https://doi.org/10.1016/j.bbapap.2014.08.006

Ghobakhlou AF, Johnston A, Harris L, Antoun H, Laberge S (2015) Microarray transcriptional profiling of Arctic *Mesorhizobium* strain N33 at low temperature provides insights into cold adaption strategies. BMC Genomics 16(1):383. https://doi.org/10.1186/s12864-015-1611-4

Goldstein RA (2007) Amino-acid interactions in psychrophiles, mesophiles, thermophiles, and hyperthermophiles: insights from the quasi-chemical approximation. Protein Sci 16 (9):1887–1895. https://doi.org/10.1110/ps.072947007

Gregson BH, Metodieva G, Metodiev MV, Golyshin PN, McKew BA (2020) Protein expression in the obligate hydrocarbon-degrading psychrophile *Oleispira antarctica* RB-8 during alkane degradation and cold tolerance. Environ Microbiol 22(5):1870–1883

Gygi SP, Rist B, Gerber SA, Turecek F, Gelb MH, Aebersold R (1999) Quantitative analysis of complex protein mixtures using isotope-coded affinity tags. Nat Biotechnol 17(10):994–999. https://doi.org/10.1038/13690

Hawwa R, Aikens J, Turner RJ, Santarsiero BD, Mesecar AD (2009) Structural basis for thermostability revealed through the identification and characterization of a highly thermostable phosphotriesterase-like lactonase from *Geobacillus stearothermophilus*. Arch Biochem Biophys 488(2):109–120. https://doi.org/10.1016/j.abb.2009.06.005

Humphery-Smith I, Cordwell SJ, Blackstock WP (1997) Proteome research: complementarity and limitations with respect to the RNA and DNA worlds. Electrophoresis 18(8):1217–1242. https://doi.org/10.1002/elps.1150180804

Jaenicke R (2000) Stability and stabilization of globular proteins in solution. J Biotechnol 79 (3):193–203. https://doi.org/10.1016/s0168-1656(00)00236-4

Kani K (2017) Quantitative proteomics using SILAC. Methods Mol Biol 1550:171–184. https://doi.org/10.1007/978-1-4939-6747-6_13

Koh HY, Park H, Lee JH, Han SJ, Sohn YC, Lee SG (2017) Proteomic and transcriptomic investigations on cold-responsive properties of the psychrophilic Antarctic bacterium *Psychrobacter* sp. PAMC 21119 at subzero temperatures. Environ Microbiol 19(2):628–644. https://doi.org/10.1111/1462-2920.13578

Kohli I, Joshi NC, Mohapatra S, Varma A (2020) Extremophile - an adaptive strategy for extreme conditions and applications. Curr Genomics 21(2):96–110. https://doi.org/10.2174/1389202921666200401105908

Koo H, Hakim JA, Morrow CD, Crowley MR, Andersen DT, Bej AK (2018) Metagenomic analysis of microbial community compositions and cold-responsive stress genes in selected antarctic lacustrine and soil ecosystems. Life 8(3):29. https://doi.org/10.3390/life8030029

Kuhn E (2012) Toward understanding life under subzero conditions: the significance of exploring psychrophilic "cold-shock" proteins. Astrobiology 12(11):1078–1086. https://doi.org/10.1089/ast.2012.0858

Kulakova L, Galkin A, Kurihara T, Yoshimura T, Esaki N (1999) Cold-active serine alkaline protease from the psychrotrophic bacterium Shewanella strain ac10: gene cloning and enzyme purification and characterization. Appl Environ Microbiol 65(2):611–617. https://doi.org/10.1128/AEM.65.2.611-617.1999

Kumar S, Suyal DC, Bhoriyal M, Goel R (2018) Plant growth promoting potential of psychrotolerant *Dyadobacter sp.* for pulses and finger millet and impact of inoculation on soil chemical properties and diazotrophic abundance. J Plant Nutr 41(8):1035–1046. https://doi.org/10.1080/01904167.2018.1433211

Kumar S, Suyal DC, Yadav A, Shouche Y, Goel R (2019) Microbial diversity and soil physiochemical characteristic of higher altitude. PLoS One 14(3):e0213844. https://doi.org/10.1371/journal.pone.0213844

Kumar S, Suyal DC, Yadav A, Shouche Y, Goel R (2020) Psychrophilic *Pseudomonas helmanticensis* proteome under simulated cold stress. Cell Stress Chaperones 25 (6):1025–1032. https://doi.org/10.1007/s12192-020-01139-4

Kumar S, Tsai CJ, Nussinov R (2000) Factors enhancing protein thermostability. Protein Eng 13 (3):179–191. https://doi.org/10.1093/protein/13.3.179

Li H, Liu X, Wang Y, Hu H, Xu X (2009) Enhanced expression of antifreeze protein genes drives the development of freeze tolerance in an Antarctica isolate of *Chlorella vulgaris*. Prog Nat Sci 19:1059–1062. https://doi.org/10.1016/j.pnsc.2008.10.0120.1007/s00792-016-0834-x

Liu C, Wang X, Wang X, Sun C (2016) Acclimation of Antarctic Chlamydomonas to the sea-ice environment: a transcriptomic analysis. Extremophiles 20(4):437–450. https://doi.org/10.1007/s00792-016-0834-x

Loman NJ, Misra RV, Dallman TJ, Constantinidou C, Gharbia SE, Wain J, Pallen MJ (2012) Performance comparison of benchtop high-throughput sequencing platforms. Nat Biotechnol 30 (5):434–439. https://doi.org/10.1038/nbt.2198

Ma R, Huang H, Bai Y, Luo H, Fan Y, Yao B (2018b) Insight into the cold adaptation and hemicellulose utilization of *Cladosporium neopsychrotolerans* from genome analysis and biochemical characterization. Sci Rep 8(1):6075. https://doi.org/10.1038/s41598-018-24443-7

Ma W, Jia J, Huang X, Xie W, Zhang X, Tang J, Lin C, Zhao L, Fang P (2018a) Stable isotope labelling by amino acids in cell culture (SILAC) applied to quantitative proteomics of *Edwardsiella tarda* ATCC 15947 under prolonged cold stress. Microb Pathog 125:12–19. https://doi.org/10.1016/j.micpath.2018.09.006

Margesin R, Zhang DC, Frasson D, Brouchkov A (2016) *Glaciimonas frigoris* sp. nov., a psychrophilic bacterium isolated from ancient Siberian permafrost sediment, and emended description of the genus *Glaciimonas*. Int J Syst Evol Microbiol 66(2):744–748. https://doi.org/10.1099/ijsem.0.000783

Martorell MM, Lannert M, Matula CV, Quartino ML, de Figueroa LIC, Cormack WM, Ruberto LAM (2021) Studies toward the comprehension of fungal-macroalgae interaction in cold marine regions from a biotechnological perspective. Fungal Biol 125(3):218–230. https://doi.org/10.1016/j.funbio.2020.11.003

McKay CP (2014) Requirements and limits for life in the context of exoplanets. Proc Natl Acad Sci U S A 111(35):12628–12633. https://doi.org/10.1073/pnas.1304212111

Medini D, Donati C, Tettelin H, Masignani V, Rappuoli R (2005) The microbial pan-genome. Curr Opin Genet Dev 15(6):589–594. https://doi.org/10.1016/j.gde.2005.09.006

Methé BA, Nelson KE, Deming JW, Momen B, Melamud E, Zhang X, Moult J, Madupu R, Nelson WC, Dodson RJ, Brinkac LM, Daugherty SC, Durkin AS, DeBoy RT, Kolonay JF, Sullivan SA, Zhou L, Davidsen TM, Wu M, Huston AL, Lewis M, Weaver B, Weidman JF, Khouri H, Utterback TR, Feldblyum TV, Fraser CM (2005) The psychrophilic lifestyle as revealed by the genome sequence of *Colwellia psychrerythraea* 34H through genomic and proteomic analyses. Proc Natl Acad Sci U S A 102(31):10913–10918. https://doi.org/10.1073/pnas.0504766102

Mock T, Otillar RP, Strauss J, McMullan M, Paajanen P, Schmutz J, Salamov A, Sanges R, Toseland A, Ward BJ, Allen AE, Dupont CL, Frickenhaus S, Maumus F, Veluchamy A, Wu T, Barry KW, Falciatore A, Ferrante MI, Fortunato AE, Glöckner G, Gruber A, Hipkin R, Janech MG, Kroth PG, Leese F, Lindquist EA, Lyon BR, Martin J, Mayer C, Parker M, Quesneville H, Raymond JA, Uhlig C, Valas RE, Valentin KU, Worden AZ, Armbrust EV, Clark MD, Bowler C, Green BR, Moulton V, van Oosterhout C, Grigoriev IV (2017) Evolutionary genomics of the cold-adapted diatom *Fragilariopsis cylindrus*. Nature 541 (7638):536–540. https://doi.org/10.1038/nature20803

Moghadam MS, Albersmeier A, Winkler A, Cimmino L, Rise K, Hohmann-Marriott MF, Kalinowski J, Rückert C, Wentzel A, Lale R (2016) Isolation and genome sequencing of four Arctic marine Psychrobacter strains exhibiting multicopper oxidase activity. BMC Genomics 17:117. https://doi.org/10.1186/s12864-016-2445-4

Moon S, Kim J, Koo J, Bae E (2019) Structural and mutational analyses of psychrophilic and mesophilic adenylate kinases highlight the role of hydrophobic interactions in protein thermal stability. Struct Dyn 6(2):024702. https://doi.org/10.1063/1.5089707

Mukhia S, Khatri A, Acharya V, Kumar R (2021) Comparative genomics and molecular adaptational analysis of Arthrobacter from Sikkim Himalaya provided insights into its survivability under multiple high-altitude stress. Genomics 113:151–158. https://doi.org/10.1016/j.ygeno.2020.12.001

Nguyen V, Wilson C, Hoemberger M, Stiller JB, Agafonov RV, Kutter S, English J, Theobald DL, Kern D (2017) Evolutionary drivers of thermoadaptation in enzyme catalysis. Science 355 (6322):289–294. https://doi.org/10.1126/science.aah3717

Nunn BL, Slattery KV, Cameron KA, Timmins-Schiffman E, Junge K (2015) Proteomics of *Colwellia psychrerythraea* at subzero temperatures - a life with limited movement, flexible membranes and vital DNA repair. Environ Microbiol 17(7):2319–2335. https://doi.org/10.1111/1462-2920.12691

Oda Y, Huang K, Cross FR, Cowburn D, Chait BT (1999) Accurate quantitation of protein expression and site-specific phosphorylation. Proc Natl Acad Sci U S A 96(12):6591–6596. https://doi.org/10.1073/pnas.96.12.6591

Ong SE, Blagoev B, Kratchmarova I, Kristensen DB, Steen H, Pandey A, Mann M (2002) Stable isotope labeling by amino acids in cell culture, SILAC, as a simple and accurate approach to expression proteomics. Mol Cell Proteomics 1(5):376–386. https://doi.org/10.1074/mcp.m200025-mcp200

Ong SE, Foster LJ, Mann M (2003) Mass spectrometric-based approaches in quantitative proteomics. Methods 29(2):124–130. https://doi.org/10.1016/s1046-2023(02)00303-1

Orellana R, Macaya C, Bravo G, Dorochesi F, Cumsille A, Valencia R, Rojas C, Seeger M (2018) Living at the frontiers of life: extremophiles in Chile and their potential for bioremediation. Front Microbiol 9:2309. https://doi.org/10.3389/fmicb.2018.02309

Park J, Kawamoto J, Esaki N, Kurihara T (2012) Identification of cold-inducible inner membrane proteins of the psychrotrophic bacterium, *Shewanella livingstonensis* Ac10, by proteomic analysis. Extremophiles 16(2):227–236. https://doi.org/10.1007/s00792-011-0422-z

Peng T, Ma L, Feng X, Tao J, Nan M, Liu Y, Li J, Shen L, Wu X, Yu R, Liu X, Qiu G, Zeng W (2017) Genomic and transcriptomic analyses reveal adaptation mechanisms of an *Acidithiobacillus ferrivorans* strain YL15 to alpine acid mine drainage. PLoS One 12(5): e0178008. https://doi.org/10.1371/journal.pone.0178008

Pereira-Medrano AG, Margesin R, Wright PC (2012) Proteome characterization of the unsequenced psychrophile *Pedobacter cryoconitis* using 15N metabolic labeling, tandem mass spectrometry, and a new bioinformatic workflow. Proteomics 12(6):775–789. https://doi.org/10.1002/pmic.201100159

Perutz MF, Raidt H (1975) Stereochemical basis of heat stability in bacterial ferredoxins and in haemoglobin A2. Nature 255(5505):256–259. https://doi.org/10.1038/255256a0

Ponder MA, Gilmour SJ, Bergholz PW, Mindock CA, Hollingsworth R, Thomashow MF, Tiedje JM (2005) Characterization of potential stress responses in ancient Siberian permafrost psychroactive bacteria. FEMS Microbiol Ecol 53(1):103–115. https://doi.org/10.1016/j.femsec.2004.12.003

Ponder MA, Thomashow MF, Tiedje JM (2008) Metabolic activity of Siberian permafrost isolates, *Psychrobacter arcticus* and *Exiguobacterium sibiricum*, at low water activities. Extremophiles 12(4):481–490. https://doi.org/10.1007/s00792-008-0151-0

Porto BA, da Silva TH, Machado MR, de Oliveira FS, Rosa CA, Rosa LH (2020) Diversity and distribution of cultivable fungi present in acid sulphate soils in chronosequence under Paraperiglacial conditions in King George Island, Antarctica. Extremophiles 24(5):797–807. https://doi.org/10.1007/s00792-020-01195-9

Rafiq M, Nadeem S, Hassan N, Hayat M, Sajjad W, Zada S, Sajjad W, Hasan F (2020) Fungal recovery and characterization from Hindu Kush mountain range, Tirich Mir glacier, and their potential for biotechnological applications. J Basic Microbiol 60(5):444–457. https://doi.org/10.1002/jobm.201900608

Rathour R, Gupta J, Mishra A, Rajeev AC, Dupont CL, Thakur IS (2020) A comparative metagenomic study reveals microbial diversity and their role in the biogeochemical cycling of Pangong lake. Sci Total Environ 731:139074. https://doi.org/10.1016/j.scitotenv.2020.139074

Raymond-Bouchard I, Tremblay J, Altshuler I, Greer CW, Whyte LG (2018) Comparative transcriptomics of cold growth and adaptive features of a eury- and steno-psychrophile. Front Microbiol 9:1565. https://doi.org/10.3389/fmicb.2018.01565

Roldán DM, Kyrpides N, Woyke T, Shapiro N, Whitman WB, Králová S, Sedláček I, Busse HJ, Menes RJ (2020) *Hymenobacter artigasi* sp. nov., isolated from air sampling in maritime Antarctica. Int J Syst Evol Microbiol 70(9):4935–4941. https://doi.org/10.1099/ijsem.0.004362

Saunders NF, Thomas T, Curmi PM, Mattick JS, Kuczek E, Slade R, Davis J, Franzmann PD, Boone D, Rusterholtz K, Feldman R, Gates C, Bench S, Sowers K, Kadner K, Aerts A, Dehal P, Detter C, Glavina T, Lucas S, Richardson P, Larimer F, Hauser L, Land M, Cavicchioli R (2003) Mechanisms of thermal adaptation revealed from the genomes of the Antarctic archaea *Methanogenium frigidum* and *Methanococcoides burtonii*. Genome Res 13(7):1580–1588. https://doi.org/10.1101/gr.1180903

Schipper MA (1967) *Mucor strictus hagem*, a psychrophilic fungus, and *Mucor falcatus sp.n.* Antonie Van Leeuwenhoek 33(2):189–195. https://doi.org/10.1007/BF02045550

Seo JB, Kim HS, Jung GY, Nam MH, Chung JH, Kim JY, Yoo JS, Kim CW, Kwon O (2004) Psychrophilicity of *Bacillus psychrosaccharolyticus*: a proteomic study. Proteomics 4 (11):3654–3659. https://doi.org/10.1002/pmic.200401025

Sharon I, Banfield JF (2013) Microbiology. Genomes from metagenomics. Science 342 (6162):1057–1058. https://doi.org/10.1126/science.1247023

Sherpa MT, Najar IN, Das S, Thakur N (2020) Distribution of antibiotic and metal resistance genes in two glaciers of North Sikkim, India. Ecotoxicol Environ Saf 203:111037. https://doi.org/10.1016/j.ecoenv.2020.111037

Soni R, Suyal DC, Agrawal K, Yadav A, Shouche Y, Goel R (2015) Differential proteomic expression of himalayan psychrotrophic diazotroph *Pseudomonas palleroniana* N26 under low temperature diazotrophic conditions. Cryo Letters 36(2):74–82

Su Y, Jiang X, Wu W, Wang M, Hamid MI, Xiang M, Liu X (2016) Genomic, transcriptomic, and proteomic analysis provide insights into the cold adaptation mechanism of the obligate psychrophilic fungus *Mrakia psychrophila*. G3 (Bethesda) 6(11):3603–3613. https://doi.org/10.1534/g3.116.033308

Suyal DC, Joshi D, Kumar S, Soni R, Goel R (2019) Differential protein profiling of soil diazotroph *Rhodococcus qingshengii* S10107 towards low-temperature and nitrogen deficiency. Sci Rep 9 (1):20378. https://doi.org/10.1038/s41598-019-56592-8

Suyal DC, Yadav A, Shouche Y, Goel R (2014) Differential proteomics in response to low temperature diazotrophy of Himalayan psychrophilic nitrogen fixing *Pseudomonas migulae* S10724 strain. Curr Microbiol 68(4):543–550. https://doi.org/10.1007/s00284-013-0508-1

Tang J, Du LM, Liang YM, Daroch M (2019) Complete genome sequence and comparative analysis of *Synechococcus* sp. CS-601 (SynAce01), a cold-adapted cyanobacterium from an oligotrophic antarctic habitat. Int J Mol Sci 20(1):152. https://doi.org/10.3390/ijms20010152

Terashima M, Ohashi K, Takasuka TE, Kojima H, Fukui M (2019) Antarctic heterotrophic bacterium *Hymenobacter nivis* $P3^T$ displays light-enhanced growth and expresses putative photoactive proteins. Environ Microbiol Rep 11(2):227–235. https://doi.org/10.1111/1758-2229.12702

Tettelin H, Riley D, Cattuto C, Medini D (2008) Comparative genomics: the bacterial pan-genome. Curr Opin Microbiol 11:472–477. https://doi.org/10.1016/j.mib.2008.09.006

Tribelli PM, Solar Venero EC, Ricardi MM, Gómez-Lozano M, Raiger Iustman LJ, Molin S, López NI (2015) Novel essential role of ethanol oxidation genes at low temperature revealed by transcriptome analysis in the Antarctic bacterium *Pseudomonas extremaustralis*. PLoS One 10(12):e0145353. https://doi.org/10.1371/journal.pone.0145353

Wang Z, Gerstein M, Snyder M (2009) RNA-Seq: a revolutionary tool for transcriptomics. Nat Rev Genet 10(1):57–63. https://doi.org/10.1038/nrg2484

Watanabe M, Kojima H, Fukui M (2020) *Labilibaculum antarcticum* sp. nov., a novel facultative anaerobic, psychrotorelant bacterium isolated from marine sediment of Antarctica. Antonie Van Leeuwenhoek 113(3):349–355. https://doi.org/10.1007/s10482-019-01345-w

Weaver LH, Kester WR, Ten Eyck LF, Matthews BW (1976) The structure and stability of thermolysin. Experientia Suppl 26:31–39. https://doi.org/10.1007/978-3-0348-7675-9_2

Williams TJ, Liao Y, Ye J, Kuchel RP, Poljak A, Raftery MJ, Cavicchioli R (2017) Cold adaptation of the Antarctic haloarchaea *Halohasta litchfieldiae* and *Halorubrum lacusprofundi*. Environ Microbiol 19(6):2210–2227. https://doi.org/10.1111/1462-2920.13705

Wu CC, Yates JR (2003) The application of mass spectrometry to membrane proteomics. Nat Biotechnol 21(3):262–267. https://doi.org/10.1038/nbt0303-262

Yu EY, Kwon MA, Lee M, Oh JY, Choi JE, Lee JY, Song BK, Hahm DH, Song JK (2011) Isolation and characterization of cold-active family VIII esterases from an arctic soil metagenome. Appl Microbiol Biotechnol 90(2):573–581. https://doi.org/10.1007/s00253-011-3132-7

Zhang C, Guo W, Wang Y, Chen X (2018) The complete genome sequence of *Colwellia* sp. NB097-1 reveals evidence for the potential genetic basis for its adaptation to cold environment. Mar Genomics 37:54–57. https://doi.org/10.1016/j.margen.2017.11.010

Zhang Z, Qu C, Yao R, Nie Y, Xu C, Miao J, Zhong B (2019) The parallel molecular adaptations to the antarctic cold environment in two psychrophilic green algae. Genome Biol Evol 11 (7):1897–1908. https://doi.org/10.1093/gbe/evz104

Zhang Z, Qu C, Zhang K, He Y, Zhao X, Yang L, Zheng Z, Ma X, Wang X, Wang W, Wang K, Li D, Zhang L, Zhang X, Su D, Chang X, Zhou M, Gao D, Jiang W, Leliaert F, Bhattacharya D, De Clerck O, Zhong B, Miao J (2020) Adaptation to extreme Antarctic environments revealed by the genome of a sea ice green alga. Curr Biol 30(17):3330–3341. https://doi.org/10.1016/j.cub.2020.06.029

Cold-Adapted Microorganisms and their Potential Role in Plant Growth

14

Arun Kumar Rai and Hemant Sharma

Abstract

Abiotic and biotic factors typically interact with each other and are the primary determinants of the agricultural yield especially for the cultivated crops growing at exotic locations. Abiotic factors naturally affect microflora of the soil therefore influencing the ecosystem responsible for the potential growth of cultivated plants. Cold-adapted microorganisms are capable of not merely enduring the harsh and adverse climatic conditions, but they are equally able to efficiently produce functionalities useful for the flourishing of plants at subzero temperatures. Several key factors are responsible for the tolerance and possible survival of the cold conditions by the specific microorganisms endowed with distinctive features like notable production of plant growth-promoting factors and typically inducing effective resistance for the plants to tolerate biotic as well as abiotic stresses. Apart from their potential use in sustainable agriculture in inhospitable conditions, microbes could be conveniently used to produce various enzymes suitable for its use in industries and pharmaceuticals. Some of the specific organisms typically possess the enhanced capability to staunchly resist and degrade hazardous pollutants with possible use in managing wastes and cleaning oil spills at low temperatures. Bioprospecting of beneficial microorganisms from unexplored niches could undoubtedly help to carefully develop formulations capable of inducing successful resistance in plants to abiotic stresses and biofertilization of arid soils inevitably leading to considerable advancement of intensive agriculture and increase productivity in arid regions with low nutrition and harsh climatic conditions along with the successful

A. K. Rai (✉) · H. Sharma
Sikkim University, Gangtok, Sikkim, India
e-mail: akrai@cus.ac.in

R. Goel et al. (eds.), *Survival Strategies in Cold-adapted Microorganisms*,
https://doi.org/10.1007/978-981-16-2625-8_14

production of various functionalities suitable for its use in various industrial sectors.

Keywords

Himalaya · Antarctica · Arctic · Himalaya

14.1 Introduction

Plants growing in extreme climatic conditions are typically presented with arduous factors which severely impact their necessary sustenance. Harsh climate naturally takes a considerable toll on the plant's survivalism especially tolerating extreme temperatures that affect the productivity of plants drastically. Planet Earth includes biospheres with varying degrees of temperatures of which 85% have seasonally or permanently cold temperatures extending below 5 °C. More than 59% of the terrestrial surface are either covered in snow or remain frozen, while others have low temperatures (Hassan et al. 2016). Around 8% of the land mass is categorized under alpine and arctic ecosystems which is much more than that classified under tropical region (Chapin and Körner 1994).

Cold environments coupled with other abiotic factors such as arid conditions and low nutrition undoubtedly have considerable impact on the survivability of organisms (Margesin and Miteva 2011) growing in cold places which are majorly classified into psychrophiles and psychrotolerants depending on their ability to withstand an extensive range of cold temperatures. Psychrophiles can grow markedly below 0 °C with optimum temperature of around 15 °C and maximum temperature below 20 °C. Psychrotolerants typically have optimum growth temperature exceeding 15 °C and can grow at temperatures beyond 20 °C. However, psychrotolerants can only grow above 0 °C (Wang et al. 2017). Despite the harsh climatic conditions, diverse group of microorganisms such as bacteria, fungi including yeasts, and microalgae thrive in this hostile environment. *Mrakia stokesii* was identified to be the most abundant species among others isolated from Scarisoara ice cave located in Bihor Mountains, Romania (Brad et al. 2018). Interactions between plants and microorganisms are essential for sustainable agriculture practices (Kumar and Verma 2018).

14.2 Possible Mechanisms Involved in Cold Tolerance in Microorganisms

Psychrophily represent the fundamental mode of life of certain microorganisms that are able to optimally grow at low temperatures and not just thrive under such harsh conditions. Microorganisms admirably adapt to such complex environments through complex array of physiological and molecular mechanisms to evade harmful effects imposed by the severe surroundings (Collins and Margesin 2019).

14.2.1 Maintaining Fluidity of the Membranes

Low temperatures undoubtedly bring about many adversities for the possible survival of the microorganisms which adversely affect the fluid nature of the flexible membranes by changing the liquid portion of the membrane into gel and eventually ceasing the functioning of the membranes. Composition of lipids in the cell membrane, depending on the temperature of the surroundings, frames the physical characteristics of the membrane. Psychrophiles are able to survive in cold temperatures by altering the nature of cellular membranes by increasing the amount of fatty acids that are either unsaturated or poly-unsaturated, and sometimes the compositions of groups of lipid heads, protein content, and carotenoid pigments that are nonpolar influence the fluidity of the membranes (Chintalapati et al. 2004; D'Amico et al. 2006; Yarzábal 2020). Different types of yeast that are cold adapted have been known to contain higher concentration of polyunsaturated fatty acids (PUFA) (Collins and Margesin 2019). Impairment in the growth of a cold-tolerant bacterium was observed due to the mutation in the gene wapH, a lipo-polysaccharide glycosyltransferase gene (Benforte et al. 2018).

14.2.2 Adaptions Related to Protein Synthesis

Proper functioning of the molecular mechanisms involved in transcription and translation of the cellular proteins is severely hampered at low temperatures. This is due to the decreased activity of the enzymes, reduction in folding of the necessary proteins due to diminished rate in the isomerization of prolyl enzyme, and maintaining the secondary structures of nucleic acids. In case of psychrophiles, certain factors and enzymes, viz., RNA polymerase, isomerases, and factors involved in isomerization, retain their functions even at 0 °C. There is a possibility that, at low temperatures, overexpression of "peptidyl–prolyl-cis–trans-isomerase" may be responsible for maintaining the rates of folding of the proteins. Overexpression of certain proteins involved in binding with nucleic acids such as CspA-related proteins from *Escherichia coli* and RNA helicases required for the destabilization of the secondary structures of nucleic acids has been observed in psychrophiles (Lim et al. 2000; D'Amico et al. 2006).

14.2.3 Responses to Cold Shock

Several proteins are overexpressed by the mesophiles when there is sudden change in temperature known as heat shock and cold shock proteins that influence the molecular physiology of the cell and maintain fluid nature of the membranes. Similarly, certain kinds of proteins such as nucleic acid binding proteins, proteins related to cold shock, and chaperons such as DnaK and GroEL have been reported from psychrophiles which are differentially regulated indicating the presence of some kind of sensory system in the membranes of cells that detect the changes in the fluidity of the membranes (D'Amico et al. 2006).

14.2.4 Cryoprotectants

Low temperature results in injury to the cells through physical damage to the cell membranes (Fonseca et al. 2016). Several kinds of proteins have been reported with an ability to interact with crystals of ice, thus lowering the temperature for the organisms to grow (Jia and Davies 2002). These proteins have been classified as antifreeze proteins and have been observed in bacteria *Marinomonas primoryensis* obtained from Antarctic lake and *Pseudomonas putida* GR12–2 with plant growth-promoting potential (Muryoi et al. 2004; Gilbert et al. 2005). Trehaloses present in the bacterial cells have been known to avert denaturation of proteins, and exopolysaccharides shield the enzymes from getting denatured, thus acting as cryoprotectant (Phadtare 2004; Nichols et al. 2005).

14.2.5 Cold-Tolerant Enzymes

Rate of enzyme reaction decreases drastically with the decrease in temperature and increasing viscosity. Psychrophilic organisms are known to secrete enzymes that are adapted to the cold temperatures due to flexibility in the structure of enzymes to reduce the effect of freezing. Flexibility in the structure is due to reduced ion pairs, reduced interactions, increase in availability of the active site, decrease in the binding of cofactors, reduced content of proline and arginine amino acids, and clustering residues of glycine amino acids (Johns and Somero 2003; Siddiqui et al. 2004; Violot et al. 2005; D'Amico et al. 2006). It was observed that small portion of mutation in amino acid chains may result in adaptation of *Halorubrum lacusprofundi* in cold environment (Laye et al. 2017).

14.3 Role of Cold-Adapted Microorganisms in Plant Growth Promotion and Plant Defenses

Agricultural land across the globe is influenced by abiotic stresses, such as drought (approximately 25%) or salt (approximately 5–7%) (Ruiz-Lozano et al. 2012), which affect the biochemical and physiological conditions in plants, thereby altering proper growth in plants (Ahmad et al. 2010). Phytohormones play an important role in regulating plant growth and metabolism along with stimulating defense mechanisms during stresses. Microorganisms residing in rhizosphere have shown to improve host tolerance to abiotic stress, especially related to crop plants (Egamberdieva et al. 2017). Fungal isolates from the soils of Indian Himalayan region have demonstrated their ability to mineralize and mobilize nutrients at low temperatures, while bacterial isolates from Antarctica have proven to promote plant growth (Yarzábal et al. 2018;

Pandey et al. 2019). Microorganisms residing in the rhizosphere region affect the growth of the plants directly or indirectly by playing a key role in improving soil fertility (Kumar et al. 2020), while some of them are able to fix nitrogen from the atmosphere or solubilize inorganic phosphates from the soil and make it accessible to the plants or produce plant hormones and influence the plant growth through direct mechanisms. Other microorganisms impact plant growth indirectly by secreting antagonistic properties such as antimicrobial molecules, siderophores, and lytic enzymes and promote establishment of beneficial mycorrhizal fungi in the rhizosphere (Pandey and Yarzábal 2019). It has been proposed that extreme environmental conditions play an important role in choosing specific symbiotic partners between plants and microorganisms that help the plants to sustain abiotic stress (Gallardo-Cerda et al. 2018). Ethylene and jasmonates are related with defense mechanisms in plants which help the plants to tolerate abiotic stresses (Kazan 2015). Plant growth-promoting microorganisms have been found to regulate hormones of plants, increase acquisition of nutrients from soil and production of siderophores, and improve tolerance to stress (Kumar and Verma 2018). A strain of cold-tolerant bacteria, *Viridibacillus arenosi* PH15, laden with various plant growth-promoting traits was isolated from a medicinal plant *Podophyllum hexandrum* growing at Sangla valley, Himachal Pradesh (Kushwaha et al. 2020b). Increasing tolerance to cold was observed in seedlings of foxtail millet inoculated with *Sphingomonas faeni*, a cold-tolerant bacterium that was able to overexpress 1-aminocyclopropane-1-carboxylate-deaminase (ACCD) gene (Srinivasan et al. 2017). ACCD activity of endophytic bacteria may help in the germination of seeds and growth of plants and tolerate biotic and abiotic stress due to increased production of IAA and other factors (Selvakumar et al. 2017).

Microorganisms may produce certain functionalities that are involved in defensive mechanisms such as antagonistic effect against the pathogens or induce systemic resistance in plants (Ongena et al. 2007). Psychrotolerant bacterial isolates belonging to *Pseudomonas* sp. and *Brevibacterium* sp. when inoculated to seedlings of *Phaseolus vulgaris* L. were able to lessen the injury caused by freezing and reduce ice-nucleating activity and other properties that improved cold resistance in bean seedlings (Tiryaki et al. 2019). Strains of *Pseudomonas* sp. isolated from Antarctic and glaciers of Venezuela were able to inhibit the growth of phytopathogenic fungi by producing siderophores and HCN molecules (Yarzábal et al. 2018; Rondón et al. 2019). Psychrotolerant fungal microorganisms such as *Candida* sp. obtained from Antarctic were able to inhibit fungal pathogens of apple through the production of volatile metabolites (Arrarte et al. 2017). Some of the microorganisms may produce organic acids with chelating potential that acts along with siderophores for inhibiting the growth of phytopathogens (Torracchi et al. 2020), while others secrete chitinase enzymes that inhibit the growth of fungi such as inhibition of *Verticillium dahlia* and *Fusarium* sps. by a strain of *Pseudomonas* obtained from Antarctica (Liu et al. 2019).

14.3.1 Cold-Adapted Fungi

Fungal species in the soil play a significant role in plant growth promotion, resist biotic and abiotic stresses, and induce systemic resistance in plant toward phytopathogens. Fungi especially belonging to the families of *Ascomycota*, *Zygomycota*, *Basidiomycota*, and few from the families of *Chytridiomycota*, *Blastomycota*, *Cryptomycota*, *Glomeromycota*, and Neocallimastigomycota have been isolated from Antarctic region. Majority of the fungal species isolated from the region belonged to the genera *Cryptococcus*, *Epicoccum*, *Cladosporium*, and few *Cadophora* species (Arenz et al. 2006; Timling et al. 2014). Endophytic fungal species, obtained from the roots of *Taxus wallichiana*, belonging to *Aspergillus* and *Penicillium* genera have demonstrated their ability to solubilize phosphates (Adhikari and Pandey 2018). Combination of secondary metabolites extracted from the species of *Trichoderma* isolated from the rhizosphere soil of potato, maize, and quinoa growing in TA mountains and Bolivian Altiplano increased the yield of quinoa grains under greenhouse conditions including the growth of radish and lettuce plants (Ortuño et al. 2017). Arbuscular mycorrhizal fungi obtained from the potatoes collected from Bolivia, Ecuador, and Peru consisted of around 40 species of fungi that were conserved. Some of the most dominating species identified included strains of *Rhizophagus* sps., *Claroideoglomus* sps., *Cetraspora* sps., and *Acaulospora* sps. That played important roles for the host plant (Senés-Guerrero and Schüßler 2016) (Table 14.1).

14.3.2 Cold-Adapted Bacteria and Plant Growth Promotion

Proteobacteria forms the most dominant bacteria in rhizosphere region and soil, while substantial numbers of *Firmicutes* and *Actinobacteria* have been isolated from tissues of plants growing in arctic region (Wu et al. 2019). Numerous factors of biocontrol agents, such as sustainability, ecologically friendlier, and cost-effectiveness, have provided scope for such agents to be utilized in an alternate approach for the control of phytopathogens rather than using chemical pesticides (Chen et al. 2020). Microorganisms belonging to the genus *Bacillus*, which are known to form durable endospores, are preferred over other forms of pesticides and fertilizers as they have better stability (Borriss 2011). Biocontrol agents that are commercially used have mostly been obtained from the genus *Bacillus* and represent around half of the products available in the market (Fravel 2005). Gram-positive bacteria belonging to the genera *Arthrobacter* and *Micrococcus* have been isolated frequently, while species belonging to *Methanococcus* and *Methanogenium* genera have been reported from waters belonging to deep sea in some areas (D'Amico et al. 2006). Gram-positive bacterial isolates produce array of antimicrobial molecules which makes it ideal for its use against phytopathogens. Strains of *Bacillus* sp. have been considered to be the best for developing into inoculants for plant growth promotion with properties that biologically control pathogens and survive under stressful conditions as they have the ability to go into dormant form due to their

Table 14.1 Significant functionalities of cold-adapted fungi useful in agriculture

Sl. No.	Name of organism	Properties	Isolated from	References
1.	*Rhodotorula* sp., *Naganishia* sp., and *Mrakia* sp.	Plant growth promotion and antifungal properties	Rhizosphere of *Arenaria* sp. and *Drab* sp. plants obtained from the crater of Xinantécatl volcano, Mexico	Tapia-Vázquez et al. (2020)
2.	Strains of fungi belonging to *Ascomycota* and *Mucoromycota*	Degradation of cellulose and reduction of nitrates	Soil and woodchip bioreactor located in cold conditions	Aldossari and Ishii (2020)
3.	*Rhodotorula mucilaginosa* sp. GUMS16	Production of exopolysaccharides with antioxidant activities	Debris of leaves from the jungle of Deylaman, Gilan, Iran	Hamidi et al. (2020)
4.	Species of *Acremonium*, *Penicillium*, and *Pseudogymnoascus*	Antifungal, nematocidal, and herbicidal properties	Marine sediments obtained from South Shetland Islands, Antarctica	Ogaki et al. (2020)
5.	*Thermobifida halotolerans* YIM 90462	β-Glucosidase production and its potential use in the hydrolysis of sugarcane bagasse	Cold-tolerant strain	Yin et al. (2020)
6.	*Pseudomonas koreensis* P2	Contains genes related to solubilization of phosphorus and cold shock proteins	Sela lake, India	Srivastava et al. (2019)
7.	*Penicillium* sp. (GBPI_P155)	Carotenoid pigment and antimicrobial activity against bacteria	Soil of Indian Himalayan region	Pandey et al. (2018)
8.	*Cladosporium herbarum* ER-25	Production of invertase enzyme and removal of color compounds from molasses	Soil samples of Erzurum city (Turkey) during winter seasons	Taskin et al. (2016a)
9.	*Lecanicillium muscarium* CCFEE 5003	Cold-tolerant chitinolytic enzymes acting against fungal pathogens	Moss samples of Victoria land (Antarctica)	Fenice (2016)
10.	*Mortierella* sp.	Insecticidal properties against housefly and waxmoth	Soil samples obtained from Signy Island, South Orkney Islands	Edgington et al. (2014)

ability of sporulation (Shafi et al. 2017). Strains of naturally occurring *Bacillus* sp. in the rhizosphere of potatoes obtained from Peruvian regions were found to inhibit *Fusarium solani* and *Rhizoctonia solani*. Similarly, many strains of *Bacillus subtilis* isolated from Indian Himalayan Region have demonstrated antimicrobial properties against array of fungal and bacterial pathogens (Pandey and Yarzábal 2019), while some of the isolates of *Bacillus* genus were found to promote plant growth and defend potato plantlets from infection by *Rhizoctonia solani* (Ghyselinck et al. 2013). Species of *Pseudomonas* genus form the diverse group of Gram-negative bacteria that are able to colonize and propagate under a broad temperature range and endure various types of stresses (Moreno and Rojo 2014). Plant growth-promoting traits were reported from the strains of *Pseudomonas* sp. and *Stenotrophomonas* sp. which were isolated from the roots of *Origanum vulgare* plant obtained from sub Himalayan belt (Bafana 2013). Bacterial species belonging to *Rahnella*, *Serratia*, and *Pseudomonas* genera were isolated from the rhizosphere region of *Physalis peruviana* L. obtained from Andean region (Ogata-Gutiérrez et al. 2016). Similarly, strains of *Bacillus* and *Pseudomonas* sp. isolated from the rhizosphere of *Lepidium meyenii* obtained from the Andean region demonstrated plant growth-promoting traits such as production of indoleacetic acid and solubilization of phosphates including promotion of germination in clover seeds (Ortiz-Ojeda et al. 2017). Seeds treated with microbial inoculum containing plant growth-promoting bacterial isolates that are cold adapted have shown significant increment in root, shoot, and biomass production and promoted seed germination along with increasing the rate of nutrient uptake (Patni et al. 2018). Rhizobacterial isolates from the roots of purple corn growing in the Peruvian Andes exhibited plant growth-promoting factors. Microbial species belonging to the genera *Bacillus*, *Paenibacillus*, *Lysinibacillus*, *Pseudomonas*, *Stenotrophomonas*, and *Achromobacter* tested positive for the production of indoleacetic acid and siderophores and solubilization of siderophores and had antagonistic effect against fungal phytopathogen such as *Fusarium oxysporum* (Castellano-Hinojosa et al. 2018) (Table 14.2).

14.4 Other Possible Applications of Cold-Adapted Microorganisms

14.4.1 Enzyme Production

Demand for different types of naturally derived environment-friendly enzymes in various industrial sectors such as detergent, paper, textile, and food are ever increasing. Enzymes that remain active under low temperatures are preferred as it helps the industrial processes to run at lower temperatures, thereby reducing expenses incurred during heating and also helps to reduce carbon footprint (Carrasco et al. 2019). Cold-adapted enzymes are cost-effective and require minimum heat for completion of the reactions, and they can be inactivated using mild heat without affecting other processes (Javed and Qazi 2016; Kuddus 2018). Bioprospecting of potential microorganisms capable of producing enzymes with suitable applications under cold temperature is gaining importance in research (Santiago et al. 2016).

Pectinases are used in industries related to food, paper, wine, fruit, and textiles (Garg et al. 2016) which are generally used for the clarification of fruit juices and wine, prevent the growth of undesirable microorganisms, and retain volatile flavors.

Table 14.2 Significant functionalities of cold-adapted bacteria useful in agriculture

Sl. No.	Name of organism	Properties	Isolated from	References
1.	*Bacillus pumilus*, *Bacillus safensis*, and *Bacillus atrophaeus*	Promotion of growth in winter wheat seedlings at 10 °C	Qinghai, Tibetan plateau	Wu et al. (2019)
2.	*Pseudomonas* sp.	Chitinolytic activity and inhibition of phytopathogens *Verticillium dahliae* and *Fusarium oxysporum*	Feces of seals and penguins, marine sediments, and soil samples obtained from Fildes peninsula, Antarctica	Liu et al. (2019)
3.	*Pseudomonas* spp.	Production of indoleacidic acid (IAA), siderophores, and hydrogen cyanide. Production of molecules with antagonistic properties against phytopathogens such as *Fusarium oxysporum*, *Phytophthora infestans*, and *Pythium ultimatum*. Promotes shoot elongation in wheat plants	Soils of Greenwich Island, Antarctic	Yarzábal et al. (2018)
4.	*Pseudomonas chlororaphis* GBPI_507	Production of siderophores, hydrogen cyanide (HCN), ammonia, lipases, and proteases and solubilization of phosphates	Rhizosphere of wheat plants obtained from the mountains of Indian Himalayan region	Jain and Pandey (2016)
5.	*Pseudomonas fredericksbergensis*	Production of chitinase which inhibits the growth of mycelia of *Botrytis cinerea*	Phyllosphere of *Deschampsia antarctica*	Melo et al. (2016)
6.	*Pseudomonas* sp. Da-bac TI-8	Solubilization of phosphates from different sources and promotion of growth of *D. antarctica*	Rhizosphere of *Deschampsia antarctica*	Berríos et al. (2013)
7.	*Pseudomonas* sp.	Positive influence on root and shoot growth in lentils along with promotion of nutrient uptake at 4 °C	Rhizosphere of different plants from North-Western Himalayas	Bisht et al. (2013)

(continued)

Table 14.2 (continued)

Sl. No.	Name of organism	Properties	Isolated from	References
8.	*Exiguobacterium acetylicum* 1P (MTCC 8707)	Production of indoleacidic acid (IAA), siderophores, and hydrogen cyanide and solubilization of phosphates at 4 °C along with promotion of growth in wheat seedlings	Rhizosphere soil of Almora, North-Western Himalayas	Selvakumar et al. (2010)
9.	*Pseudomonas* sp. strain PGERs17 (MTCC 9000)	Production of indoleacidic acid (IAA), siderophores, and hydrogen cyanide and solubilization of phosphates at 4 °C along with antifungal properties against phytopathogens	Rhizosphere of garlic growing in sub-alpine region of North-Western Himalayas	Mishra et al. (2008)
10.	*Serratia marcescens* strain SRM	Production of indoleacidic acid (IAA), siderophores, and hydrogen cyanide and solubilization of phosphates at 4 °C along with promotion of growth in wheat seedlings	Flowers of *Cucurbita pepo* growing in Almora, North-Western Himalayas	Selvakumar et al. (2008)

Pectinases occupy around 10% of the enzyme market around the globe. Cold-tolerant pectinases are preferred over normally available ones as optimum temperature for pectinase is around 50 °C and lose their activity at lower temperatures which always creates a scope for cold-tolerant pectinases with better activity (Carrasco et al. 2019). Similarly, amylases are used in different industrial applications, and search for organisms producing such enzymes with better activity is always on the rise. *Tetracladium* sp. was found to produce a novel enzyme glucoamylase that was cold adapted and useful in biofuel production (Carrasco et al. 2017). A strain of *Alteromonas* sp. ML117 obtained from the depths of Mariana Trench was able to produce β-galactosidase enzyme which was active at cold temperatures. Through the use of biotechnology, the enzyme would have possible applications in the production of milk with less lactose content (Yao et al. 2019). α-D-galactosidase from cold-adapted *Pseudoalteromonas* sp. KMM 701 could be used in biomedicines and synthesis of enzymes (Bakunina et al. 2018). A marine bacterium *Gayadomonas joobiniege* was found to produce a type of galactosidase, an agarolytic enzyme, which exhibited enzymatic activity between 7 and 15 °C (Asghar et al. 2018). A strain of yeast, *Sporobolomyces roseus* LOCK 1119, isolated from a mine in Luiza,

Zabrze, Poland, was found to be of suitable use in food industries particularly related with the production of cheese, soy sauce, and bread and tenderization of meat as well as for the production of peptides from animal proteins with antioxidant properties (Białkowska et al. 2018).

Fungal isolates from Antarctic region have shown protease properties (Duarte et al. 2018). An isolate of *Planococcus* sp. showed the production of extracellular protease with possible use in detergent industries (Chen et al. 2018). A species belonging to the genus *Lysobacter* was capable of producing serine peptidase which showed activity at low as well as high temperatures (Pereira et al. 2017). Similarly, serine protease from *Chryseobacterium* sp. was found to be active at low temperatures and tolerated salt concentrations remarkably showing its potential use in industries related to foods and meat products (Mageswari et al. 2017). Aspartic protease produced by a cold-adapted fungus *Geomyces pannorum* demonstrated features suitable for use in making of cheese (Gao et al. 2018).

Bacterial and fungal isolates have been reported to produce cold active lipases in abundance (Sahay and Chouhan 2018). An isolate of *Rhodotorula* sp. Y-23 obtained from Nella Lake in East Antarctica produced lipases that exhibited good compatibility with detergents and its possible use at low temperatures for the degradation of lipase molecules (Maharana and Singh 2018). *Rhodotorula glutinis* HL25 was able to produce lipase from waste frying olive oils through the process of immobilization (Taskin et al. 2016b). Lipases produced from a psychrotrophic bacteria, *Arthrobacter gangotriensis*, were found to be active under wide range of pH, temperature, and different types of organic solvents with good stability (Ramle and Abdul Rahim 2016). *Pseudomonas* sp. LSK25 obtained from the soil samples collected from research center located in Signy Island, Antarctica, exhibited lipase production at 10 °C (Salwoom et al. 2019). Cold active lipases produced by an isolate of *Pseudomonas palleroniana* from Himalayan region demonstrated its possible applicability in bioremediation and detergent industries (Jain et al. 2019).

Ripening of cheese involves the use of esterase enzymes that cleaves fatty acids with short chains (De Santi et al. 2016a). Species belonging to the genus *Thalassospira* isolated from the Arctic have shown the presence of esterase gene which may be useful in biotechnology industry (De Santi et al. 2016b). A mutant strain obtained from *Pseudomonas* sp. was able to hydrolyze fatty acids at cold temperatures with maximum activity observed at 0 °C showing its potential use in dairy industry for the production of aromas (Dong et al. 2017). A novel cold active esterase enzyme, EstLiu, was obtained from *Zunongwangia profunda* which retained 75% of its activity even at 0 °C (Rahman et al. 2016).

Some of the species belonging to the genera *Pseudomonas*, *Sphingomonas*, and *Hymenobacter* obtained from Antarctica were found to produce photolyase enzymes with properties that help to reverse lesions in DNA and are resistant to ultraviolet radiations (Marizcurrena et al. 2017; Bruno et al. 2019). Naturally occurring photolyase from *Sphingomonas* sp. UV9 could have possible application in cosmetic industries (Marizcurrena et al. 2020).

Cellulose is the most abundant polymeric substance produced by plants on Earth. Cellulase enzymes help in the hydrolysis of cellulose and breakdown of cell wall

(Gupta et al. 2020). Cellulases have been produced by fungal isolates obtained from soil samples collected from the Antarctic (Carrasco et al. 2016). CMCase producing cold-adapted bacteria have been isolated from the landfills in Himalayan region. The isolates were dominated by *Bacillus* sps. (Hamid et al. 2019). Endoglucanases produced by a cold-adapted strain of *Paenibacillus* sp. were found to be stable at alkaline pH and active under cold temperatures (Dhar et al. 2016).

14.4.2 Bioremediation Using Cold-Adapted Microorganisms

Fossil fuels are consumed worldwide for the generation of necessary energy and considerable risk of spilling up of fuels during necessary transportation, or storage is always prevalent (Martínez Álvarez et al. 2017). A bacterial isolate, *Rhodococcus* sp. AQ5–07, isolated from South Shetland Islands, Antarctica, fares well in degradation of oil and could help in treating wastewater and environment (Ibrahim et al. 2020). Cold-adapted consortia of microorganism was found to be suitable in promoting composting startup of food waste at temperatures as low as 10 °C (Xie et al. 2017). Differences in the density of plastics could result in the transportation and accumulation of plastics in cold seas and polar belts (Waller et al. 2017), and bacteria being the primary colonizers are responsible for biofouling of the plastic waste (Selim et al. 2017). Bacterial and fungal microorganisms have been known to degrade bioplastics. *Clonostachys rosea* and *Trichoderma* sp. isolated from Spitsbergen, Svalbard, were able to degrade bioplastics at low temperatures (Urbanek et al. 2017). A bacterial species *Ideonella sakaiensis* has shown possibility of degrading polyethylene terephthalate (PET) by the production of PETase enzyme (Austin et al. 2018).

Antarctica remains a place with less human interference and very little exploration in terms of microbial biodiversity (Fernández et al. 2017). However, Smykla et al. (2018) have reported that human activities in Scott Base, Antarctica, for around 40 years have resulted in the accumulation of heavy metals such as cadmium, copper, lead, zinc, etc. Yeast isolates from the Antarctic which could tolerate heavy metal ions and able to use phenols as carbon source could be used for the treatment of wastewater and effluents from industries and refineries (Fernández et al. 2017). Biodegradation of polychlorinated biphenyls by cold-adapted microorganisms such as *Gelidibacter* sp. has been reported by Papale et al. (2017). Indigenous bacterial species belonging to *Arthrobacter* genus isolated from Antarctic region were able to degrade diesel with the possibility of its application in degradation of Antarctic soils contaminated with diesel and heavy metals (Abdulrasheed et al. 2020).

14.4.3 Pigment Production

Fungal microorganisms are known to produce array of pigments such as carotenoid, quinone, melanin, flavin, and indigo as secondary metabolites which can be

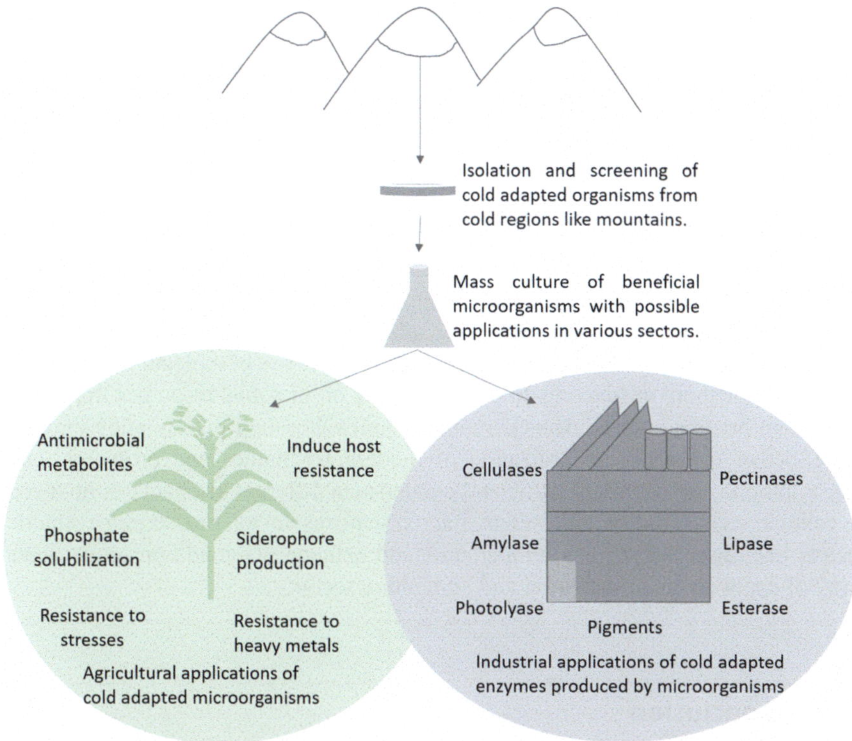

Fig. 14.1 Possible applications of cold-adapted microorganisms in various sectors

classified under terpenes, peptides, polyketides, and non-ribosomal peptides (Fig. 14.1). Some of the important genera of fungi producing pigments include *Aspergillus* sp., *Fusarium* sp., *Monascus* sp., *Neurospora* sp., and *Penicillium* sp. (Studt et al. 2012; Pandey et al. 2018). Unfavorable environmental factors stimulate the production of pigments in the cytoplasm (Pagano and Dhar 2015). Different coloration of snow can be observed depending on the type and concentration of pigments produced by microorganisms such as xanthophylls and carotenoids which impart various shades of yellow and orange (Anesio et al. 2017). Microbial isolates obtained from glaciers have been recorded to produce pigments commonly (Shen et al. 2018). Naturally occurring microbial pigments are treasured compounds with huge potential to replace pigments produced synthetically and can be obtained year-round due to its ease of production (Pandey et al. 2018). An isolate of *Planococcus maritimus* KK21 that was able to grow at broad range of temperature, i.e., from −4 to 37 °C, was able to produce C30-carotenoid (Kushwaha et al. 2020a). A carotenoid-like molecule produced by a species of *Penicillium* was able to inhibit the growth of phytopathogens belonging to *Serratia* and *Pseudomonas* genera (Pandey et al. 2018).

14.5 Future Scope and Potential

Extensive exploration of beneficial microorganisms is carried out on the fundamental basis of their molecular diversity, capability to survive and multiply in exotic environmental conditions, and unique ability to produce array of bioactive metabolites. Active investigation on cold-adapted microorganisms from unexplored niches such as Indian Himalayan Region, temperate and alpine regions, glaciers, Arctic, Antarctic regions, and cold deserts from across the globe will undoubtedly help to adequately understand specific functionalities and their profound significance to the agriculture sector to promote plant growth in hostile conditions. Further investigation on the biological activities, understanding metabolic pathways, and profiling of the active metabolites through the extensive use of molecular techniques could revel plethora of functionalities. Harnessing of microbial resources from such hostile but pristine environments certainly requires thorough investigation through the effective use and standardization of modern techniques and proteomic and transcriptomic approaches to sufficiently understand different mechanisms involved at cellular level. Ability of resilient microorganisms to tolerate extreme abiotic factors like high salinity, low temperature, and extreme pH conditions could have array of applications in industrial and agriculture sector.

14.6 Conclusion

Agriculture sector across the globe typically faces several constraints that typically involve considerable variation in climatic and environmental conditions, geographical restrictions, limited use of modern technology, and economic circumstances. A large number of households typically residing in the mountainous regions of Andes and Himalayan belt practice small-scale farming to run their sustainable livelihood. Microorganisms secreting key hormones, essential enzymes, prime factors, and active metabolites like phosphatases, siderophores, indoleacetic acid, antimicrobial metabolites of practical significance to sustainable agriculture and other plant growth-promoting traits of cold-adapted microorganisms can be properly utilized to naturally enhance the productivity of cultivated crops in these regions. Plants growing in such extreme places rely on microorganism for their growth and productivity. Most of the beneficial properties of capable microorganisms have been elucidated under in vitro conditions and undoubtedly require further investigation in the local fields to adequately understand its full potential under natural conditions. Application of symbiotic microbes could be utilized to induce the synthesis of metabolites from plants, increase root formation, promote plant growth, and tolerate stresses related to cold and arid conditions. Active partnerships between researchers and industries would help in the advancement of research in the particular filed at much faster pace.

References

Abdulrasheed M, Zakaria NN, Ahmad Roslee AF, Shukor MY, Zulkharnain A, Napis S, Convey P, Alias SA, Gonzalez-Rocha G, Ahmad SA (2020) Biodegradation of diesel oil by cold-adapted bacterial strains of Arthrobacter spp. from Antarctica. Antarct Sci 32:341–353. https://doi.org/10.1017/S0954102020000206

Adhikari P, Pandey A (2018) Phosphate solubilization potential of endophytic fungi isolated from Taxus wallichiana Zucc. Roots. Rhizosphere 9:2–9. https://doi.org/10.1016/j.rhisph.2018.11.002

Ahmad P, Jaleel CA, Salem MA, Nabi G, Sharma S (2010) Roles of enzymatic and nonenzymatic antioxidants in plants during abiotic stress. Crit Rev Biotechnol 30:161–175. https://doi.org/10.3109/07388550903524243

Aldossari N, Ishii S (2020) Isolation of cold-adapted nitrate-reducing fungi that have potential to increase nitrate removal in woodchip bioreactors. J Appl Microbiol 2020:14939. https://doi.org/10.1111/jam.14939

Anesio AM, Lutz S, Chrismas NAM, Benning LG (2017) The microbiome of glaciers and ice sheets. NPJ Biofilms Microbiom 3:10. https://doi.org/10.1038/s41522-017-0019-0

Arenz BE, Held BW, Jurgens JA, Farrell RL, Blanchette RA (2006) Fungal diversity in soils and historic wood from the Ross Sea region of Antarctica. Soil Biol Biochem 38:3057–3064. https://doi.org/10.1016/j.soilbio.2006.01.016

Arrarte E, Garmendia G, Rossini C, Wisniewski M, Vero S (2017) Volatile organic compounds produced by Antarctic strains of Candida sake play a role in the control of postharvest pathogens of apples. Biol Control 109:14–20. https://doi.org/10.1016/j.biocontrol.2017.03.002

Asghar S, Lee C-R, Park J-S, Chi W-J, Kang D-K, Hong S-K (2018) Identification and biochemical characterization of a novel cold-adapted 1,3-α-3,6-anhydro-l-galactosidase, Ahg786, from Gayadomonas joobiniege G7. Appl Microbiol Biotechnol 102:8855–8866. https://doi.org/10.1007/s00253-018-9277-x

Austin HP, Allen MD, Donohoe BS, Rorrer NA, Kearns FL, Silveira RL, Pollard BC, Dominick G, Duman R, El Omari K, Mykhaylyk V, Wagner A, Michener WE, Amore A, Skaf MS, Crowley MF, Thorne AW, Johnson CW, Lee Woodcock H, McGeehan JE, Beckham GT (2018) Characterization and engineering of a plastic-degrading aromatic polyesterase. Proc Natl Acad Sci U S A 115:E4350–E4357. https://doi.org/10.1073/pnas.1718804115

Bafana A (2013) Diversity and metabolic potential of culturable root-associated bacteria from Origanum vulgare in sub-Himalayan region. World J Microbiol Biotechnol 29:63–74. https://doi.org/10.1007/s11274-012-1158-3

Bakunina I, Slepchenko L, Anastyuk S, Isakov V, Likhatskaya G, Kim N, Tekutyeva L, Son O, Balabanova L (2018) Characterization of properties and Transglycosylation abilities of recombinant α-galactosidase from cold-adapted marine bacterium Pseudoalteromonas KMM 701 and its C494N and D451A mutants. Mar Drugs 16:349. https://doi.org/10.3390/md16100349

Benforte FC, Colonnella MA, Ricardi MM, Solar Venero EC, Lizarraga L, López NI, Tribelli PM (2018) Novel role of the LPS core glycosyltransferase WapH for cold adaptation in the Antarctic bacterium Pseudomonas extremaustralis. PLoS One 13:1–18. https://doi.org/10.1371/journal.pone.0192559

Berríos G, Cabrera G, Gidekel M, Gutiérrez-Moraga A (2013) Characterization of a novel antarctic plant growth-promoting bacterial strain and its interaction with antarctic hair grass (Deschampsia antarctica Desv). Polar Biol 36:349–362. https://doi.org/10.1007/s00300-012-1264-6

Białkowska AM, Krysiak J, Florczak T, Szulczewska KM, Wanarska M, Turkiewicz M (2018) The psychrotrophic yeast Sporobolomyces roseus LOCK 1119 as a source of a highly active aspartic protease for the in vitro production of antioxidant peptides. Biotechnol Appl Biochem 65:726–738. https://doi.org/10.1002/bab.1656

Bisht SC, Mishra PK, Joshi GK (2013) Genetic and functional diversity among root-associated psychrotrophic Pseudomonad's isolated from the Himalayan plants. Arch Microbiol 195:605–615. https://doi.org/10.1007/s00203-013-0908-4

Borriss R (2011) Bacteria in agrobiology: plant growth responses. Springer, Berlin

Brad T, Itcus C, Pascu MD, Perşoiu A, Hillebrand-Voiculescu A, Iancu L, Purcarea C (2018) Fungi in perennial ice from Scărişoara Ice Cave (Romania). Sci Rep 8:1–9. https://doi.org/10.1038/s41598-018-28401-1

Bruno S, Coppola D, di Prisco G, Giordano D, Verde C (2019) Enzymes from marine polar regions and their biotechnological applications. Mar Drugs 17:544. https://doi.org/10.3390/md17100544

Carrasco M, Alcaíno J, Cifuentes V, Baeza M (2017) Purification and characterization of a novel cold adapted fungal glucoamylase. Microb Cell Factories 16:75. https://doi.org/10.1186/s12934-017-0693-x

Carrasco M, Rozas JM, Alcaíno J, Cifuentes V, Baeza M (2019) Pectinase secreted by psychrotolerant fungi: identification, molecular characterization and heterologous expression of a cold-active polygalacturonase from Tetracladium sp. Microb Cell Factories 18:1–11. https://doi.org/10.1186/s12934-019-1092-2

Carrasco M, Villarreal P, Barahona S, Alcaíno J, Cifuentes V, Baeza M (2016) Screening and characterization of amylase and cellulase activities in psychrotolerant yeasts. BMC Microbiol 16:21. https://doi.org/10.1186/s12866-016-0640-8

Castellano-Hinojosa A, Pérez-Tapia V, Bedmar EJ, Santillana N (2018) Purple corn-associated rhizobacteria with potential for plant growth promotion. J Appl Microbiol 124:1254–1264. https://doi.org/10.1111/jam.13708

Chapin FS, Körner C (1994) Arctic and alpine biodiversity: patterns, causes and ecosystem consequences. Trends Ecol Evol 9:45–47. https://doi.org/10.1016/0169-5347(94)90266-6

Chen K, Mo Q, Liu H, Yuan F, Chai H, Lu F, Zhang H (2018) Identification and characterization of a novel cold-tolerant extracellular protease from Planococcus sp. CGMCC 8088. Extremophiles 22:473–484. https://doi.org/10.1007/s00792-018-1010-2

Chen K, Tian Z, He H, Long C, Jiang F (2020) Bacillus species as potential biocontrol agents against citrus diseases. Biol Control 151:104419. https://doi.org/10.1016/j.biocontrol.2020.104419

Chintalapati S, Kiran MD, Shivaji S (2004) Role of membrane lipid fatty acids in cold adaptation. Cell Mol Biol (Noisy-le-Grand) 50:631–642

Collins T, Margesin R (2019) Psychrophilic lifestyles: mechanisms of adaptation and biotechnological tools. Appl Microbiol Biotechnol 103:2857–2871. https://doi.org/10.1007/s00253-019-09659-5

D'Amico S, Collins T, Marx J, Feller G, Gerday C, Gerday C (2006) Psychrophilic microorganisms: challenges for life. EMBO Rep 7:385–389. https://doi.org/10.1038/sj.embor.7400662

De Santi C, Altermark B, Pierechod MM, Ambrosino L, de Pascale D, Willassen N-P (2016a) Characterization of a cold-active and salt tolerant esterase identified by functional screening of Arctic metagenomic libraries. BMC Biochem 17:1. https://doi.org/10.1186/s12858-016-0057-x

De Santi C, Leiros H-KS, Di Scala A, de Pascale D, Altermark B, Willassen N-P (2016b) Biochemical characterization and structural analysis of a new cold-active and salt-tolerant esterase from the marine bacterium Thalassospira sp. Extremophiles 20:323–336. https://doi.org/10.1007/s00792-016-0824-z

Dhar H, Swarnkar MK, Rana A, Kaushal K, Singh AK, Kasana RC, Gulati A (2016) Complete genome sequence of a low-temperature active and alkaline-stable endoglucanase-producing Paenibacillus sp. strain IHB B 3084 from the Indian trans-Himalayas. J Biotechnol 230:1–2. https://doi.org/10.1016/j.jbiotec.2016.04.037

Dong J, Gasmalla MAA, Zhao W, Sun J, Liu W, Wang M, Han L, Yang R (2017) Characterization of a cold-adapted esterase and mutants from a psychotolerant Pseudomonas sp. strain. Biotechnol Appl Biochem 64:686–699. https://doi.org/10.1002/bab.1525

Duarte AWF, dos Santos JA, Vianna MV, Vieira JMF, Mallagutti VH, Inforsato FJ, Wentzel LCP, Lario LD, Rodrigues A, Pagnocca FC, Pessoa Junior A, Durães Sette L (2018) Cold-adapted enzymes produced by fungi from terrestrial and marine Antarctic environments. Crit Rev Biotechnol 38:600–619. https://doi.org/10.1080/07388551.2017.1379468

Edgington S, Thompson E, Moore D, Hughes KA, Bridge P (2014) Investigating the insecticidal potential of Geomyces (Myxotrichaceae: Helotiales) and Mortierella (Mortierellacea: Mortierellales) isolated from Antarctica. Springerplus 3:1–8. https://doi.org/10.1186/2193-1801-3-289

Egamberdieva D, Wirth SJ, Alqarawi AA, Abd Allah EF, Hashem A (2017) Phytohormones and beneficial microbes: essential components for plants to balance stress and fitness. Front Microbiol 8:1–14. https://doi.org/10.3389/fmicb.2017.02104

Fenice M (2016) The Psychrotolerant Antarctic fungus Lecanicillium muscarium CCFEE 5003: a powerful producer of cold-tolerant Chitinolytic enzymes. Molecules 21:447. https://doi.org/10.3390/molecules21040447

Fernández PM, Martorell MM, Blaser MG, Ruberto LAM, de Figueroa LIC, Mac Cormack WP (2017) Phenol degradation and heavy metal tolerance of Antarctic yeasts. Extremophiles 21:445–457. https://doi.org/10.1007/s00792-017-0915-5

Fonseca F, Meneghel J, Cenard S, Passot S, Morris GJ (2016) Determination of intracellular Vitrification temperatures for unicellular micro organisms under conditions relevant for cryo-preservation. PLoS One 11:e0152939. https://doi.org/10.1371/journal.pone.0152939

Fravel DR (2005) Commercialization and implementation of biocontrol. Annu Rev Phytopathol 43:337–359. https://doi.org/10.1146/annurev.phyto.43.032904.092924

Gallardo-Cerda J, Levihuan J, Lavín P, Oses R, Atala C, Torres-Díaz C, Cuba-Díaz M, Barrera A, Molina-Montenegro MA (2018) Antarctic rhizobacteria improve salt tolerance and physiological performance of the Antarctic vascular plants. Polar Biol 41:1973–1982. https://doi.org/10.1007/s00300-018-2336-z

Gao B, He L, Wei D, Zhang L (2018) Identification and magnetic immobilization of a pyrophilous aspartic protease from Antarctic psychrophilic fungus. J Food Biochem 42:e12691. https://doi.org/10.1111/jfbc.12691

Garg G, Singh A, Kaur A, Singh R, Kaur J, Mahajan R (2016) Microbial pectinases: an ecofriendly tool of nature for industries. 3. Biotech 6:1–13. https://doi.org/10.1007/s13205-016-0371-4

Ghyselinck J, Velivelli SLS, Heylen K, O'Herlihy E, Franco J, Rojas M, De Vos P, Prestwich BD (2013) Bioprospecting in potato fields in the central Andean highlands: screening of rhizobacteria for plant growth-promoting properties. Syst Appl Microbiol 36:116–127. https://doi.org/10.1016/j.syapm.2012.11.007

Gilbert JA, Davies PL, Laybourn-Parry J (2005) A hyperactive, Ca2+−dependent antifreeze protein in an Antarctic bacterium. FEMS Microbiol Lett 245:67–72. https://doi.org/10.1016/j.femsle.2005.02.022

Gupta SK, Kataki S, Chatterjee S, Prasad RK, Datta S, Vairale MG, Sharma S, Dwivedi SK, Gupta DK (2020) Cold adaptation in bacteria with special focus on cellulase production and its potential application. J Clean Prod 258:1–19. https://doi.org/10.1016/j.jclepro.2020.120351

Hamid B, Jehangir A, Ahmad Baba Z, Fatima S (2019) Isolation and characterization of cold active bacterial species from municipal solid waste landfill site. Res J Environ Sci 13:1–9. https://doi.org/10.3923/rjes.2019.1.9

Hamidi M, Gholipour AR, Delattre C, Sesdighi F, Mirzaei Seveiri R, Pasdaran A, Kheirandish S, Pierre G, Safarzadeh Kozani P, Safarzadeh Kozani P, Karimitabar F (2020) Production, characterization and biological activities of exopolysaccharides from a new cold-adapted yeast: Rhodotorula mucilaginosa sp. GUMS16. Int J Biol Macromol 151:268–277. https://doi.org/10.1016/j.ijbiomac.2020.02.206

Hassan N, Rafiq M, Hayat M, Shah AA, Hasan F (2016) Psychrophilic and psychrotrophic fungi: a comprehensive review. Rev Environ Sci Bio/Technol 15:147–172. https://doi.org/10.1007/s11157-016-9395-9

Ibrahim S, Zahri KNM, Convey P, Khalil KA, Gomez-Fuentes C, Zulkarnain A, Alias SA, González-Rocha G, Ahmad SA (2020) Optimisation of biodegradation conditions for waste canola oil by cold-adapted Rhodococcus sp. AQ5-07 from Antarctica. Electron J Biotechnol 48:1–12. https://doi.org/10.1016/j.ejbt.2020.07.005

Jain R, Pandey A (2016) A phenazine-1-carboxylic acid producing polyextremophilic Pseudomonas chlororaphis (MCC2693) strain, isolated from mountain ecosystem, possesses biocontrol and plant growth promotion abilities. Microbiol Res 190:63–71. https://doi.org/10.1016/j.micres.2016.04.017

Jain R, Pandey N, Pandey A (2019) Aggregation properties of cold-active lipase produced by a psychrotolerant strain of Pseudomonas palleroniana (GBPI_508). Biocatal Biotransformation 38:1–11. https://doi.org/10.1080/10242422.2019.1666829

Javed A, Qazi JI (2016) Psychrophilic microbial enzymes implications in coming biotechnological processes. Am Sci Res J Eng Technol Sci 23:103–120

Jia Z, Davies PL (2002) Antifreeze proteins: an unusual receptor–ligand interaction. Trends Biochem Sci 27:101–106. https://doi.org/10.1016/S0968-0004(01)02028-X

Johns GC, Somero GN (2003) Evolutionary convergence in adaptation of proteins to temperature: A4-lactate dehydrogenases of Pacific damselfishes (Chromis spp.). Mol Biol Evol 21:314–320. https://doi.org/10.1093/molbev/msh021

Kazan K (2015) Diverse roles of jasmonates and ethylene in abiotic stress tolerance. Trends Plant Sci 20:219–229. https://doi.org/10.1016/j.tplants.2015.02.001

Kuddus M (2018) Cold-active enzymes in food biotechnology: an updated mini review. J Appl Biol Biotechnol 6:58–63. https://doi.org/10.7324/jabb.2018.60310

Kumar A, Verma JP (2018) Does plant—microbe interaction confer stress tolerance in plants: a review? Microbiol Res 207:41–52. https://doi.org/10.1016/j.micres.2017.11.004

Kumar N, Kumar A, Jeena N, Singh R, Singh H (2020) Factors influencing soil ecosystem and agricultural productivity at higher altitudes. In: Goel R, Soni R, Suyal D (eds) Microbiological advancements for higher altitude agro-ecosystems & sustainability. Springer, Singapore, pp 55–70

Kushwaha K, Saxena J, Agarwal MK (2020a) Identification and characterization of psychrotrophic strain of Planococcus maritimus for glucosylated C 30 carotenoid production. Indian J Exp Biol 58:190–197

Kushwaha P, Kashyap PL, Kuppusamy P (2020b) Microbes for cold stress resistance in plants: mechanism, opportunities, and challenges. In: Goel R, Soni R, Suyal D (eds) Microbiological advancements for higher altitude agro-ecosystems & sustainability. Springer, Singapore, pp 269–292

Laye VJ, Karan R, Kim J-M, Pecher WT, Das Sarma P, Das Sarma S (2017) Key amino acid residues conferring enhanced enzyme activity at cold temperatures in an Antarctic polyextremophilic β-galactosidase. Proc Natl Acad Sci 114:12530–12535. https://doi.org/10.1073/pnas.1711542114

Lim J, Thomas T, Cavicchioli R (2000) Low temperature regulated DEAD-box RNA helicase from the antarctic archaeon, Methanococcoides burtonii. J Mol Biol 297:553–567. https://doi.org/10.1006/jmbi.2000.3585

Liu K, Ding H, Yu Y, Chen B (2019) A cold-adapted chitinase-producing bacterium from Antarctica and its potential in biocontrol of plant pathogenic fungi. Mar Drugs 17(12):695. https://doi.org/10.3390/md17120695

Mageswari A, Subramanian P, Chandrasekaran S, Karthikeyan S, Gothandam KM (2017) Systematic functional analysis and application of a cold-active serine protease from a novel Chryseobacterium sp. Food Chem 217:18–27. https://doi.org/10.1016/j.foodchem.2016.08.064

Maharana AK, Singh SM (2018) A cold and organic solvent tolerant lipase produced by Antarctic strain Rhodotorula sp. Y-23. J Basic Microbiol 58:331–342. https://doi.org/10.1002/jobm.201700638

Margesin R, Miteva V (2011) Diversity and ecology of psychrophilic microorganisms. Res Microbiol 162:346–361. https://doi.org/10.1016/j.resmic.2010.12.004

Marizcurrena JJ, Acosta S, Canclini L, Hernández P, Vallés D, Lamparter T, Castro-Sowinski S (2020) A natural occurring bifunctional CPD/(6-4)-photolyase from the Antarctic bacterium Sphingomonas sp. UV9. Appl Microbiol Biotechnol 104:7037–7050. https://doi.org/10.1007/s00253-020-10734-5

Marizcurrena JJ, Morel MA, Braña V, Morales D, Martinez-López W, Castro-Sowinski S (2017) Searching for novel photolyases in UVC-resistant Antarctic bacteria. Extremophiles 21:409–418. https://doi.org/10.1007/s00792-016-0914-y

Martínez Álvarez L, Ruberto L, Lo Balbo A, Mac Cormack W (2017) Bioremediation of hydrocarbon-contaminated soils in cold regions: development of a pre-optimized biostimulation biopile-scale field assay in Antarctica. Sci Total Environ 590–591:194–203. https://doi.org/10.1016/j.scitotenv.2017.02.204

Melo I, Souza W, Silva L, Santos S, Assalin M, Zucchi T, Queiroz S (2016) Antifungal activity of Pseudomonas frederiksbergensis CMAA 1323 isolated from the Antarctic hair grass Deschampsia antarctica. Br Microbiol Res J 14:1–11. https://doi.org/10.9734/BMRJ/2016/25314

Mishra PK, Mishra S, Selvakumar G, Bisht SC, Bisht JK, Kundu S, Gupta HS (2008) Characterisation of a psychrotolerant plant growth promoting Pseudomonas sp. strain PGERs17 (MTCC 9000) isolated from North Western Indian Himalayas. Ann Microbiol 58:561–568. https://doi.org/10.1007/BF03175558

Moreno R, Rojo F (2014) Features of pseudomonads growing at low temperatures: another facet of their versatility. Environ Microbiol Rep 6:417–426. https://doi.org/10.1111/1758-2229.12150

Muryoi N, Sato M, Kaneko S, Kawahara H, Obata H, Yaish MWF, Griffith M, Glick BR (2004) Cloning and expression of afpA, a gene encoding an antifreeze protein from the Arctic plant growth-promoting Rhizobacterium Pseudomonas putida GR12-2. J Bacteriol 186:5661–5671. https://doi.org/10.1128/JB.186.17.5661-5671.2004

Nichols CAM, Guezennec J, Bowman JP (2005) Bacterial exopolysaccharides from extreme marine environments with special consideration of the Southern Ocean, sea ice, and deep-sea hydrothermal vents: a review. Mar Biotechnol 7:253–271. https://doi.org/10.1007/s10126-004-5118-2

Ogaki MB, Coelho LC, Vieira R, Neto AA, Zani CL, Alves TMA, Junior PAS, Murta SMF, Barbosa EC, Oliveira JG, Ceravolo IP, Pereira PO, Cota BB, Viana RO, Alves VS, Rosa LH (2020) Cultivable fungi present in deep-sea sediments of Antarctica: taxonomy, diversity, and bioprospecting of bioactive compounds. Extremophiles 24:227–238. https://doi.org/10.1007/s00792-019-01148-x

Ogata-Gutiérrez K, Alvarado D, Chumpitaz-Segovia C, Zúñiga-Dávila D (2016) Characterization of plant growth promoting Rhizobacteria isolated from the rhizosphere of Peruvian highlands native crops. Int J Plant Soil Sci 11:1–8. https://doi.org/10.9734/ijpss/2016/24573

Ongena M, Jourdan E, Adam A, Paquot M, Brans A, Joris B, Arpigny JL, Thonart P (2007) Surfactin and fengycin lipopeptides of Bacillus subtilis as elicitors of induced systemic resistance in plants. Environ Microbiol 9:1084–1090. https://doi.org/10.1111/j.1462-2920.2006.01202.x

Ortiz-Ojeda P, Ogata-Gutiérrez K, Zúñiga-Dávila D (2017) Evaluation of plant growth promoting activity and heavy metal tolerance of psychrotrophic bacteria associated with maca (Lepidium meyenii Walp.) rhizosphere. AIMS Microbiol 3:279–292. https://doi.org/10.3934/microbiol.2017.2.279

Ortuño N, Castillo JA, Miranda C, Claros M, Soto X (2017) The use of secondary metabolites extracted from Trichoderma for plant growth promotion in the Andean highlands. Renew Agric Food Syst 32:366–375. https://doi.org/10.1017/S1742170516000302

Pagano M, Dhar P (2015) Fungal pigments: an overview. In: Gupta V, Mach R, Sreenivasaprasad S (eds) Fungal biomolecules: sources, applications and recent developments, 1st edn. Wiley, London

Pandey A, Dhakar K, Jain R, Pandey N, Gupta VK, Kooliyottil R, Dhyani A, Malviya MK, Adhikari P (2019) Cold adapted Fungi from Indian Himalaya: untapped source for bioprospecting. Proc Natl Acad Sci India Sect B Biol Sci 89:1125–1132. https://doi.org/10.1007/s40011-018-1002-0

Pandey A, Yarzábal LA (2019) Bioprospecting cold-adapted plant growth promoting microorganisms from mountain environments. Appl Microbiol Biotechnol 103:643–657. https://doi.org/10.1007/s00253-018-9515-2

Pandey N, Jain R, Pandey A, Tamta S (2018) Optimisation and characterisation of the orange pigment produced by a cold adapted strain of Penicillium sp. (GBPI_P155) isolated from mountain ecosystem. Mycology 9:81–92. https://doi.org/10.1080/21501203.2017.1423127

Papale M, Giannarelli S, Francesconi S, Di Marco G, Mikkonen A, Conte A, Rizzo C, De Domenico E, Michaud L, Lo GA (2017) Enrichment, isolation and biodegradation potential of psychrotolerant polychlorinated-biphenyl degrading bacteria from the Kongsfjorden (Svalbard Islands, high Arctic Norway). Mar Pollut Bull 114:849–859. https://doi.org/10.1016/j.marpolbul.2016.11.011

Patni B, Panwar AS, Negi P, Joshi GK (2018) Plant growth promoting traits of psychrotolerant bacteria: a boon for agriculture in hilly terrains. Plant Sci Today 5:24–28. https://doi.org/10.14719/pst.2018.5.1.352

Pereira JQ, Ambrosini A, Passaglia LMP, Brandelli A (2017) A new cold-adapted serine peptidase from Antarctic Lysobacter sp. A03: insights about enzyme activity at low temperatures. Int J Biol Macromol 103:854–862. https://doi.org/10.1016/j.ijbiomac.2017.05.142

Phadtare S (2004) Recent developments in bacterial cold-shock response. Curr Issues Mol Biol 6:125–136. https://doi.org/10.21775/cimb.006.125

Rahman MA, Culsum U, Tang W, Zhang SW, Wu G, Liu Z (2016) Characterization of a novel cold active and salt tolerant esterase from Zunongwangia profunda. Enzym Microb Technol 85:1–11. https://doi.org/10.1016/j.enzmictec.2015.12.013

Ramle Z, Abdul Rahim R (2016) Psychrophilic lipase from Arctic bacterium. Trop Life Sci Res 27:151–157. https://doi.org/10.21315/tlsr2016.27.3.21

Rondón JJ, Ball MM, Castro LT, Yarzábal LA (2019) Eurypsychrophilic Pseudomonas spp. isolated from Venezuelan tropical glaciers as promoters of wheat growth and biocontrol agents of plant pathogens at low temperatures. Environ Sustain 2:265–275. https://doi.org/10.1007/s42398-019-00072-2

Ruiz-Lozano JM, Porcel R, Azcon C, Aroca R (2012) Regulation by arbuscular mycorrhizae of the integrated physiological response to salinity in plants: new challenges in physiological and molecular studies. J Exp Bot 63:4033–4044. https://doi.org/10.1093/jxb/ers126

Sahay S, Chouhan D (2018) Study on the potential of cold-active lipases from psychrotrophic fungi for detergent formulation. J Genet Eng Biotechnol 16:319–325. https://doi.org/10.1016/j.jgeb.2018.04.006

Salwoom L, Raja Abd Rahman R, Salleh A, Mohd Shariff F, Convey P, Pearce D, Mohamad Ali M (2019) Isolation, characterisation, and lipase production of a cold-adapted bacterial strain pseudomonas sp. LSK25 isolated from signy island, Antarctica. Molecules 24:715. https://doi.org/10.3390/molecules24040715

Santiago M, Ramírez-Sarmiento CA, Zamora RA, Parra LP (2016) Discovery, molecular mechanisms, and industrial applications of cold-active enzymes. Front Microbiol 7:1408. https://doi.org/10.3389/fmicb.2016.01408

Selim MS, Shenashen MA, El-Safty SA, Higazy SA, Selim MM, Isago H, Elmarakbi A (2017) Recent progress in marine foul-release polymeric nanocomposite coatings. Prog Mater Sci 87:1–32. https://doi.org/10.1016/j.pmatsci.2017.02.001

Selvakumar G, Kim K, Shagol CC, Joe MM, Sa T (2017) Spore associated bacteria of arbuscular mycorrhizal fungi improve maize tolerance to salinity by reducing ethylene stress level. Plant Growth Regul 81:159–165. https://doi.org/10.1007/s10725-016-0184-9

Selvakumar G, Kundu S, Joshi P, Nazim S, Gupta AD, Gupta HS (2010) Growth promotion of wheat seedlings by Exiguobacterium acetylicum 1P (MTCC 8707) a cold tolerant bacterial strain from the Uttarakhand Himalayas. Indian J Microbiol 50:50–56. https://doi.org/10.1007/s12088-009-0024-y

Selvakumar G, Mohan M, Kundu S, Gupta AD, Joshi P, Nazim S, Gupta HS (2008) Cold tolerance and plant growth promotion potential of Serratia marcescens strain SRM (MTCC 8708) isolated from flowers of summer squash (Cucurbita pepo). Lett Appl Microbiol 46:171–175. https://doi.org/10.1111/j.1472-765X.2007.02282.x

Senés-Guerrero C, Schüßler A (2016) A conserved arbuscular mycorrhizal fungal core-species community colonizes potato roots in the Andes. Fungal Divers 77:317–333. https://doi.org/10.1007/s13225-015-0328-7

Shafi J, Tian H, Ji M (2017) Bacillus species as versatile weapons for plant pathogens: a review. Biotechnol Biotechnol Equip 31:446–459. https://doi.org/10.1080/13102818.2017.1286950

Shen L, Liu Y, Wang N, Jiao N, Xu B, Liu X (2018) Variation with depth of the abundance, diversity and pigmentation of culturable bacteria in a deep ice core from the Yuzhufeng glacier, Tibetan plateau. Extremophiles 22:29–38. https://doi.org/10.1007/s00792-017-0973-8

Siddiqui KS, Bokhari SA, Afzal AJ, Singh S (2004) A novel thermodynamic relationship based on Kramers theory for studying enzyme kinetics under high viscosity. IUBMB Life 56:403–407. https://doi.org/10.1080/15216540400003052

Smykla J, Szarek-Gwiazda E, Drewnik M, Knap W, Emslie SD (2018) Natural variability of major and trace elements in non-ornithogenic Gelisoils at Edmonson point, northern Victoria land, Antarctica. Polish Polar Res 39:19–50. https://doi.org/10.24425/118737

Srinivasan R, Mageswari A, Subramanian P, Maurya VK, Sugnathi C, Amballa C, Sa T, Gothandam KM (2017) Exogenous expression of ACC deaminase gene in psychrotolerant bacteria alleviates chilling stress and promotes plant growth in millets under chilling conditions. Indian J Exp Biol 55:463–468

Srivastava AK, Saxena P, Sharma A, Srivastava R, Jamali H, Bharati AP, Yadav J, Srivastava AK, Kumar M, Chakdar H, Kashyap PL, Saxena AK (2019) Draft genome sequence of a cold-adapted phosphorous-solubilizing Pseudomonas koreensis P2 isolated from Sela Lake, India. 3 Biotech 9:256. https://doi.org/10.1007/s13205-019-1784-7

Studt L, Wiemann P, Kleigrewe K, Humpf HU, Tudzynski B (2012) Biosynthesis of fusarubins accounts for pigmentation of fusarium Fujikuroi perithecia. Appl Environ Microbiol 78:4468–4480. https://doi.org/10.1128/AEM.00823-12

Tapia-Vázquez I, Sánchez-Cruz R, Arroyo-Domínguez M, Lira-Ruan V, Sánchez-Reyes A, del Rayo S-CM, Padilla-Chacón D, Batista-García RA, Folch-Mallol JL (2020) Isolation and characterization of psychrophilic and psychrotolerant plant-growth promoting microorganisms from a high-altitude volcano crater in Mexico. Microbiol Res 232:126394. https://doi.org/10.1016/j.micres.2019.126394

Taskin M, Ortucu S, Unver Y, Tasar OC, Ozdemir M, Kaymak HC (2016a) Invertase production and molasses decolourization by cold-adapted filamentous fungus Cladosporium herbarum ER-25 in non-sterile molasses medium. Process Saf Environ Prot 103:136–143. https://doi.org/10.1016/j.psep.2016.07.006

Taskin M, Ucar MH, Unver Y, Kara AA, Ozdemir M, Ortucu S (2016b) Lipase production with free and immobilized cells of cold-adapted yeast Rhodotorula glutinis HL25. Biocatal Agric Biotechnol 8:97–103. https://doi.org/10.1016/j.bcab.2016.08.009

Timling I, Walker DA, Nusbaum C, Lennon NJ, Taylor DL (2014) Rich and cold: diversity, distribution and drivers of fungal communities in patterned-ground ecosystems of the north American Arctic. Mol Ecol 23:3258–3272. https://doi.org/10.1111/mec.12743

Tiryaki D, Aydın İ, Atıcı Ö (2019) Psychrotolerant bacteria isolated from the leaf apoplast of cold-adapted wild plants improve the cold resistance of bean (Phaseolus vulgaris L.) under low temperature. Cryobiology 86:111–119. https://doi.org/10.1016/j.cryobiol.2018.11.001

Torracchi CJE, Morel MA, Tapia-Vázquez I, Castro-Sowinski S, Batista-García RA, LA Yarzábal R (2020) Fighting plant pathogens with cold-active microorganisms: biopesticide development and agriculture intensification in cold climates. Appl Microbiol Biotechnol 104:8243–8256. https://doi.org/10.1007/s00253-020-10812-8

Urbanek AK, Rymowicz W, Strzelecki MC, Kociuba W, Franczak Ł, Mirończuk AM (2017) Isolation and characterization of Arctic microorganisms decomposing bioplastics. AMB Express 7:148. https://doi.org/10.1186/s13568-017-0448-4

Violot S, Aghajari N, Czjzek M, Feller G, Sonan GK, Gouet P, Gerday C, Haser R, Receveur-Bréchot V (2005) Structure of a full length psychrophilic Cellulase from Pseudoalteromonas haloplanktis revealed by X-ray diffraction and small angle X-ray scattering. J Mol Biol 348:1211–1224. https://doi.org/10.1016/j.jmb.2005.03.026

Waller CL, Griffiths HJ, Waluda CM, Thorpe SE, Loaiza I, Moreno B, Pacherres CO, Hughes KA (2017) Microplastics in the Antarctic marine system: an emerging area of research. Sci Total Environ 598:220–227. https://doi.org/10.1016/j.scitotenv.2017.03.283

Wang M, Tian J, Xiang M, Liu X (2017) Living strategy of cold-adapted fungi with the reference to several representative species. Mycology 8:178–188. https://doi.org/10.1080/21501203.2017.1370429

Wu H, Gu Q, Xie Y, Lou Z, Xue P, Fang L, Yu C, Jia D, Huang G, Zhu B, Schneider A, Blom J, Lasch P, Borriss R, Gao X (2019) Cold-adapted bacilli isolated from the Qinghai–Tibetan plateau are able to promote plant growth in extreme environments. Environ Microbiol 21:3505–3526. https://doi.org/10.1111/1462-2920.14722

Xie X-Y, Zhao Y, Sun Q-H, Wang X-Q, Cui H-Y, Zhang X, Li Y-J, Wei Z-M (2017) A novel method for contributing to composting start-up at low temperature by inoculating cold-adapted microbial consortium. Bioresour Technol 238:39–47. https://doi.org/10.1016/j.biortech.2017.04.036

Yao C, Sun J, Wang W, Zhuang Z, Liu J, Hao J (2019) A novel cold-adapted β-galactosidase from Alteromonas sp. ML117 cleaves milk lactose effectively at low temperature. Process Biochem 82:94–101. https://doi.org/10.1016/j.procbio.2019.04.016

Yarzábal LA (2020) Perspectives for using glacial and periglacial microorganisms for plant growth promotion at low temperatures. Appl Microbiol Biotechnol 104:3267–3278. https://doi.org/10.1007/s00253-020-10468-4

Yarzábal LA, Monserrate L, Buela L, Chica E (2018) Antarctic Pseudomonas spp. promote wheat germination and growth at low temperatures. Polar Biol 41:2343–2354. https://doi.org/10.1007/s00300-018-2374-6

Yin YR, Sang P, Xiao M, Xian WD, Dong ZY, Liu L, Yang LQ, Li WJ (2020) Expression and characterization of a cold-adapted, salt- and glucose-tolerant GH1 β-glucosidase obtained from Thermobifida halotolerans and its use in sugarcane bagasse hydrolysis. Biomass Conv Bioref:1–9. https://doi.org/10.1007/s13399-019-00556-5

Structure and Functions of Rice and Wheat Microbiome 15

Saurabh Kumar, Divya Joshi, Satish Chandra Pandey, Prasenjit Debbarma, Deep Chandra Suyal, Anand Kumar Chaubey, and Ravindra Soni

Abstract

Plant-associated microorganisms play a crucial role in functioning of agroecosystems through recycling of vital soil nutrients, thus promoting plant growth. Rice (*Oryza sativa* L.) and wheat (*Triticum aestivum* L.) are predominant cropping systems for cereal food production. Microbial communities associated with rice and wheat plants confer fitness to the host plants under normal and stressed conditions. Root-associated cold-adapted microorganisms play a crucial role in protecting rice and wheat plants under cold-stressed environment. Advancement in high-throughput sequencing technology has enhanced the

S. Kumar
Division of Crop Research, ICAR-Research Complex for Eastern Region, Patna, Bihar, India

D. Joshi
Uttarakhand Pollution Control Board, Regional Office, Kashipur, Uttarakhand, India

S. C. Pandey
Cell and Molecular Biology Laboratory, Department of Zoology, Kumaun University, SSJ Campus, Almora, Uttarakhand, India

P. Debbarma
School of Agriculture, Graphic Era Hill University, Dehradun, Uttarakhand, India

D. C. Suyal
Department of Microbiology, Akal College of Basic Sciences, Eternal University, Baru Sahib, Sirmour, Himachal Pradesh, India

A. K. Chaubey
Department of Soil Science & Agricultural Chemistry, College of Agriculture, Banda University of Agriculture and Technology, Banda, Uttar Pradesh, India

R. Soni (✉)
Department of Agricultural Microbiology, College of Agriculture, Indira Gandhi KrishiVishwaVidyalaya, Raipur, Chhatisgarh, India

R. Goel et al. (eds.), *Survival Strategies in Cold-adapted Microorganisms*,
https://doi.org/10.1007/978-981-16-2625-8_15

current understanding of microorganisms associated with rice/wheat plants and their ecological relevance. In this chapter, wheat and rice plant microbiomes are discussed with their beneficial traits and impact on plant health.

Keywords

Microbiome · PGPR · Phyllosphere · Rhizosphere

15.1 Introduction

Interaction between plants and microorganisms plays a crucial role in maintaining plant health and crop productivity. Plant-associated microorganisms play a crucial role in functioning of agroecosystems through recycling of vital soil nutrients, thus promoting plant growth. Rice (*Oryza sativa* L.) and wheat (*Triticum aestivum* L.) cropping system is a predominant cropping system for cereal food production. Rice-wheat crop rotation has been practiced for thousands of years in Asia. This crop rotation provides diverse habitats for a variety of microorganisms, which influences plant growth directly and indirectly.

Plants are habitats for microbial populations in three broadly classified categories: the rhizosphere, endosphere, and phyllosphere (Soni et al. 2017; Goel et al. 2017a). Therefore, plants have been considered as a metaorganism possessing a distinct microbiome. Phyllosphere and rhizosphere are unique habitats for microorganisms which influence plant growth. Microbial communities associated with rice and wheat plants are influenced by plant varieties, fertilization management, organic amendments, plant growth stage, water management, seasonal changes, and environmental stress (Mwajita et al. 2013). The individual and/or cumulative activities of the microbes in the rhizosphere and phyllosphere are direct factor for maintaining the plant health and vitality.

Microorganisms with tight association with a particular plant species, independent of soil nutrients and structure, management practices, and environment, are referred as the core plant microbiome. Core plant microbiome consists of keystone microbial species that are crucial for the fitness of plants (Trivedi et al. 2020). On the other hand, some microbial taxa that are present in low abundance and reduced number of sites are referred as satellite taxa. Satellite taxa are considered as drivers of some important functions for the ecosystem (Compant et al. 2019). These plant-associated microorganisms affect plant growth by inducing phosphate solubilization, nitrogen fixation, production of phytohormones, siderophores, ACC deaminase, and biocontrol agents (Kumar et al. 2019a; Kumar et al. 2018; Nag et al. 2018; Kushwah et al. 2021; Joshi et al. 2017). Abiotic stress also severely affects plant health, thus reducing crop productivity. However microbe-mediated abiotic stress tolerance in plants could be a sustainable approach of increasing crop productivity under increasing abiotic stress.

Exploring the functionality of rice-wheat plant microbe interactions and factors involved can lead to enhanced understanding of how plants can get benefit from microorganisms. Moreover, understanding succession of soil microbial communities in rice-wheat cropping system is a major goal of microbial ecology research. Root-associated cold-adapted microorganisms play crucial role in protecting rice and wheat plants under cold-stressed environment (Suyal et al. 2021a, b; Jeyakumar et al. 2020; Sahu et al. 2020). Detailed understanding of microbial communities associated with rice and wheat plants would provide the better insight for designing effective microbial inoculants. Therefore, microbial communities with their structure and function in rice and wheat rhizosphere and phyllosphere are discussed here.

15.2 Rice Microbiome

15.2.1 Structure and Function of Rice Rhizosphere Microbiome

Rice is a main crop which nourishes more than 50% of the world's population (Ge et al. 2017). Rice is different from other crops since it is cultivated over most growth stages in flooded paddy soils. Therefore, rice is colonized by different groups of microbes including aerobic, anaerobic, or facultative anaerobic microbes (Zhao et al. 2019). Oxygen is produced from the rice aerenchyma in flooded paddy soils, contributing to oxygen-rich rhizosphere enclosed by the anoxic bulk soil. For some microbial communities (methanogens and methanotrophs) that are suitable to this particular niche, the oxic-anoxic interface can be selected. Microbial populations including bacterial, archaeal, and fungal population existing in the rice rhizosphere have been extensively studied for their abundance, diversity, and composition. However the structures in the microbial communities of the rice rhizosphere, specifically the bacterial populations, are complex and dynamic. They may be mainly affected by plant- and soil-associated factors such as geographical position, type of soil, and genotype of rice (Edwards et al. 2015). Evidently, modifications in all these conditions may alter microbial growth and proliferation environments in the soil (e.g., soil pH and root exudation patterns) and eventually contribute to fluctuations in the microbial population structure of the rhizosphere. Studies have shown that *Proteobacteria* (mainly groups of *Alpha-*, *Beta-*, and *Deltaproteobacteria*), *Actinobacteria*, *Acidobacteria*, and *Chloroflexi* are primarily inhabited bacterial population in the rice rhizosphere. *Alphaproteobacteria*, *Deltaproteobacteria*, and *Actinobacteria* are regarded as a part of core rice microbiome (Edwards et al. 2015).

Not unexpectedly, in the bacterial population, structure in the rhizosphere of rice is different from the other plants' (i.e., soybean, *Populus*, potato, *Arabidopsis*, and maize) rhizosphere. This may be partly explained by the fact that a number of anaerobic microorganisms belong to the class *Deltaproteobacteria* which favors the growth and functioning of the oxic-anoxic feature of the rice root surroundings, for example, *Desulfococcus and Geobacter* genera, which contribute in ferric iron [Fe(III)] and/or sulfate reduction (Ding et al. 2015). Furthermore, in the rice rhizosphere, the *Alpha-* and *Betaproteobacteria* groups are also abundant. They are important for the functioning of the environment, including the control of nitrogen (N), carbon (C), Fe, and sulfur (S) cycles (Hernández et al. 2015), the development of phytohormones beneficial for the growth of plant (Lee et al. 2006), and the synthesis of antibiotics to conquer bacterial and fungal pathogens (Raaijmakers and Mazzola 2012). Moreover, archaebacterial population in the rhizosphere of rice is usually comprised of *Euryarchaeota*, *Thaumarchaeota*, and *Crenarchaeota* phyla. These are distinct from those generally inhabited by the *Thaumarchaeota* phylum in other crops/plants' rhizosphere (e.g., corn, soybeans, *Populus*, and potatoes). Specifically, the substantial enhancement of the *Euryarchaeota* phylum in the rhizosphere of rice could be indicative of the fact that it is associated with methanogens (e.g., genera *Methanosaeta* and *Methanosarcina*) which is responsible for the production of CH_4 in paddy soils (Lee et al. 2014). Unlike bacterial communities, the diversity of archaeabacterial populations in the rhizosphere of rice remains stable and is immune to climate conditions (Breidenbach et al. 2016).

In comparison to the bacterial and archaeal populations, the makeup of the fungal population in the rhizosphere of rice is significantly different from that of the soybean and *Populus* rhizosphere, especially with the enrichment of the selected studies based on *Chytridiomycota* phylum. Many members associated with this phylum are known to be aquatic fungi and are widespread in aquatic environments (e.g., lakes) and high-humidity ecosystems (e.g., ditches and pond banks) and hence would be preferred by paddy soils. Different members of this phylum, such as the genera *Rhizophlyctis* and *Cladochytrium* reported in the rice rhizosphere, are efficient carbohydrate decomposers such as cellulose and thus may have a role in the biogeochemical cycle of carbon in paddy soils (Gleason et al. 2011; Eichorst and Kuske 2012).

Fungi have been found to play important roles in the development of the environment, such as the solubilization of phosphorus (P) and the synthesis of indoleacetic acid (IAA) for plant growth promotion (Kumar et al. 2019; Suyal et al. 2021a, b). Fungi such *Aspergillus* can significantly accelerate P solubilization, and arbuscular mycorrhizal fungi (AMF) (e.g., *Rhizophagus* spp.) can make absorption and transfer of P easier (Mendes et al. 2013). Remarkably, the arbuscular mycorrhizal (AM) symbiosis can contribute up to 80% to the absorption of phosphorus by the rice roots, but higher level of soil soluble P negatively affects symbiosis (Mendes et al. 2013).

15.2.2 Structure and Function of Rice Phyllosphere Microbiome

The phyllosphere refers to the complete surface of the aerial plant (above-ground portions) which is colonized by a large number of microorganisms, and these microorganisms are beneficial for plant growth. Microorganisms create communities on the leaf surface that are compositionally complex. Various microbial species, including bacteria, algae, yeasts, filamentous fungi, and protozoans, constitute the rice phyllosphere (Vorholt 2012). On the leaf surface, the diversity, dispersal, and community growth of microbes are based on the host plant's physiochemistry, climate, and immunity. The colonization process is a significant event that has benefited both the microorganisms and the host plant.

On the phyllospheric environment, microbes exhibit either epiphytic or endophytic mode of life cycle, which help the plant to communicate with its surrounding. The bacteria that inhabit the rice phyllosphere and their physiological adaptations to the environment have been examined intensively (Madhaiyan et al. 2007, 2009). A large number of bacterial isolates from the rice phyllosphere have been characterized so far, and possible beneficial interactions of phyllosphere bacteria with rice plants have been examined.

So far a variety of bacterial isolates has been identified from the rice phyllosphere, and potential beneficial associations for plant growth promotion (bacterial nitrogen fixation, development of phytohormones, and protection against pathogens) of phyllosphere bacteria with rice plants have been investigated (Yang et al. 2008; Chinnadurai et al. 2009; Pedraza et al. 2009). Venkatachalam et al. (2016) reported *Alphaproteobacteria* (35%) and *Actinobacteria* (38%) as most abundant bacterial groups in rice phyllosphere, while *Pantoea*, *Exiguobacterium*, and *Bacillus* were common bacterial genera. About 34% of total bacterial bacterial isolates had higher potential for indoleacetic acid production (Venkatachalam et al. 2016). Gai-Di et al. (2014) found that 70.6–93.8% bacterial population in the rice phyllosphere is composed of *Enterobacteriaceae* family of *Gammaproteobacteria*. In another culture independent study, Yasmin et al. (2020) studied Basmati rice phyllosphere and found *Proteobacteria* (79.6%) as dominant phylum followed by *Firmicutes* (9.8%), *Bacteroidetes* (8.6%), *Chloroflexi* (4.3%), and *Actinobacteria* (0.9%). Further comparison of rhizosphere and phyllosphere bacterial communities revealed that rhizosphere is rich in bacterial diversity but *Sphingomonas*, *Pseudomonas*, *Bradyrhizobium*, GP6, and *Bacillus* were more abundant in phyllosphere (Yasmin et al. 2020). Roman-Reyna et al. (2019) reported that *Proteobacteria*, *Firmicutes*, *Actinobacteria*, *Cyanobacteria*, *Tenericutes*, and *Euryarchaeota* were the most abundant phyla in rice phyllosphere.

In the phyllosphere, pathogenic microbial interactions reduce plant health, reduce crop productivity, and threaten the protection of horticultural products for human consumption. Rice has been recognized by some pathogens, including fungi ascomycete *Pyricularia oryzae* (syn. *Magnaporthe oryzae*), for its vulnerability to infection (Zhang et al. 2016). This triggers the most devastating rice disease, the leaf blast, which contributes to significant production losses worldwide each year. Symptoms can grow throughout the entire plant and negatively affect the yield and quality. Leaf blast, in fact, is among the top ten fungal diseases that threaten worldwide food safety (Dean et al. 2012). Microbial populations associated with rice phyllosphere have been reported to act as antagonists of *Rhizoctonia solani* (De Costa et al. 2006). Phyllosphere actinomycetes reduce the growth and colonization of plant pathogens and help the plant to recover from the infection (Lindow and Brandl 2003). According to a research, the actinomycetes of the rice phyllosphere are possible biocontrol agents for fungal leaf blast disease (Harsonowati et al. 2017).

15.3 Wheat Microbiome

15.3.1 Structure and Functions of the Wheat Rhizosphere Microbiome.

Root-associated microbiomes have gained popularity due to its ability to aid in nutrition, development, and growth of the host. They have negative or positive effects on host plant, which affect production in agriculture. Therefore, more extensive understanding of the associated communities is vital for soil management practices and plant health. Several studies have depicted that *Actinobacteria*, *Bacteroidetes*, and *Proteobacteria* largely dominate wheat rhizosphere (Rascovan et al. 2016; Granzow et al. 2017). Donn et al. (2015) reported an increased population of *Actinobacteria*, *Pseudomonas*, and other oligotrophs and copiotrophs in the wheat rhizosphere in comparison to bulk soil. Moreover, reports have also revealed that communities in rhizosphere and rhizoplane were continuously altered with time, whereas the bulk soil communities remain unaffected. Rascovan et al. (2016) have reported several communities like *Pseudomonas* spp., *Paraburkholderia* spp., and *Pantoea* spp. and several others in wheat and soybean roots that depicted plant growth-promoting properties like mobilization of minerals such as phosphorous, nitrogen fixation, production of growth regulators like indoleacetic acid, regulatory behavior under stressful conditions by production of enzyme-like ACC deaminase, and increased nutrient uptake efficiency. In a field trial study, presence of *Azospirillum brasilense* strain Ab-V5 gave significant gains in grain yield in maize (30%) and wheat (16%) (Hungria et al. 2010). Similarly, GSF30T *Herbaspirillum frisingense* produces growth regulators (IAA) in the culture (Straub et al. 2013). It was also revealed post-inoculation with *Bacillus subtilis* the growth of wheat seedlings increases due to production of auxin (plant hormone). A study reported 7469 bacterial and 715 fungal unique operational taxonomic units (OTUs) in 254 analyzed wheat samples of wheat. Also the analysis of the amplicon sequence variants (ASVs) unveiled 20,061 bacterial and 891 fungal ASVs. *Pseudomonas* and *Fusarium/Gibberella* were the most common identified communities in roots and rhizosphere soils. Some other genera exclusively found in rhizosphere soils were *Adhaeribacter*, *Kitasatospora*, *Nitrosovibrio*, and *Leohumicola* (Araujo et al. 2020).

Plant-associated microbiomes whether linked to roots or shoots are some of the key factors associated with plant growth and development. It is a well-established fact that they are the futuristic assets to attain sustainable agricultural production. They aid plants in nutrient and water uptake, provide protection against pathogens, and aid in mineralization of insoluble elements, stress tolerance against abiotic stress, and degradation of harmful xenobiotics. Sugars and organic acids act as source of energy, thus making this interaction mutually beneficial (Backer et al. 2018). Some of the well-known examples are *Rhizobium* and plant pathogens, while still a lot need to be discovered, explored, and documented. There is a widely known fact that root rhizosphere is important for many host-microbe interactions. The knowledge and use of these microbial populations play key roles in modulating the cropping systems in context to the current scenario. These assemblies are believed to be highly functional and establish a dynamic interaction between the plant and environmental factors. Their variability or diversity is highly required and needs to be studied in depth so as to develop management practices that have consistent effects on soil health and thus plant growth and development.

15.3.2 Structure and Functions of the Wheat Phyllosphere Microbiome

Phyllosphere is the largest microbial habitat on Earth and refers to aboveground or total aerial plant surfaces. Unlike the much stable rhizosphere, the phyllosphere is a highly unstable habitat. Carbon and sources of nutrients are the selective factors and govern and/or limit the diversity structure on the leaf surface. Several studies have shown that the external application of nutrient source brings about the same changes as natural source in the microbial community composition and abundance (Stadler and Müller 2000). Main sources for microbial diversity in phyllosphere are soil, water, air, tree buds, and plant residues from the previous crops. The microbial populations of the phyllosphere are subjected to fluctuations in temperature, humidity, and UV light irradiation that highly influence the changing pattern of microbial growth. The diversity and development on the phyllosphere depends on the variety of factors, abiotic and biotic, which shape its structure to a great extent (Sivakumar et al. 2020) (Fig. 15.1).

The phyllosphere has diversified range of microbial populations which includes bacteria, fungi, algae, and protozoans (Verma et al. 2015, 2016). Among these, bacteria are the dominant one (Whipps et al. 2008). Generally, conventional culture-based method is used for the identification of diverse microbial communities. Moreover, the culture-independent approaches like 16S rDNA sequences of the whole microbial mass of phyllosphere can provide the structure of microbial community (Goel et al. 2017b; Suyal et al. 2019; Soni et al. 2021; Soni et al. 2016). Molecular studies have already revealed *Alpha-*, *Beta-*, and *Gammaproteobacteria* and *Firmicutes* as the dominant phyla in the phyllosphere (Sivakumar et al. 2020).

Generally, when considered the bacterial diversity, *Proteobacteria*, *Firmicutes*, *Bacteroides*, and *Actinobacteria* (Durand et al. 2018) are the four major predominant phyla in wheat phyllosphere. Studies have also reported some methylotrophic bacteria in phyllosphere which belongs to major genera like *Methylobacterium*, *Methylophilus*, *Methylibium*, and *Methylocystis* (Kwak et al. 2014; Krishnamoorthy et al. 2018). *Methylobacterium* and *Sphingomonas* (class *Alphaproteobacteria*) are the main inhabitants in some plant phyllospheres (Kumar et al. 2019b). *Massilia*, *Flavobacterium*, *Pseudomonas*, and *Rathayibacter* are also abundant bacterial community found in *Arabidopsis thaliana* apart from *Proteobacteria*, *Actinobacteria*, and *Bacteroidetes* and *Deinococcus-Thermus* on trees, and *Bacillus* and *Pantoea* are main dominating phyla on lettuce (Sivakumar et al. 2020).

Kembel et al. (2014) reported approximately 400 bacterial taxa on leaves of tropical trees, mainly *Actinobacteria*; *Alpha-*, *Beta-*, and *Gammaproteobacteria*; and *Sphingobacteria*. Steven et al. (2018) reported and documented *Pseudomonas*

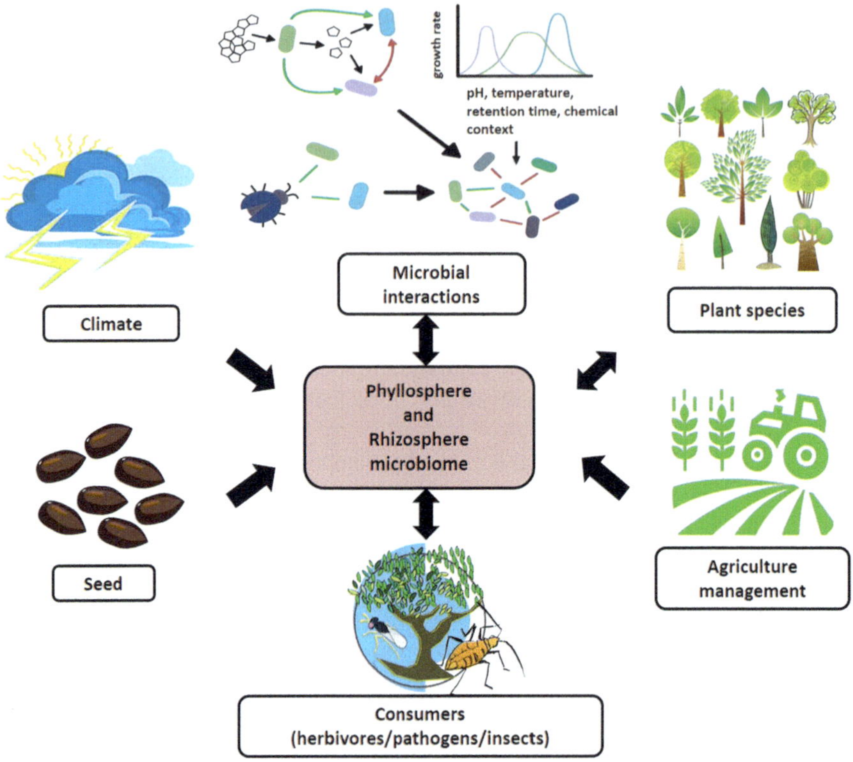

Fig. 15.1 Factors affecting the phyllosphere and rhizosphere microbiome

and *Enterobacteriaceae* as main taxa from apple phyllosphere. In terms of negative interaction, the most common pathogen is *Pseudomonas syringae* which infect many economically important plants (Burch et al. 2014). *Rhizobiales*, *Clostridiales*, *Pseudomonadales*, *Burkholderiales*, *Sphingomonadales*, *Lactobacillales*, and *Bacillales* are reported to be the inhabitants of *Cinnamomum camphora* (L.) Presl. leaves (Hamd Elmagzob et al. 2019). Some diazotrophs like *Beijerinckia*, *Azotobacter*, *Klebsiella*, and *Cyanobacteria* (e.g., *Nostoc*, *Scytonema*, and *Stigonema*) are also reported inhabitants of phyllosphere (Vacher et al. 2016). *Proteobacteria*, *Firmicutes*, *Bacteroidetes*, *Actinobacteria*, *Gemmatimonadetes*, *Acidobacteria*, *Planctomycetes*, *Chloroflexi*, and *Fusobacteria* are also present on leaves of *Cinnamomum camphora* (L.) as revealed by the 16S rRNA gene metagenomics study (Elmagzob et al. 2019).

Some fungal species have also been reported from herbs to woody plants. *Aureobasidium pullulans* are most abundant fungal species to be found in phyllosphere (Abdelfattah et al. 2015). Several filamentous fungi have also been reported from healthy and infected leaves. Ripa et al. (2019) isolated several fungal species like *Aspergillus niger*, *Fusarium oxysporum*, *Penicillium aurantiogriseum*,

Fusarium incarnatum, *Aspergillus flavus*, *Trichoderma aureoviride*, *Trichoderma harzianum*, *Penicillium janthinellum*, *Fusarium proliferatum*, *Fusarium equiseti*, and *Aspergillus stellatus* from wheat plant using the culture-dependent method. Osono (2008) reported that *Camellia japonica* is inhabited by *Colletotrichum gloeosporioides* and *C. acutatum* (endophytes) and *Pestalotiopsis* sp., *Aureobasidium pullulans*, *Phoma* sp., and *Ramichloridium* sp. (epiphytes). However, seasonal variations and leaf age also alter phyllosphere fungal assembly.

A large number of *Actinobacteria* are reported in plants with medicinal value, crop plants, and some other terrestrial plants (Dinesh et al. 2017; Nalini and Prakash 2017). Some of the species of *Actinobacteria* are reported in wheat, lupin, lobelia, northern black wattle, agar wood, rice, cannonball mangrove, and wild olive from various environments. These species include *Actinoplane missouriensis*, *Amycolatopsis tolypomycina*, *Jishengella endophytica*, *Kribbella* sp., *Microbispora* sp., *Micromonospora* sp., *Nocardioides* sp., *Nonomuraea rubr Micromonospora* sp. *Nonomuraea* sp., *Pseudonocardia* sp., *Planotetraspora* sp., *Pseudonocardia endophytica*, *Pseudonocardia halophobica*, *Streptomyces* sp., and *Streptomyces javensis* (Yadav 2017; Yadav et al. 2018). These primarily are from arid, semiarid, and swampy areas. The diversity is higher in tropical and temperate region (Yadav and Yadav 2019).

Phyllosphere diversity plays a key role in absorption of minerals and their recycling process. This specifically relates to carbon, nitrogen, and phosphorus recycling in the forest ecosystem. For instance, the phyllospheric fungi and its physiology with host plant are well explored for the purpose. Generally, fungi can help plant in its growth as well as provide resistance to biotic (pathogens) and abiotic (drought and salinity) stresses (Yadav et al. 2018). Phyllosphere bacteria are also involved in nutrient cycling like carbon and nitrogen and include nitrogen fixation and nitrification, thus affecting the plant health (Gupta and Patil 2020). Some *Actinobacteria* colonizing the plant tissues (symbiotic association) are of great importance to the host and its environment due to their novel metabolites (Sivakumar et al. 2020). The phyllosphere is a microniche for diverse microbial communities that affect the ecosystem. Reports have suggested that plants with altered genetic networks have a different microbiome and develop dysbiosis in the phyllosphere. This highlights plant genetics as a key factor in assembly of phyllosphere microbiome and links it to plant health. Dysbiosis refers to a microbial imbalance that occurs due to overgrowth of pathogenic species (pathobionts). This results in loss of diversity, which in turn links to potential health issues of animals (Liu et al. 2020). This suggested that an impaired genetic network can alter leaf microbial communities. The resultant microbial variation links to leaf phenotypes. This diversity is the key factor in maintaining plant health, as high diversity supports more positive or beneficiary microbial interactions to a substantial degree which further helps in controlling the pathobionts.

15.4 Conclusion and Future Prospects

A systematic investigation for examining the microbiomes of rice and wheat plants is desired to identify the associated core and hub microbiota and their biological functions. Study of plant-associated microbiome provides good insight of plant health. Considering the potential of microorganisms in plant growth promotion and stress adaptation to plants, several bio-inoculants have been developed, but majority of them underperform under field conditions. Before applications of microbial inoculants, detailed understating is required on how these inoculants change the resident microbiome associated with wheat and rice plant. Future study on rice and wheat microbiome providing more insight into the other residents of the plant microbiome including viruses, archaeal, and protists would unravel the complete picture of rice and wheat-associated microbiome.

References

Abdelfattah A, Li Destri Nicosia MG, Cacciola SO, Droby S, Schena L (2015) Metabarcoding analysis of fungal diversity in the phyllosphere and carposphere of olive (Oleaeuropaea). PLoS One 10(7):e0131069

Araujo R, Dunlap C, Franco CM (2020) Analogous wheat root rhizosphere microbial successions in field and greenhouse trials in the presence of biocontrol agents Paenibacillus peoriae SP9 and Streptomyces fulvissimus FU14. Mol Plant Pathol 21(5):622–635

Backer R, Rokem JS, Ilangumaran G, Lamont J, Praslickova D, Ricci E et al (2018) Plant growth-promoting rhizobacteria: context, mechanisms of action, and roadmap to commercialization of biostimulants for sustainable agriculture. Front Plant Sci 9:1473

Breidenbach B, Pump J, Dumont MG (2016) Microbial community structure in the rhizosphere of rice plants. Front Microbiol 6:1537

Burch AY, Zeisler V, Yokota K, Schreiber L, Lindow SE (2014) The hygroscopic biosurfactant syringafactin produced by Pseudomonas syringae enhances fitness on leaf surfaces during fluctuating humidity. Environ Microbiol 16(7):2086–2098

Chinnadurai C, Balachandar D, Sundaram SP (2009) Characterization of 1-aminocyclopropane-1-carboxylate deaminase producing methylobacteria from phyllosphere of rice and their role in ethylene regulation. World J Microbiol Biotechnol 25(8):1403–1411

Compant S, Samad A, Faist H, Sessitsch A (2019) A review on the plant microbiome: ecology, functions, and emerging trends in microbial application. J Adv Res 19:29–37

De Costa DM, Rathnayake RMPS, De Costa WAJM, Kumari WMD, Dissanayake DMN (2006) Variation of phyllosphere microflora of different rice varieties in Sri Lanka and its relationship to leaf anatomical and physiological characters. J Agron Crop Sci 192(3):209–220

Dean R, Van Kan JA, Pretorius ZA, Hammond-Kosack KE, Di Pietro A, Spanu PD et al (2012) The top 10 fungal pathogens in molecular plant pathology. Mol Plant Pathol 13(4):414–430

Dinesh R, Srinivasan V, TE, S., Anandaraj, M., & Srambikkal, H. (2017) Endophytic actinobacteria: diversity, secondary metabolism and mechanisms to unsilence biosynthetic gene clusters. Crit Rev Microbiol 43(5):546–566

Ding LJ, Su JQ, Xu HJ, Jia ZJ, Zhu YG (2015) Long-term nitrogen fertilization of paddy soil shifts iron-reducing microbial community revealed by RNA-13 C-acetate probing coupled with pyrosequencing. ISME J 9(3):721–734

Donn S, Kirkegaard JA, Perera G, Richardson AE, Watt M (2015) Evolution of bacterial communities in the wheat crop rhizosphere. Environ Microbiol 17(3):610–621

Durand A, Maillard F, Alvarez-Lopez V, Guinchard S, Bertheau C, Valot B et al (2018) Bacterial diversity associated with poplar trees grown on a Hg-contaminated site: community characterization and isolation of Hg-resistant plant growth-promoting bacteria. Sci Total Environ 622: 1165–1177

Edwards J, Johnson C, Santos-Medellín C, Lurie E, Podishetty NK, Bhatnagar S et al (2015) Structure, variation, and assembly of the root-associated microbiomes of rice. Proc Natl Acad Sci 112(8):E911–E920

Eichorst SA, Kuske CR (2012) Identification of cellulose-responsive bacterial and fungal communities in geographically and edaphically different soils by using stable isotope probing. Appl Environ Microbiol 78(7):2316–2327

Elmagzob AAH, Ibrahim MM, Zhang GF (2019) Seasonal diversity of endophytic Bacteria associated with Cinnamomumcamphora (L.). Presl Divers 11(7):112

Gai-Di REN, Jian-Guo ZHU, Zhong-Jun JIA (2014) Contrasting response patterns of rice phyllosphere bacterial taxa to elevated CO_2. Pedosphere 24(4):544–552

Ge T, Li B, Zhu Z, Hu Y, Yuan H, Dorodnikov M et al (2017) Rice rhizodeposition and its utilization by microbial groups depends on N fertilization. Biol Fertil Soils 53(1):37–48

Gleason FH, Marano AV, Digby AL, Al-Shugairan N, Lilje O, Steciow MM et al (2011) Patterns of utilization of different carbon sources by Chytridiomycota. Hydrobiologia 659(1):55–64

Goel R, Kumar V, Suyal DK, Dash B, Kumar P, Soni R (2017a) Root-associated bacteria: rhizoplane and endosphere. In: Singh D, Singh H, Prabha R (eds) Plant-microbe interactions in agro-ecological perspectives. Springer, Singapore. https://doi.org/10.1007/978-981-10-5813-4_9

Goel R, Suyal DC, Narayan DB, Soni R (2017b) Soil metagenomics: a tool for sustainable agriculture. In: Kalia V, Shouche Y, Purohit H, Rahi P (eds) Mining of microbial wealth and metagenomics. Springer, Singapore. https://doi.org/10.1007/978-981-10-5708-3_13

Granzow S, Kaiser K, Wemheuer B, Pfeiffer B, Daniel R, Vidal S, Wemheuer F (2017) The effects of cropping regimes on fungal and bacterial communities of wheat and faba bean in a greenhouse pot experiment differ between plant species and compartment. Front Microbiol 8:902

Gupta R, Patil R (2020) Phyllospheric microbes: diversity, functions, interaction, and applications in agriculture. In: Current trends in microbial biotechnology for sustainable agriculture. Springer, Cham, pp 301–323

Harsonowati W, Astuti RI, Wahyudi AT (2017) Leaf blast disease reduction by rice-phyllosphere actinomycetes producing bioactive compounds. J Gen Plant Pathol 83(2):98–108

Hernández M, Dumont MG, Yuan Q, Conrad R (2015) Different bacterial populations associated with the roots and rhizosphere of rice incorporate plant-derived carbon. Appl Environ Microbiol 81(6):2244–2253

Hungria M, Campo RJ, Souza EM, Pedrosa FO (2010) Inoculation with selected strains of Azospirillum brasilense and A. lipoferum improves yields of maize and wheat in Brazil. Plant Soil 331(1-2):413–425

Jeyakumar SP, Dash B, Singh AK, Suyal DC, Soni R (2020) Nutrient cycling at higher altitudes. In: Goel R, Soni R, Suyal D (eds) Microbiological advancements for higher altitude agro-ecosystems & sustainability, Rhizosphere biology. Springer, Singapore. https://doi.org/10.1007/978-981-15-1902-4_15

Joshi D, Kumar S, Suyal DC, Goel R (2017) The microbiome of the Himalayan ecosystem. In: Mining of microbial wealth and metagenomics. Springer, Singapore, pp 101–116

Kembel SW, O'Connor TK, Arnold HK, Hubbell SP, Wright SJ, Green JL (2014) Relationships between phyllosphere bacterial communities and plant functional traits in a neotropical forest. Proc Natl Acad Sci 111(38):13715–13720

Krishnamoorthy R, Kwon SW, Kumutha K, Senthilkumar M, Ahmed S, Sa T, Anandham R (2018) Diversity of culturable methylotrophic bacteria in different genotypes of groundnut and their potential for plant growth promotion. 3 Biotech 8(6):275

Kumar M, Kour D, Yadav AN, Saxena R, Rai PK, Jyoti A, Tomar RS (2019b) Biodiversity of methylotrophic microbial communities and their potential role in mitigation of abiotic stresses in plants. Biologia 74(3):287–308

Kumar S, Suyal DC, Bhoriyal M, Goel R (2018) Plant growth promoting potential of psychrotolerant Dyadobacter sp. for pulses and finger millet and impact of inoculation on soil chemical properties and diazotrophic abundance. J Plant Nutr 41(8):1035–1046

Kumar V, Soni R, Jain L, Dash B, Goel R (2019) Endophytic fungi: recent advances in identification and explorations. In: Singh B (ed) Advances in endophytic fungal research, Fungal biology. Springer, Cham. https://doi.org/10.1007/978-3-030-03589-1_13

Kumar S, Suyal DC, Yadav A, Shouche Y, Goel R (2019a) Microbial diversity and soil physiochemical characteristic of higher altitude. PLoS One 14(3):e0213844

Kushwah S, Singh AK, Chowdhury T, Soni R (2021) Comparative effect of selected bacterial consortia on yield and yield attributes of Chickpea (Cicer arietinum L.). Asian Jr Microbiol Biotech Env Sc 23(1):72–74

Kwak MJ, Jeong H, Madhaiyan M, Lee Y, Sa TM, Oh TK, Kim JF (2014) Genome information of Methylobacterium oryzae, a plant-probiotic methylotroph in the phyllosphere. PLoS One 9(9): e106704

Lee HJ, Kim SY, Kim PJ, Madsen EL, Jeon CO (2014) Methane emission and dynamics of methanotrophic and methanogenic communities in a flooded rice field ecosystem. FEMS Microbiol Ecol 88(1):195–212

Lee HS, Madhaiyan M, Kim CW, Choi SJ, Chung KY, Sa TM (2006) Physiological enhancement of early growth of rice seedlings (Oryzasativa L.) by production of phytohormone of N 2-fixing methylotrophic isolates. Biol Fertil Soils 42(5):402–408

Lindow SE, Brandl MT (2003) Microbiology of the phyllosphere. Appl Environ Microbiol 69(4):1875–1883

Liu H, Brettell LE, Singh B (2020) Linking the phyllosphere microbiome to plant health. Trends Plant Sci 25(9):841–844

Madhaiyan M, Poonguzhali S, Kwon SW, Sa TM (2009) Methylobacterium phyllosphaerae sp. nov., a pink-pigmented, facultative methylotroph from the phyllosphere of rice. Int J Syst Evol Microbiol 59(1):22–27

Madhaiyan M, Poonguzhali S, Sa TM (2007) Influence of plant species and environmental conditions on epiphytic and endophytic pink-pigmented facultative methylotrophic bacterial populations associated with field-grown rice cultivars. J Microbiol Biotechnol 17(10):1645–1654

Mendes R, Garbeva P, Raaijmakers JM (2013) The rhizosphere microbiome: significance of plant beneficial, plant pathogenic, and human pathogenic microorganisms. FEMS Microbiol Rev 37(5):634–663

Mwajita MR, Murage H, Tani A, Kahangi EM (2013) Evaluation of rhizosphere, rhizoplane and phyllosphere bacteria and fungi isolated from rice in Kenya for plant growth promoters. Springerplus 2(1):606

Nag J, Dash B, Singh AK, Chowdhury T, Gupta SB, Soni R (2018) Effect of plant growth-promoting rhizobacterial composite culture on the growth of chickpea seedlings. Biologija 64(2):129–136

Nalini MS, Prakash HS (2017) Diversity and bioprospecting of actinomycete endophytes from the medicinal plants. Lett Appl Microbiol 64(4):261–270

Osono T (2008) Endophytic and epiphytic phyllosphere fungi of Camellia japonica: seasonal and leaf age-dependent variations. Mycologia 100(3):387–391

Pedraza RO, Bellone CH, de Bellone SC, Sorte PMFB, dos Santos Teixeira KR (2009) Azospirillum inoculation and nitrogen fertilization effect on grain yield and on the diversity of endophytic bacteria in the phyllosphere of rice rainfed crop. *european*. Journal of soil Biology 45(1):36–43

Raaijmakers JM, Mazzola M (2012) Diversity and natural functions of antibiotics produced by beneficial and plant pathogenic bacteria. Annu Rev Phytopathol 50:403–424

Rascovan N, Carbonetto B, Perrig D, Díaz M, Canciani W, Abalo M et al (2016) Integrated analysis of root microbiomes of soybean and wheat from agricultural fields. Sci Rep 6:28084
Ripa FA, Cao WD, Tong S, Sun JG (2019) Assessment of plant growth promoting and abiotic stress tolerance properties of wheat endophytic fungi. Biomed Res Int 2019:6105865
Roman-Reyna V, Pinili D, Borjaa FN, Quibod I, Groen SC, Mulyaningsih ES, et al. (2019). The rice leaf microbiome has a conserved community structure controlled by complex host-microbe interactions. CELL-HOST-MICROBE-D-19-00340
Sahu B, Singh AK, Chaubey AK, Soni R (2020) Effect of rhizobium and phosphate solubilizing bacteria inoculation on growth and yield performance of Lathyrus (Lathyrus sativus L.) in Chhattisgarh plains. Curr J Appl Sci Technol 39(48):237–247
Sivakumar N, Sathishkumar R, Selvakumar G, Shyamkumar R, Arjunekumar K (2020) Phyllospheric microbiomes: diversity, ecological significance, and biotechnological applications. In: Plant microbiomes for sustainable agriculture. Springer, Cham, pp 113–172
Soni R, Suyal DC, Sai S, Goel R (2016) Exploration of nifH gene through soil metagenomes of the western Indian Himalayas. 3 Biotech 6:25
Soni R, Kumar V, Suyal DC, Jain L, Goel R (2017) Metagenomics of plant rhizosphere microbiome. In: Singh R, Kothari R, Koringa P, Singh S (eds) Understanding host-microbiome interactions - an omics approach. Springer, Singapore. https://doi.org/10.1007/978-981-10-5050-3_12
Soni R, Suyal DC, Sahu B, Phulara SC (2021) Metagenomics. In: Verma A, Saini JK, Hesham AE-L, Singh HB (eds) Phytomicrobiome interactions and sustainable agriculture. https://doi.org/10.1002/9781119644798.ch3
Stadler B, Müller T (2000) Effects of aphids and moth caterpillars on epiphytic microorganisms in canopies of forest trees. Can J For Res 30(4):631–638
Steven B, Huntley RB, Zeng Q (2018) The influence of flower anatomy and apple cultivar on the apple flower phytobiome. Phytobiomes 2(3):171–179
Straub D, Rothballer M, Hartmann A, Ludewig U (2013) The genome of the endophytic bacterium H. frisingense GSF30T identifies diverse strategies in the Herbaspirillum genus to interact with plants. Front Microbiol 4:168
Suyal DC, Joshi D, Debbarma P, Soni R, Das B, Goel R (2019) Soil metagenomics: unculturable microbial diversity and its function. In: Varma A, Choudhary D (eds) Mycorrhizosphere and Pedogenesis. Springer, Singapore. https://doi.org/10.1007/978-981-13-6480-8_20
Suyal DC et al (2021a) Phosphate-solubilizing fungi: current perspective and future need for agricultural sustainability. In: Yadav AN (ed) Recent trends in mycological research, Fungal biology. Springer, Cham. https://doi.org/10.1007/978-3-030-60659-6_5
Suyal DC, Prasad P, Sahu B, Soni R, Goel R (2021b) Rhizosphere fungi and their plant association: current and future prospects. In: Sharma VK, Shah MP, Parmar S, Kumar A (eds) Fungi bio-prospects in sustainable agriculture, environment and nano-technology. Academic Press, pp 339–356
Trivedi P, Leach JE, Tringe SG, Sa T, Singh BK (2020) Plant–microbiome interactions: from community assembly to plant health. Nat Rev Microbiol 18(11):607–621
Vacher C, Hampe A, Porté AJ, Sauer U, Compant S, Morris CE (2016) The phyllosphere: microbial jungle at the plant–climate interface. Annu Rev Ecol Evol Syst 47:1–24
Venkatachalam S, Ranjan K, Prasanna R, Ramakrishnan B, Thapa S, Kanchan A (2016) Diversity and functional traits of culturable microbiome members, including cyanobacteria in the rice phyllosphere. Plant Biol 18(4):627–637
Verma P, Yadav AN, Khannam KS, Kumar S, Saxena AK, Suman A (2016) Molecular diversity and multifarious plant growth promoting attributes of bacilli associated with wheat (Triticum aestivum L.) rhizosphere from six diverse agro-ecological zones of India. J Basic Microbiol 56(1):44–58
Verma P, Yadav AN, Khannam KS, Panjiar N, Kumar S, Saxena AK, Suman A (2015) Assessment of genetic diversity and plant growth promoting attributes of psychrotolerant bacteria allied with

wheat (*Triticum aestivum*) from the northern hills zone of India. Ann Microbiol 65(4):1885–1899
Vorholt JA (2012) Microbial life in the phyllosphere. Nat Rev Microbiol 10(12):828–840
Whipps J, Hand P, Pink D, Bending GD (2008) Phyllosphere microbiology with special reference to diversity and plant genotype. J Appl Microbiol 105(6):1744–1755
Yadav AN (2017) Agriculturally important microbiomes: biodiversity and multifarious PGP attributes for amelioration of diverse abiotic stresses in crops for sustainable agriculture. Biomed J Sci Tech Res 1:1–4
Yadav AN, Verma P, Kumar V, Sangwan P, Mishra S, Panjiar N et al (2018) Biodiversity of the genus Penicillium in different habitats. In: New and future developments in microbial biotechnology and bioengineering. Elsevier, Amsterdam, pp 3–18
Yadav N, Yadav AN (2019) Actinobacteria for sustainable agriculture. J Appl Biotechnol Bioeng 6: 38–41
Yang JH, Liu HX, Zhu GM, Pan YL, Xu LP, Guo JH (2008) Diversity analysis of antagonists from rice-associated bacteria and their application in biocontrol of rice diseases. J Appl Microbiol 104(1):91–104
Yasmin S, Hakim S, Zaheer A, Mirza B, Mirza MS (2020) Metagenomic analysis of bacterial community associated with rhizosphere and Phyllosphere of basmati Rice. bioRxiv
Zhang N, Luo J, Rossman AY, Aoki T, Chuma I, Crous PW et al (2016) Generic names in Magnaporthales. IMA Fungus 7(1):155–159
Zhao Z, Ge T, Gunina A, Li Y, Zhu Z, Peng P et al (2019) Carbon and nitrogen availability in paddy soil affects rice photosynthate allocation, microbial community composition, and priming: combining continuous 13 C labeling with PLFA analysis. Plant Soil 445(1):137–152

Cold-Adapted Microorganisms: Survival Strategies and Biotechnological Significance

16

Hemant Dasila, Damini Maithani, Deep Chandra Suyal, and Prasenjit Debbarma

Abstract

Thermal stress either cold or heat stress is one of the major factors that influence microorganisms, and to survive from these adverse conditions, microorganisms have to adapt different survival strategies. Some major survival strategies adapted by the microorganisms in response to cold stress are metabolic adaptations, change in cell membrane structure and functions such as membrane fluidity, molecular adaptations that includes change in gene expression, production of cold-adaptive enzymes, and the production of compounds like cryoprotectants that protects microorganisms from these adverse effects. These survival strategies represent bacterial adaptations, and those microorganisms that have potential to adapt better in these survival conditions are most likely to survive in cold stress. Cold-adaptive microorganisms possess high survival instincts and simultaneously offer numerous advantages in pursuits of biotechnological advances. This advancement includes production of cold-adaptive enzymes and proteins, and these proteins and enzymes are very important for commercial purpose that contribute to Indian economy too.

Hemant Dasila and Damini Maithani contributed equally with all other contributors.

H. Dasila · D. Maithani
Department of Microbiology, C.B.S.H., G. B. Pant University of Agriculture and Technology, Pantnagar, Uttarakhand, India

D. C. Suyal
Department of Microbiology, Akal College of Basic Sciences, Eternal University, Baru Sahib, Sirmaur, Himachal Pradesh, India

P. Debbarma (✉)
School of Agriculture, Graphic Era Hill University, Dehradun, Uttarakhand, India

R. Goel et al. (eds.), *Survival Strategies in Cold-adapted Microorganisms*,
https://doi.org/10.1007/978-981-16-2625-8_16

Keywords

Thermal stress · Metabolic · Cryoprotectants · Biotechnological

16.1 Introduction

To survive in different conditions, microorganisms have to adapt different mechanisms to counter any environmental stress. These stresses may include thermal stress, osmotic stress, and nutrient stress. Temperature plays an important role in bacterial growth, and their survival can severely affect functioning and productivity of an ecosystem. Temperature plays a very important role in regulating the growth of microorganisms as it directly affects the cellular constituents, for instance, macromolecules like proteins and enzymes, which are the key players in determining the rate of a metabolic reactions. Arrhenius equation provides the significant relationship between temperature and rate of reaction:

$$k = Ae^{-Ea/RT}$$

where K = rate constant, Ea = activation energy, R = universal gas constant, A = constant of steric collision and frequency, and T = absolute temperature.

Generally, for most of the enzymes, activation energy is about 420 KJ mol^{-1}, so when temperature drops from 20 °C to 0 °C, there is fourfold decrease in enzymatic activity. Thus, temperature plays a very important role in regulating cell overall biochemistry (Russell 1984). Freezing temperatures can cause decreased growth rate, reduced enzymatic activity, and altered physiological and cellular properties of microorganisms.

In most part of the earth, cold environment is predominant in nature contributing about 80% of total aquatic and terrestrial area (Yadav et al. 2017). A vast amount of the information regarding bacterial adaptation to low temperatures can be generated by analyzing bacterial population with molecular tools and techniques such as MALDI-TOF, metagenomics, genome mining, etc. (Pandey et al. 2019). In cold stress several modifications occur in microorganisms that include changes at cellular, physiological, metabolic, and molecular level (Phadtare 2004). To thrive in such inhospitable conditions, cold-adapted microorganisms use a vast array of strategies that include maintenance of membrane fluidity, accumulation and synthesis of compatible solutes, antifreezing compounds, differential gene expression, change in protein as well as total soluble protein within the cell, RNA degradasomes, and ice nucleator proteins (Mishra et al. 2010). Microorganisms that are mostly predominant in cold environment are categorized as psychrophiles and psychrotolerant. According to Schmidt-Nielsen (1902), term psychrophile is used for those microorganisms which are able to grow at 0 °C. Stokes, in 1963, termed psychrophiles as those microorganisms which are able to grow at 0 °C and macroscopically visible within 1 week. Psychrophiles are those which can grow at temperatures ranging from −20 to 25 °C. These groups of microorganisms however are unable to grow at temperatures above 15 °C, whereas psychrotolerant microorganisms grow optimally at 20–25 °C and are metabolically active even at 0 °C (Santiago et al. 2016). Terms like *Stenopsychrophiles* and *Eurypsychrophiles* are often used to indicate narrow and wide range of temperature tolerance of microorganisms, respectively (Atlas and Bartha 1998). Obligate psychrophiles can be classified as stenothermal psychrophiles, whereas facultative psychrophiles, often termed as psychrotrophs, are generally classified as eurythermal psychrophiles (Feller and Gerday 2003).

Due to their broad range of temperature tolerance, psychrotolerant microorganisms are more abundant than psychrophilic microorganisms (Yadav et al. 2017). Extreme environments like temperature regions are hotspot for microbial diversity. Cold-tolerant microbial communities include archaea, fungi, algae, and different phyla of domain bacteria (Zachariah and Das 2017). While bacteria dominate and are present in greater diversity than archaea in polar environments, archaea are widespread in cold, deep ocean waters (Mishra et al. 2010). *Flavobacterium, Janthinobacterium, Kocuria, Lysinibacillus, Methylobacterium, Microbacterium, Pseudomonas, Psychroflexus, Paenibacillus, Halorubrum, Arthrobacter, Providencia, Brevundimonas, Serratia, Citricoccus, Azotobacter, Clostridium, Exiguobacterium, Hydrogenophaga, Burkholderia, Enterobacter, Azospirillum, Pseudoalteromonas, Moraxella, Psychrobacter, Polaromonas, Polaribacter, Moritella, Vibrio, Bacillus* and *Micrococcus, Methanogenium, Methanococcoides, Candida, Cryptococcus, Penicillium, Cladosporium,* and *Chloromonas* species are some of the microorganisms that are well adapted to colder regions (Feller and Gerday 2003; Bhandari et al. 2020). Yeasts have been reported to be better adapted to low temperatures in comparison to bacteria (Buzzini and Margesin 2014). These cold-adapted microorganisms are gaining attention as they are a treasure trove of bioactive compounds and enzymes of biotechnological and industrial significance (Santiago et al. 2016).

16.2 How Does Psychrophiles and Psychrotrophs Grow at Lower Temperature?

Adaptive changes in cellular constituents, viz., proteins and lipids, are key determinants that are used by psychrophiles and psychrotrophs. These adaptations may be phenotypic or genotypic. Genotypic adaptations are those adaptations that arise over an evolutionary time scale which are easily observed in interspecies, and phenotypic adaptations are those that occur within the lifetime of organisms which may be seasonal or not. Genotypic adaptations of organisms are the results of Darwinian selection that are adapted by organisms in continued stress. Genotypic adaptation of psychrophiles and psychrotrophs represents sum of end points in phenotypic adaptations also, and different organisms reach to this end point, viz., combination of phenotypic as well as genotypic adaptation in different time and

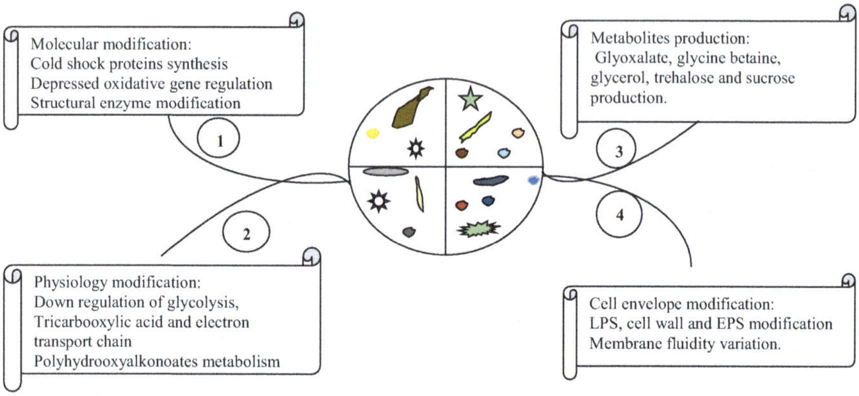

Fig. 16.1 Survival strategies adopted by cold native microorganisms

different ways, and their extent level is also different. However phenotypic adaptations provide an advantage by providing the mechanisms that are involved (Hochachka and Somero 1984). Various changes at structural, physiological, and molecular level occur such as regulation of metabolism in response to the low temperature conditions, maintenance of cellular integrity, etc. An increased expression of enzymes and pigments has been reported in microorganisms native to cold climates. Low temperatures also induce a shift in carbon source utilization as well increase their susceptibility toward antibiotics (Mishra et al. 2010). *Cryptococcus* genus has been reported to be dominantly present in cold regions due to its ability to produce a polysaccharide capsule which serves as an additional survival strategy for the organism (Buzzini and Margesin 2014) (Fig. 16.1).

16.3 Role of Metabolic Pathway in Cold Adaptation

With decrease in temperature, growth rate decreases as lag phase of bacteria extends before growth which finally results in decrease in cell number. During lag phase some major physiological changes occur that include decrease in saturation of fatty acid and synthesis of DNA, RNA, and proteins stops (Berry and Foegeding 1997). With decrease in temperature, major pathway like glycolysis, tricarboxylic acid (TCA), and electron transport chain get depressed. Analysis of psychrophile *Psychrobacter arcticus* 273–4 revealed that it can grow at temperature of 10 °C and lacks major genes that are responsible for glycolysis and lacks phosphotransferase system but possesses enzymes for gluconeogenesis pathway. Phenotypic analysis of this bacteria revealed that major energy metabolism pathway depends upon acetate pathway as acetate can easily diffuse through the membrane. Metabolic activity at given temperature is a function of two variables, viz., temperature function (Q_{10}) and temperature characteristics (μ). Most of the substrates have temperature coefficient value lower for psychrophiles as compared to mesophiles

(Ingraham and Bailey 1959). Microorganisms have been isolated from cold stress condition, and their metabolic activity has been checked to find out basic and advanced changes that are occurring during thermal adaptability. In a given research, a strain has been isolated from Arctic sea, e.g., *Psychromonas ingrahamii*, and its generation time was measured and was found to be 24° h at −12 °C which is much higher when compared to generation time at 5 °C that was 12 h, and this was the first case of growth analysis at −12 °C (Breezee et al. 2004). Using cell-free system study, it was found that rate of transcription and translation is relatively slow when compared to its optimum temperature.

16.4 Polyhydroxyalkonoate Metabolisms

Several microorganisms seem to synthesize polyhydroxyalkonoates (PHA) that are considered to be a reserve polymer under cold stress. This PHA seems to play a very important role in survival strategy adapted by microorganisms under cold stress. Genomic analysis of *C. psychrerythraea* revealed that there is a duplication of gene like enoyl-CoA hydratase, and acyl CoA dehydrogenase occurs that shows number of versatile compounds that can be synthesized by PHAs (Methe et al. 2005). Proteome analysis of *S. alaskensis* suggests that there is increase in number of enzymes at low temperature that are related to PHA metabolism, and these enzymes seem to compensate the reduced enzymatic activity or poor nutritional transport as a survival strategy acquired by *S. alaskensis* under cold stress (Ting et al. 2010). Among all the proteins, the major protein was found to be phasin which is a granule-associated protein that plays an important role in stress protection (Mezzina and Pettinari 2016). This finding suggests that cold adaptation in *S. alaskensis* not only involves de novo synthesis of PHA but also involves enzyme secretion for scavenging extracellular PHA (Ting et al. 2010).

16.5 Temperature-Induced Modifications in Cell Membrane

Cell membrane consists of phospholipids that are arranged in the form of bilayer with polar head groups which are at periphery and thus can interact with aqueous phase inside and outside the cell (Neidhart et al. 1990). It is a very well-established fact that microorganisms are able to adjust their lipid composition within membrane in response to fluctuation in temperature. This helps bacteria to maintain the solute transport rate and enzymatic activity in such extreme conditions (Brown and Minnikin 1973). At low temperatures, changes tend to occur in fatty acid and glycolipid compositions that lead to change in membrane fluidity (Berry and Foegeding 1997). These changes cause decreased membrane permeability and impaired membrane functions due to low temperature-induced gel-phase transitions; therefore, in order to counter these effects, the lipid bilayer must have proper fluidity under freezing conditions (Feller and Gerday 2003). Generally, at low temperatures, unsaturation of fatty acid within the cell membrane and reduced fatty acid chain length are observed which result in an increased membrane fluidity as increased unsaturation destabilizes the steric constraints and improves their mobility by decreasing packing density. In *Clostridium botulinum* with drop in temperature from 37 °C to 8 °C, level of unsaturation in fatty acid chain occurs from 27% to 40% with the help of desaturase enzyme that is localized within the membrane itself (Berry and Foegeding 1997). Average fatty acid chain length is also shortened that results in fewer carbon to carbon interaction which allow membrane fluidity to increase in the cell membrane (Russell 1990). Increased rate of unsaturation of fatty acids in cold stress has been reported in one of the Gram-positive strain, e.g., *Micrococcus roseus*, and a Gram-negative bacterial strain, e.g., *Sphingobacterium antarcticus* (Chattopadhyay and Jagannadham 2001).

Another mechanism for cold adaptation was observed in *Salmonella* spp. and *C. botulinum*, where an increase in branched chain fatty acids and decreased numbers of cyclic fatty acids were observed (Russell 1984; Evans et al. 1998). To maintain the content of fatty acids in response to cold stress, increase in one fatty acid content may result in decrease of other fatty acid contents and vice versa. For example, as there is a decreased amount of laurate in the cell in response to cold stress, an increase in palmitoleate content was observed (Carty et al. 1999). Since palmitoleate is an unsaturated fatty acid, this increase clearly indicates that unsaturation in fatty acid content is directly linked to membrane fluidity as it lowers the phase transition. Certain research reported that in response to cold stress, acyltransferase Lpx P gets activated, and this seems to be the responsible for attaching palmitoleate to lipid A (Vorachek-Warren et al. 2002). In *Bacillus subtilis* desaturation of membrane fatty acid occurs in already existing phospholipids, and this was done by desaturase (Des) which is regulated by DesK sensor kinase and DesR response regulators (Aguilar and De Mendoza 2006). When there is a shift toward cold temperature, there is activation of DesK kinase; DesK phosphorylates the Des R which is a transcriptional activator which further activates des gene promoter, and in return this des gene promoter expresses D5-desaturase. This D5-desaturase catalyzes the addition of double bond in already existing fatty acid chain inside the cell (Aguilar and De Mendoza 2006; Albanesi et al. 2004).

Microorganisms (*Psychrotolerant* and *Psychrophiles*) that are adapted to low temperature tend to have reduced number of branched chain lipid and increased number of unsaturated fatty acid as compared to mesophiles and thermophiles (Morita 1975). In thermophiles, there is high level of saturated and branched chain fatty acid to make membrane stiff at higher temperatures. Membrane transport proteins also play an important role during survival of microorganisms at different temperatures, and *Psychrophiles* thus have much more efficient mechanisms of adaptation than mesophiles (Jay 1986). Envelope of Gram-negative bacteria consists of inner and outer membrane separated via periplasmic space. Outer membrane of Gram-negative consists of lipopolysaccharide, proteins, and phospholipids. LPS contain lipid A anchored to the membrane, external O-polysaccharide, and intermediate oligosaccharide component. In case of *P. haloplanktis*, high percentage of cell envelope gene is observed in the genome and concluded that it might somehow help in countering the effect of cold (Médigue et al. 2005). A similar result was observed when *S. alakensis* drops to temperature of 10 °C. Its proteome analysis revealed that there is upregulation severalfold of genes that are related to envelope biogenesis and exopolysachharide biosynthesis (Ting et al. 2010).

In the case of Gram-positive bacteria, the cell envelope bacteria consist of cell wall and inner membrane. Under cold stress thickening of cell wall occurs like increase in cell wall biosynthesis gene observed in *E. sibiricum* at 2.5 °C, which suggests that thickening of cell wall prevents the cell from disruption during ice formation (Rodrigues et al. 2008). *Planococcus halocryophilus* has an unusual cell envelope which consists of characteristic surrounding encrustation when put under cold stress. Transcriptomic analysis of this microorganisms also revealed that genes that are relevant to peptidoglycan synthesis and its precursor compounds were upregulated. Microscopic analysis results show that there is increase in hydrophobicity content with 20% calcium carbonate, 29% choline, and 50% increase in peptidoglycan along with increase of several copies of gene encoding carbonic anhydrase which is responsible for mineralization of calcium carbonate. There is another surprising feature of *P. halocryophilus* at zero temperature which is an increase in fatty acid saturation percentage with decrease in temperature (Mykytczuk et al. 2013). It has been postulated that *P. halocryophilus* adapts a different mechanism for preserving membrane fluidity (Ronholm et al. 2015). EPS (exopolysaccharide) is a constituent of extracellular polymers surrounding bacterial cells and also one of the important factors that are required to cope with cold environments (Mancuso Nichols et al. 2004). *Pseudoalteromonas* sp. that is isolated from Arctic marine regions produces a complex EPS that contains mannose as one of the main components. This finding suggests that EPS production along with some modification in it is the key feature shared by cold-adaptive microorganisms (Corsaro et al. 2004).

16.6 Enzyme Adaptations in Cold Stress

Enzyme plays a very crucial role in all the metabolic and molecular activities that are occurring inside the cells under low temperature conditions. Cold stress hampers the normal metabolic functioning and chemical reactions in microorganisms; thus it becomes important for the psychrophiles and psychrotolerant microorganisms to maintain an optimally working enzymatic machinery for catalyzing the reactions at normal rate. The microorganisms occupying these places express cold-active enzymes that are heat labile and maintain high activity, i.e., ten times higher than their mesophilic homologues under low temperature conditions (Feller and Gerday 2003). For instance, a cold-active superoxide dismutase was isolated from *Deschampsia antarctica.* The enzyme was found to be highly thermostable, showing its optimal activity at 20 °C. However, it retains 80% of its activity when temperature is reduced to 0 °C (Rojas-Conteras et al. 2015). High activity is generally attained in these enzymes by destabilization of their active sites which increases the flexibility of its catalytic center under freezing conditions. Thus, particular enzymes for a particular physiological process are expressed under cold stress, and ribonuclease

(Reddy et al. 2004), alkaline phosphatase (Chattopadhyay et al. 1995), and DNA-dependent RNA polymerase (Uma et al. 1999) have been identified already from Antarctic regions. The nature of both thermal adaptive enzymes and cold-adaptive enzymes is different, viz., generally thermophilic enzymes are poor catalyst, whereas cold stress enzymes require greater structural flexibility (Gerday et al. 2000). Attempts have been made to find out the difference between nature and functionality of enzymes that are isolated from mesophilic bacteria and cold stress bacteria. Amino acid profiles of cold-adapted enzymes indicate that there is a lower content of proline and arginine that cause restrictions in free backbone movements. In addition to that, there is a decreased hydrophobic amino acid content in comparison to polar amino acid with fewer disulfide bond residues (D′ Amico et al. 2002). All the weak interactions are minimized, and protein interior is generally less compact due to decreased hydrophobicity of the nonpolar core (Feller and Gerday 2003). β-Galactosidase isolated from psychrotolerant bacteria, e.g., *Arthrobacter*, retain a 50% activity when it is fed from 18 °C to 0 °C, and this activity was surprisingly 5.0 times higher than mesophilic bacteria, e.g., *E.coli* at 20 °C and 10 °C, respectively (Coker et al. 2003). This shows the adaptability of microorganisms with respect to cold stress. Besides, in comparison to their mesophilic counterparts, ΔG and ΔH values are lower in the case of reactions catalyzed by enzymes that are active under cold stress (Feller and Gerday 2003). Research postulates that there is a cooperative as well as synergistic role of intramolecular interaction that somehow promotes the thermal adaptability in microorganisms (Wintrode et al. 2000; Zartler et al. 2001). Low stability of cold-active enzymes is indicated by the fact that they show cooperativity during unfolding of the enzyme. Besides, these show high degree of unfolding reversibility as well (Feller and Gerday 2003).

To maintain the demand supply of cold-active enzymes for biotechnological and industrial purposes, various methods are being used for production such as use of molecular chaperones, stimulation of cold-active promoters, optimized heterologous host, etc. (Santiago et al. 2016). Cold-active enzymes that are active at moderate as well as high temperatures are beneficial for economic purposes and thus are helpful in various industrial processes. For instance, adding enzymes in detergents suitable for washing clothes at low temperature in order to protect their color, texture, and quality is more profitable as mesophilic enzymes are not optimally active at low temperatures. Similarly, using cold-active proteases helps to perform the process of peeling in leather industry by using tap water instead of heating at 37 °C which is required while working with mesophilic proteases (Feller and Gerday 2003).

16.7 Compounds Involved in Cold Adaptation

Many microorganisms that are adapted to cold stress tend to have accumulation of compatible solutes. These compatible solutes not only play an important role in cryoprotection and osmoprotection but also serve as nitrogen, carbon, and energy sources (Methe et al. 2005). Some important compatible solutes include trehalose, sucrose, glycine, betaine, glycerol, sorbitol, and mannitol. These compatible solutes tend to decrease freezing point of the cytoplasm and also prevent macromolecular aggregation and stabilize cellular membrane under cold stress (Collins and Deming 2013). Glycine betaine was detected in *L. monocytogenes*, a food-borne pathogen which survives at low temperatures (Angelidis and Smith 2003). Glycine betaine acts as a molecular chaperone that prevents protein aggregation and also helps in maintaining membrane fluidity at low temperatures (Chattopadhyay 2002a). Trehalose, another cryoprotectant used by microorganisms native to low temperature regions, is a nonreducing sugar that stabilizes the membrane and proteins. Presence of exogenous trehalose is also helpful in providing protection against freezing conditions. Enzymes, trehalose-6-phosphate synthases, and trehalose-6-phosphatase are key players in trehalose biosynthesis which are encoded by genes *ots*A and *ots*B (Mishra et al. 2010). Cryoprotectants, viz., chemical substances, also protect the microorganisms from cold stress. These substances include sugars (fructose, glucose), amino acids (proline, alanine), and sugar alcohol (glycerol, mannitol). Cryoprotective role of glycine betaine has been observed in bacteria (Chattopadhyay 2002b). Glycine betaine seems to prevent aggregations of proteins that otherwise accumulate in cold temperature and promotes membrane fluidity. Two strains isolated from Antarctica, viz., *Pseudomonas haloplanktis* TAC125 and *Pseudoalteromonas* sp. TB 41, show remarkable difference in glutathione metabolism in cold stress, and this glutathione seems to promote cryotolerance in *Pseudomonas haloplanktis* TAC125 at much higher efficient rate by serving as compatible solute (Mocali et al. 2017). Strain N33 also accumulates large amount of compatible solutes like valine, threonine, and sarcosine which seems to work as cryoprotectant (Ghobakhlou et al. 2015).

16.8 Carotenoids' Role in Cold Adaptations

Carotenoids have been reported to occur in plants and some bacteria as well as fungi. Most unique feature of these molecules is presence of conjugated double bonds in the backbone structure. Due to the presence of highly delocalized π-electrons, these compounds are able to absorb the radiation in the wavelength region ranging from 400 to 500 nm. Major role of carotenoids in microorganisms is to prevent oxidative damage from reactive oxygen species (ROS) (Moliné et al. 2014). Role of carotenoids is also one of the important mechanisms adapted by microorganisms to counter cold stress. Strains like *Sphingobacterium antarcticus* and *Micrococcus roseus* isolated from Antarctic regions when subjected to fractionation seem to have carotenoids attached to membrane and promote the membrane fluidity. There is also increase in the polar carotenoids content, but amount of nonpolar carotenoid decreases. Role of polar carotenoids is somehow not clear, but it was postulated that in response to increase in unsaturated fatty acid content to increase membrane fluidity in cold stress, polar carotenoid amount increases to rigidify membrane (Chattopadhyay and Jagannadham 2001). *Phaffia rhodozyma*, a cold-adapted yeast, has been reported to be the only known species to produce astaxanthin,

Table 16.1 Some important survival strategy adapted by microorganisms with representations of microorganisms

Cold microorganism adaptations	Example	References
1. Enzyme adaptations. (a) Amino acid composition change. (b) Production of cold adaptive enzyme	*Arthrobacter psychrolactophilus* • aminopeptidase • β-galactosidase • ribonuclease • protein-tyrosine phosphatase • alkaline phosphatase • DNA-dependent RNA polymerase	Coker et al. (2003) Huston (2008) Groudieva et al. (2004) Reddy et al. (2004) Tsuruta et al. (2004) Chattopadhyay et al. (1995) Uma et al. (1999)
2. Fatty acid composition. (a) Increase in unsaturation of fatty acid in the membrane. (b) Shortening of fatty acid chain length	Gram-positive *Micrococcus roseus* Gram-negative *Sphingobacterium antarcticus* *Listeria* sp.	Chattopadhyay and Jagannadham (2001) Russell (2002)
3. Structural change. (a) Cell size increases. (b) Filament formation	*Candida utilis* *Pseudomonas putida*	Herbert (1986) Phillips et al. (1998)
4. Carotenoid production	Dihydroxycarotenoids Zeaxanthin Violaxanthin	Subczynski et al. (1992)
5. Cold shock protein. (a) CspA. (b) Hsc 25. (c) CspC. (d) CspD. (e) CspE	*E. coli* *Pantoea ananas* *C. Botulinum* *E. coli* *E. coli*	Ray et al. (1994) Kawahara et al. (2000) Derman et al. (2015) Uppal et al. (2008) Feng et al. (2001)
6. Cryoprotectant production. Glycine betaine	*L. monocytogenes*	Koo et al. (2016)
7. Antifreeze protein production (AFP)	*Marinomonas primoryensis*	Gilbert et al. (2005)
8. Heat shock protein. ClpB Htp G	*Synechococcus* PCC 7942	Porankiewicz and Clarke (1997) Hossain and Nakamoto (2003)

representing 80% of the total carotenoids produced by it (Moliné et al. 2014) (Table 16.1).

16.9 Potential Role of Cold Microorganisms in Biotechnology

Several studies are ongoing to find out the important role of cold-adaptive microorganisms in industry sector. Cold-adaptive microorganisms offer so many advantages to different industry sectors, and one of the main sectors is biotechnology sector. Cold-adaptive microorganisms are employed in different biotechnology sectors to produce some important cold-adaptive proteins, some important biotechnology vectors that are associated with bioremediation and disease control. This property of cold-adaptive microorganisms is also very important for agriculture sector which directly contributes to countries' economy too. Some important properties of cold-adaptive microorganisms are as follows.

16.9.1 Saving Energy

Microorganisms that are adapted to low temperature (0–20 °C) are able to grow and perform various enzymatic activities at this range only; this turns out to be an advantage because other microorganisms are not able to interfere; as a result source of contamination is less. This offers an advantage to various biotechnology processes by reducing process time in removing contamination source and thus plays a very important role.

16.9.2 Promoting Plant Growth in Cold Regions

Due to poor soil nutrient availability, frost and freezing conditions, low soil fertility, and moisture content, productivity and yield of agriculturally important crops in cold regions are hampered. Plants suffer chilling, wilting chlorosis, etc. due to low temperatures. Some cold-adapted microorganisms colonize the plant rhizosphere under cold stress conditions and help the plant to combat the stress in direct or indirect manner like other plant growth-promoting microorganisms. Thus, microbial inoculants from these extreme conditions possessing beneficial attributes can be efficiently utilized as biofertilizers or biocontrol agents for promoting growth and yield of crops grown in cold regions (Bhandari et al. 2020). These microorganisms promote plant health by either frost injury protection, growth stimulation, or protection from pests and diseases. For instance, Sinorhizobium meliloti was reported to improve growth of under cold and anaerobic conditions, i.e., ice encasement (Prévost et al. 2003). Similarly, Pseudomonas lurida isolated from rhizosphere Himalayan plants was found to protect plant from chilling stress (Bisht et al. 2014). Some members of genus Pseudomonas can cause frost damage to crops, and this could be prevented by inoculating ice mutant strains, as these strains are further prevented from ice-positive strains, viz., Pseudomonas syringae (Herbert 1992).

16.9.3 Enzyme Production

Psychrophiles or psychrotrophs seemed to have very potential role in having higher number of lipase and protease activity. Lipases are involved in recovery of silver from X-ray film and integral part of food industry. Proteases are used in beer treatment, in bakeries, in production and maturation of cheese, and in production of fermented foods. Production of both enzymes can be enhanced by cold-adaptive microorganisms; thus they became a very important part of food industry and are also used as detergent additives. The first cold-adapted enzyme was isolated from Antarctic bacteria which are further sequenced, cloned, and expressed in a recombinant host for production of subtilisins, α-amylase, and lipases, and all these enzymes are well characterized as true representative of industrial enzymes (Hoag 2008).

16.9.4 Biodegradation and Bioremediation

Biodegradation of organic contaminant in cold environment is stimulated by indigenous psychrophiles. These Psychrophiles can degrade organic contaminant to less harmful substances which are on later stage integrated to biogeochemical cycles. In most cases in cold environments, petroleum hydrocarbon bioremediation is focused because these cold regions are very much exposed to petroleum transportation and its production. These environments include alpine, permafrost, sea, sediments, and polar regions where temperature rarely exceed 10 °C. There are large number of bacteria and fungi strains that have been reported to have a potential role to bioremediate petroleum hydrocarbon (Yeargeau et al. 2009). Microorganisms that are adapted to these cold environments face many challenges like increased viscosity of liquid hydrocarbon, reduced rate of enzyme catalyzed reaction, limited bioavailability of contaminants, poor nutrient supply, and extreme pH and salinity (Aislabie et al. 2006). The most common widely used method of bioremediation in cold soil is biostimulation of indigenous microorganisms with good supply of appropriate nutrients (Walworth et al. 2007). Bioaugmentation is another method to treat petroleum contaminant where addition of cultured microorganisms is done at subsurface of soil or ground water to remove contaminants, and several studies have been done in Alaska, Greenland, and Canada, but this strategy is not well established as it gives no better results than fertilization. Bioaugmentation with nonindigenous microorganisms or genetically engineered microorganisms has already been banned in Sweden, Norway, Iceland, and Antarctica (Filler et al. 2009). Biodegradation under low temperature of another major pollutant (phenol is one of the major representatives of toxic aromatic compound in water bodies) removed by Psychrophilic bacteria (Rhodococcus spp.) and yeast (Rhodotorula psychrophenolica). They are able to remove 12.5–15 mM of phenol at 10 °C (Margesin et al. 2005).

16.9.5 Recombinant Protein System

Recombinant protein secretion system at low temperature is absolutely necessary for facilitating biotechnological application of psychrophiles. Production of cold-active enzymes by wild-type strain is usually very low. The first recombinant protein isolated was α-amylase from Antarctic Pseudomonas haloplanktis. This method of cold gene expression system was further optimized and developed for extracellular secretion of heterologous protein in P. haloplanktis (Cusano et al. 2006). Another recombinant protein system that works at low temperature was developed by using Shewanella sp. strain (Miyake et al. 2007) by selecting a suitable promoter and a plasmid having broad host range. Much higher production of β-lactamase was observed when strain is grown at 4 °C around 64% more than that when strain is grown at 18 °C. Expression vectors are also designed in such a manner that transcript sequences which are of microorganism interest, for example, enable microorganisms to survive in response to cold stress. A number of experiments have been designed to build a series of expression vectors which are based on cold shock protein transcript sequences, e.g., CspA in E. coli as a host to produce an improved quality of proteins and also in some research CspA transcript sequence used as a promoter sequence too (Vasina and Baneyx 1996).

16.9.6 Psychrophiles and Enzymes at Industry Level

At industry level in polar regions, the most well-characterized microorganism is Candida antarctica. This microorganism is known to produce two lipases A and B. Lipase B reported to be involved in large number of organosynthesis application which seems to be very important for pharmaceutical, food processing, or cosmetics (Babu et al. 2008). In a survey it was established that in most of the patents that are registered related to Antarctica, patents of lipases produced by C. antarctica by far are most of them in product base or processed based. This is a very good example of potential biocatalyst that is isolated from genetic resources in polar regions. The market for enzymes that are commonly used as detergents comprises of 30–40% worldwide. Among all the enzymes used in detergents, subtilisins are by far large in number. Bacillus spp. for the first time reported to produce to psychrophilic subtilisins in Antarctica (Narinx et al. 1997). Pseudoalteromonas haloplanktis reported to produce xylanases, and it is a classic example of successful biotechnological transfer of academic research to application-based industry production. Xylanases are enzymes that are responsible of breaking down β-1,4 xylan to hemicelluloses which is one of the major components in plant cell wall. It is also the most important ingredient in dough conditioner and thus seems to improve bread quality. Further careful optimization of this psychrophilic xylanase results in production of highest known psychrophilic enzyme up to date, and its products are now sold by Puratos (Belgium) (Collins et al. 2003).

16.10 Molecular Adaptation in Response to Cold Stress

Jones and others (1987) for the first time reported a change in gene expression pattern in response to cold stress in *E. coli*, and this leads to induction of those proteins that are very crucial for cold adaptation in microorganisms, and these proteins are said to be cold shock proteins (Jones et al. 1992a). Cold shock proteins can be defined as set of proteins that are involved in various important cellular process especially made at 10 °C rather than at 27 °C. Both machinery (transcription and translation) regulated in such manner that microorganisms that are exposed to low temperature get enough amount of cold shock proteins. Although the role of cold shock proteins is not well understood, the major role of cold shock protein is to recover the microorganisms from the partial block in protein synthesis during cold stress, and therefore it increases the rate of protein synthesis of those proteins that are engaged in cold stress adaptation. Mihoub et al. (2003) also reported a major shift in gene expression in *E. coli* under cold stress by using proteomics analysis. Nature of polypeptide chain in cold shock proteins can be same or different in different microorganisms. For example, cold shock proteins CspA identified in *E. coli* have a sequence similarity with two different microorganisms *Streptomyces clavuligerus* and *Bacillus subtilis*. Sequence homology of *E. coli* CspA with *B. subtilis* cold shock protein, viz., CspB, was found to be 61% (Willimsky et al. 1992), and isolated protein from *S. clavuligerus* of around 7.0 KDa has the sequence similarity of around 56% (Jones and Inouye 1994). In response to shift in temperature, many components (transcription and translation factors) of molecular machinery are still synthesized which normally block in case of those microorganisms that are not able to adapt with cold temperature stress. During lag phase when *E. coli* was subjected to a temperature of 10 °C, synthesis of cold shock proteins gets affected by regulators like guanosine 5'diphosphate 3'diphosphate (ppGpp) and 5'triphosphate 3′ diphosphate (pppGpp). Variation in the level of these regulators has been observed in low temperature. In *E.coli* trace amount of pppGpp has been found when *E. coli* is subjected to low temperature and variation in this pppGpp is directly proportional to extent of low temperature can proceed. Mutant experiment analysis provides the evidence that mutant microorganisms that do not contain a detectable amount of pppGpp when exposed to cold stress (10 °C) seem to promote induction of many cold shock proteins and increase level of many transcriptional and translational proteins. Jones et al. (1992b) suggested that there is an inverse relationship between pppGpp and cold shock proteins, viz., downregulation of pppGpp somehow upregulates the cold shock proteins at 10 °C, and it is somehow related to adaptation to cold stress. Not only cold stress but many chemicals like tetracycline and chloramphenicol which are known to inhibit bacterial protein synthesis also seem to induce many cold shock proteins. These compounds are responsible for downregulating the pppGpp level, and this reduction in pppGpp level upregulates the cold shock proteins (Van Bogelen and Neidhardt 1990).

Cold shock proteins have been found in many genera of microorganisms including *Bacillus cereus* (Berry and Foegeding 1996), *Pseudomonas fragi* (Berry and Foegeding 1997), *Trichosporon pullulans* (Juleseth and Inniss 1990), and *Vibrio vulnificus* (McGovern and Oliver 1995). More than 50 different cold shock proteins have been identified, and their number and their level of expression vary from species to species. The number of cold shock proteins and level of their expression depend upon the extent of temperature drop. For example, *Pseudomonas fragii* is known to produce 15 cold shock proteins when temperature drops from 20 °C to 5 °C (Russell 2002). Cold shock responses of psychrotrophic bacteria when fed to temperature close to freezing are different from the cold shock responses of mesophilic bacteria (Hebraud and Potier 1999). The second group of cold-induced proteins was identified as cold acclimation proteins (caps) that seem to be very similar with Csps (cold shock proteins) and continuously synthesized during prolonged growth at low temperature which can be easily differentiated from mesophiles and psychrophiles. At DNA level when subjecting *E. coli* to cold stress, DNA gets more negatively super coiled, and negative supercoiling is also somehow related to environmental factors such as change in osmolarity (Higgins et al. 1988). In cold stress stabilization of secondary structure is necessary other than it can affect the process of transcription and translation. A main function of these cold shock proteins is to prevent any such secondary structure and also facilitates the degradation of structured RNA. Csps family includes nucleic acid chaperones which target any unnecessary secondary structures. Helicase DeaD also falls under Csps family, and they are also responsible for removing any such secondary structure with the help of exoribonuclease RNase R and PNPase. RNase R have been isolated from *E. coli* that contain 3–59 exonuclease that efficiently removes double-stranded RNA (Matos et al. 2009; Phadtre 2004). The role of these cold shock proteins can be similar or different from each other in terms of their chaperone activity, and chaperone activity of these Csps is very crucial for maintaining mRNA stability. CspsA is one of the major cold shock proteins as it destabilizes secondary structure. CspE acts in an opposite manner as it binds to poly-A tail of mRNA which prevents its degradation by exoribonuclease like PNPase and RNase E (Feng et al. 2001). Thus, CspA and CspE act antagonistically but at last promote microorganism's survival under cold stress.

16.11 Bacterial Adaptations to Cold Stress

Microbiomes of the extreme environment impart important information about the critical limits for survival and adaptability of microorganisms. Bacterial communities are also adapted to cold regions in such manner that there is a functional abundance of those genes that are very important for survival in cold stress. Metagenomics analysis of cyanobacterial mats from the Antarctic and Arctic revealed that there is an abundance of genes that are involved in membrane modification, cold shock protein synthesis, and exopolysachharide synthesis (Varin et al. 2012). Several metagenomics datasets that are taken from ice-covered regions of Antarctic Lake Joyce showed that there is an induction of antifreeze proteins, cold shock proteins (CspA, CspB, CspC, CspD, CspE, and CspG), cold shock DEAD-box protein A, trehalose synthase, and ice nuclear protein (Koo et al. 2016). Although there are many research in progress to discover the variation in microbial community and adaptation in cold stress, coexistence of many metabolic groups that are yet to be identified results in complexity of understanding microbial communities. Recent study also revealed permafrost communities typically use highly diverse and complex type of biochemical process that are involved in organic matter decomposition, carbon processing, methane generation, and nitrogen cycling.

This is also one of the big reasons of complexity that arise in microbial communities which are adapted to cold stress (MacKelprang et al. 2017).

16.11.1 Heat Shock Protein Role in Cold Stress

Heat shock proteins (HSP) are ubiquitously expressed protein in microorganisms which promote growth of microorganisms against thermal stress, but also these heat shock proteins promote survival of microorganisms in cold stress. Some known heat shock proteins are expressed in *E. coli* during heating of these microorganisms at 42 °C before putting it to −80 °C, and this heating somehow promotes survival of these microorganisms at such extreme cold temperature because there is an accumulation of heat shock proteins (Chow and Tung 1998). Another HSP called Clp B tends to increase in *Synechococcus* PCC 7942, a cyanobacterial strain, when this strain is transferred from 37 °C to 25 °C. In mutant experiment where Clp B is knocked out, there is a repression in photosynthetic activity of these microorganisms observed (Porankiewicz and Clarke 1997). Another HSP, viz., Htp G from *Synechococcus* PCC 7942, is also under cold stress, and this was too confirmed by mutation experiments. This protein helps microorganism in photosynthetic activity, and this was confirmed by Western blotting (Hossain and Nakamoto 2003). Although these HSP help microorganisms to counter cold stress, these HSP not only induce cold stress condition, but also other environmental factors can initiate these HSP. These HSP seem to promote correct protein folding which gets distorted during thermal stress.

Other compounds termed as cold acclimation proteins (caps) are a set of 20 proteins that are present in cold-adapted microorganisms. These proteins ensure improved protein synthesis, continued growth, and cell cycle regulation at low temperatures (Margesin et al. 2005). Hsc 25, an example of Caps, is produced by *Pantoea ananas* KUIN-3, an ice-nucleating bacterium, that was found to be useful in refolding of cold-denatured enzymes (Kawahara et al. 2000). Cold shock proteins (Csps) are expressed in microorganisms when they are exposed to a downshift in temperature. This cold shock response is not only specific to psychrophiles or psychrotolerant microorganisms but all the microorganisms that are exposed to such temperature shifts (Mishra et al. 2010). Csps are involved in RNA folding and stabilization of secondary structures of macromolecules and regulate protein synthesis in low temperature stress. Regulation of Csps occurs at the level of transcription and translation (Horn et al. 2007). Ice nucleator proteins prevent supercooling at temperatures below 0 °C by forming crystal-like arrangements on water molecules, thus reducing the energy required for ice formation (Zachariassen and Kristiansen 2000). Ice-nucleating agents perform cellular protection by either establishing extracellular freezing instead of intracellular freezing that is lethal toward the cell or releasing the heat of fusion (Mishra et al. 2010). Based on presence of Ina protein, on bacterial cell wall microorganisms are categorized into ice-plus and ice-minus. Ina acts as a nucleating center for the formation of ice crystals.

Erwinia herbicola, *Pseudomonas*, and *Xanthomonas* have been reported as potent ice nucleators (Lindow et al. 1978; Maki et al. 1974; Obata et al. 1990).

16.11.2 Antifreeze Protein Role in Cold Stress

Antifreeze proteins (AFPs) get accumulated in cold stress and promote survival of microorganisms in cold stress. The role of (AFPs) has been already observed in blood of fishes. Not only in fishes but also in plants and in insects role of AFPs has been identified. The role of thermal hysteresis for the first time in bacteria is observed in strain of *Moraxella* sp., and not only that, the first AFPs were also found in this strain (Duman et al. 2004). Thermal hysteresis value of AFPs that are isolated from microorganisms is lower as compared to thermal value of AFPs isolated from animals. Calcium-dependent AFPs were also found in isolates of *Marimonas primoryensis* that are isolated from Antarctic regions (Gilbert et al. 2005).

16.11.3 Ribonuclease Role in Cold Stress

Degradosome is a multisubunit protein complex which is responsible for maintaining stability of RNA. This stability of RNA is somehow regulatory mechanisms in protein synthesis and in protein degradations. Degradosome protein isolated from *Pseudomonas syringae* of Antarctic regions has similar kind of degradosome that has been isolated from *E. coli*. But in addition to that, degradosome protein complex of *P. syringae* has an addition of another endoribonucleases, e.g., RNase E, RNase R, and RNA helicase. The role of RNase E is somehow not clear, but role of RNase R is to degrade any protein secondary structure that can be seen during cold stress (Purusharth et al. 2005).

16.12 Conclusion

Low temperature conditions strongly influence growth and metabolism of organisms that limit their growth. But metabolic and molecular adaptations by the microorganisms provide a very significant prospectus in terms of modification in metabolic pathway and switch on and off of particular gene through operon system or just by regulating gene at post-transcription and post-translation modification. Cold-adaptive microorganisms possess different patterns of gene expression and accumulation of cold-adaptive proteins which offer so many advantages to biotechnology sectors that result into ongoing research to identify more and more cold-accumulated proteins. Different numbers of cold-adaptive protein are yet to be identified that will unlock role of various mysterious players in cold adaptation strategy.

References

Aguilar PS, De Mendoza D (2006) Control of fatty acid desaturation: a mechanism conserved from bacteria to humans. Mol Microbiol 62(6):1507–1514

Aislabie J, Saul DJ, Foght JM (2006) Bioremediation of hydrocarbon-contaminated polar soils. Extremophiles 10:171–179

Albanesi D, Mansilla MC, De Mendoza D (2004) The membrane fluidity sensor DesK of Bacillus subtilis controls the signal decay of its cognate response regulator. J Bacteriol 186 (9):2655–2663

Angelidis AS, Smith GM (2003) Role of glycine betaine and carnitine transporters in adaptation of Listeria monocytogenes to chill stress in defined medium. Appl Environ Microbiol 69:7492–7498

Atlas RM, Bartha R (1998) Microbial ecology: fundamentals and applications, 4th edn. Benjamin/ Cummings Science Publishers, Menlo Park, CA, pp 1–704

Babu E, Ramteke PW, Thomas G (2008) Cold active microbial lipases: some hot issues and recent developments. Biotechnol Adv 26:457–470

Berry ED, Foegeding PM (1996) Cold shock proteins and cold shock domains in Bacillus cereus. In Proceedings of the 96th General Meeting of American Society of Microbiology, Washington, DC, pp 19–23

Berry ED, Foegeding PM (1997) Cold temperature adaptation and growth of microorganisms. J Food Prot 60(12):1583–1594

Bhandari G, Nautiyal N, Sharma M (2020) Assessment of genes and enzymes of microorganisms of high altitudes and their application in agriculture. In: Microbiological advancements for higher altitude agro-ecosystems & sustainability. Springer, Singapore, pp 307–326

Bisht SC, Joshi GK, Mishra PK (2014) CspA encodes a major cold shock protein in Himalayan psychrotolerant Pseudomonas strains. Interdiscip Sci Comput Life Sci 6(2):140–148

Breezee JN, Cady JT, Staley (2004) Subfreezing growth of the sea ice bacterium "Psychromonas ingrahamii". Microbial Ecol 47:300–304

Brown CM, Minnikin DE (1973) Effect of growth temperature on fatty-acid composition of some psychrophilic marine pseudomonads. J General Microbiol 75:R9

Buzzini P, Margesin R (2014) Cold-adapted yeasts. In: Miscellaneous cold-active yeast enzymes of industrial importance. New York, Springer

Carty SM, Sreekumar KR, Raetz CR (1999) Effect of cold shock on lipid a biosynthesis in Escherichia coli induction at 12° C of an acyltransferase specific for palmitoleoyl-acyl carrier protein. J Biol Chem 274(14):9677–9685

Chattopadhyay MK (2002a) Glycine betaine is a bacterial cryoprotectant and is believed to act by stabilizing cellular proteins and membranes at low temperature. Resonance 7:59–63

Chattopadhyay MK (2002b) Bacterial cryoprotectants. Resonance 7(11):59–63

Chattopadhyay MK, Devi KU, Gopishankar Y, Shivaji S (1995) Thermolable alkaline phosphatase from *Sphingobacterium antarcticus*, a psychrotrophic bacterium from Antarctica. Polar Biol 15:215–219

Chattopadhyay MK, Jagannadham MV (2001) Maintenance of membrane fluidity in Antarctic bacteria. Polar Biol 24:386–388

Chow KC, Tung WL (1998) Overexpression of dnaK/dnaJ and groEL confers freeze tolerance to Escherichia coli. Biochem Biophys Res Commun 253:502–505

Coker JA, Sheridan PP, Loveland-Curtze J, Gutshall KR, Auman AJ, Brenchley JE (2003) Biochemical characterization of a β-galactosidase with a low temperature optimum obtained from an Antarctic *Arthrobacter* isolate. J Bacteriol 185:5473–5482

Collins RE, Deming JW (2013) An inter-order horizontal gene transfer event enables the catabolism of compatible solutes by Colwellia psychrerythraea 34H. Extremophiles 17:601–610

Collins T, Meuwis MA, Gerday C, Feller G (2003) Activity, stability and flexibility in glycosidases adapted to extreme thermal environments. J Mol Biol 328:419–428

Corsaro MM, Lanzetta R, Parrilli E, Parrilli M, Tutino ML, Ummarino S (2004) Influence of growth temperature on lipid and phosphate contents of surface polysaccharides from the antarctic bacterium Pseudoalteromonas haloplanktis TAC 125. J Bacteriol 186:29–34

Cusano AM, Parrilli E, Duilio A, Sannia G, Marino G, Tutino ML (2006) Secretion of psychrophilic alpha-amylase deletion mutants in Pseudoalteromonas haloplanktis TAC125. FEMS Microbiol Lett 258:67–71

D'Amico S, Claverie P, Collins T, Georlette D, Gratia E, Hoyoux A, Meuwis M-A, Feller G, Gerday C (2002) Molecular basis of cold adaptation. Philos Trans R Soc London B Biol Sci 357:917–925

Derman Y, Söderholm H, Lindström M, Korkeala H (2015) Role of csp genes in NaCl, pH, and ethanol stress response and motility in *Clostridium botulinum* ATCC3502. Food Microbiol 46:463–470. https://doi.org/10.1016/j.fm.2014.09.004

Duman JG, Bennett V, Sformo T, Hochstrasser R, Barnes BM (2004) Antifreeze proteins in Alaskan insects and spiders. J Insect Physiol 50(4):259–266

Evans RI, McClure PJ, Gould GW, Russell NJ (1998) The effect of growth temperature on the phospholipid and fatty acyl compositions of non-proteolytic Clostridium botulinum. Int J Food Microbiol 40(3):159–167

Feller G, Gerday C (2003) Psychrophilic enzymes: hot topics in cold adaptation. Nat Rev Microbiol 1(3):200–208

Feng P, Everly DN, Read GS (2001) mRNA decay during herpesvirus infections: interaction between a putative viral nuclease and a cellular translation factor. J Virol 75(21):10272–10280

Filler DM, van Stempvoort DR, Leigh MB (2009) Remediation of frozen ground contaminants with petroleum hydrocarbons: feasibility and limits. In: Margesin R (ed) Permafrost soils. Springer, Berlin, pp 279–301

Gerday C, Aittaleb M, Bentahir M, Chessa J-P, Claverie P, Collins T, D'Amico S, Dumont J, Garsoux G, Georlette D, Hoyoux A, Lonhienne T, Meuwis M-A, Feller G (2000) Cold-adapted enzymes:from fundamentals to biotechnology. Trends Biotechnol 18:103–107

Ghobakhlou AF, Johnston A, Harris L, Antoun H, Laberge S (2015) Microarray transcriptional profiling of Arctic Mesorhizobium strain N33 at low temperature provides insights into cold adaption strategies. BMC Genomics 16:383

Gilbert JA, Davies PL, Laybourn-Parry J (2005) A hyperactive, Ca2+—dependent antifreeze protein in an Antarctic bacterium. FEMS Microbiol Lett 245(1):67–72

Groudieva T, Kambourova M, Yusef H, Royter M, Grote R, Trinks H, Antranikian G (2004) Diversity and cold-active hydrolytic enzymes of culturable bacteria associated with Arctic sea ice, Spitzbergen. Extremophiles 8(6):475–488

Hébraud M, Potier P (1999) Cold shock response and low temperature adaptation in psychrotrophic bacteria. J Mol Microbiol Biotechnol 1(2):211–219

Herbert RA (1986) The ecology and physiology of psychrophilic microorganisms. In: Herbert RA, Codd GA (eds) Microbes in extreme environments. The Society for General Microbiology, Academic Press, London, pp 1–24

Herbert RA (1992) A perspective on the biotechnological potential of extremophiles. Trends Biotechnol 10:395–401

Higgins CF, Dorman CJ, Stirling DA, Waddell L, Booth IR, May G, Bremer E (1988) A physiological role for DNA supercoiling in the osmotic regulation of gene expression in S. typhimurium and E. coli. Cell 52(4):569–584

Hoag H (2008) Polar biotech. Nature Biotechnol 26:1204

Hochachka PW, Somero GN (1984) Biochemical adaptation. Princeton University Press, Princeton, NJ, p 538

Horn G, Hofweber R, Kremer W, Kalbitzer HR (2007) Structure and function of bacterial cold shock proteins. Cell Mol Life Sci 64(12):1457–1470

Hossain MM, Nakamoto H (2003) Role for the cyanobacterial HtpG in protection from oxidative stress. Curr Microbiol 46(1):0070–0076

Huston AL (2008) Biotechnological aspects of cold-adapted enzymes. In: Psychrophiles: from biodiversity to biotechnology, pp 347–363
Ingraham JL, Bailey GF (1959) Comparative study of effect of temperature on metabolism of psychrophilic and mesophilic bacteria. J Bacteriol 77(5):609
Jay JM (1986) Modern food microbiology, 3rd edn. Van Nostrand Reinhold, Wayne, UK, p 214
Jones PG, Cashel M, Glaser G, Neidhardt FC (1992a) Function of a relaxed-like state following temperature downshifts in Escherichia coli. J Bacteriol 174(12):3903–3914
Jones PG, Inouye M (1994) The cold-shock response—a hot topic. Mol Microbiol 11(5):811–818
Jones PG, Krah R, Tafuri SR, Wolffe AP (1992b) DNA gyrase, CS7. 4, and the cold shock response in Escherichia coli. J Bacteriol 174(18):5798–5802
Julseth CR, Inniss WE (1990) Induction of protein synthesis in response to cold shock in the psychrotrophic yeast Trichosporon pullulans. Can J Microbiol 36(8):519–524
Kawahara H, Koda N, Oshio M, Obata H (2000) A cold acclimation protein with refolding activity on frozen denatured enzymes. Biosci Biotechnol Biochem 64(12):2668–2674
Koo H, Hakim JA, Fisher PR, Grueneberg A, Andersen DT, Bej AK (2016) Distribution of cold adaptation proteins in microbial mats in Lake Joyce, Antarctica: analysis of metagenomic data by using two bioinformatics tools. J Microbiol Methods 120:23–28
Lindow SE, Arny DC, Upper CD (1978) Erwinia herbicola: a bacterial ice nucleus active in increasing frost injury to corn. Phytopathology 68:523–527
Mackelprang R, Burkert A, Haw M, Mahendrarajah T, Conaway CH, Douglas TA, Waldrop MP (2017) Microbial survival strategies in ancient permafrost: insights from metagenomics. ISME J 11(10):2305–2318
Maki IR, Galyon EL, Chang-Chien M, Cald WDR (1974) Ice nucleation induced by Pseudomonas syringae. Appl Microbiol 28:456–460
Mancuso Nichols CA, Garon S, Bowman JP, Raguénès G, Guézennec J (2004) Production of exopolysaccharides by Antarctic marine bacterial isolates. J Appl Microbiol 96:1057–1066
Margesin R, Fonteyne PA, Redl B (2005) Low-temperature biodegradation of high amounts of phenol by Rhodococcus spp. and basidiomycetous yeasts. Res Microbiol 156:68–75
Matos RG, Barbas A, Arraiano CM (2009) RNase R mutants elucidate the catalysis of structured RNA: RNA-binding domains select the RNAs targeted for degradation. Biochem J 423:291–301
McGovern VP, Oliver JD (1995) Induction of cold-responsive proteins in Vibrio vulnificus. J Bacteriol 177(14):4131–4133
Médigue C, Krin E, Pascal G, Barbe V, Bernsel A, Bertin PN, Cheung F, Cruveiller S, D'Amico S, Duilio A et al (2005) Coping with cold: the genome of the versatile marine Antarctica bacterium Pseudoalteromonas haloplanktis TAC125. Genome Res 15:1325–1335
Methe BA, Nelson KE, Deming JW, Momen B, Melamud E, Zhang X, Moult J, Madupu R, Nelson WC, Dodson RJ et al (2005) The psychrophilic lifestyle as revealed by the genome sequence of Colwellia psychrerythraea 34H through genomic and proteomic analyses. Proc Natl Acad Sci U S A 102:10913–10918
Mezzina MP, Pettinari MJ (2016) Phasins, multifaceted polyhydroxyalkanoate granule-associated proteins. Appl Environ Microbiol 82(17):5060–5067
Mihoub F, Mistou MY, Guillot A, Leveau JY, Boubetra A, Billaux F (2003) Cold adaptation of Escherichia coli: microbiological and proteomic approaches. Int J Food Microbiol 89 (2–3):171–184
Mishra PK, Joshi P, Bisht SC, Bisht JK, Selvakumar G (2010) Cold-tolerant agriculturally important microorganisms. In: Plant growth and health promoting bacteria. Springer, Berlin, pp 273–296
Miyake R, Kawamoto J, Wei YL, Kitagawa M, Kato I, Kurihara T, Esaki N (2007) Construction of a low-temperature protein expression system using a coldadapted bacterium, Shewanella sp. strain Ac10, as the host. Appl Environ Microbiol 73(15):4849–4856

Mocali S, Chiellini C, Fabiani A, Decuzzi S, Pascale D, Parrilli E, Tutino ML, Perrin E, Bosi E, Fondi M et al (2017) Ecology of cold environments: new insights of bacterial metabolic adaptation through an integrated genomic-phenomic approach. Sci Rep 7:839

Moliné M, Libkind D, de Garcia V, Giraudo MR (2014) Production of pigments and photo-protective compounds by cold-adapted yeasts. In: Cold-adapted Yeasts. Springer, Berlin, pp 193–224

Morita RY (1975) Psychrophilic Bacteria. Bacteriol Rev 39:144–167

Mykytczuk NCS, Foote SJ, Omelon CR, Southam G, Greer CW, Whyte LG (2013) Bacterial growth at 15 °C; molecular insights from the permafrost bacterium Planococcus halocryophilus Or1. ISME J 7:1211–1226

Narinx E, Baise E, Gerday C (1997) Subtilisin from psychrophilic Antarctic bacteria: characterization and site-directed mutagenesis of residues possibly involved in the adaptation to cold. Protein Eng 10:1271–1279

Neidhardt FC, Ingraham JL, Schaechter M (1990) Physiology of the bacterial cell: a molecular approach, vol 20. Sinauer Associates, Sunderland, MA

Obata H, Kakinami K, Tanishita J, Hasegawa Y (1990) Identification of new ice-nucleating bacterium and its ice nucleation properties. Agric Biol Chem 54:725–730

Pandey A, Jain R, Sharma A, Dhakar K, Kaira GS, Rahi P, Shouche YS (2019) 16S rRNA gene sequencing and MALDI-TOF mass spectrometry based comparative assessment and bioprospection of psychrotolerant bacteria isolated from high altitudes under mountain ecosystem. SN Appl Sci 1(3):1–12

Phadtare S (2004) Recent developments in bacterial cold-shock response. Curr Issues Mol Biol 6:125–136

Phillips LE, Humphrey TJ, Lappin-Scott HM (1998) Chilling invokes different morphologies in two Salmonella enteritidis PT4 strains. J Appl Microbiol 84(5):820–826

Porankiewicz J, Clarke AK (1997) Induction of a heat shock protein, ClpB, affects cold acclimation in the cyanobacterium Synechococcus sp. strain PCC 7942. J Bacteriol 179:5111–5117

Prevost D, Drouin P, Laberge S, Bertrand A, Cloutier J, Levesque G (2003) Cold-adapted rhizobia for nitrogen fixation in temperate regions. Can J Bot 81:1153–1161

Purusharth RI, Klein F, Sulthana S, Jäger S, Jagannadham MV, Evguenieva-Hackenberg E, Klug G (2005) Exoribonuclease R interacts with endoribonuclease E and an RNA helicase in the psychrotrophic bacterium Pseudomonas syringae Lz4W. J Biol Chem 280(15):14572–14578

Ray MK, Sitaramamma T, Ghandhi S, Shivaji S (1994) Occurrence and expression of cspA, a cold shock gene, in Antarctic psychrotropic bacteria. FEMS Microbiol Lett 116(1):55–60

Reddy GS, Matsumoto GI, Schumann P, Stackebrandt E, Shivaji S (2004) Psychrophilic pseudomonads from Antarctica: *Pseudomonas antarctica* sp nov., *Pseudomonas meridiana* sp. nov. and *Pseudomonas proteolytica* sp. nov. Int J Syst Evol Microbiol 54:713–719

Rodrigues DF, Ivanova N, He Z, Huebner M, Zhou J, Tiedje JM (2008) Architecture of thermal adaptation in an Exiguobacterium sibiricum strain isolated from 3 million year old permafrost: a genome and transcriptome approach. BMC Genomics 9:547

Rojas-Contreras JA, de la Rosa APB, De León-Rodríguez A (2015) Expression and characterization of a recombinant psychrophilic Cu/Zn superoxide dismutase from Deschampsia antarctica E. Desv.[Poaceae]. Appl Biochem Biotechnol 175(7):3287–3296

Ronholm J, Raymond-Bouchard I, Creskey M, Cyr T, Cloutis EA, Whyte LG (2015) Characterizing the surface-exposed proteome of Planococcus halocryophilus during cryophilic growth. Extremophiles 19:619–629

Russell NJ (1984) Mechanisms of thermal adaptation in bacteria: blueprints for survival. Trends Biochem Sci 9(3):108–112

Russell NJ (1990) Cold adaptation of microorganisms. Philosophical transactions of the royal society of London. B Biol Sci 326(1237):595–611

Russell NJ (2002) Bacterial membranes: the effects of chill storage and food processing. An overview. Int J Food Microbiol 79(1–2):27–34

Santiago M, Ramírez-Sarmiento CA, Zamora RA, Parra LP (2016) Discovery, molecular mechanisms, and industrial applications of cold-active enzymes. Front Microbiol 7:1408

Schmidt-Nielsen S (1902) Ueber einige psychrophile Mikroorganismen und ihr Vorkommen. Zentr Bakteriol Parasitenkd Infektionsk Hyg Abt II 9:145–147

Subczynski WK, Markowska E, Gruszecki WI, Sielewiesiuk J (1992) Effects of polar carotenoids on dimyristoylphosphatidylcholine membranes: a spin-label study. Biochimica et Biophysica Acta (BBA)-Biomembranes 1105(1):97–108

Ting L, Williams TJ, Cowley MJ, Lauro FM, Guilhaus M, Raftery MJ, Cavicchioli R (2010) Cold adaptation in the marine bacterium, Sphingopyxis alaskensis, assessed using quantitative proteomics. Environ Microbiol 12:2658–2676

Tsuruta H, Tamura J, Yamagata H, Aizono Y (2004) Specification of amino acid residues essential for the catalytic reaction of cold-active protein-tyrosine phosphatase of a psychrophile, *Shewanella* sp. Biosci Biotechnol Biochem 68:440–443

Uma S, Jadhav RS, Kumar GS, Shivaji S, Ray MK (1999) A RNA polymerase with transcriptional activity at 0°C from the Ant-arctic bacterium *Pseudomonas syringae*. FEBS Lett 453:313–317

Uppal S, Akkipeddi VS, Jawali N (2008) Posttranscriptional regulation of cspE in *Escherichia coli*: involvement of the short 5′-untranslated region. *FEMS Microbiol*. Lett 279:83–91. https://doi.org/10.1111/j.1574-6968.2007.01009

VanBogelen RA, Neidhardt FC (1990) Ribosomes as sensors of heat and cold shock in Escherichia coli. Proc Natl Acad Sci 87(15):5589–5593

Varin T, Lovejoy C, Jungblut AD, Vincent WF, Corbeil J (2012) Metagenomic analysis of stress genes in microbial mat communities from Antarctica and the high Arctic. Appl Environ Microbiol 78(2):549–559

Vasina JA, Baneyx F (1996) Recombinant protein expression at low temperatures under the transcriptional control of the major Escherichia coli cold shock promoter CspA. Appl Environ Microbiol 62:1444–1447

Vorachek-Warren MK, Carty SM, Lin S, Cotter RJ, Raetz CR (2002) An Escherichia coli mutant lacking the cold shock-induced Palmitoleoyltransferase of lipid a biosynthesis absence of unsaturated acyl chains and antibiotic hypersensitivity at 12° C. J Biol Chem 277 (16):14186–14193

Walworth JL, Pond A, Snape I, Rayner J, Ferguson S, Harvey P (2007) Nitrogen requirements for maximizing petroleum bioremediation in a sub-Antarctic soil. Cold Regions Sci Tech 48:84–91

Willimsky G, Bang H, Fischer G, Marahiel MA (1992) Characterization of cspB, a Bacillus subtilis inducible cold shock gene affecting cell viability at low temperatures. J Bacteriol 174 (20):6326–6335

Wintrode PL, Miyazaki K, Arnold FH (2000) Cold-adaptation of a mesophilic subtilisin-like protease by laboratory evolution. J Biol Chem 275:31635–31640

Yadav AN, Verma P, Kumar V, Sachan SG, Saxena AK (2017) Extreme cold environments: a suitable niche for selection of novel psychrotrophic microbes for biotechnological applications. Adv Biotechnol Microbiol 2:1–4

Yergeau E, Arbour M, Brousseau R, Juck D, Lawrence JR, Masson L, Whyte LG, Greer CW (2009) Microarray and real-time PCR analyses of the responses of high-arctic soil bacteria to hydrocarbon pollution and bioremediation treatments. Appl Environ Microbiol 75:6258–6267

Zachariassen KE, Kristiansen E (2000) Ice nucleation and anti-nucleation in nature. Cryobiology 41:257–279

Zachariah S, Das SK (2017) Idiomarina andamanensis sp. nov., an alkalitolerant bacterium isolated from Andaman Sea water. Antonie Van Leeuwenhoek 110(12):1581–1592

Zartler ER, Jenney FE Jr, Terrell M, Eidsness MK, Adams MW, Prestegard JH (2001) Structural basis for thermostability in aporubredoxins from *Pyrococcus furiosus* and *Clostridium pasteurianum*. Biochemistry 40:7279–7290

An Insight to Cold-Adapted Microorganisms and their Importance in Agriculture

17

Shriniketan Puranik, Sandeep Kumar Singh, and Livleen Shukla

Abstract

A major part of earth experiences less than 5 °C temperature, adversely affecting agricultural productivity. Water, the most important factor for existence of living organisms nearly freezes at such temperatures and becomes unavailable for utilization. Frozen soils and chilled/ frozen water challenge the survivability of plants, thus reducing crop yields to a large extent. Some parts of the world experience subzero temperature, reduced kinetics, and scarcity of nutrient and water conditions, thereby challenging the survival of autochthonous biota too. But, even under icy conditions, soil microorganisms playing a vital role in agriculture, are well adapted due to evolution of diverse mechanisms to overcome and perform better, making them a potent source to be tapped for increasing productivity under cold stress. They employ various methods like altering cell envelope, energy metabolism and membrane fluidity, quenching reactive oxygen species, production of compatible solutes, cold shock proteins, and exopolysaccharides in alleviation of cold stress in plants. However, such bioinoculants in cold regions are yet to be fully mined for their capability.

Keywords

Cold-adapted microorganisms · PGPR · Stress management · Biocontrol agents

S. Puranik · S. K. Singh · L. Shukla (✉)
Division of Microbiology, Indian Agricultural Research Institute, New Delhi, India

R. Goel et al. (eds.), *Survival Strategies in Cold-adapted Microorganisms*,
https://doi.org/10.1007/978-981-16-2625-8_17

17.1 Introduction

More than 85% of the Earth's biosphere experiences temperature lower than 5 °C all across the year (Hoshino and Matsumoto 2012). Some parts of the world experience subzero temperature, reduced kinetics, and scarcity of nutrient and water conditions, thereby challenging the survival of autochthonous biota (Nikrad et al. 2016; Altshuler et al. 2017). In such areas, soils remain frozen for a long time, challenging agricultural activities, their production and productivity. Microorganisms play a vital role in agriculture. Even under icy conditions, soil microorganisms are well adapted due to evolution of diverse mechanisms to overcome and perform better, making them a potent source to be tapped for increasing productivity (Margesin and Schinner 1999). Depending upon their cardinal temperatures, they can be classified into psychrophiles (cold-loving) or psychrotolerants (cold-bearing). Psychrophiles can grow optimally below 15 °C. They are found mainly in permanent cold habitats, like polar ice caps, very high altitudes, or very deep in oceans. Psychrotolerants grow optimally over 20 °C but have an ability to tolerate lower temperatures. They are found in areas with periodic temperature fluctuations (Margesin et al. 2007). These microorganisms survived and performed better under cold conditions by evolving distinct strategies. Both psychrotolerant and psychrophilic microorganisms belong to different groups like *Archaea*, bacteria, and fungi (Li et al. 2020a; Margesin et al. 2016; Zachariah et al. 2016; Ciobanu et al. 2014). In fact, Forster (1887) was the first to report on the high proliferation rate of some microorganisms even under cold conditions. There are several reports on microbial diversity of microorganisms inhabiting these cold areas (Miteva and Brenchley 2005; Pradhan et al. 2010; Sahay et al. 2013; Prasad et al. 2014). Some extreme environments like water bodies under ice sheets of Antarctica and permafrost regions have found to harbor a wide range of chemotrophs (Mikucki et al. 2015). These regions are characterized by very high salinity to keep water at liquid state under such freezing temperature or extreme anhydrous conditions, making the organisms inhabiting them either halophilic or xerophiles (Robinson and Mikucki 2016). It is reported that at least an a_w (water activity) of 0.61 is required to support life and that some brines have a_w as low as 0.45, making it difficult for microorganisms to thrive (Bakermans 2017).

Even after these hurdles, exploration of low-temperature environments has led to the discovery of psychrotrophic microbiomes that can be tapped for agricultural applications under cold conditions. Various techniques like phospholipid fatty acid (PLFA) analysis, nucleic acid techniques, gene clone library method, DNA fingerprinting techniques, DNA microarray, transcriptomics, proteotmics, CRISPR-cas spacer studies, stable isotope probing (SIP), and metagenomic approaches have paved the way for isolation, characterization, and functional understanding of these cold-adapted microorganisms (Lopatina et al. 2016; Mackelprang et al. 2016; Yadav et al. 2015a, b; Yadav et al. 2017a, b; Koh et al. 2017; Raymond-Bouchard et al. 2017; Koch et al. 2018). Also, various studies state that such bacteria belong to major genera like *Azospirillum*, *Sinorhizobium*, *Burkholderia*, *Pantoea*, *Serratia*, *Rahnella*, *Pseudomonas*, *Polymorphobacter*, *Pedomicrobium*, *Phormidium*, *Lyngbya*, *Nostoc*, *Phormidesmis*, *Anaerococcus*, *Viridibacillus*, and *Pseudochrobactrum* (Okon and Labandera-Gonzalez 1994; Katiyar and Goel 2004;

Barka et al. 2006; Selvakumar et al. 2008a, b; Vyas et al. 2010; Bisht et al. 2014; Gokul et al. 2016; Qin et al. 2017; Chaya et al. 2019; Gautam et al. 2019; Pavankumar et al. 2020). Some bacteria like *Thiobacillus* not only thrive but also metabolize chemotrophically in cold conditions, thus proving themselves to be well adapted (Harrold et al. 2016). Cold-dwelling fungal genera belong to phyla *Ascomycota*, *Deuteromycota*, *Zygomycota*, and Basidiomycota, including *Aspergillus*, *Paecilomyces*, *Penicillium*, *Trichoderma*, *Cladosporium*, *Mucor*, *Alternaria*, *Botrytis*, *Geomyces*, *Arbuscular*, *Mycorrhiza*, and *Phialophora* (Pandey et al. 2008; Kostadinova et al. 2009; Rinu et al. 2013; Rinu et al. 2014; Dhakar and Pandey 2016; Pedranzani et al. 2016). Many genera of *Archaea* are also found to be cold-adapted, among which *Methanolobus*, *Methanococcoides*, *Methanosarcina*, *Methanogenium*, and *Methanosaeta* are well studied (Taha et al. 2016; Chaya et al. 2019; Dev et al. 2019; Li et al. 2020b; Saralov 2019). However, *Archaea* and fungi are 200–1000-folds lower in abundance than bacteria (Frey et al. 2016). Other eukaryotes like algae are known to harbor snow too (Brown et al. 2016). It has been well established that the structure of soil microbial communities is governed by soil organic matter, varying along the altitudinal gradient (Kumar et al. 2019). Such organisms are reported to use organic matter for their growth (Edwards and Cameron 2017). A study reported that total microbial biomass was significantly low with no microbial activity in regions at higher elevation as compared to regions at lower elevation. This was attributed to lower solute concentrations which hindered film formation of water for availability to microorganisms (Goordial et al. 2016).

At such cold temperatures, it is very difficult for metabolism to occur. Various authors have reported different vital metabolic activities by microorganisms even at subzero temperature which are well-compiled (Bakermans 2008). Some of them are production of carbon dioxide and methane, cell division, nucleic acid synthesis, protein synthesis, photosynthesis, and respiration. There is a report indicating possibilities of cellular reproduction by microorganisms at temperature as low as −20 °C (Bakermans 2017). A study showed a lower temperature limit for active metabolizing life, T_g (also called glass transition temperature) ranging between −12 °C and − 26 °C in five microorganisms, with lowest recorded to be *Arthrobacter arilaitensis* (−26 °C) (Clarke et al. 2013). Even though the lowest T_g is yet to be reported, a study has stated that T_g of *Lactobacillus delbrueckii* ssp. bulgaricus CFL1 reduced to −51 °C from −19 °C, due to addition of selected cryoprotectants (Fonseca et al. 2016). This paves way for new possibilities of exploiting cold-adapted microorganisms.

17.2 Ecological Diversity of Cold-Adapted Microorganisms

According to general perseverance, the more the existence of extreme environmental conditions, the lower the diversity of organisms. Bacterial growth, viability, and metabolism are greatly temperature dependent (Tripathi and Klingmuller 1992). However, a number of microorganisms dominate most cold, inhospitable habitats and make them the most challenging mode of life. The lowest temperature limit for life appears to be approximately −20 °C, which is the recorded value for bacteria

(*Pseudomonas* sp., *Serratia marcescens*, *Pantoea* sp., *Azospirillum* sp., *Acinetobacter* sp., *Meyerozyma* sp.) that are living in permafrost soil and sea ice. Microbial development at this temperature is confined within the permafrost soil or ice and brine channels to a small amount of unfrozen water. This involves high concentrations of salts, particulate matter, exopolymeric compounds, and temperature gradients that sustain the fluid flow (D'Amico et al. 2006).

Even in the refrigerated environment, cold-tolerant microorganisms have been a significant cause for concern in the food production and storage industries. Greenland and Losleben (2001) mentioned that in the alpine soil conditions, physical and biochemical properties are distinguished by seasonal changes due to the occasional cover of snow and changing freezing temperatures in winter and heavy drying sunlight, marked by rare rains in summer. It is not rare for a wide range of cold-tolerant microorganisms to be present in alpine and subalpine ecosystems. In cold condition most of the eukaryotes, Archaea, and bacteria occur. Although in polar ecosystems bacteria prevail more diversely than Archaea, the latter is widespread in cold, deep waters of the ocean (Deming 2002; Karner et al. 2001). In cold environments, morphological forms observed include filamentous bacteria, non-spore-formers, and some spore-formers. Together they represent a broad variety of metabolic types from aerobes to anaerobes and include both heterotrophs and autotrophs.

17.3 Effect of Cold Stress on Plants and Microorganisms

Generally, temperature plays an important role in the vibration of molecules. Temperature is an important factor that determines integrity of protein structures, enzymatic activities, and enthalpy of a system (Bakermans 2008). Below subzero temperatures, water freezes to lose its liquid property and becomes unavailable for plants and microorganisms. Living organisms greatly depend on temperature for their growth and development. Plants need an ideal range of temperature to develop and attain maximum growth. Any anomaly results in structural, functional, and molecular changes in plants causing severe metabolic dysfunction and retardness in growth (Fitter and Hay 1981). Cold stress in plants hampers regular growth and development (Patni et al. 2018; Rihan et al. 2017). Several important crop species like paddy, cotton, tomato, and some cucurbits are highly affected by cold stress (Hussain et al. 2018; Ghadirnezhad and Fallah 2014). It can cause certain visible symptoms in leaves like decreased enlargement, wilting, yellowing, and/or necrosis. The development of reproductive parts at the flower opening stage in rice was seriously affected leading to sterility in inflorescence due to cold stress (Jiang et al. 2002). Also, reduction in plant attributes like germination, seedling vigor, growth, tillering, withering of leaves, and pollen sterility due to cold stress causes reduction in crop yields (Imin et al. 2004; Suzuki et al. 2008; Wang et al. 2016a, b). Chickpea also experiences higher rates of floral abortion under cold stress (Sharma and Nayyar 2014). Cold temperature is found to reduce root metabolic activity, relative water content, and chlorophyll content (Lootens et al. 2004; Burchett et al. 2006; Sun et al. 2017). Apart from phenotypic and physiological damages, cold stress imparting injuries at cellular level is also observed. Cold stress causes dehydration, causing injuries to the plasma membrane. Also, due to comparatively lesser solute concentration in apoplastic space, ice formation occurs, leading to dehydration (Steponkus et al. 1993; Olien and Smith 1997). Cold stress is reported to cause modifications in lipid constituents in membranes, thus hindering fluidity (Welti et al. 2002). It altercates membrane stability, resulting in leakage of cellular components. The compartments holding cellular organelles are disturbed by cold temperatures. They also cause feedback inhibition of photosynthesis due to accumulation of sucrose in leaves (Ruelland et al. 2009), impairment of protein associations, and cellular metabolism. In maize, cold stress reduces levels of pectin and imparts thickening of leaf cell walls (Bilska-Kos et al. 2017). Increase in disease incidence in soybean and corn due to exposure to cold stress is reported (Bradley 2008; Robertson and Munkvold 2012; Serrano and Robertson 2016).

Like plants, the survival of microorganisms is greatly affected by cold temperature too (Margesin and Miteva 2011). The lag period stretches as the temperature decreases, leading to a decline in the growth rate and the overall number of cells. Some face reduced rate of growth, enzymatic activity, cellular structure alterations, and varying nutritional needs under cold conditions. At the cellular level, the functional phase of lipid bilayer changes to gel phase from liquid to crystalline phase due to cold stress, thereby making alterations in membrane attributes. Temperatures below −10 °C increase solute concentration significantly due to conversion of water to ice crystals, creating severe desiccation. This also causes rigidity in membranes, thus hampering their fluidity and osmosis, accompanied by inefficient nutrient uptake and transportation. Formation of ice crystals greatly suppresses protein activity of cells (Raymond et al. 2007; Chandler 2018). Under such nutrient-deficit, desiccated, impaired growth conditions, the proliferation and diversity of microorganisms are greatly altered in addition to anthropogenic reasons (Lopatina et al. 2013; Hauptmann et al. 2014). The reduction in rate of physiological processes may further prohibit the transmission of signals required for chemotaxis and escalated viscosity reducing diffusion (Stocker and Seymour 2012; Showalter and Deming 2018).

17.4 Cold-Adapted PGPR as Bioinoculants for Stress Management

The developmental cycle of most of the crops around the globe is subjected to cold temperature variations. Even under such conditions, application of plant growth-promoting bacteria is found to be an agronomically excellent strategy. They metabolize in cold states and develop metabolites like plant growth regulators that stimulate growth directly and encourage plants' nutrient intake. PGPRs increase the growth of plants while increasing their stress tolerance. Plant growth-promoting rhizobacteria (PGPRs) can promote host plant development; disrupt the organization of plant pathogens; contribute to systemic pathogen resistance; affect the production of phytohormones; and enhance the control of nutrients and water management

(Gupta et al. 2020; Kumar et al. 2016a, 2017; Barka et al. 2006). Mechanism of PGPR and their role to combat cold stress are described in Fig. 17.1.

17.4.1 Cold-Adapted Plant Growth-Promoting Rhizobacteria

Soil is considered a complicated living entity that is important for the development and maintenance of the majority of lifecycles, not only in farming but also in food production (Mishra et al. 2019). Soil is a microbial storage site, and it is estimated that less than 5% of the total area is filled by living microorganisms. The soil around the roots is biologically, physically, and chemically influenced by plant roots, commonly known as the "rhizosphere." The rhizospheric soil containing residential microorganisms is considered to influence fertility of soil and the health of plants. Rhizospheric region containing microbial population, in which PGPR is one of the crucial components, was first described by Kloepper and Schroth in 1978. Many strains that are not found in the rhizospheric zone, yet exhibiting PGP activities, are nowadays classified by the name PGPB (Singh et al. 2020b; Andrews and Harris 2003). Rhizospheric microorganisms in cold regions largely depend on the temperature of the root zone as metabolic activities influencing plant growth seize at cold temperature. Under such circumstances, rhizospheric microorganisms must retain their activity so as to produce intermediates/end products to promote plant growth (Singh et al. 2019a).

The suboptimal temperature that is being interspersed by growing seasons prevailing in temperate agroecosystem slows down or deteriorates microbial activities having a negative effect on productivity. This shows that temperature and time deeply affect growth of crops and growth of microbes. Microorganisms play a keen role during development and productivity of agroecosystems via various functions including fixation of N_2, solubilization of nutrients, PGP, translocation, and inhibition of metabolic cycle of toxic pathogens and insects. Cold-tolerant microbes maintain their function in the suboptimal conditions, thereby playing an important role in case of nutrient transition. Microbes play a crucial role in nutrient cycling under the cold environment, one of the integral parts of any ecosystem's environment (Haggblom and Margesin 2005).

Unfortunately, there have been little attempts to consider the existence and property of cold-adapted microbes, and there are scanty details available concerning their use in cold agroecosystems. The development of stimulating phytohormones through PGPR/PGPBs in root areas is one of the main facilities in plant growth promotion. Such hormones stimulate root hair density and length, resulting in increased water and mineral intake through soil (Volkmar and Bremer 1998). Besides the development of phytohormones, promotion of plant growth is supposed to be enhanced via processes including siderophore production, deamination of ethylene's precursor molecule (whose aggregated tissue found in roots is supposed to influence growth and development of root), induction of systemic resistance to plant pathogens, and antagonism toward deleterious root microorganisms (Singh

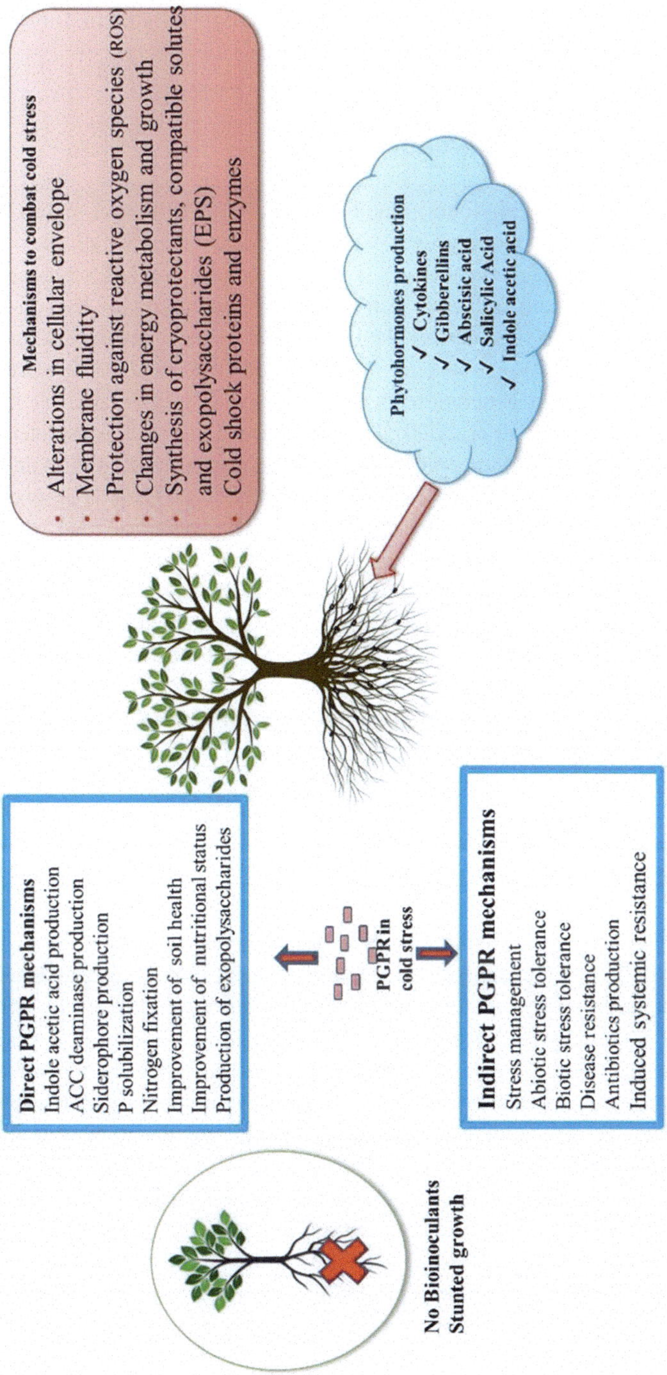

Fig. 17.1 Schematic descriptive mechanism of several traits shown by PGPR and their mechanisms to combat cold stress

et al. 2010, Katiyar and Goel 2004; Glick et al. 1998; Misaghi et al. 1982). Various cold-adapted microbes and their niches of isolation are mentioned in Table 17.1.

17.4.2 Indoleacetic Acid Producers

The release of stimulating phytohormone by PGPRs within the root region is a key mechanism for promoting plant growth, stimulating density and length of root hairs. The rise of the root surface area allows a significant quantity of soil to increase plant absorption of water and mineral nutrients. The potential of IAA-producing microorganisms helps to distinguish them and also gives them a vital marker to research the IAA's physio-persistence functions or ecological value (Bric et al. 1991). Concerning auxin generation, pentose phosphate is regulated by bacteria-regulating proline (McCue et al. 2000). Selvakumar et al. (2008a, b) isolated PGPB which could adapt to cold temperature from North-western Indian Himalayas and named them *Serratia marcescens* SRM and *Pantoea* 1A. Their IAA production capacity sustained at 4 and 15 °C by these strains. The bacterial strain performing seed bacterization has improved biomass of plant and absorption of nutrients in wheat seedlings during cold temperature. Mishra et al. (2008, 2009) reported that *Pseudomonas* sp. could successfully adapt to cold temperature and produce indoleacetic acid. At cold temperature, bacterization of seed by strains PGERs17 and NARs9 played a significant role in germination and root and shoot length of low-temperature wheat seedlings. In the future, a brand new cold-resistant genus of PGPB will be discovered in the light of the metabolic flexibility of pseudomonads.

17.4.3 ACC-Deaminase Producers

As several other abiotic and biotic factors, researchers have recorded a wide variety of accelerated ethylene development in the rhizosphere, plant tissue, and microbial species. Plant subjected to negative impact by stress hormone ethylene can be partly alleviated by bacterial strains containing ACC-deaminases. 1-Aminocyclopropane-1-carboxylate deaminase (ACC deaminase) is a crucial enzyme in pathways for PGPBs to promote plant development. The plant hormone ethylene, thereby impacting plant growth and development, is greatly regulated by this enzyme. ACC deaminase plants can deal with this adverse condition by reducing their ethylene level in the same manner as other environmental stresses. Under salt stress condition and low temperature, canola plant growth is mediated by psychrotolerant *Pseudomonas putida* UW4, an ACC deaminase producing bacteria (Saleem et al. 2007; Cheng et al. 2007). With respect to ethylene production in stress conditions, more work is required to decode the role of ACC deaminase producing PGPR.

Table 17.1 Cold-adapted microbes and their role in different plants with isolation source

Plant	Microbes	Effect/result	Temperature and location	References
Vitis vinifera L. cv.	*Burkholderia phytofirmans* strain PsJN	Increased grapevine growth and physiological activity, increased levels of starch, proline, and phenolics	4–26 °C	Barka et al. (2006)
Cicer arietinum, *Vigna mungo*, *Vigna radiata*, *Cajanus cajan* and *Eleusine coracana*	*Dyadobacter* sp.	Promote plant growth by fixing atmospheric N_2, making it available to plant; enhancements in agronomical parameters, leaf nitrate reductase activity, and total chlorophyll content	10 °C Bhowali, Western Indian Himalaya	Kumar et al. (2018)
Triticum aestivum L.	*Pseudomonas* sp.	Enhanced root/shoot biomass and nutrient uptake, significantly improved level of cellular metabolites like chlorophyll, anthocyanin, free proline, total phenolics, starch content, physiologically available iron, proteins, and amino acids	4–8 °C Uttarakhand state located in the NW Indian Himalayan Region (IHR)	Mishra et al. (2011)
Phaseolus vulgaris L.	*Pseudomonas fragi*, *P. chlororaphis*, *P. fluorescens*, *P. proteolytica* and *Brevibacterium frigoritolerans*	Improved the cold resistance of bean, exhibit ACC deaminase activity, increase tolerance of cold-sensitive crops	4 °C, Faculty of Agriculture, Atatürk University, Erzurum (Turkey)	Tiryaki et al. (2019)
Amaranthus sp.	*Pseudomonas* sp. NARs9 (MTCC9002)	Enhanced germination, shoot and root lengths	4 °C Nainital District (29°30' N and 79°30′E) in the North Western Indian Himalayas	Mishra et al. (2009)
Betula utilis Don and *Rhododendron campanulatum* Don	*Pseudomonas putida* (B0)	Increase in plant biomass; ß-1,3-glucanase, salicylic acid,	4 °C and 28 °C Indian central Himalaya, (Phurkia, district Pithoragarh, Uttaranchal	Pandey et al. (2006)

(continued)

Table 17.1 (continued)

Plant	Microbes	Effect/result	Temperature and location	References
		siderophore, produce chitinase and hydrogen cyanide		
Solanum lycopersicum cv Mil.	*Flavobacterium* sp. OR306 and *Pseudomonas frederiksbergensis* OS211	1-Aminocyclopropane-1-Carboxylate deaminase (ACCD) gene; ethylene release, ACC content, and ACC oxidase activity increased in tomato under chilling	12/10 °C, University of Waterloo, Canada	Subramanian et al. (2015)
Triticum aestivum spp. *vulgare* cv Bezostiya and *Hordeum vulgare* cv Tokak	*Bacillus megaterium* M3, *Bacillus subtilis* OSU142, *Azospirillum brasilense* Sp245, and *Raoultella terrigena*	Increased root and shoot dry weight of wheat and barley plant	2–20 °C Bioengineering Faculty of Engineering and Architecture at Yeditepe university Istanbul, Turkey	Turan et al. (2013)
Solanum lycopersicum cv Mil	Specific strains of *Arthrobacter*, *Flavimonas*, *Flavobacterium*, *Massilia*, *Pedobacter*, and *Pseudomonas*	Improved germination and plant growth and induce anti oxidant capacity	15 °C Chungbuk agricultural research and extension services, Ochang-eup, South Korea	Subramanian et al. (2016)
Eleusine coracana (L.), *Cicer arietinum* (L.), *Vigna radiata* (L.) *Vigna mungo* (L.) and *Cajanus cajan* (L.)	*Pseudomonas jesenii* MP1	Stimulated growth of shoot length, root length, plant fresh weight, and plant dry weight; significant increase in chlorophyll content, nitrate reductase activity, and phosphorous content	5–15 °C western Indian Himalayas	Kumar et al. (2014)
Lens culinaris (L.)	*Pseudomonas* sp.	Increased lentil shoot length, root length, root biomass, and shoot biomass	4 °C north western Himalayas	Bisht et al. (2013)

17.4.4 Siderophore Producers

Iron, as a plant micronutrient, serves as a cofactor in various enzymes performing redox reaction. Ferric hydroxide, an insoluble form of iron that exists in a higher proportion in the soil, functions as a limiting factor during plant growth. Oxidation of ferrous to ferric concerning supply of iron to organisms is found to be low. Ferric ion, an insoluble form, is one of the obstacles during possession via organisms under physiological conditions (Neilands 1995). Specialized mechanisms for assimilating iron have been developed by microorganisms, including iron-chelating, low molecular weight compounds called siderophores, that assist during its transport into the cell. Siderophores provide advantage in plant and bacterial survival by reducing the supply of iron for pathogen survival, so they mediate the struggle leading to the exclusion of fungal infections and other microbial predators in the rhizosphere. A cold-tolerant mutant, *Pseudomonas fluorescens*, boosted the colonization of mung bean plant rhizosphere at 10 and 25 °C and increased siderophore concentration by 17-folds (Katiyar and Goel 2004). Research on the development of psychrotolerant bacteria by siderophore-mediated growth remains at a beginning stage and requires further research.

17.4.5 Role of Microbes in Nitrogen Fixation

Fixation of nitrogen is one of the key sources for sustaining life on Earth by using certain free-living, associative, and symbiotic bacterial genera (Singh et al. 2020a). However, cold temperature stress has a major effect on this process. Low temperature affects rhizobial behaviors including depression of the competitiveness and functioning of nodules. McKay and Djordjevic (1993) found that *Rhizobium leguminosarum* bv. trifolii decreased production of Nod metabolites under low temperature, which affected host legume yield and nodulation. Similarly, from many experiments, Lynch and Smith (1994) found that suboptimum temperature affected competition for nodulation, inhibition of nodule activity, and infection of root. In temperate condition, a successful symbiosis can double N_2 fixation in crop, hence improving the production of legume crops (Sprent 1979). To resolve the cold-induced stress, it is therefore necessary to choose cold-adapted/cold-resistant strains of *Rhizobia*. Prevost et al. (1999) selected cold-adapted rhizobia from Canadian soils, in order to improve the production of legumes subjected to cool conditions in the growing season. *Rhizobia* in arctic and subarctic regions associated with legume species were used. Isolation of different rhizobial strains were done from *Oxytropis*, *Lathryrus* and *Astragalus*, namely, *R. leguminosarum* and *Mesorhizobium* sp.; because of their capacity to expand at 0 °C, these rhizobia were declared psychrotrophs.

Temperate forage legume sainfoin defined the benefits of cold-adapted arctic *Mesorhizobium* in enhancing legume symbiosis. It was found that arctic rhizobia was comparatively better than temperate rhizobia in nodule development and overall growth of sainfoin. In cold-adapted rhizobia, biochemical experiment results showed higher synthesis of cold shock proteins than their mesophilic counterparts. Nodulation genes and bacterial signals are defined as the initial step in changing the specificity of nodulation in host, since the arctic *Mesorhizobium* could not agronomically nodulate essential legumes. Zhang et al. (2003) found that the rhizobia from colder climates in North America had a favorable impact on the N_2 fixation and nodulation in *Glycine max* compared to southern warmer climates.

Superior strain of *Sinorhizobium meliloti* was picked and found effective in field as well as laboratory conditions to enhance the growth of alfalfa by effective nodulation. Rhizobia increased growth at low temperatures and at anaerobic stress, which means that in different abiotic stresses inherent in temperate climates, selected rhizobia can be modified. Therefore, a high level of competition for nodule and nitrogen-fixing capability combined with cold-tolerant properties will need specific rhizobia for temperate legumes (Prevost et al. 2003). To allow synthesis of membrane-associated nod factors and activity playing an important role in nodulation and host specificity, rhizobia must maintain their membrane fluidity at low temperature.

Primarily associated with tropical grass and cereals crops, *Azospirillum* is a well-known PGPB (Kumar et al. 2020a). *Azospirillum* inoculation in a large-scale use discouraged its impact on winter crops (Kaushik et al. 2001). A study involved a mutant, i.e., Tn5:lacZ, isogenic to wild-type *A. brasilense* and confirmed its growth at a cool temperature (Kaushik et al. 2000). It could affect wheat growth at low temperatures in field studies. This was an unusual study on *Azospirillum*'s field success under suboptimal temperatures. *Azospirillum* is a bacterium that gives a chance to explore under cold stress because of its agronomic importance.

17.4.6 Role of Phosphate Solubilizers

Phosphate solubilization is vital for various geochemical processes as well as plant nutrition. P is present in the soil in a fixed form, therefore requiring transformation. For promoting plant growth, one of the most suitable paths is solubilization of P by rhizosphere microflora. Glucose dehydrogenase, an enzyme bound to the membrane, is involved in solubilization of phosphate by direct oxidation of glucose to gluconic acid. Subsequently, gluconic acid is enzymatically converted to 2-ketogluconic and 2,5-diketogluconic acid. Compared to gluconic acid, 2-ketogluconic acid is more effective in P solubilization (Goldstein 1995; Kim et al. 2002). Das et al. (2003) used cold-tolerant *Pseudomonas* mutants for solubilization of phosphate at 10 °C. They found that phosphate solubility was more effective in cold-tolerant mutant (10 °C) than the respective wild type (25 °C). Similar results were reported by Trivedi and Sa (2008).

The Indian Himalayan region has been a potent niche and has made the most progress in this direction. Pandey et al. (2006) isolated antagonist and cold-tolerant P-solubilizing strains from central Indian Himalayas. Strains of *P. putida* and *P. fragi* solubilized phosphate at temperatures ranging between 4 and 28 °C (Selvakumar et al. 2009a). Besides, rate of germination, plant biomass, and nutrient intake of wheat seedlings under cold-temperature conditions was improved. Gulati et al. (2009) isolated a rhizosphere competent phosphate-solubilizing strain from Himalayan cold deserts, namely, *Acinetobacter rhizosphaerae*. Screening for tolerance against salinity, temperature, calcium salts, alkalinity, and desiccation-induced stress, *Pseudomonas fluorescens* isolates from cold deserts of Himalayas also were efficient P solubilizers (Vyas et al. 2009). The above studies are mainly exploratory, while the fundamental necessity is developing commercially viable cold-tolerant PSB inoculants in temperate farming.

17.5 Mechanisms to Combat Cold Stress

As discussed earlier, cold stress affects plant growth and development severely by limiting their metabolism and reducing water availability. Most biochemical and physiological functions concerning plant growth are impaired at these cold temperatures, causing shortening of cropping cycles and greatly influencing crop yield (Mishra et al. 2010). Even though plants use various mechanisms to outgrow cold stress, they are hit hard and subsequently yield lesser than normal conditions (Haldiman 1998). Compared to plants, cold-adapted microorganisms carry out metabolism more actively under cold conditions by exhibiting a variety of adaptation strategies (D'Amico et al. 2006) that are as follows.

17.5.1 Alterations in Cellular Envelope and Movement

Cold-adapted microorganisms reduce their cell size and envelope enclosing cell membrane. To retain the cell viability and metabolism, especially under low-temperature stress, the consistency of the bacterial cell envelope is crucial. Microorganisms dynamically modify membrane fluidity through adjustments in lipid composition, preferring shorter chain lengths and decreased saturated fatty acids (Mykytczuk et al. 2007). In some bacteria like *Planococcus*, an increase in rate of peptidoglycan synthesis was observed, which contributes to negative charge, thereby forming a crust by mineralization of $CaCl_2$ around the cell wall (Mykytczuk et al. 2016). Others are found to remodel their cell envelope under cold conditions (Williams et al. 2017). Furthermore, during low availability of water, lipid composition changes appear to support overall increase of negative charge that provides a poor permeability to Na^+ ions and regulates osmotic effects like swelling or shrinkage. Increase in cellular Na^+/K^+ ratio hampers the metabolic activity of many enzymes. This is important as membrane stability is an attribute of thermotolerance (Thiemann and Imhoff 1991; Mykytczuk et al. 2016). Apart from permeability, cold temperature also influences chemotaxis of several microorganisms. Some cold-adapted bacteria have shown to place themselves within the ice matrix found in the sea via chemotaxis in order to attain optimum energy and as response to chemosignals. By production of *molecular fishing hook* and exopolysaccharides under cold conditions, bacteria such as *Marinomonas* sp. and *Colwellia* sp. are reported to bind to ice particles, thus exhibiting biofilm character (Bar Dolev et al. 2016; Carillo et al. 2015; Casillo et al. 2017; Showalter and Deming 2018).

17.5.2 Membrane Fluidity

External environment is faced by cell membrane that acts as the initial defense. Thus, proper functioning of the membrane is important for any cell. Made up of phospholipid bilayer and special protein domains, fluidity of membrane easily gets disturbed due to changes in external conditions (Strancar et al. 2000). Cold temperatures freeze and dehydrate the cells by altering membrane lipid constitution, changing it to gel phase from liquid phase, thereby hampering its function. In order to adapt themselves to such conditions, cold-adapted microorganisms modify the lipid structure to induce cold resistance (Mykytczuk et al. 2007; Arvizu-Gomez et al. 2013). Some fungi belonging to genera *Cadophora*, *Geomyces*, and *Mortierella* isolated from the Antarctic synthesize unsaturated fatty acids like arachidonic and linoleic acids. Similarly, a study demonstrated production of stearidonic acid in *Mortierella* under cold stress (Hassan et al. 2016). Psychrotolerant yeast like *Rhodosporidium diobovatum* also undergoes similar changes, thus proving the importance of unsaturated fatty acids in maintenance of cell membrane integrity at low temperature conditions.

17.5.3 Protection Against Reactive Oxygen Species (ROS)

Many studies confirm the existence of oxidative stress at low temperatures because of high gas solubility and iron uptake and enhanced rates of certain enzymatic activities (Chattopadhyay et al. 2011; De Maayer et al. 2014). In addition to these factors, iron acquisition (which influences production of oxygen radicals) and activity of Fe-S cofactors also were found to increase at −5 °C in *Rhodococcus* sp. JG3 and *Polaromonas* sp. Eur3 1.2.1 as compared to higher temperature (25 °C). Some studies confirmed the reduction of reactive oxygen species produced during oxidative stress by production of detoxifying enzymes like superoxide dismutase and catalase. To counteract the activities, *Polaromonas* sp. Eur3 1.2.1 boosted the production of catalases and cytochrome peroxidase by twofold. Similarly, *Rhodococcus* sp. JG3 enhanced the production of catalase by three times. It also produced superoxide dismutase that transforms superoxide radical to O_2 or H_2O_2 (Raymond-Bouchard et al. 2018). Upregulation of catalase coding genes was found in *Pseudomonas syringae* pv phaseolicola under cold conditions (Arvizu-Gomez et al. 2013). When freeze-thawed, activation of oxidative stress pathways was found in *Arthrobacter* sp., *Dioszegia hungarica*, and *P. syringae* (Joly et al. 2015). There are a few key antioxidants like glutathione and carnitine that are related to oxidative stress under cold temperature conditions. Glutathione constitutes cysteine, glutamate, and glycine that form a tripeptide, while carnitine is important for lipid metabolism. The concentrations of both glutathione and carnitine were reported to be increased at 5 °C in *P. syringae* (Jousse et al. 2018). Production of glutathione increased oxidative stress response in *Lactobacillus sanfranciscensis* (Zhang et al. 2012).

17.5.4 Changes in Energy Metabolism and Growth

The changes in metabolism of cold-adapted microorganisms are variable. Cold stress often decelerates enzymatic activities and molecular transportation in few psychrophiles (Tribelli et al. 2015), while some are metabolically highly active (Chong et al. 2011). Maintenance of metabolic activity is a challenge and is tackled by microorganisms by making adjustments in energy metabolism in order to increase yields of ATP (Amato et al. 2015). *Pseudomonas syringae* PDD-32b-74 degraded glycogen to enhance the concentrations of ATP, glucose, and trehalose under cold conditions, which indicated activation of carbon metabolism and TCA (Jousse et al. 2018). A psychrophilic *Propionibacterium* sp. metabolized excess lactate due to promoted expression of lactate dehydrogenase enzyme under low temperature (Dalmasso et al. 2012). A similar kind of overexpression of such enzymes is reported in *Psychrobacter* (Bakermans et al. 2007). *Planococcus halocryophilus* showed a varied response by enhancing activity of many enzymes like FMN reductase, NADPH dehydrogenase, and glutamate-1-semialdehyde aminotransferase, except for phosphoglycerate mutase (in EMP), UTP-glucose-1-phosphate uridylyltransferase (in carbon metabolism), and ATP synthase which had a reduced activity (Raymond-Bouchard et al. 2018). Proteomic study in *Aurantiochytrium* under cold stress revealed that enzymes pertaining to glycolysis and TCA were downregulated. However, the energy needs of the cell were met via pentose phosphate pathway (Ma et al. 2017a).

17.5.5 Synthesis of Cryoprotectants, Compatible Solutes, and Exopolysaccharides (EPS)

Cold stress causes formation of ice crystals in cytoplasm that leads to osmotic instability. Microorganisms are known to produce low molecular weight molecules known as "cryoprotectants" or "compatible solutes" to adapt such conditions (Shivaji and Prakash 2010). Several studies report the induction of biosynthetic genes and transportation of these compounds in various microorganisms due to fall in temperatures (Rodrigues et al. 2008; Campanaro et al. 2011; Mykytczuk et al. 2013, 2016). However, microorganisms usually prefer to uptake the compounds from their surroundings instead of newly synthesizing them, thus conserving energy (Raymond-Bouchard et al. 2018). Some of these important compounds belong to carbohydrates, derivatives of amino acids, polyols like glycine, trehalose, glycerol, mannitol, proline, betaine, carnitine, etc. (Table 17.2). Apart from stabilization of plasma membrane and maintenance of cell integrity, they also are known to inhibit denaturation and/or aggregation of proteins. They are found to decrease the limits of

Table 17.2 List of different compounds (cryoprotectants, compatible solutes, and EPS) produced by microorganisms during cold stress

Cryoprotectant/compatible solutes/EPS	Microorganism	References
Spermidine and putrescine	*Pseudomonas helmanticensis*	Kumar et al. (2020b)
Glycine betaine, choline, EPS	*Rhodococcus* sp.	Raymond-Bouchard et al. (2018)
Unknown hydrophilic solutes, EPS	*Polaromonas* sp.	Raymond-Bouchard et al. (2018)
EPS	*Winogradskyella* sp., *Colwellia* sp., *Shewanella* sp.	Caruso et al. (2017a)
Ectoine	*Vibrio anguillarum*	Ma et al. (2017b)
Glutathione, ornithine, proline, EPS, and mannose	*Pseudoalteromonas haloplanktis*	Mocali et al. (2017), Nichols et al. (2005b), Caruso et al. (2017b)
Trehalose	*Burkholderia pseudomallei*	Vanaporn et al. (2017)
EPS	*Pseudomonas* sp.	Carrión et al. (2015)
Sarcosine, threonine, and valine	*Mesorhizobium* sp.	Ghobakhlou et al. (2015)
Glycine betaine	*Colwellia psychrerythraea*	Collins and Deming (2013)
Betaine, carnitine, choline, EPS	*Colwellia psychrerythraea*	Methé et al. (2005)
EPS	*P. nigrifaciens*	Mancuso Nichols et al. (2004)
Glycine betaine	*Listeria monocytogenes*	Mendum and Smith (2002)
Glycine betaine	*Synechococcus* sp.	Deshnium et al. (1997)

water activity (a_w) in some microorganisms (Stevenson et al. 2015). They help in quenching reactive oxygen species. Exopolysaccharide (EPS) production is yet another effective way of managing cold stress because they drop the freezing point and temperature needed for ice nucleation of water due to their high polyhydroxyl structure. They also increase water holding capacity, nutrient availability, metal ion chelation, surface adhesion, aggregation, biofilm formation, and protection of enzymes against cold stress (De los Ríos et al. 2004; Nichols et al. 2005a; Qin et al. 2007). EPS also has shown to increase disorder and pore density in ice crystals, thereby modifying the microstructure and salt content of ice crystals. This gives such microorganisms a chance to survive and colonize icy environments by altering their salt content (Ewert and Deming 2011; Krembs et al. 2011; De Maayer et al. 2014).

17.5.6 Cold Shock Proteins and Enzymes

Enzymes play a vital role in metabolic processes. They are greatly influenced by temperatures. Cold stress greatly affects enzymatic systems, for which cold-adapted microorganisms produce special proteins, "cold shock proteins" (CSP). They help in proper folding of polypeptides and prevent the loss of their functionality (Williams et al. 2017). Protein folding is coupled with and enhanced production of chaperone elements in psychrophiles (Cipolla et al. 2012; Williams et al. 2017). Most of the CSPs have high specific activity and low thermal stability because they contain more hydrophobic residues than uncharged or aromatic residues (Goldstein 2007; Marx et al. 2007). They bind to nucleic acids and regulate mRNA synthesis, protein synthesis, folding, and degradation of mRNA. Some RNA helicases are found to function as chaperones to inhibit secondary RNAs. These have been increasingly synthesized and help in stabilization of cold-adapted processes like transcription and translation (De Maayer et al. 2014). Overexpression of RNA helicases, DNA helicases, and chaperones in many folds is a proof that cold-adapted microorganisms efficiently carry out metabolic activities under cold stress (Raymond-Bouchard et al. 2018). Cold-adapted enzymes exhibit high turnover and catalytic efficiency, even at very low temperatures. They are characterized by pliable structures and inherit low activation enthalpy and negative entropy of activation in comparison with their homologs in mesophiles. Some studies confirm that cold-adapted enzymes are longer than homologs found in mesophiles and contain loops in their structures that aid in adjusting to cold stress (Miao et al. 2016). Their pliability is attributed to features like high hydrophobicity, longer loops, reduced number of disulfide bridges, weak subunit interaction, etc. (Gerday et al. 2000; Siddiqui and Cavicchioli 2006; Ramli et al. 2012, 2013). Some cold-adapted microorganisms also produce anti-freeze or ice-binding proteins (AFP) which are known to control formation of ice crystals by altering freezing point (Kawahara 2002).

17.6 Cold-Adapted Microbes as Biocontrol Agents

In a natural ecosystem, the key players involving ecological interactions are cold-adapted microorganisms. Some microorganisms colonizing rhizospheric region of plants prove to be lethal to plants, while some microorganisms, through indirect and direct mechanisms, protect plants and play a role in the development and growth of plants. These "protective" microbes are referred to as biocontrol agents (BCA) (Singh et al. 2019b). By the secretion of a significant number of secondary metabolites, BCA either limit or impair the development of pathogens (including siderophore, soluble and volatile antibiotics, quorum sensing interfering agents). Directly, BCA produce enzymes or metabolites or involve in a hyperparasitic behavior that interferes with the growth of pathogens. Production and release of secondary metabolites of microbes are related to BCA and plant pathogenic microorganisms (i.e., low-molecular weight compounds, chemically diverse, not required for cell functioning, during late exponential phase growth secretion and production) (Karlovsky 2008). Indirectly, BCA controls plant pathogens via nutrient and space competence or by interfering with chemical communication. To control plant disease and pests, BCA are used because they are eco-friendly living cells and can be included as formulations in commercial biopesticides. In the current scenario, proven technology is the development of biopesticides from mesophilic microorganisms. At low temperature, these biopesticides are not active. BCA have been one of the most effective strategic tools in recent decades for the sustainable intensification of farming, particularly in the mountainous areas of the developing world without posing a significant environmental threat. As suggested by Swaminathan (2006), green revolution includes BCA-containing biopesticides. The second green revolution relies on sustainable agricultural practices to conserve the fragile ecological equilibrium of agroecosystems without further interference or deterioration of the soil environment. The purpose is to increase crop productivity, by adding necessary environmental dimensions, thereby intensifying agriculture (Swaminathan 2006).

The ability of a biocontrol agent to survive and proliferate is a crucial factor for its competitiveness. However, the activity of a specific microbe in the soil is often challenging to predict, as a bacterium's soil survival can be affected by soil temperature and other different variables. It is logical to assume that biocontrol agents are cold resistant, since certain fungal phytopathogens are most devastating when the soil temperature is very low. Many *Trichoderma* sp. strains were isolated by McBeath (1995), serving as low-temperature biocontrol agents (i.e., 4–10 °C) for a variety of fungal pathogens. In the Garhwal region of Indian Himalayas, *Pseudomonas* community was identified by Negi et al. (2005), which was cold tolerant in nature. At temperatures ranging between 4 and 25 °C, these strains produced siderophore and exhibited PGP activity. They suppressed root disease of garden pea upon seed treatment.

Moreno and Rojo (2014) identified isolates of *Pseudomonas* from cold environments which are most studied BCA and PGPR. Microorganisms are known to produce antimicrobial compounds of volatile nature, e.g., iron-chelating compounds, HCN, or biosurfactants (Tapia Vázquez et al. 2020). To alleviate stress caused by cold temperature and other biotic stresses in plants, induction of plant systemic resistance was done by some isolates belonging to genus *Pseudomonas* that induce plant systemic resistance (Ogata-Gutiérrez et al. 2017; Ortiz-Ojeda et al. 2017). *Pseudomonas* strain is isolated from cold Antarctic regions which produce both volatile metabolites like HCN and soluble-like siderophores that have potential to inhibit growth of plant pathogenic fungi and oomycetes. The important plant pathogen biocontrol method which is by gluconic acid production by bacteria was suggested. To cure the root disease in wheat which was caused by *Gaeumannomyces graminis* var. tritici, gluconic acid-producing *Pseudomonas* are found effective. *Gluconacetobacter diazotrophicus* is also related to the generation of gluconic acid in the broad spectrum of antimicrobial activity. From the glaciers of Venezuela, psychrotolerant and psychrophilic bacteria are isolated that have capacity to generate large amount of gluconic acid (Rondón et al. 2019). At the temperature 15 °C, the microorganism *Lactobacilli* has a good impact in controlling fruit and food spoilage. In recent time Daranas and coworker reported broad-scale ability of *Lactobacillus plantarum* to protect fruits, namely, strawberry, kiwifruit, and prunes, against pathogens like *Xanthomonas arboricola* pv. *pruni*, *P. syringae* pv. *actinidiae*, and *X. fragariae* (Daranas et al. 2019). In various environmental conditions, biological control has been recorded using in vivo and in vitro techniques which is also related to the generation of acidification through accumulation of lactic acid. *Trichoderma atroviride*, a hyper-mycoparasitic and psychrotolerant organism, has shown to hamper growth of pathogen, causing severe damages to aerial and terrestrial organs of plants in cold climate (e.g., snow mold pathogens like *Myriosclerotinia (Sclerotinia) borealis*, *Coprinus psychromorbidus*, *Microdochium nivale*, and several species of the *Typhula* genus). *Trichoderma gamsii* isolated from western part of Himalayas has shown antagonizing effect against *F. oxysporum* f. sp. *lycopersici*, thereby resisting *Fusarium* wilts in greenhouse tomatoes at 18–20 °C. *Rhodotorula mucilaginosa*, a Tibetan strain, is shown to inhibit at a temperature around 4 °C in density-dependent manner growth of *Penicillium expansum* that causes blue mold in pear fruits. In addition to active hydrolase secretion, yeast was found to outgrow pathogenic fungus by consuming available nutrients left in the medium. Cold-adapted *Meyerozyma guilliermondii* strain has shown to protect pear fruits from *P. expansum* (Yan et al. 2018).

Selvakumar et al. (2009b) isolated *Exiguobacterium acetylicum* 1P from North-west Indian Himalayan soils that produced siderophores at 4 °C in pot-cultural conditions. It was also found to suppress growth and development of *F. oxysporum*, *Phythium*, *Rhizoctonia solani*, and *Sclerotium rolfsii*. The strain PGERs17, a cold-tolerant species of *Pseudomonas*, could generate siderophores and HCN at 4 °C. It demonstrated inhibitory activity in three separate bioassays against many phytopathogens (Singh et al. 2020c). Mishra et al. (2008) reported that *S. rolfsii* and *R. solani* showed maximum relative growth, i.e., 100%, whereas *Pythium* sp. is 73.1% followed by *F. oxysporum* which is 19.7% respectively in volatile compound assays. Psychrotolerant *Streptomyces* strains were isolated by Malviya et al. (2009) from Indian Himalayan glacial sites that exhibited antagonism against many plant-pathogenic fungi. In a modern-day scenario that demands non-pesticide food products, many research efforts are required to develop cold-tolerant strains as biocontrol agents in temperate agriculture.

17.7 Ice Bacteria for Frost Management

Every year, agriculture is being badly affected by frost-damaged crops. Due to the nonuniform nature of different plant parts (stem, leaf, bud, and flowers) injury caused due to freezing is found out to be complex. Lindow (1983) found that the ice nucleation in different parts of plants is not endogenous but is induced by catalytic sites that are present in microbial parasites. Frost settling on the surface of plants in cold weather is harmful. Frost damage causes huge losses in the agriculture industry annually. Ice nucleation activities (INAs), similar to ice crystals in structure, induce freezing below zero temperature, thus restricting supercooling. They reduce the energy required for formation of ice by forming an arrangement similar to ice on water molecules that are in contact with the outer part of bacteria. There are broadly two types of bacteria in this aspect. “Ice plus” bacteria contain INA proteins that mimic the nucleating center of ice crystals (Lee et al. 1995). This influences formation of ice at low temperature. On the contrary, “ice minus” bacteria reduce ice nucleation temperature due to absence of INAs, thereby resisting ice formation (Zachariassen and Kristiansen 2000). Therefore, foliar application of "ice minus" bacteria might reduce risks of frost damage.

Pseudomonas syringae ice-nucleating strains improve the frost vulnerability of tomatoes and soybean when applied on leaves before low temperature stress and additionally impart pathogenesis to these plants. Recognition of the ice nucleation gene in *P. syringae* initially contributes to the synthesis of a mutant ice minus, which has been detected inactive for the promotion of ice nucleation in plant leaves (Anderson et al. 1982; Xu et al. 1998). A classic example of how a biocontrol agent eliminates a bacterial pathogen from the bacteria is an efficient and environmentally friendly way of managing freezing damages in plants, minimizing the number of ice-nucleating bacteria by various methods. Bacteria which are ice-minus are the mutated strains of the common wild-type *P. syringae* bacteria. The *P. syringae* wild type is known as ice plus due to its outer cell wall containing a surface protein, which helps to form a freeze. With the mutant *P. syringae*, a frost-facilitating surface protein fails, and therefore bacteria are not recognized as ice-minus bacteria and cannot facilitate frost formation. In the natural condition, both ice-plus and ice-minus bacteria are found. For spraying on crops, the ice-minus bacteria are manufactured by rDNA technology on a large scale.

17.8 Tolerance or Management of Chilling Resistance by Cold-Tolerant PGPRs

Cold-temperature stress has several impacts on the metabolism of plants and significantly decreases yield (Gupta et al. 2019). Several exploratory studies have been performed to overcome this with microbial strains. *Vitis vinifera* cv chardonnay explant in vitro inoculation using a PGPR *Burkholderia phytofirmans* strain PsJN at low temperature increases the growth of grapevines and physiological activities (Kumar et al. 2020c). There was a connection between the bacterial endophytic colonization, growth, and chill sensitivity at both temperatures, i.e., ambient (26 °C) and low (4 °C), of the grapevine plantlets. The key advantages of bacterization for root growth, i.e., 11.8-fold increase, were observed at 26 °C, while 10.7-fold was reported at 4 °C. Plant biomass increased by sixfold at 26 °C, whereas at 4 °C it was observed to increase by 2.2-fold. PsJN greatly increased cold resistance of plants relative to non-bacterized control inoculation. Furthermore, bacterized plantlets were substantially improved with starch, proline, and phenolics in contrast with the non-inoculated controls (Kumar et al. 2016b). These changes were associated with the rise in cold resistance of the grapevine plantlets (Barka et al. 2006).

17.9 Conclusion

High-altitude agriculture is a challenge due to significantly very low temperature conditions. Cold-adapted microorganisms are always under limelight when it comes to agriculture in cold, inhospitable environments, hills, and mountainous agroecosystems. Bioinoculants such as biofertilizers or biocontrol agents withstanding such low temperature are required in order to promote plant growth and development in cold regions. Selecting the indigenous microflora as bioinoculants has immense advantages because they are well adapted to cold conditions and may drive metabolic processes in plants efficiently. They are known to promote growth by N-fixation, P solubilization, siderophore production, ACC deaminase production, biocontrol, etc., thereby alleviating cold stress in plants. By employing various mechanisms to combat cold stress like altering cell envelope, energy metabolism and membrane fluidity, quenching reactive oxygen species, and production of compatible solutes, cold shock proteins, and EPS, they make peace with subzero temperatures too. With diverse applications of these cold-adapted bioinoculants, agriculture in cold regions can turn more productive. However, microbial communities in cold regions are yet to be fully tapped for their potential.

References

Altshuler I, Goordial J, Whyte LG (2017) Microbial life in permafrost. In: Margesin R (ed) Psychrophiles: from biodiversity to biotechnology. Springer, Cham. https://doi.org/10.1007/978-3-319-57057-0_8

Amato P, Joly M, Schaupp C, Attard E, Möhler O, Morris CE, Brunet Y, Delort AM (2015) Survival and ice nucleation activity of bacteria as aerosols in a cloud simulation chamber. Atmos Chem Phys 15(11):6455–6465. https://doi.org/10.5194/acp-15-1-2015

Anderson RM, Gordon DM, Crawley MJ, Hassell MP (1982) Variability in the abundance of animal and plant species. Nature 296:243–248. https://doi.org/10.1038/296245a0

Andrews JH, Harris RF (2003) The ecology and biogeography of microorganisms on plant surfaces. Annu Rev Phytopathol 38:145–180. https://doi.org/10.1146/annurev.phyto.38.1.145

Arvizu-Gomez JL, Hernadez-Morales A, Agilar JRP, Alavrez-Morales A (2013) Transcriptional profile of *P. syringae* pv *phaesicola* NPS3121 at low temperature: physiology of phytopathogenic bacteria. BMC Microbiol 13:81. https://doi.org/10.1186/1471-2180-13-81

Bakermans C (2008) Limits for microbial life at subzero temperatures. In: Psychrophiles: from biodiversity to biotechnology. Springer, Cham, pp 17–28. https://doi.org/10.1007/978-3-540-74335-4_2

Bakermans C (2017) Determining the limits of microbial life at subzero temperatures. In: Psychrophiles: from biodiversity to biotechnology. Springer, Cham, pp 21–38. https://doi.org/10.1007/978-3-319-57057-0_2

Bakermans C, Tollaksen SL, Giometti CS, Wilkerson C, Tiedje JM, Thomashow MF (2007) Proteomic analysis of *Psychrobacter cryohalolentis* K5 during growth at subzero temperatures. Extremophiles 11:343–354. https://doi.org/10.1007/s00792-006-0042-1

Bar Dolev M, Bernheim R, Guo S, Davies PL, Braslavsky I (2016) Putting life on ice: bacteria that bind to frozen water. J R Soc Interface 13:20160210. https://doi.org/10.1098/rsif.2016.0210

Barka EA, Nowak J, Clément C (2006) Enhancement of chilling resistance of inoculated grapevine plantlets with a plant growth-promoting rhizobacterium, *Burkholderia phytofirmans* strain PsJN. Appl Environ Microbiol 72:7246–7252. https://doi.org/10.1128/AEM.01047-06

Bilska-Kos A, Solecka D, Dziewulska A, Ochodzki P, Jończyk M, Bilski H, Sowiński P (2017) Low temperature caused modifications in the arrangement of cell wall pectins due to changes of osmotic potential of cells of maize leaves (*Zea mays* L.). Protoplasma 254(2):713–724. https://doi.org/10.1007/s00709-016-0982-y

Bisht SC, Joshi GK, Mishra PK (2014) CspA encodes a major cold shock protein in Himalayan psychrotolerant *Pseudomonas* strains. Interdiscip Sci: Comput Life Sci 6:140–148. https://doi.org/10.1007/s12539-013-0015-x

Bisht SC, Mishra PK, Joshi GK (2013) Genetic and functional diversity among root-associated psychrotrophic Pseudomonad's isolated from the Himalayan plants. Arch Microbiol 195 (9):605–615. https://doi.org/10.1007/s00203-013-0908-4

Bradley CA (2008) Effect of fungicide seed treatments on stand establishment, seedling disease, and yield of soybean in North Dakota. Plant Dis 92(1):120–125. https://doi.org/10.1094/PDIS-92-1-0120

Bric JM, Bostock RM, Silverstone SR (1991) Rapid in situ assay for indole acetic acid production by bacteria immobilized on a nitrocellulose membrane. Appl Microbiol 57:535–538

Brown SP, Ungerer MC, Jumpponen A, Graham LE (2016) A community of clones: snow algae are diverse communities of spatially structured clones. Int J Plant Sci 177(5):432–439. https://doi.org/10.1086/686019

Burchett S, Niven S, Fuller MP (2006) The effect of cold acclimation on the water relations and freezing tolerance of *Hordeum vulgare* L. Cryo Letters 27:295–303

Campanaro S, Williams TJ, Burg DW, De Francisci D, Treu L, Lauro FM, Cavicchioli R (2011) Temperature-dependent global gene expression in the Antarctic archaeon *Methanococcoides burtonii*. Environ Microbiol 13:2018–2038. https://doi.org/10.1111/j.1462-2920.2010.02367.x

Carillo S, Casillo A, Pieretti G, Parrilli E, Sannino F, Bayer-Giraldi M, Cosconati S, Novellino E, Ewert M, Deming JW, Lanzetta R (2015) A unique capsular polysaccharide structure from the psychrophilic marine bacterium *Colwellia psychrerythraea* 34H that mimics antifreeze (glyco) proteins. J Am Chem Soc 137(1):179–189. https://doi.org/10.1021/ja5075954

Carrión O, Delgado L, Mercade E (2015) New emulsifying and cryoprotective exopolysaccharide from Antarctic *Pseudomonas* sp. ID1. Carbohydr Polym 6(117):1028–1034. https://doi.org/10.1016/j.carbpol.2014.08.060

Caruso C, Rizzo C, Mangano S, Poli A, Di Donato P, Finore I, Nicolaus B, Di Marco G, Michaud L, Lo Giudice A (2017a) Production and biotechnological potentialities of extracellular polymeric substances from sponge-associated Antarctic bacteria. Appl Environ Microbiol 84(4):e01624. https://doi.org/10.1128/AEM.01624-17

Caruso C, Rizzo C, Mangano S, Poli A, Di Donato P, Nicolaus B, Di Marco G, Michaud L, Giudice AL (2017b) Extracellular polymeric substances with metal adsorption capacity produced by *Pseudoalteromonas* sp. MER144 from Antarctic seawater. Environ Sci Pollut Res Int 25:4667–4677. https://doi.org/10.1007/s11356-017-0851-z

Casillo A, Parrilli E, Sannino F, Mitchell DE, Gibson MI, Marino G, Lanzetta R, Parrilli M, Cosconati S, Novellino E, Randazzo A (2017) Structure-activity relationship of the exopolysaccharide from a psychrophilic bacterium: a strategy for cryoprotection. Carbohydrate Polym 156:364–371. https://doi.org/10.1016/j.carbpol.2016.09.037

Chandler S (2018) The effect of temperature on cell membranes. Symp Soc Exp Biol 42:237–258

Chattopadhyay MK, Raghu G, Sharma YV, Biju AR, Rajasekharan MV, Shivaji S (2011) Increase in oxidative stress at low temperature in an antarctic bacterium. Curr Microbiol 62:544–546. https://doi.org/10.1007/s00284-010-9742-y

Chaya A, Kurosawa N, Kawamata A, Kosugi M, Imura S (2019) Community structures of Bacteria, archaea, and eukaryotic microbes in the freshwater glacier Lake Yukidori-Ike in Langhovde, East Antarctica. Diversity 11(7):105. https://doi.org/10.3390/d11070105

Cheng Z, Park E, Glick BR (2007) 1-Aminocyclopropane-1-carboxylate (ACC) deaminase from *Pseudomonas putida* UW4 facilitates the growth of canola in the presence of salt. Can J Microbiol 53:912–918. https://doi.org/10.1139/W07-050

Chong GL, Chu WL, Othman RY, Phang SM (2011) Differential gene expression of an Antarctic Chlorella in response to temperature stress. Polar Biol 34(5):637–645. https://doi.org/10.1007/s00300-010-0918-5

Ciobanu MC, Burgaud G, Dufresne A, Breuker A, Rédou V, Maamar SB, Gaboyer F, Vandenabeele-Trambouze O, Lipp JS, Schippers A, Vandenkoornhuyse P (2014)

Microorganisms persist at record depths in the subseafloor of the Canterbury Basin. ISME J 8 (7):1370–1380. https://doi.org/10.1038/ismej.2013.250

Cipolla A, Delbrassine F, Da Lage JL, Feller G (2012) Temperature adaptations in psychrophilic, mesophilic and thermophilic chloride-dependent alpha-amylases. Biochimie 94(9):1943–1950. https://doi.org/10.1016/j.biochi.2012.05.013

Clarke A, Morris GJ, Fonseca F, Murray BJ, Acton E, Price HC (2013) A low temperature limit for life on earth. PLoS One 8(6):e66207. https://doi.org/10.1371/journal.pone.0066207

Collins RE, Deming JW (2013) An inter-order horizontal gene transfer event enables the catabolism of compatible solutes by *Colwellia psychrerythraea* 34H. Extremophiles 17(4):601–610. https://doi.org/10.1007/s00792-013-0543-7

D'Amico S, Colins T, Marx JC, Feller G, Gerday C (2006) Psychrophilic microorganisms: a challenge for life. EMBO Rep 7:385–389. https://doi.org/10.1038/sj.embor.7400662

Dalmasso M, Aubert J, Briard-Bion V, Chuat V, Deutsch SM (2012) A temporal -omic study of *Propionibacterium freudenreichii* CIRM-BIA1T adaptation strategies in conditions mimicking cheese ripening in the cold. PLoS One 7(1):e29083. https://doi.org/10.1371/journal.pone.0029083

Daranas N, Roselló G, Cabrefiga J, Donati I, Francés J, Badosa E, Spinelli F, Montesinos E, Bonaterra A (2019) Biological control of bacterial plant diseases with Lactobacillus plantarum strains selected for their broad-spectrum activity. Ann Appl Biol 174:92–105. https://doi.org/10.1111/aab.12476

Das K, Katiyar V, Goel R (2003) 'P' solubilization potential of plant growth promoting Pseudomonas mutants at low temperature. Microbiol Res 158(4):359–362. https://doi.org/10.1078/0944-5013-00217

De los Ríos A, Wiezchos J, Sancho LG, Ascaso C (2004) Exploring the physiological state of continental Antarctic endolithic microorganisms by microscopy. FEMS Microbiol Ecol 50:143–152. https://doi.org/10.1016/j.femsec.2004.06.010

De Maayer P, Anderson D, Cary C, Cowan DA (2014) Some like it cold: understanding the survival strategies of psychrophiles. EMBO Rep 15:508–517. https://doi.org/10.1002/embr.201338170

Deming JW (2002) Psychrophiles and polar regions. Curr Opin Microbiol 5:301–309. https://doi.org/10.1016/s1369-5274(02)00329-6

Deshnium P, Gombos Z, Nishiyama Y, Murata N (1997) The action in vivo of glycine betaine in enhancement of tolerance of *Synechococcus* sp. strain PCC 7942 to low temperature. J Bacteriol 179:339–344. https://doi.org/10.1128/jb.179.2.339-344.1997

Dev S, Saha S, Kurade MB, Salama ES, El-Dalatony MM, Ha GS, Chang SW, Jeon BH (2019) Perspective on anaerobic digestion for biomethanation in cold environments. Renew Sust Energ Rev 103:85–95. https://doi.org/10.1016/j.rser.2018.12.034

Dhakar K, Pandey A (2016) Wide pH range tolerance in extremophiles: towards understanding an important phenomenon for future biotechnology. Appl Microbiol Biotechnol 100:2499–2510. https://doi.org/10.1007/s00253-016-7285-2

Edwards A, Cameron KA (2017) Microbial life in supraglacial environments. In: Psychrophiles: from biodiversity to biotechnology. Springer, Cham, pp 57–81. https://doi.org/10.1007/978-3-319-57057-0_4

Ewert M, Deming JW (2011) Selective retention in saline ice of extracellular polysaccharides produced by the cold-adapted marine bacterium *Colwellia psychrerythraea* strain 34H. Ann Glaciol 52:111–117. https://doi.org/10.3189/172756411795931868

Fitter AH, Hay RKM (1981) Environmental physiology of plants. Academic Press, New York

Fonseca F, Meneghel J, Cenard S, Passot S, Morris GJ (2016) Determination of intracellular vitrification temperatures for unicellular microorganisms under conditions relevant for cryopreservation. PLoS One 11(4):e0152939. https://doi.org/10.1371/journal.pone.0152939

Forster J (1887) Über eihnigh Eigenschafter leuchtender Bakterien. Centr Bakteriol Rev Parasitenk 2:337–340

Frey B, Rime T, Phillips M, Stierli B, Hajdas I, Widmer F, Hartmann M (2016) Microbial diversity in European alpine permafrost and active layers. FEMS Microbiol Ecol 92(3):fiw018. https://doi.org/10.1093/femsec/fiw018

Gautam N, Sharma P, Rana JC, Singh M (2019) Plant growth promoting traits of a novel psychrotrophic bacterium *Virdibacillus arenosi* PH15 isolated from rhizosphere of *Podophyllum hexandrum* Royle. Acad J Med Plants 7(1):013–019. https://doi.org/10.15413/ajmp.2018.0165

Gerday C, Aittaleb M, Bentahir M, Chessa JP, Claverie P, Collins T, D'Amico S, Dumont J, Garsoux G, Georlette D, Hoyoux A (2000) Cold-adapted enzymes: from fundamentals to biotechnology. Trends Biotechnol 18(3):103–107. https://doi.org/10.1016/s0167-7799(99)01413-4

Ghadirnezhad R, Fallah A (2014) Temperature effect on yield and yield components of different rice cultivars in flowering stage. Int J Agron 2014:846707. https://doi.org/10.1155/2014/846707

Ghobakhlou AF, Johnston A, Harris L, Antoun H, Laberge S (2015) Microarray transcriptional profiling of Arctic *Mesorhizobium* strain N33 at low temperature provides insights into cold adaption strategies. BMC Genomics 16:383. https://doi.org/10.1186/s12864-015-1611-4

Glick BR, Penrose DM, Jiping L (1998) A model for the lowering plant ethylene concentrations by plant growth promoting bacteria. J Theor Biol 190:63–68. https://doi.org/10.1006/jtbi.1997.0532

Gokul JK, Hodson AJ, Saetnan ER, Irvine-Fynn TD, Westall PJ, Detheridge AP, Takeuchi N, Bussell J, Mur LA, Edwards A (2016) Taxon interactions control the distributions of cryoconite bacteria colonizing a high Arctic ice cap. Mol Ecol 25:3752–3767. https://doi.org/10.1111/mec.13715

Goldstein A (1995) Recent progress in understanding the molecular genetics and biochemistry of calcium phosphate solubilization by gram-negative bacteria. Biol Agric Hortic 12:185–193. https://doi.org/10.1080/01448765.1995.9754736

Goldstein RA (2007) Amino-acid interactions in psychrophiles, mesophiles, thermophiles, and hyperthermophiles: insights from the quasi-chemical approximation. Protein Sci 16(9):1887–1895. https://doi.org/10.1110/ps.072947007

Goordial J, Davila A, Lacelle D, Pollard W, Marinova MM, Greer CW, DiRuggiero J, McKay CP, Whyte LG (2016) Nearing the cold-arid limits of microbial life in permafrost of an upper dry valley, Antarctica. ISME J 10:1613–1624. https://doi.org/10.1038/ismej.2015.239

Greenland D, Losleben M (2001) Structure and function of an alpine ecosystem. In: Bowman WD, Seastedt TR (eds) Climate. Oxford University Press, Niwot Ridge, CO, pp 15–31

Gulati A, Vyas P, Rai P, Kasana RC (2009) Plant growth promoting and rhizosphere-competent *Acinetobacter rhizosphaerae* strain BIHB 723 from the cold deserts of the Himalayas. Curr Microbiol 58:371–377. https://doi.org/10.1007/s00284-008-9339-x

Gupta A, Singh PP, Singh P, Singh K, Singh AV, Singh SK, Kumar A (2019) Medicinal plants under climate change: impacts on pharmaceutical properties of plants. In: Climate change and agricultural ecosystems. Woodhead Publishing, Cambridge, UK, pp 181–209. https://doi.org/10.1016/B978-0-12-816483-9.00008-6

Gupta A, Singh SK, Singh MK, Singh VK, Modi A, Singh PK, Kumar A (2020) Plant growth–promoting rhizobacteria and their functional role in salinity stress management. In: Abatement of environmental pollutants. Elsevier, Amsterdam, pp 151–160. https://doi.org/10.1016/B978-0-12-818095-2.00007-2

Haggblom M, Margesin R (2005) Microbial life in cold ecosystems. FEMS Microbiol Ecol 53:186–188. https://doi.org/10.1016/j.femsec.2005.03.006

Haldiman P (1998) Low growth temperature induced changes to pigment composition and photosynthesis in *Zea mays* genotypes differing in chilling sensitivity. Plant Cell Environ 21:200–208. https://doi.org/10.1046/j.1365-3040.1998.00260.x

Harrold ZR, Skidmore ML, Hamilton TL, Desch L, Amada K, Van Gelder W, Glover K, Roden EE, Boyd ES (2016) Aerobic and anaerobic thiosulfate oxidation by a cold-adapted, subglacial chemoautotroph. Appl Environ Microbiol 82(5):1486–1495. https://doi.org/10.1128/aem.03398-15

Hassan N, Rafiq M, Hayat M, Shah AA, Hasan F (2016) Psychrophilic and psychrotrophic fungi: a comprehensive review. Rev Environ Sci Biotechnol 15(2):147–172. https://doi.org/10.1007/s11157-016-9395-9

Hauptmann AL, Stibal M, Bælum J, Sicheritz-Pontén T, Brunak S, Bowman JS, Hansen LH, Jacobsen CS, Blom N (2014) Bacterial diversity in snow on north pole ice floes. Extremophiles 18(6):945–951. https://doi.org/10.1007/s00792-014-0660-y

Hoshino T, Matsumoto N (2012) Cryophilic fungi to denote fungi in the cryosphere. Fungal Biol Rev 26:102–105. https://doi.org/10.1016/j.fbr.2012.08.003

Hussain HA, Hussain S, Khaliq A, Ashraf U, Anjum SA, Men S, Wang L (2018) Chilling and drought stresses in crop plants: implications, cross talk, and potential management opportunities. Front Plant Sci 9:393. https://doi.org/10.3389/fpls.2018.00393

Imin N, Kerim T, Rolfe BG, Weinman JJ (2004) Effect of early cold stress on the maturation of rice anthers. Proteomics 4(7):1873–1882. https://doi.org/10.1002/pmic.200300738

Jiang QW, Kiyoharu O, Ryozo I (2002) Two novel mitogen-activated protein signaling components, OsMEK1 and OsMAP1, are involved in a moderate low-temperature signaling pathway in rice. Plant Physiol 129:1880–1891. https://doi.org/10.1104/pp.006072

Joly M, Amato P, Sancelme M, Vinatier V, Abrantes M, Deguillaume L, Delort AM (2015) Survival of microbial isolates from clouds toward simulated atmospheric stress factors. Atmos Environ 07:009

Jousse C, Dalle C, Canet I, Lagrée M, Traïkia M, Lyan B, Mendes C, Sancelme M, Amato P, Delort AM (2018) Metabolomic study of the response to cold shock in a strain of *Pseudomonas syringae* isolated from cloud water. Metabolomics 14(1):1–3. https://doi.org/10.1007/s11306-017-1295-7

Karlovsky P (2008) Secondary metabolites in soil ecology. In: Secondary metabolites in soil ecology. Springer, Berlin, pp 1–19

Karner MB, Delong EF, Karl DM (2001) Archeal dominance in the mesophilic zone of Pacific Ocean. Nature 409:507–510. https://doi.org/10.1038/35054051

Katiyar V, Goel R (2004) Siderophore mediated plant growth promotion at low temperature by mutant of fluorescent pseudomonad. Plant Growth Regul 42:239–244. https://doi.org/10.1023/B:GROW.0000026477.10681.d2

Kaushik R, Saxena AK, Tilak KVBR (2000) Selection of Tn5::lac Z mutants isogenic to wild type *Azospirillum brasilense* strains capable of growing at sub-optimal temperature. World J Microbiol Biotechnol 16:567–570. https://doi.org/10.1023/A:1008901331991

Kaushik R, Saxena AK, Tilak KVBR (2001) Selection and evaluation of *Azospirillum brasilense* strains capable of growing at sub-optimal temperature in rhizocoenosis with wheat. Folia Microbiol 46:327–332. https://doi.org/10.1007/BF02815622

Kawahara H (2002) The structures and functions of ice crystal controlling proteins from bacteria. J Biosci Bioeng 94:492–496. https://doi.org/10.1016/S1389-1723(02)80185-2

Kim KY, Hwangbo H, Kim YW, Kim HJ, Park KH, Kim YC, Seoung KY (2002) Organic acid production and phosphate solubilization by *Enterobacter intermedium* 60-2 G. Kor J Soil Sci Fertil 35:59–67

Kloepper JW, Schroth MN (1978) Plant growth-promoting rhizobacteria on radishes. In: proceedings of the 4th international conference on plant pathogenic bacteria. Gilbert-Clarey, Tours, France 879–882

Koch BJ, McHugh TA, Hayer M, Schwartz E, Blazewicz SJ, Dijkstra P, van Gestel N, Marks JC, Mau RL, Morrissey EM, Pett-Ridge J, Hungate BA (2018) Estimating taxon-specific population dynamics in diverse microbial communities. Ecosphere 9(1):e02090. https://doi.org/10.1002/ecs2.2090

Koh HY, Park H, Lee JH, Han SJ, Sohn YC, Lee SG (2017) Proteomic and transcriptomic investigations on cold-responsive properties of the psychrophilic Antarctic bacterium

Psychrobacter sp PAMC 21119 at subzero temperatures. Environ Microbiol 19(2):628–644. https://doi.org/10.1111/1462-2920.13578

Kostadinova M, Krumova E, Tosi S, Pashova AM (2009) Isolation and identification of filamentous fungi from island Livingston, Antarctica. Biotech Biotechnol Equip 23:267–270. https://doi.org/10.1080/13102818.2009.10818416

Krembs C, Eicken H, Deming JW (2011) Exopolymer alteration of physical properties of sea ice and implications for ice habitability and biogeochemistry in a warmer Arctic. Proc Natl Acad Sci U S A 108:3653–3658. https://doi.org/10.1073/pnas.1100701108

Kumar A, Droby S, Singh VK, Singh SK, White JF (2020a) Entry, colonization, and distribution of endophytic microorganisms in plants. In: Microbial endophytes. Woodhead Publishing, Cambridge, UK, pp 1–33. https://doi.org/10.1016/B978-0-12-819654-0.00001-6

Kumar A, Droby S, White JF, Singh VK, Singh SK, Zhimo VY, Biasi A (2020b) Endophytes and seed priming: agricultural applications and future prospects. In: Microbial endophytes. Woodhead Publishing, Cambridge, UK, pp 107–124. https://doi.org/10.1016/B978-0-12-819654-0.00005-3

Kumar A, Singh M, Singh PP, Singh SK, Raj P, Pandey KD (2016a) Antioxidant efficacy and curcumin content of turmeric (*Curcuma longa* L.) flower. Int J Curr Pharm Res 8:112–114

Kumar A, Singh M, Singh PP, Singh SK, Singh PK, Pandey KD (2016b) Isolation of plant growth promoting rhizobacteria and their impact on growth and curcumin content in *Curcuma longa* L. Biocatal Agric Biotechnol 8:1–7. https://doi.org/10.1016/j.bcab.2016.07.002

Kumar A, Verma H, Yadav A, Ansari WA, Singh PP, Singh SK, Singh PK, Pandey KD (2017) Plant growth promoting rhizobacteria of *Curcuma amada* (mango ginger). J Pure Appl Microbiol 11(1):513–520. https://doi.org/10.22207/JPAM.11.1.67

Kumar S, Suyal DC, Bhoriyal M, Goel R (2018) Plant growth promoting potential of psychrotolerant *Dyadobacter* sp. for pulses and finger millet and impact of inoculation on soil chemical properties and diazotrophic abundance. J Plant Nutr 41(8):1035–1046. https://doi.org/10.1080/01904167.2018.1433211

Kumar S, Suyal DC, Dhauni N, Bhoriyal M, Goel R (2014) Relative plant growth promoting potential of Himalayan Psychrotolerant *Pseudomonas jesenii* strain MP1 against native *Cicer arietinum* (L.)., *Vigna mungo* (L.) Hepper; *Vigna radiata* (L.) Wilczek., *Cajanus cajan* (L.) Millsp. and *Eleusine coracana* (L.) Gaertn. Afr J Microbiol Res 8(50):3931–3943. https://doi.org/10.5897/AJMR2014.7035

Kumar S, Suyal DC, Yadav A, Shouche Y, Goel R (2019) Microbial diversity and soil physiochemical characteristic of higher altitude. PLoS One 14(3):e0213844. https://doi.org/10.1371/journal.pone.0213844

Kumar S, Suyal DC, Yadav A, Shouche Y, Goel R (2020c) Psychrophilic *Pseudomonas helmanticensis* proteome under simulated cold stress. Cell Stress Chaperones 25:1025–1032. https://doi.org/10.1007/s12192-020-01139-4

Lee RE, Warren GJ, Gusta LV (1995) Biochemistry of bacterial ice nuclei. In: Ray F, Paul WK (eds) Biological ice nucleation and its application. APS Press, St Paul, MI, pp. 63–83

Li L, Ren M, Xu Y, Jin C, Zhang W, Dong X (2020a) Enhanced glycosylation of an S-layer protein enables a psychrophilic methanogenic archaeon to adapt to elevated temperatures in abundant substrates. FEBS Lett 594(4):665–677. https://doi.org/10.1002/1873-3468.13650

Li WL, Wu YZ, Gw Z (2020b) Hui Huang & Yong Wang Metabolic diversification of anaerobic methanotrophic archaea in a deep-sea cold seep. Mar Life Sci Technol 2:431–441. https://doi.org/10.1007/s42995-020-00057-9

Lindow SE (1983) The role of bacterial ice nucleation in frost injury to plants. Rev Phytopathol 21:363–384. https://doi.org/10.1146/annurev.py.21.090183.002051

Lootens P, Van Waes J, Carlier L (2004) Effect of a short photoinhibition stress on photosynthesis, chlorophyll a fluorescence, and pigment contents of different maize cultivars. Can a rapid and objective stress indicator be found? Photosynthetica 42(2):187–192. https://doi.org/10.1023/B:PHOT.0000040589.09614.a0

Lopatina A, Krylenkov V, Severinov K (2013) Activity and bacterial diversity of snow around Russian Antarctic stations. Res Microbiol 164:949–958. https://doi.org/10.1016/j.resmic.2013.08.005

Lopatina A, Medvedeva S, Shmakov S, Logacheva MD, Krylenkov V, Severinov K (2016) Metagenomic analysis of bacterial communities of Antarctic surface snow. Front Microbiol 7:398. https://doi.org/10.3389/fmicb.2016.00398

Lynch DH, Smith DL (1994) The effects of low temperature stress on two soybean (*Glycine max*) genotypes when combined with *Bradyrhizobium* strains of varying geographic origin. Physiol Plant 90:105–113. https://doi.org/10.1111/j.1399-3054.1994.tb02199.x

Ma Y, Wang Q, Xu W, Liu X, Gao X, Zhang Y (2017b) Stationary phase-dependent accumulation of ectoine is an efficient adaptation strategy in *Vibrio anguillarum* against cold stress. Microbiol Res 205:8–18. https://doi.org/10.1016/j.micres.2017.08.005

Ma Z, Tian M, Tan Y, Cui G, Feng Y, Cui Q, Song X (2017a) Response mechanism of the docosahexaenoic acid producer *Aurantiochytrium* under cold stress. Algal Res 1(25):191–199. https://doi.org/10.1016/j.algal.2017.05.021

Mackelprang R, Saleska SR, Jacobsen CS, Jansson JK, Taş N (2016) Permafrost meta-omics and climate change. Annu Rev Earth Planet Sci 44(1):439–462. https://doi.org/10.1146/annurev-earth-060614-105126

Malviya MK, Pandey A, Trivedi P, Gupta G, Kumar B (2009) Chitinolytic activity of cold tolerant antagonistic species of *Streptomyces* isolated from glacial sites of Indian Himalaya. Curr Microbiol 59:502–508. https://doi.org/10.1007/s00284-009-9466-z

Mancuso Nichols CA, Garon S, Bowman JP, Raguénès G, Guezennec J (2004) Production of exopolysaccharides by Antarctic marine bacterial isolates. J Appl Microbiol 96:1057–1066. https://doi.org/10.1111/j.1365-2672.2004.02216.x

Margesin R, Miteva V (2011) Diversity and ecology of psychrophilic microorganisms. Res Microbiol 162:346–361. https://doi.org/10.1016/j.resmic.2010.12.004

Margesin R, Neuner G, Storey KB (2007) Cold-loving microbes, plants, and animals—fundamental and applied aspects. Naturwissenschaften 94:77–99. https://doi.org/10.1007/s00114-006-0162-6

Margesin R, Schinner F (1999) Cold-adapted organisms. Ecology, physiology, enzymology and molecular biology. Springer, Berlin. https://doi.org/10.1007/978-3-662-06285-2

Margesin R, Zhang DC, Frasson D, Brouchkov A (2016) *Glaciimonas frigoris* sp. nov., a psychrophilic bacterium isolated from ancient Siberian permafrost sediment, and emended description of the genus *Glaciimonas*. Int J Syst Evol Microbiol 66(2):744–748. https://doi.org/10.1099/ijsem.0.000783

Marx JC, Collins T, D'Amico S, Feller G, Gerday C (2007) Cold-adapted enzymes from marine Antarctic microorganisms. Mar Biotechnol 9(3):293–304. https://doi.org/10.1007/s10126-006-6103-8

McBeath J (1995) Cold tolerant Trichoderma. US patent #5,418,165

McCue P, Zheng Z, Pinkham JL, Shetty K (2000) A model for enhanced pea seedling vigour following low pH and salicylic acid treatments. Proc Biochem 35:603–613. https://doi.org/10.1016/S0032-9592(99)00111-9

McKay IA, Djordjevic MA (1993) Production and excretion of nod metabolites by *Rhizobium leguminosarum* bv. *trifolii* are disrupted by the same environmental factors that reduce nodulation in the field. Appl Environ Microbiol 59:3385–3392

Mendum ML, Smith LT (2002) Characterization of glycine betaine porter I from listeria monocytogenes and its roles in salt and chill tolerance. Appl Environ Microbiol 68 (2):813–819. https://doi.org/10.1128/AEM.68.2.813-819.2002

Methé BA, Nelson KE, Deming JW, Momen B, Melamud E, Zhang X, Moult J, Madupu R, Nelson WC, Dodson RJ, Brinkac LM (2005) The psychrophilic lifestyle as revealed by the genome sequence of *Colwellia psychrerythraea* 34H through genomic and proteomic analyses. Proc Natl Acad Sci U S A 102:10913–10918. https://doi.org/10.1073/pnas.0504766102

Miao LL, Hou YJ, Fan HX, Qu J, Qi C, Liu Y, Li DF, Liu ZP (2016) Molecular structural basis for the cold adaptedness of the psychrophilic beta-glucosidase BglU in *Micrococcus antarcticus*. Appl Environ Microbiol 82:2021–2030. https://doi.org/10.1128/aem.03158-15
Mikucki JA, Auken E, Tulaczyk S, Virginia RA, Schamper C, Sørensen KI, Doran PT, Dugan H, Foley N (2015) Deep groundwater and potential subsurface habitats beneath an Antarctic dry valley. Nat Commun 6(1):1–9. https://doi.org/10.1038/ncomms7831
Misaghi IJ, Stowell LJ, Grogan RG, Spearman LC (1982) Fungistatic activity of water-soluble fluorescent pigments of fluorescent pseudomonads. Phytopathology 72:33–36. https://doi.org/10.1094/Phyto-72-33
Mishra P, Singh PP, Singh SK, Verma H (2019) Sustainable agriculture and benefits of organic farming to special emphasis on PGPR. In: Role of plant growth promoting microorganisms in sustainable agriculture and nanotechnology. Woodhead Publishing, Cambridge, UK, pp 75–87. https://doi.org/10.1016/B978-0-12-817004-5.00005-1
Mishra PK, Bisht SC, Ruwari P, Selvakumar G, Joshi GK, Bisht JK, Bhatt JC, Gupta HS (2011) Alleviation of cold stress in inoculated wheat (*Triticum aestivum* L.) seedlings with psychrotolerant pseudomonads from NW Himalayas. Arch Microbiol 193(7):497–513. https://doi.org/10.1007/s00203-011-0693-x
Mishra PK, Joshi P, Bisht SC, Bisht JK, Selvakumar G (2010) Cold-tolerant agriculturally important microorganisms. In: Plant growth and health promoting bacteria. Springer, Berlin, pp 273–296
Mishra PK, Mishra S, Bisht SC, Selvakumar G, Kundu S, Bisht JK, Gupta HS (2009) Isolation, molecular characterization and growth-promotion activities of a cold tolerant bacterium *Pseudomonas* sp. NARs9 (MTCC9002) from the Indian Himalayas. Biol Res 42(3):305–313. https://doi.org/10.4067/S0716-97602009000300005
Mishra PK, Mishra S, Selvakumar G, Bisht SC, Bisht JK, Gupta HS (2008) Characterization of a psychrotolerant plant growth promoting *Pseudomonas* sp. strain PGERs17 (MTCC 9000) isolated from North Western Indian Himalayas. Ann Microbiol 58:561–568. https://doi.org/10.1007/BF03175558
Miteva VI, Brenchley JE (2005) Detection and isolation of ultrasmall microorganisms from a 120,000-year-old Greenland glacier ice core. Appl Environ Microbiol 71:7806–7818. https://doi.org/10.1128/AEM.71.12.7806-7818.2005
Mocali S, Chiellini C, Fabiani A, Decuzzi S, de Pascale D, Parrilli E, Tutino ML, Perrin E, Bosi E, Fondi M, Giudice AL (2017) Ecology of cold environments: new insights of bacterial metabolic adaptation through an integrated genomic-phenomic approach. Sci Rep 7(1):839. https://doi.org/10.1038/s41598-017-00876-4
Moreno R, Rojo F (2014) Features of pseudomonads growing at low temperatures: another facet of their versatility. Environ Microbiol Rep 6:417–426. https://doi.org/10.1111/1758-2229.12150
Mykytczuk NCS, Lawrence JR, Omelon CR, Southam G, Whyte LG (2016) Microscopic characterization of the bacterial cell envelope of *Planococcus halocryophilus* Or1 during subzero growth at -15 °C. Polar Biol 39(4):701–712. https://doi.org/10.1007/S00300-015-1826-5
Mykytczuk NCS, Trevors JT, Leduc LG, Ferroni GD (2007) Bacterial cytoplasmic membrane polarization under environmental stress. Prog Biophys Mol Biol 95:60–82
Mykytczuk NC, Foote SJ, Omelon CR, Southam G, Greer CW, Whyte LG (2013) Bacterial growth at −15 C; molecular insights from the permafrost bacterium Planococcus halocryophilus Or1. ISME J 7:1211–1226. https://doi.org/10.1038/ismej.2013.8
Negi YK, Kumar J, Garg SK (2005) Cold-tolerant fluorescent Pseudomonas isolates from Garhwal Himalayas as potential plant growth promoting and biocontrol agents in pea. Curr Sci 89:2151–2156
Neilands JD (1995) Siderophores: structure and function of microbial iron transport compounds. J Biol Chem 270:26723–26726. https://doi.org/10.1074/jbc.270.45.26723
Nichols CA, Guezzenec J, Bowman JP (2005a) Bacterial exopolysaccharides from extreme environments with special consideration of the southern ocean, sea ice, and deep-sea hydrothermal vents: a review. Mar Biotechnol 7:253–271. https://doi.org/10.1007/s10126-004-5118-2

Nichols CM, Lardière SG, Bowman JP, Nichols PD, Gibson JA, Guézennec J (2005b) Chemical characterization of exopolysaccharides from Antarctic marine bacteria. Microb Ecol 49:578–589. https://doi.org/10.1007/s00248-004-0093-8

Nikrad MP, Kerkhof LJ, Häggblom MM (2016) The subzero microbiome: microbial activity in frozen and thawing soils. FEMS Microbiol Ecol 92:fiw081. https://doi.org/10.1093/femsec/fiw081

Ogata-Gutiérrez K, Chumpitaz-Segovia C, Lirio-Paredes J, Finetti-Sialer MM, Zúñiga-Dávila D (2017) Characterization and potentiajl of plant growth promoting rhizobacteria isolated from native Andean crops. World J Microbiol Biotechnol 33:1–3. https://doi.org/10.1007/s11274-017-2369-4

Ortiz-Ojeda P, Ogata-Gutiérrez K, Zúñiga-Dávila D (2017) Evaluation of plant growth promoting activity and heavy metal tolerance of psychrotrophic bacteria associated with maca (Lepidium meyenii Walp.) rhizosphere. AIMS Microbiol 3:279. https://doi.org/10.3934/microbiol.2017.2.279

Okon Y, Labandera-Gonzalez CA (1994) Agronomic applications of *Azospirillum*: an evaluation of 20 years worldwide field inoculation. Soil Biol Biochem 26:1591–1601. https://doi.org/10.1016/0038-0717(94)90311-5

Olien CR, Smith MN (1997) Ice adhesions in relation to freeze stress. Plant Physiol 60:499–503

Pandey A, Das N, Kumar B, Rinu K, Trivedi P (2008) Phosphate solubilization by *Penicillium* spp. isolated from soil samples of Indian Himalayan region. World J Microbiol Biotechnol 24:97–102. https://doi.org/10.1007/s11274-007-9444-1

Pandey A, Trivedi P, Kumar B, Palni LM (2006) Characterization of a phosphate solubilizing and antagonistic strain of *Pseudomonas putida* (B0) isolated from a sub-alpine location in the Indian central Himalaya. Curr Microbiol 53(2):102–107. https://doi.org/10.1007/s00284-006-4590-5

Patni B, Panwar AS, Negi P, Joshi GK (2018) Plant growth promoting traits of psychrotolerant bacteria: a boon for agriculture in hilly terrains. Plant Sci Today 5(1):24–28. https://doi.org/10.14719/pst.2018.5.1.352

Pavankumar TL, Mittal P, Hallsworth JE (2020) Molecular insights into the ecology of a psychrotolerant Pseudomonas syringae. Environ Microbiol. https://doi.org/10.1111/1462-2920.15304. Epub ahead of print. PMID: 33169927

Pedranzani H, Rodríguez-Rivera M, Gutiérrez M, Porcel R (2016) Arbuscular mycorrhizal symbiosis regulates physiology and performance of *Digitaria eriantha* plants subjected to abiotic stresses by modulating antioxidant and jasmonate levels. Mycorrhiza 26(2):141–152. https://doi.org/10.1007/s00572-015-0653-4

Pradhan S, Srinivas T, Pindi PK, Kishore KH, Begum Z, Singh PK, Singh AK, Pratibha M, Yasala AK, Reddy G (2010) Bacterial biodiversity from Roopkund glacier, Himalayan mountain ranges. India Extrem 14:377–395. https://doi.org/10.1007/s00792-010-0318-3

Prasad S, Manasa P, Buddhi S, Tirunagari P, Begum Z, Rajan S, Shivaji S (2014) Diversity and bioprospective potential (cold-active enzymes) of cultivable marine bacteria from the subarctic glacial fjord, Kongsfjorden. Curr Microbiol 68:233–238. https://doi.org/10.1007/s00284-013-0467-6

Prevost D, Drouin P, Antoun H (1999) The potential use of cold adapted rhizobia to improve nitrogen fixation in legumes cultivated in temperate regions. In: Margesin R, Schinner F (eds) Biotechnological application of cold-adapted organisms. Springer, Berlin, pp 161–176. https://doi.org/10.1007/978-3-642-58607-1_11

Prevost D, Drouin P, Laberge S, Bertrand A, Cloutier J, Levesque G (2003) Cold-adapted rhizobia for nitrogen fixation in temperate regions. Can J Bot 81:1153–1161. https://doi.org/10.1139/b03-113

Qin G, Zhu L, Chen Z, Wang PG, Zhang Y (2007) Structural characterization and ecological roles of a novel exopolysaccharide from the deep sea psychrotolerant bacterium *Pseudomonas* sp. SM9913. Microbiology 153:1566–1572. https://doi.org/10.1099/mic.0.2006/003327-0

Qin Y, Fu Y, Kanga W, Li H, Gao H, Vitalievitch KS, Liu H (2017) Isolation and identification of a cold-adapted bacterium and its characterization for biocontrol and plant growth-promoting activity. Ecol Eng 105:362–369. https://doi.org/10.1016/j.ecoleng.2017.04.045

Ramli ANM, Azhar MA, Shamsir MS, Rabu A, Murad AM, Mahadi NM, Illias RM (2013) Sequence and structural investigation of a novel psychrophilic α-amylase from *Glaciozyma antarctica* PI12 for cold adaptation analysis. J Mol Model 19:3369–3383. https://doi.org/10.1007/s00894-013-1861-5

Ramli ANM, Mahadi NM, Shamsir MS, Rabu A, Joyce-Tan KH, Murad AM, Illias RM (2012) Structural prediction of a novel chitinase from the psychrophilic *Glaciozyma antarctica* PI12 and an analysis of its structural properties and function. J Comput Aided Mol Des 26:947–961. https://doi.org/10.1007/s10822-012-9585-7

Raymond J, Fritsen C, Shen K (2007) An ice-binding protein from an Antarctic Sea ice bacterium. FEMS Microbiol Ecol 61:214–221. https://doi.org/10.1111/j.1574-6941.2007.00345.x

Raymond-Bouchard I, Chourey K, Altshuler I, Iyer R, Hettich RL, Whyte LG (2017) Mechanisms of subzero growth in the cryophile *Planococcus halocryophilus* determined through proteomic analysis. Environ Microbiol 19(11):4460–4479. https://doi.org/10.1111/1462-2920.13893

Raymond-Bouchard I, Tremblay J, Altshuler I, Greer CW, Whyte LG (2018) Comparative transcriptomics of cold growth and adaptive features of a eury- and steno-psychrophile. Front Microbiol 9:1565. https://doi.org/10.3389/fmicb.2018.01565

Rihan HZ, Al-Issawi M, Fuller MP (2017) Advances in physiological and molecular aspects of plant cold tolerance. J Plant Interact 12(1):143–157. https://doi.org/10.1080/17429145.2017.1308568

Rinu K, Malviya MK, Sati P, Tiwari SC, Pandey A (2013) Response of cold tolerant *Aspergillus* spp. to solubilization of Fe and Al phosphate in presence of different nutritional sources. ISRN Soil Sci 2013:59854. https://doi.org/10.1155/2013/598541

Rinu K, Sati P, Pandey A (2014) *Trichoderma gamsii* (NFCCI 2177): a newly isolated endophytic, psychrotolerant, plant growth promoting, and antagonistic fungal strain. J Basic Microbiol 54:408–417. https://doi.org/10.1002/jobm.201200579

Robertson AE, Munkvold GP (2012) Seedling diseases reported in corn and soybean, vol 193. Iowa State University, Ames, IA

Robinson JM, Mikucki JA (2016) Occupied and empty regions of the space of extremophile parameters. In: Habitability of the universe before earth. Elsevier, Amsterdam, pp 199–230. https://doi.org/10.1016/b978-0-12-811940-2.00009-5

Rodrigues DF, Ivanova N, He Z, Huebner M, Zhou J, Tiedje JM (2008) Architecture of thermal adaptation in an *Exiguobacterium sibiricum* strain isolated from 3 million year old permafrost: a genome and transcriptome approach. BMC Genomics 9:547. https://doi.org/10.1186/1471-2164-9-547

Rondón JJ, Ball MM, Castro LT, Yarzábal LA (2019) Eurypsychrophilic Pseudomonas spp. isolated from Venezuelan tropical glaciers as promoters of wheat growth and biocontrol agents of plant pathogens at low temperatures. Environ Sustain 2:265–275. https://doi.org/10.1007/s42398-019-00072-2

Ruelland E, Vaultier MN, Zachowski A, Hurry V (2009) Cold signalling and cold acclimation in plants. Adv Bot Res 49:35–150. https://doi.org/10.1016/S0065-2296(08)00602-2

Sahay H, Babu BK, Singh S, Kaushik R, Saxena AK, Arora DK (2013) Cold-active hydrolases producing bacteria from two different sub-glacial Himalayan lakes. J Basic Microbiol 53:703–714. https://doi.org/10.1002/jobm.201200126

Saleem M, Arshad M, Hussain S, Bhatti AS (2007) Perspective of plant growth promoting rhizobacteria (PGPR) containing ACC deaminase in stress agriculture. J Ind Microbiol Biotechnol 34:635–648. https://doi.org/10.1007/s10295-007-0240-6

Saralov (2019) Adaptivity of archaeal and bacterial extremophiles. Microbiology 88:379–401. https://doi.org/10.1134/S0026261719040106

Selvakumar G, Joshi P, Nazim S, Mishra PK, Bisht JK, Gupta (2009a) Phosphate solubilization and growth promotion by *Pseudomonas fragi* CS11RH1 (MTCC 8984) a psychrotolerant bacterium isolated from a high altitude Himalayan rhizosphere. Biologia 64:239–245. https://doi.org/10.2478/s11756-009-0041-7

Selvakumar G, Joshi P, Nazim S, Mishra PK, Kundu S, Gupta HS (2009b) *Exiguobacterium acetylicum* strain 1P (MTCC 8707) a novel bacterial antagonist from the North Western Indian Himalayas. World J Microbiol Biotechnol 25:131–137. https://doi.org/10.1007/s11274-008-9874-4

Selvakumar G, Kundu S, Joshi P, Gupta AD, Nazim S, Mishra PK, Gupta HS (2008a) Characterization of a cold-tolerant plant growth-promoting bacterium *Pantoea dispersa* 1A isolated from a sub-alpine soil in the North Western Indian Himalayas. World J Microbiol Biotechnol 24:955–960. https://doi.org/10.1007/s11274-007-9558-5

Selvakumar G, Mohan M, Kundu S, Gupta AD, Joshi P, Nazim S, Gupta HS (2008b) Cold tolerance and plant growth promotion potential of *Serratia marcescens* strain SRM (MTCC 8708) isolated from flowers of summer squash (*Cucurbita pepo*). Lett Appl Microbiol 46:171–175. https://doi.org/10.1111/j.1472-765X.2007.02282.x

Serrano M, Robertson A (2016) Cold stress at planting increased susceptibility to damping-off caused by *Pythium sylvaticum*. (Abstr.). Phytopathology 106:S4–S40

Sharma KD, Nayyar H (2014) Cold stress alters transcription in meiotic anthers of cold tolerant chickpea (*Cicer arietinum* L.). BMC Res Notes 7(1):1–3. https://doi.org/10.1186/1756-0500-7-717

Shivaji S, Prakash JS (2010) How do bacteria sense and respond to low temperature? Arch Microbiol 192:85–95. https://doi.org/10.1007/s00203-009-0539-y

Showalter GM, Deming JW (2018) Low-temperature chemotaxis, halotaxis and chemohalotaxis by the psychrophilic marine bacterium *Colwellia psychrerythraea* 34H. Environ Microbiol Rep 10 (1):92–101. https://doi.org/10.1111/1758-2229.12610

Siddiqui KS, Cavicchioli R (2006) Cold-adapted enzymes. In: Pandey A, Webb C, Soccol CR et al (eds) Enzyme technology. Springer, New York, pp 615–638

Singh D, Singh SK, Modi A, Singh PK, Zhimo VY, Kumar A (2020c) Impacts of agrochemicals on soil microbiology and food quality. In: Agrochemicals detection, treatment and remediation. Butterworth-Heinemann, Oxford, UK, pp 101–116. https://doi.org/10.1016/B978-0-08-103017-2.00004-0

Singh D, Singh SK, Singh VK, Gupta A, Aamir M, Kumar A (2020a) Plant growth–promoting bacteria and their role in environmental management. In: Abatement of environmental pollutants. Elsevier, Amsterdam, pp 161–175. https://doi.org/10.1016/B978-0-12-818095-2.00008-4

Singh N, Kumar S, Bajpai VK, Dubey RC, Maheashwari DK, Kang SC (2010) Biological control of *Macrophomina phaseolina* by chemotactic fluorescent *Pseudomonas aeruginosa* PN1 and its plant growth promontory activity in chir-pine. Crop Protect 29:1142–1147. https://doi.org/10.1016/j.cropro.2010.04.008

Singh SK, Singh PP, Gupta A, Singh AK, Keshri J (2019a) Tolerance of heavy metal toxicity using PGPR strains of Pseudomonas species. In: PGPR amelioration in sustainable agriculture. Woodhead Publishing, Cambridge, UK, pp 239–252. https://doi.org/10.1016/B978-0-12-815879-1.00012-4

Singh VK, Singh M, Singh SK, Kumar C, Kumar A (2019b) Sustainable agricultural practices using beneficial fungi under changing climate scenario. In: Climate change and agricultural ecosystems. Woodhead Publishing, Cambridge, UK, pp 25–42. https://doi.org/10.1016/B978-0-12-816483-9.00002-5

Singh VK, Singh SK, Singh PK, Verma H, Pandey KD, Singh PK, Kumar A (2020b) Impact of pesticides applications on the growth and function of cyanobacteria. In: Advances in cyanobacterial biology. Academic Press, Cambridge, MA, pp 151–162. https://doi.org/10.1016/B978-0-12-819311-2.00010-3

Sprent JI (1979) The biology of nitrogen-fixing organisms. McGraw-Hill Book, New York

Steponkus PL, Uemura M, Webb MS (1993) A contrast of the cryostability of the plasma membrane of winter rye and spring oat-two species that widely differ in their freezing tolerance and plasma membrane lipid composition. In: Steponkus PL (ed) Advances in low-temperature biology, vol 2. JAI Press, London, pp 211–312

Stevenson A, Cray JA, Williams JP, Santos R, Sahay R, Neuenkirchen N, McClure CD, Grant IR, Houghton JD, Quinn JP, Timson DJ (2015) Is there a common water-activity limit for the three domains of life? ISME J 9:1333–1351. https://doi.org/10.1038/ismej.2014.219

Stocker R, Seymour JR (2012) Ecology and physics of bacterial chemotaxis in the ocean. Microbiol Mol Bio Rev 76:792–812. https://doi.org/10.1128/MMBR.00029-12

Strancar J, Sentjurc M, Shara M (2000) Fast and accurate characterization of biological membranes by EPR spectral simulations of nitroxides. J Magn Reson 142:254–265. https://doi.org/10.1006/jmre.1999.1919

Subramanian P, Kim K, Krishnamoorthy R, Mageswari A, Selvakumar G, Sa T (2016) Cold stress tolerance in psychrotolerant soil bacteria and their conferred chilling resistance in tomato (*Solanum lycopersicum* Mill.) under low temperatures. PloS One 11(8):e0161592. https://doi.org/10.1371/journal.pone.0161592

Subramanian P, Krishnamoorthy R, Chanratana M, Kim K, Sa T (2015) Expression of an exogenous 1-aminocyclopropane-1-carboxylate deaminase gene in psychrotolerant bacteria modulates ethylene metabolism and cold induced genes in tomato under chilling stress. Plant Physiol Biochem 89:18–23. https://doi.org/10.1016/j.plaphy.2015.02.003

Sun B, Liu GL, Phan TT, Yang LT, Li YR, Xing YX (2017) Effects of cold stress on root growth and physiological metabolisms in seedlings of different sugarcane varieties. Sugar Tech 19:165–175. https://doi.org/10.1007/s12355-016-0452-z

Suzuki K, Nagasuga K, Okada M (2008) The chilling injury induced by high root temperature in the leaves of rice seedlings. Plant Cell Physiol 49:433–442. https://doi.org/10.1093/pcp/pcn020

Swaminathan MS (2006) An evergreen revolution. Crop Sci 46:2293–2303. https://doi.org/10.2135/cropsci2006.9999

Tapia-Vázquez I, Sánchez-Cruz R, Arroyo-Domínguez M, Lira-Ruan V, Sánchez-Reyes A, del Rayo Sánchez-Carbente M, Padilla-Chacón D, Batista-García RA, Folch-Mallol JL (2020) Isolation and characterization of psychrophilic and psychrotolerant plantgrowth promoting microorganisms from a high-altitude volcano crater in Mexico. Microbiol Res 1:232–126394. https://doi.org/10.1016/j.micres.2019.126394

Taha, Siddiqui KS, Campanaro S, Najnin T, Deshpande N, Williams TJ, Aldrich-Wright J, Wilkins M, Curmi PM, Cavicchioli R (2016) Single TRAM domain RNA-binding proteins in Archaea: functional insight from Ctr3 from the Antarctic methanogen Methanococcoides burtonii. Environ Microbiol 18:2810–2824. https://doi.org/10.1111/1462-2920.13229

Thiemann B, Imhoff JF (1991) The effect of salt on the lipid composition of *Ectothiorhodospira*. Arch Microbiol 156:376–384. https://doi.org/10.1007/BF00248714

Tiryaki D, Aydın İ, Atıcı Ö (2019) Psychrotolerant bacteria isolated from the leaf apoplast of cold-adapted wild plants improve the cold resistance of bean (*Phaseolus vulgaris* L.) under low temperature. Cryobiology 86:111–119. https://doi.org/10.1016/j.cryobiol.2018.11.001

Tribelli PM, Solar Venero EC, Ricardi MM, Gómez-Lozano M, Raiger Iustman LJ, Molin S, López NI (2015) Novel essential role of ethanol oxidation genes at low temperature revealed by transcriptome analysis in the antarctic bacterium *Pseudomonas extremaustralis*. PLoS One 10 (12):e0145353. https://doi.org/10.1371/journal.pone.0145353

Tripathi AK, Klingmuller W (1992) Temperature sensitivity of nitrogen fixation in *Azospirillum* sp. Can J Microbiol 38:1238–1241. https://doi.org/10.1139/m92-204

Trivedi P, Sa T (2008) *Pseudomonas corrugata* (NRRL B-30409) mutants increased phosphate solubilization, organic acid production, and plant growth at low temperatures. Curr Microbiol 56:140–144. https://doi.org/10.1007/s00284-007-9058-8

Turan M, Güllüce M, Çakmak R, Şahin F (2013) Effect of plant growth-promoting rhizobacteria strain on freezing injury and antioxidant enzyme activity of wheat and barley. J Plant Nutr 36 (5):731–748. https://doi.org/10.1080/01904167.2012.754038

Vanaporn M, Sarkar-Tyson M, Kovacs-Simon A, Ireland PM, Pumirat P, Korbsrisate S, Titball RW, Butt A (2017) Trehalase plays a role in macrophage colonization and virulence of *Burkholderia pseudomallei* in insect and mammalian hosts. Virulence 8:30–40. https://doi.org/10.1080/21505594.2016.1199316

Volkmar KM, Bremer E (1998) Effects of seed inoculation with a strain of *Pseudomonas fluorescens* on root growth and activity of wheat in well-watered and drought-stressed glass-fronted rhizotrons. Can J Plant Sci 78:545–551. https://doi.org/10.4141/P97-129

Vyas P, Joshi R, Sharma KC, Rahi P, Gulati A (2010) Cold-adapted and rhizosphere competent strain of *Rahnella* sp. with broad-spectrum plant growth-promotion potential. J Microbiol Biotechnol 20:1724–1734

Vyas P, Rahi P, Gulati A (2009) Stress tolerance and genetic variability of phosphate-solubilizing fluorescent *Pseudomonas* from the cold deserts of the trans-Himalayas. Microb Ecol 58:425–434. https://doi.org/10.1007/s00248-009-9511-2

Wang C, Wang C, Gao YL, Wang YP, Guo JH (2016a) A consortium of three plant growth promoting rhizobacterium strains acclimates *Lycopersicon esculentum* and confers a better tolerance to chilling stress. J Plant Growth Regul 35(1):54–64. https://doi.org/10.1007/s00344-015-9506-9

Wang W, Chen Q, Hussain S, Mei J, Dong H, Peng S (2016b) Pre-sowing seed treatments in direct seeded early rice: consequences for emergence, seedling growth and associated metabolic events under chilling stress. Sci Rep 6:19637. https://doi.org/10.1038/srep19637

Welti R, Li W, Li M, Sang Y, Biesiada H, Zhou HE, Rajashekar CB, Williams TD, Wang X (2002) Profiling membrane lipids in plant stress responses: role of phospholipase Dα in freezing-induced lipid changes in Arabidopsis. J Biol Chem 277(35):31994–32002. https://doi.org/10.1074/jbc.M205375200

Williams TJ, Liao Y, Ye J, Kuchel RP, Poljak A, Raftery MJ, Cavicchioli R (2017) Cold adaptation of the Antarctic haloarchaea *Halohasta litchfieldiae* and *Halorubrum lacusprofundi*. Environ Microbiol 19(6):2210–2227. https://doi.org/10.1111/1462-2920.13705

Xu H, Griffith M, Patten CL, Glick BR (1998) Isolation and characterization of an antifreeze protein with ice-nucleation activity from the plant growth promoting rhizobacterium *Pseudomonas putida* GR12-2. Can J Microbiol 44:64–73. https://doi.org/10.1139/cjm-44-1-64

Yadav AN, Sachan SG, Verma P, Saxena AK (2015b) Prospecting cold deserts of North-Western Himalayas for microbial diversity and plant growth promoting attributes. J Biosci Bioeng 119:683–693. https://doi.org/10.1016/j.jbiosc.2014.11.006

Yadav AN, Sachan SG, Verma P, Tyagi SP, Kaushik R, Saxena AK (2015a) Culturable diversity and functional annotation of psychrotrophic bacteria from cold desert of Leh Ladakh (India). World J Microbiol Biotechnol 31:95–108. https://doi.org/10.1007/s11274-014-1768-z

Yadav AN, Verma P, Kumar V, Sachan SG, Saxena AK (2017b) Extreme cold environments: a suitable niche for selection of novel psychrotrophic microbes for biotechnological applications. Adv Biotechnol Microbiol 2:1–4. https://doi.org/10.19080/AIBM.2017.02.555584

Yadav AN, Verma P, Sachan SG, Saxena AK (2017a) Biodiversity and biotechnological applications of psychrotrophic microbes isolated from Indian Himalayan regions. EC Microbiol ECO 01:48–54

Yan Y, Zhang X, Zheng X, Apaliya MT, Yang Q, Zhao L, Gu X, Zhang H (2018) Control of postharvest blue mold decay in pears by Meyerozyma guilliermondii and it's effects on the protein expression profile of pears. Postharvest Biol Technol 1:124–131. https://doi.org/10.1016/j.postharvbio.2017.10.016

Zachariah S, Kumari P, Das SK (2016) *Psychrobacter pocilloporae* sp. nov., isolated from a coral, *Pocillopora eydouxi*. Int J Syst Evol Microbiol 66(12):5091–5098. https://doi.org/10.1099/ijsem.0.001476

Zachariassen KE, Kristiansen E (2000) Ice nucleation and anti-nucleation in nature. Cryobiology 41:257–279. https://doi.org/10.1006/cryo.2000.2289

Zhang H, Prithiviraj B, Charles TC, Driscoll BT, Smith DL (2003) Low temperature tolerant *Bradyrhizobium japonicum* strains allowing improved nodulation and nitrogen fixation of soybean in a short season (cool spring) area. Eur J Agron 19:205–213. https://doi.org/10.1016/S1161-0301(02)00038-2

Zhang J, Li Y, Chen W, Du GC, Chen J (2012) Glutathione improves the cold resistance of *Lacotococcus sanfranciscensis* by physiological regulation. Food Microbiol 31:285–292. https://doi.org/10.1016/j.fm.2012.04.006

Nanotechnology for Agricultural and Environmental Sustainability

18

Rajni Kant Thakur, Pramod Prasad, Siddanna Savadi, S. C. Bhardwaj, O. P. Gangwar, and Subodh Kumar

Abstract

Nanotechnology provides answer for sustainable agriculture by enhancing nutrient utilization efficacy, improving efficiency of pest control, mitigating impact of climate change, and reducing harmful environmental impacts of agriculture food production. A lot of auspicious nanotechnologies have been anticipated and needed to be checked for their beneficial role. Here we explore nanotechnology in relation to agriculture and environmental aspects. We have discussed how nanotechnology can be applied to enhance plant growth and development, and provided comprehensive overview about nanofertilizers, nano-pesticide, and applications in field of food sector.

Keywords

Nanotechnology · Agriculture · Nanoparticles · Soil fertility

18.1 Introduction

Nanotechnology is the alteration of material at range of 1–100 nm in one dimension. Definition of nanotechnology varies across the globe; hence there is no universal definition of nanotechnology (Thakur and Shirkot 2017). Nanotechnology has been utilized to increase plant growth and protection including nanocarriers for fertilizers

R. K. Thakur · P. Prasad (✉) · S. C. Bhardwaj · O. P. Gangwar · S. Kumar
ICAR-Indian Institute of Wheat and Barley Research, Regional Station, Shimla, Himachal Pradesh, India
e-mail: pramod.prasad@icar.gov.in

S. Savadi
ICAR-Directorate of Cashew Research, Puttur, Karnataka, India

R. Goel et al. (eds.), *Survival Strategies in Cold-adapted Microorganisms*,
https://doi.org/10.1007/978-981-16-2625-8_18

and nanopesticides and help to develop plants with enhanced photosynthetic capability and sensors for plant health monitoring. In vivo and in vitro studies show tremendous promises of nanotechnology in making agriculture more sustainable, efficient, and resilient. Agriculture puts huge pressure on earth's environment and is a big contributor for biodiversity loss. The current agriculture practices are not meant for meeting production goal of 2050 (Prasad et al. 2017). Crop yield plateau is getting flattened and race is increasing for natural resources such as water, energy, and arable land. Before the formal start of agriculture by *Homo sapiens*, the predators like lifestyle supported about four million people globally. Today agriculture provide food for up to six million people. Total cereal production of the world has increased significantly in the past 40 years. Major contributors for yield increase are fertilizers, water, pesticides, and new variety with disease resistance power. This has led to reduced hunger, improving nutrition, and preventing ecosystems from conversion. In 2050, world population is estimated to be 50% greater than at present, and world grain production is projected to increase by 2.4 fold. If we have to increase production of cereals and grains, we need much more precise and pinpointed target delivery of fertilizers, pesticides, and insecticides and require super green revolution variety (Kah and Tufenkji 2019). Half of the world's land is already used for intensive agriculture, it causes loss of ecosystems, and adds significant and environmentally damaging amounts of nitrogen and phosphorous to land ecosystem. A threefold increase in nitrogen and phosphorous quantity is required if we have to meet challenges to achieve 2050 or 2030 UN goals. Sustainable agriculture are methods or techniques that help us to meet present and future needs of people for food and fiber for ecosystem service and for healthy lives. Nanotechnology can up to some extent fulfill some needs in the future through nano-enabled pesticides and by converting plants into sensors. Only 0.1% of pesticides applied to the field reach their target, that is, target locations of desired product. Hence nanocarriers and nanoformulations have potential to improve the efficacy of existing pesticides by improving the accuracy of preexisting pesticides by accurate delivery (Giraldo 2014). The concept of sustainable release of an active compound from nanocarriers has existed since the 1960s: nanotechnology can produce smart pesticides which can release active compounds when plants are in stress like pathogen, water, temperature, pH, redox conditions, light, and plant biochemicals (Fig. 18.1).

18.2 Nanotechnology in Agriculture

There were many efforts to use nanotechnology in agriculture which started after the advent of new field of science, namely, nanotechnology. Age-old farming practices are neither able to increase productivity nor will restore damage caused by existing technology back to Darwin time. Nanotechnology became popular with plenty of public funding, but the growth is not like as desired from this technology (Thakur and Shirkot 2017). Developing countries and least developed countries lacked foresight; infrastructure-wise, they are unwilling to spent money on innovation

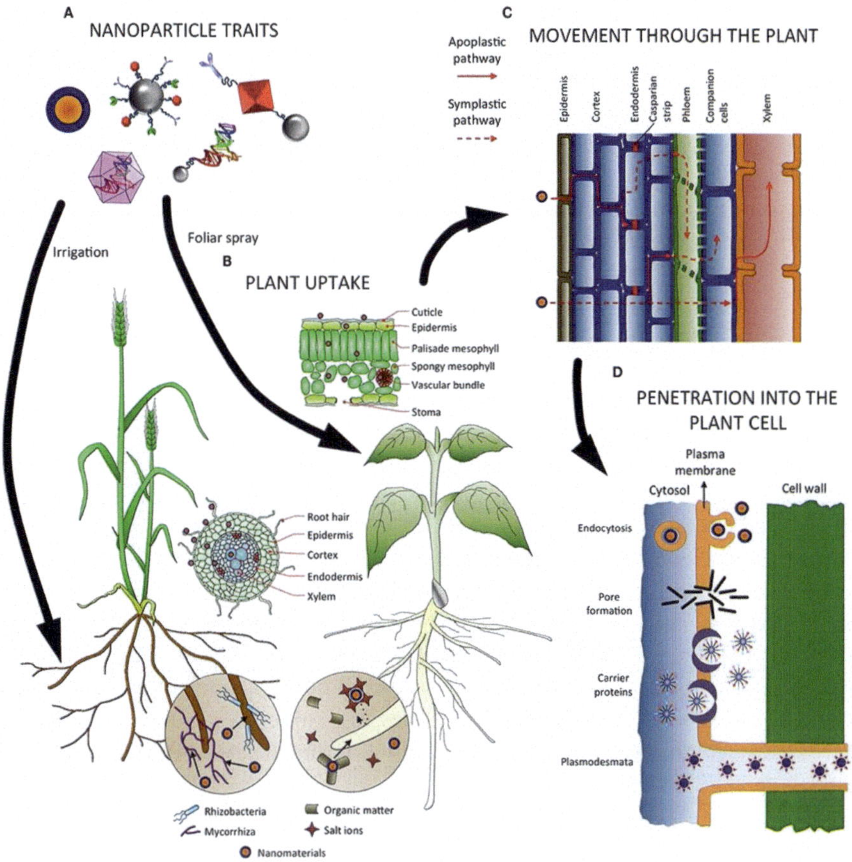

Fig. 18.1 (**a**) Picture represent how nanoparticles are translocated and their application. (**b**) represents nanoparticle interaction with microorganisms and other chemicals in soil, (**c**) represents nanoparticle traffic pathways (apoplastic/symplastic), (**d**) mechanisms for the transport of nanoparticles within cell

(research and development project). With advances in technology and increased awareness about production and quality, nowadays consumers not only want to eat food but also want a nutrient-balanced diet. Nanotechnology certainly can reduce millions of tons of chemical which are poured into the soil in the name of fertilizers and pesticides and can reduce about 70–80% chemical fertilizer which are used currently by farmers. Nanotechnology in farming has optimistic future for increasing the efficiency of nutrient use through nanoformulations of fertilizers, removing yield barriers using high-throughput phenotyping and control of pest and diseases (Thakur et al. 2018). Knowing more about mechanisms of host-pathogen interaction at molecular level will greatly help to increase production for increasing populations. Nanotechnology gets much awaited hype because of its large numbers of applications in field of nano-pesticide and nano-insecticide and in sustainable delivery of chemicals in field/soil for long-term effect of these chemicals. Biosensor is an area which needs miniaturizations, and nanotechnology can greatly benefit through small and smart devices (Fraceto et al. 2016) such as nanosensors and other nanosystems that are very important in biochemical analysis (Viswanathan and Radecki 2008). Nanosensors can detect mycotoxin present in various foods and their products (Sertova 2015). Agriculture sector always benefitted from various techniques and tools which are developed by different branches of sciences, like hybrid varieties, fertilizers, and pesticides.

18.2.1 Soil Fertility

To achieve high production in various crops like cereals and others, we have to supply synthetic fertilizer. Ancient methods of applying fertilizer cause losses of macronutrients and micronutrients, due to which very minute concentration reach to target sites and about 40–70% N, 80–90% P, and 50–90% potassium of conventionally applied fertilizers are lost in the environment (Tilman et al. 2002). Nanomaterial supplies single or various nutrients to plants ensuring better yield of crops and reduced environmental degradation (Liu and Lal 2015) (Chinnamuthu and Boopathi 2009). Encapsulation of nanofertilizer is ensured by three ways: (1) nanoporous materials, (2) film polymer, and (3) nanoparticle or nanoemulsions. Nanomaterial trapping on fertilizer sticks more strongly material of high surface area (Brady and Weil 1999). Nanofertilizer possesses qualities like solubility, stability, sustainable release, less toxicity with ease, and safe distribution and disposal (Torney et al. 2007 and Green and Beestman 2007).

18.2.2 Plant Growth

Nanoparticles of various metals have shown good potential for increasing crop yield and health by better nutrient uptake (Khot et al. 2012). True effectiveness of nanomaterials only depends on nanomaterial size, concentration, surface, and chemical properties other than susceptibility of the plant species (Ma et al. 2010). Use of advanced technology largely helps to understand interaction between plants and nanomaterials (Table 18.1).

18.2.3 Plant Growth and Development

Nanomaterial interplay causes many physiological changes in the plants depending upon the property of nanoparticles. Efficiency of nanomaterials is determined through numerous factors most importantly the dose at which these are applied (Khodakovskaya et al. 2012). Nanomaterials induce different effects on plant growth and development; their impact on plants is controlled by nanomaterial composition, size, shape, other properties, and plant species (Ma et al. 2010). Khodakovskaya

Table 18.1 Effects of nanomaterials on plant growth and development

Active element	Effects	Particle size and concentration	References
Calcium	Helps in management of insect pests in *Z. mauritiana*	26%	Hua et al. (2015)
Magnesium	Increased growth and yield in *T. aestivum* and *V. unguiculata*	0.25–0.5 g L^{-1}	Delfani et al. (2014)
Zinc	Increased carbohydrate, fat, and fiber content in *S. oleracea*	10 ppm	Burman et al. (2013)
Iron	Increased carbohydrate content in *O. basilicum*	1–3 mg L^{-1}	Elfeky et al. (2013)
Silicon	Increased protein and chlorophyll content in pea	15 mg L^{-1}	Suriyaprabha et al. (2014)
Titanium	Increased nitrogen and protein content in *S. oleracea*	0.25%	Yang et al. (2007)
Silver	Increased growth and development of *B. juncea* and *V. unguiculata*	50–75 ppm	Pallavi et al. (2016)
Gold	Enhanced growth in *Brassica juncea*	10–25 ppm	Arora et al. (2012)
Carbon	Reduction of oxidative stress in *B. vulgaris*	0.01 and 0.001 nmol	Kaphle et al. (2018)

et al. (2012) reported that application of gold nanoparticles enhances the redox status of the treated plants. Germination of *Brassica. juncea* seedlings improves when inoculated with 25 ppm gold nanoparticles; 19% improvement has been found when 10 ppm of gold nanoparticles were inoculated.

Christou et al. (1988) transferred a gene into soybean genome through DNA-coated gold particles. The optimistic results of gold nanoparticles were identified. Shah and Belozerova (2009) studied effects of gold nanoparticles within plants which inhibited aquaporin function, a group of transmembrane proteins that help in the transportation of wide range of molecules including water. Krishnaraj et al. (2012) investigated effect of silver nanoparticles on *Bacopa monnieri*. Morphological study showed minimal reduction in root and shoot lengths. Seed germination rate has been increased in case of lettuce and cucumber (Barrena et al. 2009), *Brassica juncea* (Arora et al. 2012), and *Gloriosa superba* Gopinath et al. (2014). It has been noted that increase in number of leaves, leaf area, plant height, and chlorophyll content led to better crop yield. Kumar et al. (2010) found that 24 nm-sized gold nanoparticles increased the seed yield three times than control in case of *Arabidopsis thaliana.* It was further noted that inoculum of 10–80μg/ml concentrations increases shoot growth and free radical scavenging activity. Gunjan et al. (2014) observed that exposure of gold nanoparticles led to increase in shoot length from 6.81 to 7.36 cm in seedlings of *Brassica juncea*, and at 100 ppm average root length increased up to 1.62 cm due to gold nanoparticle exposure. As gold nanoparticle concentration increased, there was an increase in antioxidative enzyme activity, H_2O_2, and proline content. Siddiqi and Husen (2016) reported that plants exposed to gold nanoparticles showed mixed effects on growth and yield of different crops. Toxicity of gold nanoparticles depends on the size and shape of nanomaterials. It has been observed through several studies that lower concentration of nanomaterials helps in increase in growth and yield of fruit/seed.

18.2.4 Fertilizer Release

In the past few decades, the application of nanoparticles in the agricultural sector has increased, accompanied by the creation of fertilizers, which have improved product performance and quality. The current trend is to incorporate nutrients through nanostructure in order to facilitate their availability to plants. Different encapsulation techniques including nutrient encapsulated inside nanomaterials such as nanotubes or nanoporous materials have been developed to improve fertiliaer avilability for crop plants. The nutrients can be coated with a thin protective polymer film and delivered as particles or emulsions of nanoscale dimensions. Several studies focus on the use of artificial fertilizers such as nitrogen, phosphorus, and potassium, so-called NPK fertilizers for various crops and under various growing conditions (Solanki et al. 2015). Therefore, studies are underway for encapsulations, in which NPK fertilizers are entrapped within chitosan nanoparticles (Hasannen et al. 2014). On the other hand, urea is utilized as a model fertilizer to access fertilizer loading and the controlled release behavior of nanoparticles (Wanyika et al. 2012). Castro-Enriquez et al. (2012) reported use of nanomembranes of wheat gluten to release urea. These membranes provided a potential solution to the problem of loss by leaching of these fertilizer into agricultural crops. A similar objective was obtained by employing gold nanoparticles supported on TiO_2 or titanate nanotubes. Thus, nano fertilizer gives high surface area relative to the number of nanomaterials.

18.2.5 Pesticide, Herbicide, and Insecticide Release

Various plant pathogens and pest reduce the crop production by 20–40% globally per year. Currently farmers employ various chemical formulations to manage pest and insect problem. There are various reasons that lead to vast application of these chemicals into the field, but keeping in mind environmental safety and sustainability of agriculture, we need to shift toward nanoformulation choice. It has been observed that 90% of applied pesticide are degraded after application. So, we have to move toward development of eco-friendly nanopesticides and insecticides.

Nanotechnology is being explored toward delivery of agrichemicals (insecticide, pesticides), nanobiosensors, and others. Technology enabled scientist and researchers to craft nanomaterials with desired shape, size, and surface property, so that they can be utilized as surface-coating materials for efficient delivery of agrochemicals. There are two different mechanisms by which nanomaterial can protect plant or crop from diseases, pest, insect, and other predators. Nanomaterials by their own provide crop protection. Nanomaterials can be used as delivery agent for agrochemicals or other active bioagents such as dsRNA. Nanomaterials as delivery agents provide benefits like large shelf life, more solubility, less toxicity, and precise delivery of chemicals. Nanopesticide is much more stable under environmental conditions such as rain and UV. De-Oliveira et al. (2015) reported that nanomaterials can be used as a delivery agent for combination of atrazine and simazine to control vegetation. Herbicidal activity was determined through pre- and post-emergent treatment of *Raphanus raphanistrum.* The formulation showed growth inhibition at concentrations ten times lower than those of the commercial formulation. Atrazine was also studied in gel beads of chitosan, obtaining an extended-release period of the compound up to 7 months. Nanoemulsions are useful for the formulation of pesticides, and the latter can be effective against various insect pests in agriculture.

18.3 Antimicrobial Activity of Nanoparticles

Microorganisms play an important role in agriculture; however some bacteria and fungi are playing the major role to crop contamination and degradation, which leads to huge economic losses worldwide. Nanotechnology provides a wide avenue to antimicrobial compounds. Gold, zinc oxide, and silver nanoparticles are the prime metals on considering antimicrobial activity. Nanotechnology is emerging as a new tool to mitigate plant diseases and helps in disease management. Nanoparticles are found to be effective for inhibiting growth of broad range of pathogenic microorganisms such as bacteria, fungi, viruses, and yeasts (Nirmala and Pandian 2007).

18.4 Psychrophiles' Role in Agriculture

Psychrophiles are microorganisms which live in extremely low temperature (−20 to 8^0C) conditions. About 70% of earth area comes under temperature zone of 1–5°C, which is permanently cold (Feller and Gerday 2003). Psychrophilic enzyme offers lucrative application for different industries such as textile, brewing food, and dairy industry. Psychrophiles are the excellent source of polyunsaturated fatty acids, which are used by pharma industry for the development of novel therapeutic agents. Mukhopadhyay et al. (2015) studied the effect of nanoparticles on the stability of pectate lyase at a temperature 4 °C; they have found 70% activity after repeated freezing and thawing at 25 °C.

18.5 In Food Sector

Applications of nanotechnology in the food sector offer great benefits in biosensing, detection of food pathogens and toxins, food packaging, delivery systems, delivery of bioactive compounds, and protection of functional ingredients. This technology may completely revolutionize the food sector because the application of nanotechnology improves the safety and the nutritional value of food products. In recent years, there has been an increasing interest in nanotechnology in the food industry, in which nanotechnology has grown enormously, as well as the creation of new food products to satisfy the needs for food quality, sensory appeal, texture, taste, improving supplements, other sensory attributes, coloring, strength, processability, stability during shelf life, and safety while simultaneously being a good source of nutrients. Prakash (2012) reported that nanotechnology could be used in quality control, food additives, and the detection of bacterial and fungal contamination in food products. It was reported that the application of nanoparticles in food is primarily focused on optimizing the use of dispersion systems and release of the bioactive compounds of liposomes and micelles, thereby increasing the bioavailability of food. Lopes and colleagues conducted experiments on other functional food ingredients in order to develop new functional material and the design of methods such as nanofiltration and instrumentation (nanosensors). The different applications of nanoparticles in the food sector will be described in the following sections.

The development of nanostructured food ingredients and of delivery systems for nutrients and supplements are the main focus of nanotechnology applications in foods. Numerous bioactive agents intended for oral ingestion are nonpolar compounds with high melting points, low water solubility, and poor oral bioavailability. Thus certain bioactive compounds are difficult to incorporate into commercial products, such as functional foods and beverages; therefore it is necessary to incorporate these into particles that facilitate their bioavailability. Moreover, micelles, liposomes, and nanoemulsions can be good options due to their high stability under moderate conditions, such as pH value, temperature, or salt concentration. Popov et al. (2010) have proposed the use of nanoemulsion technology in order to obtain aromatized beverages, juices, and milk enriched with controllably released vitamins, minerals, and functional components. The use of nanoemulsion technology has been reported for the manufacture of encapsulating systems for functional compounds, in order to prevent their degradation and to improve their bioavailability, such as nutraceuticals, drugs, flavors, antioxidants, and antimicrobial agents. Nutraceuticals are utilized in food to provide health benefits. The efficacy of nutraceuticals in disease prevention depends on the preservation of bioactive bioavailable ingredients until their release at target sites, taking into account that nanoparticles in food can be used as additives or supplements, for example, zinc oxide nanoparticles have been employed in nutritional supplements such as multivitamins. Lopes et al. (2013) used liposomes containing functional food ingredients to protect these from oxygen and water. Nanoencapsulation aids in solving certain difficulties such as loss of functionality during processing or storage and the loss of activity of enzymes. Nanoencapsulation comprises a promising technique to protect bioactive compounds from environmental damage and to mask their unpleasant properties. Microorganisms such as *Lactobacillus acidophilus*, *Lactobacillus casei*, *Lactobacillus rhamnosus*, and *Bifidobacterium* spp. have been nanoencapsulated for the protection and controlled release of these beneficial, live probiotic species to promote healthy gut function.

Table 18.2 Products developed through nanotechnology

Product	Application	Institution
Nano-biosensors	Used in checking of contamination of packaged food	Nestle, USA
Precision farming	Nano sensors give real-time access to monitoring of crop growth and soil health	US Department of Agriculture, USA
Livestock and fisheries	Nano-veterinary medicine used for nano-vaccines, drug delivery, smart herds	NanoVic, Dingley, Australia
Use of agricultural waste	Cotton waste used for improving strength of cloth through nanofiber	Cornell University, USA
Nanoparticles	*Campylobacter jejuni* removed from poultry through nanoparticles	Clemson University, USA
Buckyball fertilizer	Ammonia from buckyballs	Kyoto University, Japan
Nanocides	Nanoparticles were used to encapsulated pesticides	BASF, Germany

Nanotechnology in the food industry exerts a great impact on the development of novel food packaging materials that can help to control the oxidation of foodstuffs and to prevent the formation of off flavors and undesirable-textured foods. For example, edible films, edible coating, and polymer nanocomposites are effective, but their application depends on the level of adherence between the materials involved. Campos et al. (2011) reported that edible coatings or films could be used as a vehicle for incorporating functional ingredients such as antioxidants, flavor colors, antimicrobial agents, and nutraceuticals. Natamycin-loaded poly (*N*-isopropylacrylamide) nanohydrogels are used to aid controlled release of active compounds. It was reported that this compound does not cause changes in the main properties. The most important characteristic of food packaging compounds is that they should maintain their bioactivity. In case of nanocomposites, these exhibit good characteristics, such as augmented barrier properties, increased mechanical strength, and improved heat resistance. Cellulose nanocomposites have been employed in mango pulp to improve tensile properties and water vapor permeability. Moreover, nanoparticles are also being currently used in edible coatings and films in a wide variety of foods, including fruits, vegetables, meats, and seafood. In food packaging, nanotechnology offers huge opportunities that can benefit both consumers and the food industry (Table 18.2).

18.6 Conclusion and Future Prospects

Progress of nanotechnology is impressive in agriculture; it has been noted that it takes 20 years for any technology to move from lab to field. Surely nanotechnology or nanobiotechnology will boom in the near future or may be in the next decade. It is well known that applying new technologies to agriculture sector could bring big breakthrough in improving our current terrible nutrient usage effectively by use of nanoformulation of fertilizers, breaking yield, and nutritional quality barrier through bionanotechnology, surveillance, and control of pest and diseases. Presently we are better enabled by technology to know more about host parasite interactions at molecular level. These technologies has helped in generation of safe nanopesticides, safe carriers, better preserving and packing materials for food and food additives. Nanotechnology can improve shelf life of vegetables and flowers and also help in better water management practices, restoring soil fertility, retrieval of salinity of soil problem. Implementation of theories like theory of chaos and string theory opens a Pandora box for agriculture production system. Nanotechnologist needs more understanding in material technology, in conjunction with knowledge of the agriculture production system.

References

Arora S, Sharma P, Kumar S, Nayan R, Khanna PK, Zaidi MGH (2012) Gold-nanoparticle induced enhancement in growth and seed yield of *Brassica juncea*. Plant Growth Regul 66:303–310

Barrena R, Casals E, Colon J, Font X, Sanchez A, Puntes V (2009) Evaluation of the ecotoxicity of model nanoparticles. Chemosphere 75:850–857

Brady NR, Weil RR (1999) In: Brady NR, Weil RR (eds) The nature and properties of soils. Prentice Hall, Upper Saddle River, NJ, pp 415–473

Burman U, Saini M, Kumar P (2013) Effect of zinc oxide nanoparticles on growth and antioxidant system of chickpea seedlings. Toxicol Environ Chem 95:605–612

Campos RP, Kwiatkowski A, Clemente E (2011) Post-harvest conservation of organic strawberries coated with cassava starch and chitosan. Rev Ceres Viçosa 58:54–560

Castro-Enriquez DD, Rodriguez-Felix F, Ramirez-Wong B, Torres-Chavez PI, Castillo-Ortega MM, Rodriguez-Felix DE, Armenta-Villegas L, Ledesma-Osuna AI (2012) Preparation, characterization and release of urea from wheat gluten electrospun membranes. Materials 5:2903–2916

Chinnamuthu CR, Boopathi PM (2009) Nanotechnology and agroecosystem. Madras Agric J 96(1–6):17–31

Christou P, McCabe DE, Swain WF (1988) Stable transformation of soybean callus by DNA coated gold particles. Plant Physiol 87:671–674

Delfani M, Baradarn-Firouzabadi M, Farrokhi N, Makarian H (2014) Some physiological responses of black-eyed pea to iron and magnesium nanofertilizers. Commun Soil Sci Plant Anal 45:530–540

De-Oliveira JL, Campos EVR, Da-Silva CMG, Pasquoto T, Lima R, Fraceto LF (2015) Solid lipid nanoparticles co-loaded with simazine and atrazine:preparation, characterization, and evaluation of herbicidal activity. J Agric Food Chem 63:422–432

Elfeky SA, Mohammed MA, Khater MS, Osman YA, Elsherbini E (2013) Effect of magnetite nano-fertilizer on growth and yield of Ocimum basilicum L. Int J Indigenous Med Plants 46:1286–1293

Feller G, Gerday C (2003) Psychrophilic enzymes: hot topics in cold adaptation. Nat Rev Microbiol 1:200–208. https://doi.org/10.1038/nrmicro773

Fraceto LF, Grillo R, de Medeiros GA, Scognamiglio V, Rea G, Bartolucci C (2016) Nanotechnology in agriculture: which innovation potential does it have? Front Environ Sci 4:20. https://doi.org/10.3389/fenvs.2016.00020

Giraldo JP et al (2014) Plant nanobionics approach to augment photosynthesis and biochemical sensing. Nat Mater 13:400–408

Gopinath K, Gowri S, Karthika V, Arumugam A (2014) Green synthesis of gold nanoparticles from fruit extract of *Terminalia arjuna*, for the enhanced seed germination activity of *Gloriosa superba*. J Nanostruct Chem 4:1–11

Green JM, Beestman GB (2007) Recently patented and commercialized formulation and adjuvant technology. Crop Prot 26(3):320–327

Gunjan B, Zaidi MGH, Sandeep A (2014) Impact of gold nanoparticles on physiological and biochemical characteristics of *Brassica juncea*. Plant Biochem Physiol 2:133–139

Hasannen MNA, Abdel-Aziz HMM, El-Bialy DMA, Omer M (2014) Preparation of chitosan nanoparticles for loading with NPK fertilizer. Am J Nanobiotechnol 13:3158–3164

Hua KH, Wang HC, Chung RS, Hsu JC (2015) Calcium carbonate nanoparticles can enhance plant nutrition and insect pest tolerance. J Pestic Sci 40:208–213

Kah M, Tufenkji N, White JC (2019) Nano-enabled strategies to enhance crop nutrition and protection. Nat Nanotechnol 14:532–540

Kaphle A, Navya PN, Umapathi A, Daima HK (2018) Nanomaterials for agriculture, food and environment: applications, toxicity and regulation. Environ Chem Lett 16:43

Khodakovskaya M, Dervishi E, Mahmood M, Xu Y, Li Z, Watanabe F, Biris AS (2012) Carbon nanotubes are able to penetrate plant seed coat and dramatically affect seed germination and plant growth. ACS Nano 3:3221–3227

Khot LR, Sankaran S, Maja JM, Ehsani R, Schuster EW (2012) Applications of nanomaterials in agricultural production and crop protection: a review. Crop Prot 35:64–70. https://doi.org/10.1016/j.cropro.2012.01.007

Krishnaraj C, Jagan EG, Ramachandran R, Abirami SM, Mohan N, Kalaichelvan PT (2012) Effect of biologically synthesized silver nanoparticles on *Bacopa monnieri*. (Linn.) Wettst. plant growth metabolism. Process Biochem 47:51–658

Kumar CG, Mamidyala SK, Das B, Sridhar B, Devi GS, Karuna MS (2010) Synthesis of biosurfactant-based silver nanoparticles with purified Rhamnolipids isolated from *Pseudomonas aeruginosa* BS-161R. J Microbiol Biotechnol 20:1061–1068

Liu R, Lal R (2015) Potentials of engineered nanoparticles as fertilizers for increasing agronomic productions. Sci Total Environ 514:131–139. https://doi.org/10.1016/j.scitotenv.2015.01.104

Lopes CM, Fernandes JR, Martins-Lopes P (2013) Application of nanotechnology in the agro-food sector. Food Technol Biotechnol 51:183–197

Ma X, Geiser-Lee J, Deng Y, Kolmakov A (2010) Interactions between engineered nanoparticles (ENPs) and plants: phytotoxicity, uptake and accumulation. Sci Total Environ 408:3053–3061. https://doi.org/10.1016/j.scitotenv.2010.03.031

Mukhopadhyay A, Bhattacharyya T, Dasgupta AK, Chakrabarti K (2015) Nanotechnology based activation-immobilization of psychrophilic pectate lyase: A novel approach towards enzyme stabilization and enhanced activity. J Mol Catal B Enzym 119:54–63

Nirmala GA, Pandian K (2007) Antibacterial efficacy of aminoglycosidic antibiotics protected gold nanoparticles a brief study. Colloids Surf A Physicochem Eng 297:63–70

Pallavi, Mehta CM, Srivastava R, Arora S, Sharma AK (2016) Impact assessment of silver nanoparticles on plant growth and soil bacterial diversity. 3 Biotech 6:254

Popov KI, Filippov AN, Khurshudyan SA (2010) Food nanotechnologies. Russ J Gen Chem 80:630–642

Prakash A, Sen S, Dixit R (2012) The emerging usage and applications of nanotechnology in food processing industries: The new age of nanofood. Int J Pharma Sci Rev Res 22:107–111

Prasad R, Bhattacharyya A, Nguyen QD (2017) Nanotechnology in sustainable agriculture: recent developments, challenges, and perspectives. Front Microbiol 8:1014

Sertova NM (2015) Application of nanotechnology in detection of mycotoxins and in agricultural sector. J Cent Eur Agric 16:117–130. https://doi.org/10.5513/JCEA01/16.2.1597

Shah V, Belozerova I (2009) Influence of metal nanoparticles on the soil microbial community and germination of lettuce seeds. Water Air Soil Pollut 197:143–148

Siddiqi KS, Husen A (2016) Engineered gold nanoparticles and plant adaptation potential. Nanoscale Res Lett 11:400–410

Solanki PS, Bhargava A, Chhipa H, Panwar J (2015) Nano-fertilizers and Their smart delivery system. In: Rai M, Mattoso CRL, Duran N (eds) Nanotechnologies in food and agriculture. Springer, Switzerland, pp 81–101

Suriyaprabha R, Karunakaran G, Yuvakkumar R, Rajendran V, Kannan N (2014) Foliar application of silica nanoparticles on the phytochemical responses of maize (Zea mays L.) and its toxicological behavior. Synth React Inorg Met Org Nano Met Chem 44:1128–1131

Thakur RK, Dhirta B, Shirkot P (2018) Studies on nano toxicity effect of gold nanoparticles on *M. incognita* and tomato plants growth and development. Ann Nanosci Nanotechnol 2(1):1005

Thakur RK, Shirkot P (2017) Nanoparticles: how they are synthesized and their applications. Res J Biotechnol 14(2):92–102

Tilman D, Knops J, Wedin D, Reich P (2002) Plant diversity and composition: effects on productivity and nutrient dynamics of experimental grasslands. In: Loreau M, Naeem S, Inchausti P (eds) Biodiversity and ecosystem functioning. Oxford University Press, Oxford, pp 21–35

Torney F, Trewyn BG, Lin VS-Y, Wang K (2007) Mesoporous silica nanoparticles deliver DNA and chemicals into plants. Nat Nanotechnol 2:295–300

Viswanathan S, Radecki J (2008) Nanomaterials in electrochemical biosensors for food analysis- a review. Pol J Food Nutr Sci 58:157–164

Wanyika H, Gatebe E, Kioni P, Tang Z, Gao Y (2012) Mesoporous silica nanoparticles carrier for urea: potential applications in agrochemical delivery systems. J Nanosci Nanotechnol 12:2221–2228

Yang F, Liu C, Gao F, Su M, Wu X, Zheng L, Hong F, Yang P (2007) The improvement of spinach growth by nano-anatase TiO2 treatment is related to nitrogen photoreduction. Biol Trace Elem Res 119:77–88

Recent Trends and Advancements for Agro-Environmental Sustainability at Higher Altitudes

19

Parul Chaudhary, Anuj Chaudhary, Upasana Agri, Hina Khatoon, and Alka Singh

Abstract

Himalaya is one of the coldest environments on Earth and characterizes stressful conditions due to lack of nutrients and freezing conditions. More refined and scientific farming is crucially needed for sustainable agricultural involving uncultivable lands to boost the agricultural productivity and soil fertility without any detrimental response on soil. Hence it is essential to improve the soil beneficial population by applying bio-inoculants, rhizosphere engineering, and nanoparticles in high-altitude cold lands which helps to alter the microbial community to enhance the plant growth by uptake of nutrients and soil health improvement as a substitute in agricultural tradition. Use of culture-dependent and culture-independent technology revealed the complete information on microbial diversity and structural and functional potential of rhizospheric microbes. Nanotechnology spreads out a broader opportunity to achieve better crop production in agricultural fields because of their unique properties. The properties like uniform particle size distribution and large internal porosity of nanoparticles make them desirable for improving characteristics of soil and crops. This chapter

P. Chaudhary (✉)
Department of Animal Biotechnology, National Dairy Research Institute, Karnal, Haryana, India

A. Chaudhary
School of Agriculture and Environmental Sciences, Shobhit University, Gangoh, Uttar Pradesh, India

U. Agri
G.B.Pant University of Agriculure & Technology, Pantnagar, Uttarakhand, India

H. Khatoon
Wildlife Institute of India, Dehradun, Uttarakhand, India

A. Singh
Gurukul Kangri University, Haridwar, Uttarakhand, India

R. Goel et al. (eds.), *Survival Strategies in Cold-adapted Microorganisms*,
https://doi.org/10.1007/978-981-16-2625-8_19

provides the information to improve the rhizospheric microbiome using rhizosphere engineering and omics techniques for the better soil health and improvement in plant growth for sustainable agriculture in higher altitude.

Keywords

Nanotechnology · Rhizosphere · Agriculture · Microbes

19.1 Introduction

Higher-altitude ecosystem is commonly characterized by low temperature and has soil nutrient stress which has major impact on microbial diversity. Himalaya soil has cold-adapted microbes which are able to survive at cold temperature (Tejeda et al. 2013). These regions are characterized as more stressful conditions because of heavy snowfall, lack of nutrition, and freezing due to seasonal deviation in climate. Extreme climatic conditions of Himalaya affect the soil microbial diversity. Majority of the microbes at high altitude are psychrophiles (cold loving) and psychrotolerants (cold tolerant) which possess structural and functional adaptation to perform normal life processes under associated stresses. These microorganisms have the ability to cope with stress conditions by production of cold-active enzymes and metabolites which work better at lower temperature (Yadav et al. 2016). Food production nowadays is a demanding issue due to growing population and inadequate accessible resources through progressive climate alteration all over the world. Hence, the main concern is to facilitate accelerated adaption of plants exclusive of threatening the soil microbial population to cope with environmental stresses. Different efforts are needed by researchers to widen the technologies for sustainable agriculture.

Rhizosphere is the largely composite system for microbial territory. Rhizosphere consists of plant roots, soil, and a site of higher microbial activity of different microbial population of bacteria, fungi, archaea, and microeukaryotes, which play an important function in organic material decomposition and plant nutrient cycles. Considerate predicting and protecting the formation and role of the rhizosphere activities will permit to exploit the plant microbe communications and rhizospheric actions to boost the plant productivity (Ahkami et al. 2017). Soil quality depends on microbial population, soil enzyme activities, and micro/macronutrient status of soil (Chaudhary et al. 2021b). Soil physicochemical properties display deviation according to season and can accordingly alter the composition of bacterial communities. Overall food requirement is putting a remarkable demand on earth agriculture, so it is needed for sustainable agricultural strategy. In this situation, application of bioinoculants and nanoparticles in higher-altitude lands possibly will be a promising advancement to support mountain agro-ecosystems.

Change in climatic conditions that leads to modification in rainfall pattern and causes incidence of severe climate incidents, like low and high temperatures, can negatively influence crop productivity in diverse region of the world (Sunoj et al. 2016). Recently the extreme climate conditions like chilling affects the production of crops. To mitigate the effect of stress conditions, new low-cost advanced technologies are needed for farmers. Agriusable nanoparticles have an enormous prospective in agriculture field and can be a valuable solution for farmers to improve the productivity and develop stress tolerance in plants (Guo et al. 2019; Chaudhary et al. 2021a). Microbial diversity of Himalayan soil has been widely studied using omics techniques such as metagenomic sequencing which provides a complete picture of associated microbes. Plants are exposed to a variety of stress conditions in nature especially cold stress which affected the growth, production, and soil microbial flora. Enhancing crop production is the primary goal for researchers to nourish rising population globally in a sustainable way. Nowadays, application of indigenous beneficial plant microbes under stress conditions is gaining momentum. Omics techniques like Illumina sequencing, proteomics, gene editing, and bioinformatics allow us to study the interactions of microbiomes to augment crop nutrient attainment and stress tolerance. In this chapter we focused on rhizospheric microbiome and advanced technologies for the exploitation of genes in crops for improving plant and soil health.

19.2 Rhizosphere Engineering

Rhizosphere of plants is compactly populated by innumerable microbes, many of which allow their host to grow better, known as plant growth-promoting microorganism, involved in mobilization of nutrients and nitrogen fixation, which protects plants from pathogens and helps in abiotic stress alleviation (Glick 2012). Also it enhances the multiplication of beneficial microbes in the roots. Engineering of these microbes can be done through the plant parts which help in the production and multiplication of the microbes. Thus, selection of the right crops is a must. These crops are specific to the production and multiplication of microbes and helps in the suppression of the phyto-pathogens which thrive in the soil. To take advantage of the microbiome functions, it is imperative to recognize the biochemical and molecular determinant in the roots that governs the selective microbial enrichment.

19.2.1 Rhizosphere Engineering Using Microbes

It is well recognized that soil microbes facilitate the soil structure formation and solubilization of different minerals necessary for plant growth and repression of pathogens. The most efficient approach is to manipulate the microbiome through bioinoculation (Agri et al. 2021; Chaudhary et al. 2021c, d). Different products were available in market formulated with beneficial consortium of microorganism such as PGPR and plant growth-promoting fungi (Hashem et al. 2018; Khan et al. 2019). Some bioinoculants are unsuccessful in field environment as these are attacked by numerous predators. Valuable bioinoculants should have possibility to form association with close to microbiome. So it is imperative to add beneficial microbes from indigenous population of host plant so that isolated population survive and show

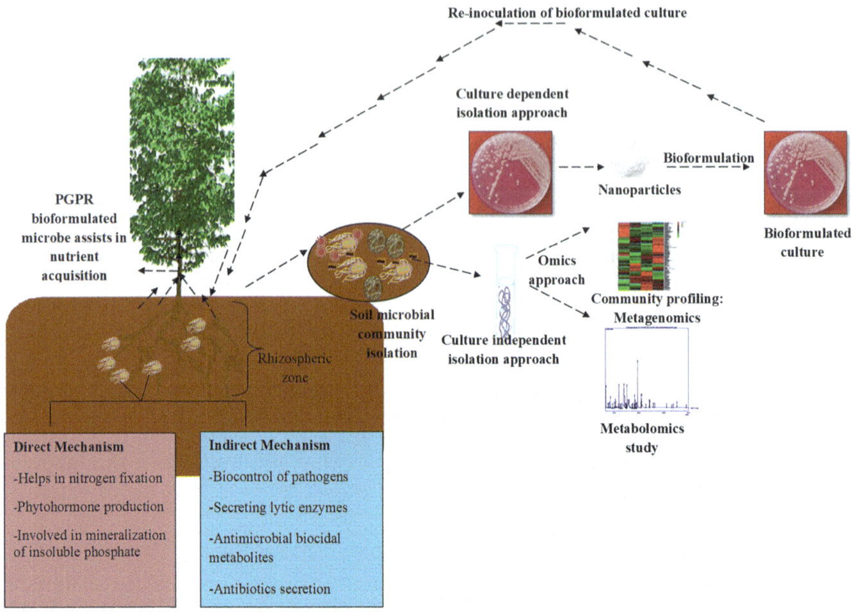

Fig. 19.1 Rhizosphere engineering using microbes and molecular technology to study rhizomicrobiome

resistance to plants toward stress conditions (Khare et al. 2018). Engineering of rhizosphere requires culturing of indigenous microbes which augment the cultivability of microbes in the rhizosphere to recognize the function and perseverance of bacteria isolates (Fig. 19.1). This helps in collection of data, so that bioformulations can be utilized in the agricultural field. There are different species of *Rhizobium*, *Sinorhizobium*, *Pseudomonas*, *Azospirillum*, and *Bacillus* which fix nitrogen; mineralize phosphorus, iron, potassium, and zinc; produce antibiotic compound like HCN, phenazine, and pyoluteorin; and are used as biocontrol agent involved in production of plant growth-promoting hormones such as IAA, gibberellins, and cytokinin. Joshi et al. (2019); Rajwar et al. (2018); Suyal et al. (2017) studied the different bacteria involved in nitrogen fixation which contain Nif gene in higher altitudes. *Pseudomonas migulae*, *Dyadobacter psychrophilus migulae*, and *Pseudomonas palleroniana* have been isolated from hilly area which involved in plant growth promotion (Suyal et al. 2014; Suyal et al. 2018). Tomer et al. (2017) observed that phosphate-solubilizing bacteria *Lysinibacillus macroides* enhanced the seed germination of chickpea and have an importance in hill agriculture system.

19.2.2 Plant-Based Methods of Rhizosphere Engineering

This method involves two approaches: plant breeding and genetic engineering. For plant breeding selection of a specific bacterial population is an attractive advancement to enhance the crop productivity and provide the resistance toward stress conditions (Ryan et al. 2009). Koyama et al. (2000) observed that transgenic plants contain better skill to conceal citrate from the roots which survive on phosphate-restricted soil as compared to natural type. This study recommended to facilitate the different crop plants with an improved capacity to nurture in acidic soils and tolerance to aluminum. Mazzola (2002) reported that wheat cultivars have the capacity to repress disease with increasing *Pseudomonas* population, anatomized towards *Rhizoctonia solani*. There is a change in the pH level of the soil by incorporating the engineered rhizosphere into the root zone or the genes incorporated in the plants. It enhances the activity of the microbes that benefits the plant, and in return the microbes obtain energy from the plant. Gevaudant et al. (2007) reported that by manipulating the pH of plant rhizospheric soil through transgenic lines of *Nicotiana tabacum* and *Arabidopsis* which altered the expression of H+ ATPase proteins reducing phenotypes, increase of H+ efflux from plant roots generates acidic situation in the rhizosphere and results in growth increment at lower pH which helps in phosphate mineralization and enhanced resistance toward stress conditions.

19.2.3 Plant-Microbe-Mediated Rhizosphere Engineering

Microbes and plants are inter-reliant to each other and identified as secondary genome of plant and subsequently may perhaps function as a metaorganism (Lakshmanan et al. 2014). This process helps in management of soil microbial diversity with stimulation of exploitive soils and helped in recovering the physico-chemical properties of soil and increase in organic carbon and different nutrient cycling using crop rotation (Kumar et al. 2015). Application of mineral and organic fertilizers such as composts and animal manure ultimately enhances biological activity of the soil by rising the soil organic matter, crop deposit, and biological activity. Opine concept concerned toward host plants to secreted fastidious root exudates concurrently through the inoculation of microbes so as to engineer to degrade this substrate regularly results colonization of the specific type of rhizospheric microbial population. For this reason, opines formed using transgenic plants direct the choice of host-specific microbial population that can preserve themselves at exceptionally high concentration, still later than transgenic plant is separated. These methods employ precise metabolic resources. Gan et al. (2015) reported that chickpea and yellow pea direct the selection of specific microbial population in the rhizosphere and enhance the wheat production under field condition. Berendsen et al. (2018) reported that plants alter their root microbiome ahead of

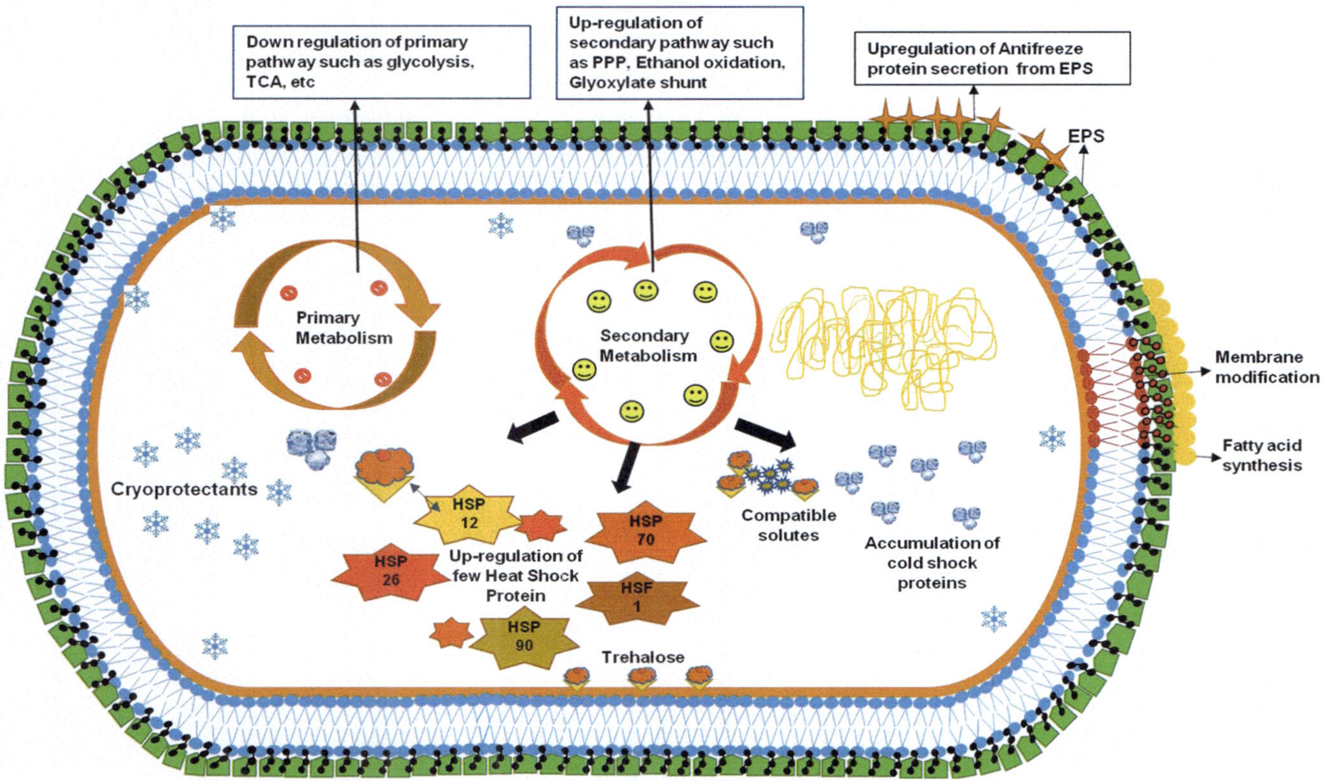

infection by pathogens and particularly employ a group of disease resistance and growth-promoting microbes for getting better chance of survival.

19.3 Culture-Dependent and Culture-Independent Techniques to Study Microbial Diversity in Hilly Agro-Ecosystem

Earlier it was thought that there is no life at colder regions; however using advanced technologies and science revealed that a huge and complicated world of tiny warriors inhabiting in the cold environment play an important role in environmental processing. Himalaya is solitary of the geographic locations that are known as a distinctive low-temperature environment as it holds the utmost glaciers after the polar regions. It is known with highly diverse regions in terms of geographical and biological aspects such as evergreen forests, cold deserts, grass lands, and glaciers (Khan et al. 2019). The extreme cold conditions have several factors which cause stress for animals and plants. Cold-adapted microbes survive under such cold environment and alleviate the effect of low temperature by alternating at cellular and biochemical levels (Margesin and Collins 2019). These microorganisms change their lipid composition to maintain integrity of membrane. Bore et al. (2019) revealed that microorganism switches among different metabolic pathways in response to low temperature to obtain energy.

Omics approaches have brought huge information at the molecular level, providing information on the microbial diversity and their interactions in specific environment. Use of LC-MS and GC-MS permits for untargeted methods known as metabolomics which helps in qualitative and quantitative analysis of the chemical composition of plant rhizosphere (Zhang et al. 2012). Himalaya region diversity has been studied by various researchers using traditional and metagenomic approaches which show the complicated composition of microbial communities. Prokaryotes were highly targeted due to their involvement in diverse ecological processes. Investigation of cold microbial diversity is now continuously increasing due to their values in agriculture production. High dominance of *Proteobacteria* was observed in Pangong lake of Himalaya using metagenomic sequencing. Microbial diversity plays an important role in nutrient cycling such as nitrogen, carbon, and phosphorus. More research is required on microbial diversity of soil, targeting functional attributes that can help to understand the microbial communities. It is widely known that the microbial diversity at low temperature is highly sensitive and can serve as indicators of climate change. All the cold-adapted microbial diversity requires more research regarding their isolation, preservation, and characterization.

Low temperature of Himalaya regions causes oxidative stress which is cope with antioxidants for normal functioning. Different cold shock proteins have been reported for the survival of microorganism at severe conditions such as chaperons (csp and rbps). These proteins bind to the RNA to destabilize the secondary structures and facilitate appropriate performance of the biochemical reactions. Ice-binding proteins have been reported in microorganisms, plants, and animals which help in survival at low temperature and facilitate microbial growth by inhibiting the formation of ice crystals. A cold-adapted microbe has pigments which protect them from harmful radiations. Nash et al. (2018) used the metagenomic approach to study the Arctic glacier and revealed their metabolic diversity including sulfur cycling bacteria and halophiles. Symbiotic nitrogen fixation efficiency in *Bradyrhizobium elkanii* and *Sinorhizobium meliloti* is reported by Marx et al. (2016). Further, Estibaliz and Stefanie (2017) reviewed a proteomic outlook on the function of legume symbiotic interactions, agreed on the need to develop a potential biomarker for symbiosis and signaling in pathogens, formation of nodules, and nitrogen fixation effectiveness, with improved stress tolerance. Different novel bacterial populations were reported in hilly areas using metagenomics and culturable techniques such as *Planococcus stackebrandtii*, *Dietzia kunjamensis*, *Actinoalloteichus spitiensis*, *Bacillus cecembensis*, *Cryobacterium roopkundense*, *Ferribacterium*, and *Wautersiella* (Mayilraj et al. 2005; Mayilraj et al. 2006; Singla et al. 2005; Reddy et al. 2008, 2010). Kumar et al. (2019) reported that bacterial population of *Acidobacteria*, *Actinobacteria*, and *Proteobacteria* were higher at higher altitude area. *Cytophaga* and *Chloroflexi* were positively found in cold deserts reported by Yang et al. (2016).

Different bioinformatics software such as omeSOM, MetGen-MAP, and PRIme plant are used in bioinformatics to analyze the data generated from different omics approaches (Milone et al. 2011; Sakurai et al. 2013). Illumina sequencing examination generates a vast amount of data which further requires analysis to get significant results by using QIIME, MEGAN, and other for data filtering and microbial diversity analysis (Gerlach et al. 2009).

19.4 Role of Nanotechnology for Plant Growth, Development, and Soil Health

Nanocompounds have relevance in agriculture field due to their unique properties. There are various nanoparticles used in agricultural field such as nanoclays, nanozeolite, nanochitosan, and nanogypsum which boost water-holding capacity of soil that improved plant growth and productivity (Chaudhary and Sharma 2019; Khati et al. 2019a; Kumari et al. 2020; Chaudhary et al. 2021a). Nanoparticles interact with plants and may lead to many physiological and morphological changes in plants which depend on the nature and concentration of nanocompounds (Siddiqui et al. 2015; Kukreti et al. 2020). Literature indicates both positive and toxic effects of nanoparticles on plants. Nanocompounds can significantly alter soil bacterial richness (Khati et al. 2017; Khati et al. 2019b). Silver nanoparticles and nanozeolite have been reported to alter soil microbial biomass, soil enzymes, diversity, and soil structure formation (Khati et al. 2018; Kumari et al. 2021). Kumar et al. (2015) reported that *Rhizobium* population was decreased in presence of silver nanoparticles (660 mg/Kg) in treated arctic soil. SiO_2 nanoparticles (300 ug/L) improved the growth of sugarcane under chilling stress by enhancing the photosynthesis and photoprotection mechanism (Elsheery et al. 2020).

19.5 Conclusion and Future Prospects

It is crucial to explore the cold area of Himalaya to develop our perspective on the metabolic performance, microbial ecology, interaction, and functional ecology of microbes at low-temperature environments using culture-dependent and culture-independent techniques. This would make possible to recognize ecological trouble linked with mountain system, leading to the valuable use of natural resources beside with the conservation of the microbial biodiversity of Himalaya. In agriculture field, introducing beneficial microbes and nanoparticles used to increase crop production by improving the soil quality for sustainable agriculture. Nanocompounds encourage an innovative green revolution with reduced farming threats. The advancement of the technology in the molecular biochemistry has evolved and gave a pavement to new era of increasing the soil rhizosphere in different agro-climatic regions. These technologies not only help in the engineering but also the closer study of these biochemical involved for further future purposes. On the other hand, there are still enormous gaps in the uptake ability, concentration, and ecotoxicity of diverse nanoparticles. Consequently, advance research is immediately essential to unravel the behavior and outcome of altered agriculture inputs and their communication with bio-macromolecules present in the living system and surroundings. More knowledge at field level is highly needed for large-scale implementation of nanocompounds strategies.

References

Agri U, Chaudhary P, Sharma A (2021) In vitro compatibility evaluation of agriusable nanochitosan on beneficial plant growth-promoting rhizobacteria and maize plant. Natl Acad Sci Lett. https://doi.org/10.1007/s40009-021-01047-w

Ahkami A, Allen White R, Handakumbura PP, Jansson C (2017) Rhizosphere engineering: enhancing sustainable plant ecosystem productivity in a challenging climate. Rhizosphere 3:233–243

Berendsen RL, Vismans G, Yu K, Song Y, De Jonge R, Burgman WP (2018) Disease-induced assemblage of a plant-beneficial bacterial consortium. ISME J 12(6):1496–1507

Bore EK, Halicki S, Kuzyakov Y, Dippold MA (2019) Structural and physiological adaptations of soil microorganisms to freezing revealed by position-specific labeling and compound-specific 13C analysis. Biogeochemistry 143:207–219

Chaudhary P, Sharma A (2019) Response of Nanogypsum on the performance of plant growth Promotory Bacteria recovered from Nanocompound infested agriculture field. Environ and Ecol 37:363–372

Chaudhary P, Khati P, Chaudhary A, Gangola S, Kumar R, Sharma A (2021a) Bioinoculation using indigenous *Bacillus* spp. improves growth and yield of *Zea mays* under the influence of nanozeolite. 3 Biotech 11:11

Chaudhary P, Sharma A, Chaudhary A, Sharma P, Gangola S, Maithani D (2021b) Illumina based high throughput analysis of microbial diversity of rhizospheric soil of maize infested with nanocompounds and *Bacillus* sp. Appl Soil Ecol 159:103836

Chaudhary P, Khati P, Gangola S, Kumar A, Kumar R, Sharma A (2021c) Impact of nanochitosan and Bacillus spp. on health, productivity and defence response in Zea mays under field condition. 3 Biotech 11:237

Chaudhary P, Khati P, Chaudhary A, Maithani D, Kumar G, Sharma A (2021d) Cultivable and metagenomic approach to study the combined impact of nanogypsum and Pseudomonas taiwanensis on maize plant health and its rhizospheric microbiome. PLoS One 16:e0250574

Elsheery NI, Sunoj VSJ, Wen Y, Zhu JJ, Muralidharan G, Cao KF (2020) Foliar application of nanoparticles mitigates the chilling effect on photosynthesis and photoprotection in sugarcane. Plant Physiol Biochem 149:50–60

Estibaliz L, Stefanie W (2017) A proteomic view on the role of legume symbiotic interactions. Front Plant Sci 8:1267

Gan Y, Hamel C, O'Donovan JT, Cutforth H, Zentner RP, Campbell CA (2015) Diversifying crop rotations with pulses enhances system productivity. Sci Rep 5:14625

Gerlach W, Jünemann S, Tille F, Goesmann A, Stoye J, WebCARMA (2009) A web application for the functional and taxonomic classification of unassembled metagenomic reads. BMC Bioinf 10:430

Gevaudant F, Duby G, von Stedingk E, Zhao R, Morsomme P, Boutry M (2007) Expression of a constitutively activated plasma membrane H+-ATPase alters plant development and increases salt tolerance. Plant Physiol 144:1763–1776

Glick BR (2012) Plant growth-promoting bacteria: mechanisms and applications. Scientifica 2012:963401

Guo K, Hu A, Wang K, Wang L, Fu D, Hao Y, Wang Y, Ali A, Adeel M, Rui Y, Tan W (2019) Effects of spraying nano-materials on the absorption of metal(loid)s in cucumber. IET Nanobiotechnol 13:712–719

Hashem A, Kumar A, Al-dbass AM, Alqarawi AA, Al-arjani AF, Singh G (2018) Arbuscular mycorrhizal fungi and biochar improves drought tolerance in chickpea. Saudi J Biol Sci 26 (3):614–624

Joshi D, Chandra R, Suyal DC, Kumar S, Goel R (2019) Impact of bioinoculants *Pseudomonas jesenii* MP1 and *Rhodococcus qingshengii* S10107 on *Cicer arietinum* yield and soil nitrogen status. Pedosphere 29(3):388–399

Khan N, Bano A, Rahman MA, Guo J, Kang Z, Babar MA (2019) Comparative physiological and metabolic analysis reveals a complex mechanism involved in drought tolerance in chickpea (Cicer arietinum L.) induced by PGPR and PGRs. Sci Rep 9:2097

Khare E, Mishra J, Arora NK (2018) Multifaceted interactions between endophytes and plant: developments and prospects. Front Microbiol 9:2732

Khati P, Chaudhary P, Gangola S, Bhatt P, Sharma A (2017) Nanochitosan supports growth of *Zea mays* and also maintains soil health following growth. 3 Biotech 7:81

Khati P, Chaudhary P, Gangola S, Sharma P (2019a) Influence of nanozeolite on plant growth promotory bacterial isolates recovered from nanocompound infested agriculture field. Envi Ecol 37:521–527

Khati P, Parul, Bhatt P, Nisha Kumar R, Sharma A (2018) Effect of nanozeolite and plant growth promoting rhizobacteria on maize. 3 Biotech 8:141

Khati P, Sharma A, Chaudhary P, Singh AK, Gangola S, Kumar R (2019b) High-throughput sequencing approach to access the impact of nanozeolite treatment on species richness and evens of soil metagenome. Biocatal Agri Biotechnol 20:101249

Koyama H, Kawamura A, Kihara T, Hara T, Takita E, Shibata D (2000) Overexpression of mitochondrial citrate synthase in *Arabidopsis thaliana* improved growth on a phosphorus-limited soil. Plant Cell Physiol 41:1030–1037

Kukreti B, Sharma A, Chaudhary P, Agri U, Maithani D (2020) Influence of nanosilicon dioxide along with bioinoculants on *Zea mays* and its rhizospheric soil. 3 Biotech 10:345

Kumar A, Dames JF, Gupta A, Sharma S, Gilbert JA, Ahmad P (2015) Current developments in arbuscular mycorrhizal fungi research and its role in salinity stress alleviation: a biotechnological perspective. Crit Rev Biotechnol 35:461–474

Kumar S, Suyal DC, Yadav A, Shouche Y, Goel R (2019) Microbial diversity and soil physiochemical characteristic of higher altitude. PLoS One 14(3):e0213844

Kumari S, Sharma A, Chaudhary P, Khati P (2020) Management of plant vigor and soil health using two agriusable nanocompounds and plant growth promotory rhizobacteria in fenugreek. 3 Biotech 10:461

Kumari H, Khati P, Gangola S, Chaudhary P, Sharma A (2021) Performance of plant growth promotory rhizobacteria on maize and soil characteristics under the influence of TiO_2 nanoparticles. Pantnagar J Res 19:28–39

Lakshmanan V, Selvaraj G, Bais HP (2014) Functional soil microbiome: belowground solutions to an aboveground problem. Plant Physiol 166:689–700

Margesin R, Collins T (2019) Microbial ecology of the cryosphere (glacial and permafrost habitats): current knowledge. Appl Microbiol Biotechnol 103:2537–2549

Marx H, Minogue CE, Jayaraman D, Richards AL, Kwiecien NW, Siahpirani AF, Rajasekar S, Maeda J, Garcia K, Valle-Echevarria ARD (2016) A proteomicatlas of the legume Medicago truncatula and its nitrogen-fixing endosymbiont Sinorhizobium meliloti. Nat Biotechnol 34 (11):1198–1205

Mayilraj S, Prasad GS, Suresh K, Saini HS, Shivaji S, Chakrabarti T (2005) *Planococcus stackebrandtii* sp. nov., isolated from a cold desert of the Himalayas, India. Int J Syst Evol Microbiol 55(Pt 1):91–94

Mayilraj S, Suresh K, Kroppenstedt RM, Saini HS (2006) Dietzia kunjamensis sp. nov., isolated from the Indian Himalayas. Int J Syst Evol Microbiol 56(Pt 7):1667–1671

Mazzola M (2002) Mechanisms of natural soil suppressiveness to soilborne diseases. Int J General Mol Microbiol 81:557–564

Milone DH, Stegmayer GS, Kamenetzky L, López M, Lee JM, Giovannoni JJ (2011) omeSOM: a software for clustering and visualization of transcriptional and metabolite data mined from interspecific crosses of crop plants. BMC Bioinf 11:438

Nash MV, Anesio AM, Barker G, Tranter M, Varliero G, Eloe-Fadrosh EA, Nielsen T, Turpin-Jelfs T, Benning LG, Sánchez-Baracaldo P (2018) Metagenomic insights into diazotrophic communities across arctic glacier forefields. FEMS Microbiol Ecol 94:114

Rajwar J, Chandra R, Suyal DC, Tomer S, Kumar S, Goel R (2018) Comparative phosphate solubilizing efficiency of psychrotolerant *Pseudomonas jesenii* MP1 and *Acinetobacter* sp. ST02 against chickpea for sustainable hill agriculture. Biologia 73(8):793–802

Reddy GS, Pradhan S, Manorama R, Shivaji S (2010) *Cryobacterium roopkundense* sp. nov., a psychrophilic bacterium isolated from glacial soil. Int J Syst Evol Microbiol 60(Pt 4):866–870

Reddy GS, Uttam A, Shivaji S (2008) *Bacillus cecembensis* sp. nov., isolated from the Pindari glacier of the Indian Himalayas. Int J Syst Evol Microbiol 58(Pt 10):2330–2335

Ryan PR, Dessaux Y, Thomashow LS, Weller DM (2009) Rhizosphere engineering and management for sustainable agriculture. Plant Soil 321:363–383

Sakurai T, Yamada Y, Sawada Y, Matsuda F, Akiyama K, Shinozaki K (2013) PRIMe update: innovative content for plant metabolomics and integration of gene expression and metabolite accumulation. Plant Cell Physiol 54:e5

Siddiqui MH, Al-Whaibi MH, Firoz M, Al-Khaishany MY (2015) Role of nanoparticles in plants. In: Nanotechnology and plant sciences. Springer, Cham, pp 19–35

Singla AK, Mayilraj S, Kudo T, Krishnamurthi S, Prasad GS, Vohra RM (2005) *Actinoalloteichus spitiensis* sp. nov., a novel actinobacterium isolated from a cold desert of the Indian Himalayas. Int J Syst Evol Microbiol 55(Pt 6):2561–2564

Sunoj VSJ, Shroyer KJ, Krishna J, Prasad V (2016) Diurnal temperature amplitude alters physiological and biochemical response of maize (*Zea mays*) during the vegetative stage. Environ Exp Bot 130:113–121

Suyal DC, Kumar S, Joshi D, Soni R, Goel R (2018) Quantitative proteomics of psychotrophic diazotroph in response to nitrogen deficiency and cold stress. J Proteome 187:235–242

Suyal DC, Kumar S, Yadav A, Shouche Y, Goel R (2017) Cold stress and nitrogen deficiency affected protein expression of psychrotrophic *Dyadobacter psychrophilus* B2 and *Pseudomonas jessenii* MP1. Front Microbiol 8(430):1–6

Suyal DC, Yadav A, Shouche Y, Goel R (2014) Differential proteomics in response to low temperature diazotrophy of Himalayan psychrophilic nitrogen fixing *Pseudomonas migulae* S10724 strain. Curr Microbiol 68:543–550

Tejeda M, Moreno-JI EL, Beniston M (2013) The changing roles of temperature and precipitation on snowpack variability in Switzerland as a function of altitude. Geophys Res Lett 40:2131–2136

Tomer S, Suyal DC, Rajwar J, Yadav A, Shouche Y, Goel R (2017) Isolation and characterization of phosphate solubilizing bacteria from Western Indian Himalayan soils. 3 Biotech 7(2):95

Yadav AN, Sachan SG, Verma P, Kaushik R, Saxena AK (2016) Cold active hydrolytic enzymes production by psychrotrophic *Bacilli* isolated from three sub-glacial lakes of NW Indian Himalayas. J Basic Microbiol 56:94–307

Yang GL, Hou SG, Le Baoge R, Li ZG, Xu H, Liu YP (2016) Differences in bacterial diversity and communities between glacial snow and glacial soil on the Chongce ice cap, west Kunlun Mountains. Sci Rep 6:36548

Zhang A, Sun H, Wang P, Han Y, Wang X (2012) Modern analytical techniques in metabolomics analysis. Analyst 137:293–300

The manufacturer's authorised representative in the EU is Springer Nature Customer Service Centre GmbH, Europaplatz 3, 69115 Heidelberg, Germany. If you have any concerns regarding our products, please contact ProductSafety@springernature.com

Printed and bound by CPI Group (UK) Ltd, Croydon, CR0 4YY
30/09/2026
02225247-0015